Nanostructured Semiconductors

Nanostructured Semiconductors

Amorphization and Thermal Properties

edited by

Konstantinos Termentzidis

PAN STANFORD PUBLISHING

Published by

Pan Stanford Publishing Pte. Ltd.
Penthouse Level, Suntec Tower 3
8 Temasek Boulevard
Singapore 038988

Email: editorial@panstanford.com
Web: www.panstanford.com

British Library Cataloguing-in-Publication Data
A catalogue record for this book is available from the British Library.

Nanostructured Semiconductors: Amorphization and Thermal Properties

ISBN 978-981-4745-64-2 (Hardcover)
ISBN 978-1-315-36445-2 (eBook)

Printed in the USA

Dedicated to my parents

Anastasia and Christos
and in memory of Georgios Kanellis

Contents

Preface

This book is the result of a collective work of 42 authors through 19 chapters on nanostructures and nanostructured materials containing both amorphous and crystalline phases with a particular focus on their thermal properties. There are two distinct parts: The first combines theory and simulations methods with specific examples, and the second part discusses methods to fabricate nanomaterials with crystalline and amorphous phases and experimental techniques to measure the thermal conductivity of such materials.

Chapter 1 details the structural properties of the crystalline and the amorphous phases, highlighting their differences in the density, radial, and angular distribution functions, and analyzing the different amorphous models used in the literature. This chapter introduces and explains a lot of the terms used herein. In Chapter 2, a general theory for the lattice thermal conductivity is developed, for both crystals and disorder systems. This analysis is used for ab initio simulations of the thermal conductivity of bulk materials. The taxonomy of vibration modes in amorphous materials is fully detailed. In Chapter 3, the differences in thermal properties between bulk materials and nanostructures are discussed. The transition from ballistic to diffusive regimes and the underlying scattering processes occurring at the nanoscale are explained. Furthermore, in this chapter, there are sections dedicated to the phonon coherence effects, the phonon rectification, and the amorphous limit of the thermal conductivity. In Chapter 4, the importance of heat transport at the micro/nanoscale for several applications is treated for a wide range of domains such as electronics, bio-engineering, and energy harvesting. In Chapter 5, the Monte Carlo methodology to solve the Boltzmann transport equation is given, with case studies such as nanowires and nanomaterials. The latter methodology is the most adequate simulation method for micro/nanostructures with length scales greater than some tenths of nanometers.

For the prediction of the thermal properties in the ballistic regime, atomistic approaches are preferred. Chapters 6, 7, and 8 present three different molecular dynamics (MD) methodologies: equilibrium MD (EMD) in Chapter 6, non-equilibrium MD (NEMD) in Chapter 7, and the approach to equilibrium MD (AEMD) in Chapter 8. In Chapter 6, an introduction to MD is given and the EMD methodology based on the Green–Kubo formula is derived to estimate the thermal conductivity, thermal conductance, and resistance with an example: the thermal conductance of the argon/heavy argon interface. In Chapter 7, the NEMD methodology is described to calculate the thermal conductivity of nanomaterials and applied in two examples: superlattices and nanowires which incorporate both crystalline and amorphous phases. Striking results show that an amorphous fraction of only 20% in certain geometries of nanostructures is enough to obtain sub-amorphous thermal conductivity. In Chapter 8, the AEMD method is analyzed, which is the only method among the three MD methods appropriate to describe transient regimes. The thermal conductivities of amorphous silica and α-quartz thin films are calculated with this latter method. In Chapter 9, numerical methods to treat the vibrational properties of isolated crystalline and amorphous nanoparticles and nanoinclusions are analyzed: resonant modes analysis, vibrational density of states, and dynamical structural factors. The last chapter of the first part of the book is dedicated to the III-nitrides group and their structural properties and defects as extended defects, polar and non-polar interfaces, diffusion of defects, and surface reconstruction.

The second part of the book starts with Chapter 11, in which a deep analysis of the role of the heavy ion irradiation for the amorphization of porous silicon nanostructures is proposed. It explains why the heavy ion irradiation influences only porous materials while remains unaffected on the bulk ones. The quantification of the amorphous fraction as a function of the ion fluence is also given. The infrared absorbance and the Raman spectra are used to appraise the influence of different irradiation rates on porous silica. In Chapter 12, another crystalline/amorphous porous material is treated for germanium compounds, with a porosity induced by electrochemical etching. A review of the possible geometries of the pores is given. It is also explained how these geometries can be realized in the case of Ge by changing two key parameters during the electrochemical etching: the anodic current density and the hydrogen passivation degree. In Chapter 13, an experimental protocol to achieve amorphous/crystalline composites is described—more precisely, the methodology to obtain crystalline nanoinclusions embedded into an amorphous matrix. Starting from amorphous materials, the strategy is to interrupt the crystallization process before the crystallization peak is completed. Chapter 14 deals with the ball milling technique, which is used to amorphize and to nanocrystallize materials with or without chemical transformation. Pedagogic case studies are given for each category with explanations of each key parameter and their influence on the mechanical properties of the elaborated nanostructures. In Chapter 15, the tin-induced crystallization of amorphous silicon manufactured by physical vapor deposition and the possibility of the crystallization control during laser annealing are analyzed. These methods can advance the wide-scale commercialization of nanocrystalline silicon for the third and next generations of solar cells and other optoelectronic devices. Chapter 16 contains both the experimental methodology and simulation evidence for the crystallization of silicon nanoparticles utilizing magnetron sputtering. Two methods are proposed: first, the use of inert gas temperature control during magnetron sputtering, in which the inert gas pressure and composition are the key parameters to control the size distribution of the nanocrystals, and second, the use of silver atoms for the heterogeneous crystallization of silicon nanoparticles.

The last three chapters of the book—Chapter 17 (scanning thermal microscopy, SThM), Chapter 18 (thermal wave methods), and Chapter 19 (X-rays and neutron spectroscopy)—are focused on experimental methods for the measurement of thermal properties appropriate for structures that contains both crystalline and amorphous phases. In Chapter 17, a review of SThM in the active mode is given, which enables nanometer-scale heat flow measurements and materials' thermal characterization. Then, the approaches to modeling and calibrating SThM probes are explained and finally the method is used to measure the thermal conductivity of nanoporous silicon with amorphous shells around the pores. In Chapter 18, a review of experimental thermal wave methods is presented. With the ac-current ($2\omega/3\omega$) method, the thermal conductivity components of thin films (in-plane/cross-plane) can be estimated, using thin metallic layers acting as a heater and/or a thermometer. The photothermal gas-microphone technique uses light to generate a heat source while the measured pressure variations can be directly related to the thermal properties. The method is applied on a partially amorphous porous silicon. The last chapter also focuses on experimental techniques: inelastic neutron scattering and X-ray scattering to measure phonon velocities and lifetimes. The authors explain the measurement principles and the experimental limitations while in the second part of Chapter 19, the atomic dynamics differences of an amorphous and a crystalline system are detailed.

Foreword

Nanostructured Semiconductors: Amorphization and Thermal Properties aims to expose the reader to the state of the art in several areas of modern science of materials. The most prominent focus of this book is on theory and simulations of thermal conduction in all major types of materials microstructures, including bulk crystals, nanocrystalline and nanoporous materials, and fully amorphous materials. Chapters 2 to 9 provide a comprehensive description of all major theoretical and simulation methods used to understand, model, and predict thermal conduction characteristics of materials and their interfaces. What is of particular value in this book is the presentation of a number of topics from multiple perspectives. For example, the rigorous theoretical foundations of thermal conductivity formulas derived from the equilibrium-based, fluctuation-dissipation theory presented in Chapter 2 are complemented by structure/problem-specific perspectives presented in Chapter 3, the application of the theory to molecular dynamics simulations presented Chapter 6, and to the phonon-level description of thermal conduction based on the Boltzmann transport equation in Chapter 5. For molecular-level simulation practitioners, Chapters 7 and 8 provide an introduction to major non-equilibrium methods, i.e., methods employing macroscopic heat flux and temperature gradients and relying directly on Fourier's law of diffusive heat conduction. Chapter 9 completes the presentation of atomic-level modeling and calculation tools with a focus on direct phonon lifetime determination and vibrational analysis of complex, non-periodic systems. For those practicing or contemplating research in modeling and simulations of phonon-based thermal transport in complex structure materials, the first part of the book provides a one-stop store for all major state-of-the-art tools and concepts they need.

The second part of the book focuses on experimental synthesis, the processing and characterization of complex nanostructure, and porous and amorphous material and their composites. Amorphization via ion irradiation is the subject of Chapter 11, while electrochemical etching–based synthesis of similar materials is described in Chapter 12. The following three chapters focus on different processing methods such as annealing-based recrystallization, mechanical-based ball milling–induced nanostructuring and amorphization, and laser processing. Finally, Chapters 17 and 18 introduce modern techniques for thermal transport characterization, allowing for the completion of a concept of structure—thermal property relationship determination with the synergistic approach based on theory, modeling, simulations, and experiments as described comprehensively throughout this book.

Dr. Pawel Keblinski
Materials Science and Engineering Department
Rensselaer Polytechnic Institute
Troy, New York, USA

Part I

THEORY-SIMULATIONS

Chapter 1

Structural Properties: Crystalline versus Amorphous Phases

Laurent Pizzagalli[a] **and Andreas Pedersen**[b]
[a] *Department of Physics and Mechanics of materials, Institut Pprime CNRS—Université de Poitiers, Bd Pierre et Marie Curie, BP30179, F86962 Futuroscope Chasseneuil Cedex, France*
[b] *Integrated Systems Laboratory, ETHZ, Gloriastrasse 35, 8092 Zurich, Switzerland*
laurent.pizzagalli@univ-poitiers.fr

1.1 Introduction

According to the usual definition, a material is classified as amorphous when it lacks the long-range structural order characteristic of a crystal. The latter is indeed distinctive by having atoms (or molecules) located in ordered and well-defined sites, forming an infinite periodic lattice. Such a translational order is lost in an amorphous phase, because the elementary components are arranged according to patterns without periodicity. Nevertheless, an amorphous material can exhibit a certain degree of order at short range, due to local structural features of the chemical bonding alike in the corresponding crystal. Short range typically refers to distances that are shorter than about 10 Å, while long range is attributed to

Nanostructured Semiconductors: Amorphization and Thermal Properties
Edited by Konstantinos Termentzidis

ISBN 978-981-4745-64-2 (Hardcover), 978-1-315-36445-2 (eBook)
www.panstanford.com

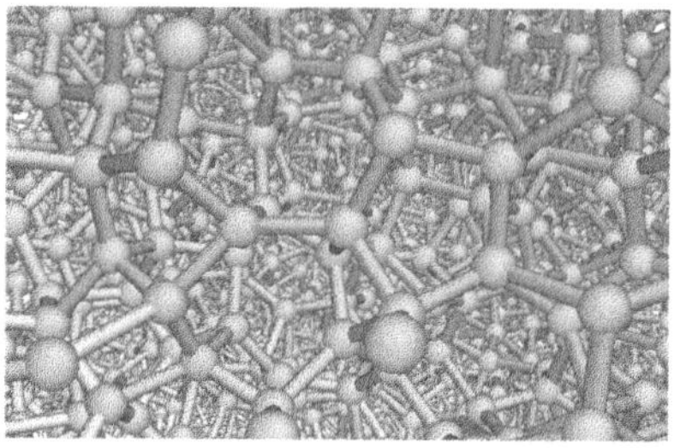

Figure 1.1 Ball and stick representation of amorphous silicon, exhibiting a local tetrahedral order due to the character of the chemical bonds (model obtained by quenching the liquid phase).

larger values (note that a medium range from 10 to about 30 Å is also employed in specific cases). Structurally, amorphous materials often resemble liquids, which also lack symmetry, but they are always in a solid phase. Figure 1.1 shows a model structure of a tetrahedrally bonded amorphous semiconductor.

Several structural quantities are useful for characterizing amorphous systems. One trivial is the local density, i.e., the number of atoms per volume unit (Section 1.2.1). Then, the distances between atoms are also important features, which can be experimentally determined with high accuracy (Section 1.2.2). Assuming that chemical bonds link neighboring atoms, one can also examine the angles spanned by adjacent bonds (bond angles for two successive bonds, and dihedral or torsion angles for three consecutive bonds) (Section 1.2.3). Finally, the topology of the network formed by the atoms and the bonds joining them is also an important structural fingerprint of the amorphous system, which is usually analyzed in terms of sizes of rings formed by connected atoms (Section 1.2.4).

Note that because of an inherent disorder, the amorphous phase is characterized by distributions of structural properties, rather than single values as for a crystal. Due to this fact, all experimentally determined values are obtained within an average range that depends on the spatial resolution of the experiment. This is an important singularity of disordered materials (and also one of the main difficulties in the study of these).

An important class of amorphous materials is semiconductors, which include tetrahedrally bonded systems such as group IV (a-Si, a-Ge), or III-V (a-GaAs) materials. Other amorphous materials are chalcogenide glasses (a-S, a-Se,...) based on group V and VI elements, and amorphous organic semiconductors. In the remaining of this chapter, we focus on tetrahedrally bonded materials, and more specifically on amorphous silicon. The latter has been thoroughly studied and is therefore a good system for discussing the various quantities used in the structural characterization of a solid amorphous material.

1.2 Structural Indicators

1.2.1 Density

The number density is probably the simplest structural quantity that can be used to differentiate amorphous and crystalline phases. It is the number of atoms per volume unit, typically expressed in nm^{-3} or $Å^{-3}$ units, and its inverse is the volume associated to a single atom (partial densities can be defined when the material consists of several species). This property can be determined directly from measurements on the whole sample (weighting and volume calibration) [6], or indirectly using diffraction experiments [11].

An important limitation of such measurements is the obtained spatial resolution, which typically belongs to the large-scale regime. Then, only averaged values of the densities are obtained, and local variations are hardly measurable. This might be critical because such density variations could exist at short range, at the scale of the nanometer, if it is assumed that the concept of point defects in crystals can be extended to disordered systems. In that case, a local excess (an interstitial) or lack (a vacancy) of atoms would occur, which is equivalent to a local inhomogeneity of the density. At a larger scale, density measurements will be affected if the presence of voids in a porous sample cannot be resolved [15].

1.2.2 Radial Distribution Functions

Another structural quantity is the radial distribution function (RDF) $g(r)$ (also called pair distribution function), that is often used as

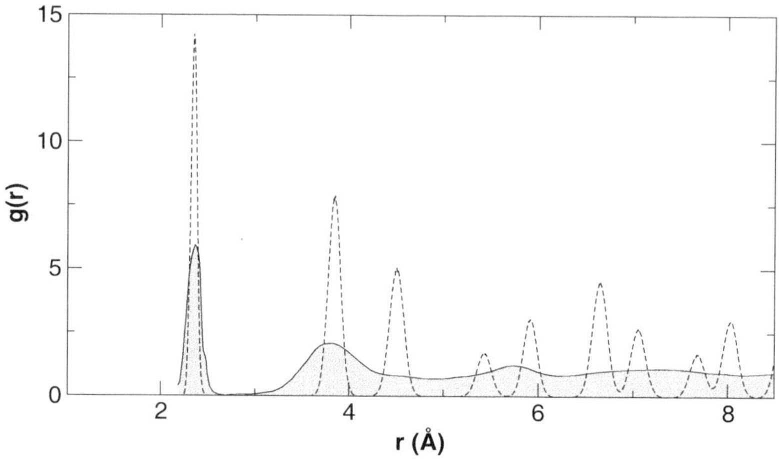

Figure 1.2 Radial distribution function of silicon at 300 K, in amorphous phase (solid line, exp. data from [11]), and computed for the crystalline phase (dashed line).

the main property for characterizing the disorder of a system. $g(r)$ is related to the probability of finding another atom at the distance r from an arbitrary reference atom, and is customarily normalized with respect to the ideal gas. Figure 1.2 shows the RDF resulting from amorphous and crystalline silicon. Two well-defined peaks can be seen at short distances in the disordered case, whereas the peaks at larger pair separations quickly vanish for the amorphous phase but remain in the data for the crystalline material. The two strong peaks result from the first- and second-neighbor shells in the crystalline structure, and their occurrence for the amorphous phase confirms that a-Si bears some similarities with c-Si at short range. However, since no additional peaks can be clearly identified at separations larger than the third-neighbor shell, the a-Si RDF becomes structureless while the c-Si RDF still clearly shows peaks that are characteristics for a configuration with a general order [7]. The RDF also allows the determination of the average coordination number, which can be obtained by integrating the area under peaks for a properly normalized RDF. This can be easily done for the first-neighbors [11], since the associated peak is well separated from others. Note that in the case of the crystal, the observed dispersion of distances to neighbor shells solely originate from thermal disorder (at 0 K the RDF would be a collection of Dirac delta functions). On the contrary, static disorder is the main source of dispersion for the amorphous system. It should also be noted that the limited experimental resolution yields an additional dispersion contribution in both cases. Finally, it is also possible to extract first-neighbor separations and coordination numbers using several experimental techniques such as EXAFS [8].

The RDF can be easily determined from the structure factor $S(k)$, which can be obtained from diffraction experiments. Since computing the RDF of amorphous models is straightforward, this quantity is often the first used to assess the validity of a proposed model. However, one has to be cautious that the RDF is a volume averaged quantity by definition and only carries a limited amount of information. Hence, it has been shown that different disordered silicon models yield a similar RDF in agreement with experiments, which indicates that other properties at higher order of correlations also should be considered. For instance, it has recently been proposed to consider four-body spatial correlations, which can be determined using fluctuation electron microscopy [23]. These experiments suggest a possible ordering beyond short range (typically 1–3 nm).

1.2.3 Angular Distributions

The angular distributions are other quantities that are useful when characterizing a material in its amorphous phase. The simplest one, the bond angle distribution, concerns the angles between two bonds sharing a common atom. In the crystalline state of silicon, all bond angles are equal to $180^\circ - \arccos(1/3) = 109.47^\circ$ (with a small dispersion associated to thermal broadening). A much wider range of values is obtained for the amorphous state, as seen in Fig. 1.3, leading to a distribution centered on a slightly lower value than for the crystalline phase [7, 10, 11]. A significant dispersion is also observed, which directly relates to the bond distortion caused by disorder.

These two characteristics of the bond angle distribution, in particular the dispersion, can be determined from diffraction and

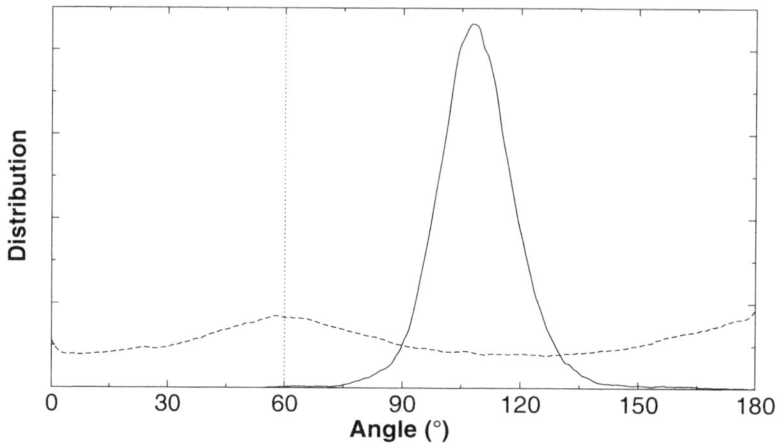

Figure 1.3 Bond (solid line) and dihedral (dashed line) angles distributions for amorphous silicon, calculated at 0 K.

Raman spectroscopy experiments. Like the RDF, the bond angle distribution is often used to validate amorphous models since it is easy to compute. However, the same restriction as pointed out above regarding the average nature of the quantity also applies for angular distributions. It is also noteworthy that the relationship between measurements and the angular dispersion is not straightforward [18].

The distributions for the higher order angle terms, such as the dihedral angle (also known as the torsion angle) distribution, also carry valuable information. By considering four atoms, three bonds can be identified and the dihedral angle is defined as the angle between the two planes that are spanned by respectively (i) the first and the shared bond and (ii) the shared and the last bond. For a crystalline cubic diamond structure, the dihedral angles are equal to 60° or 180°, in equal proportions. In the case of the amorphous phase, the two values correspond to maxima of the distribution; however, these are not as well defined and other dihedral angles also have a high probability (see Fig. 1.3). This behavior suggests that ordering is progressively being reduced beyond the short range.

1.2.4 Network Topology

Another option for comparing the crystalline and the amorphous phases is by investigating the connectivity of the network formed by the chemically bonded atoms. The quantity that is usually determined is the statistics of n-fold rings, which are defined as the closed paths formed by bonds connecting n atoms. Although there are several possibilities, a shortest path criterion is often used when identifying rings.

In crystalline silicon, there are only 6-fold rings. Those also constitute the majority in the disordered state. However, 5-fold rings represent a sizeable proportion, which is followed by much smaller amounts of 7-fold and 4-fold rings. Figure 1.4 shows examples of such rings in an amorphous silicon model.

Ring statistics provide a wealth of information on the structure of amorphous semiconductors. However, an experimental determination is still lacking, which prohibits comparison to the proposed models for the disordered state.

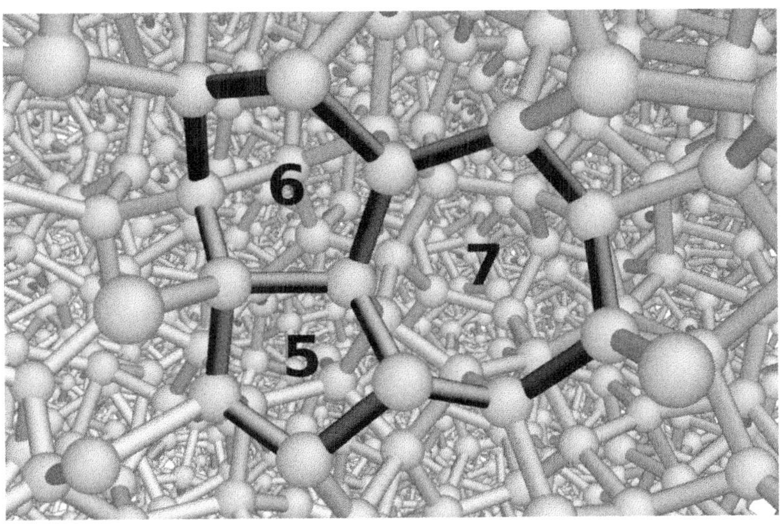

Figure 1.4 Same model as in Fig. 1.1, showing 5-fold, 6-fold, and 7-fold topological rings (black sticks).

1.3 Amorphous Models

The determination of the structure of amorphous semiconductors remains an unfinished quest. This is the case despite decades of dedicated efforts (read, for instance, [26] for a short history) and is partly due to the lack of knowledge. In fact, while the investigation of crystalline phases can be done using many experimental tools, only a few of these are suited for probing the amorphous phase. For the disordered state, an alternative approach is to build models for which the aforementioned structural indicators can be computed and compared to available experimental findings [22].

For tetrahedrally coordinated amorphous semiconductors like silicon, the vast majority of experimental data suggest a very high degree of local order, with a coordination in accordance with chemical valence. Complying with these requirements, several models of the disordered infinite solid have been proposed in the past, which roughly can be classified in the following two categories.

1.3.1 CRN-Like Models

Historically attributed to Zachariasen [26], the continuous random network (CRN) model strictly requires that all atoms are fully coordinated (i.e., like in the corresponding pristine bulk system, without any dangling bonds). This means that disorder only originates from small bond angle and length distortions relative to the crystal values, and there is order at short range, quickly vanishing for increasing distances.

The earliest CRN model for a-Si was hand-built by Polk in 1971 [17]. A significant step forward was later accomplished when Wooten, Winer, and Weaire proposed a computer-based procedure (commonly called the WWW method) to generate a CRN model, which is also known as the sillium [25]. Additional refinements to this method have later been proposed [1]. Another option is to lift the restriction about perfect coordination, thus allowing the presence of over- and under-coordinated atoms, which makes it possible to build CRN-like amorphous state models.

The more advanced CRN (CRN-like) models are able to reproduce most of available structural data, such as RDF and the angular

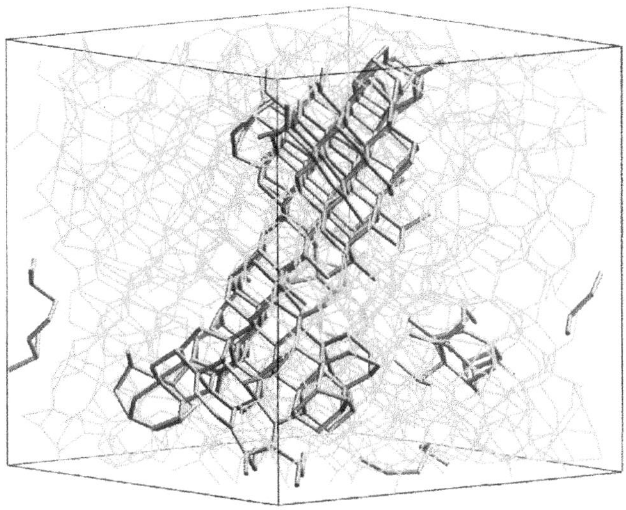

Figure 1.5 Paracrystalline model generated by combining interatomic potential relaxation and experimental data fitting. Thin light gray lines represent the CRN network, while thick dark gray lines connect atoms in crystalline-like environment. Courtesy of M. M. J. Treacy and K. B. Borisenko.

distributions, with great accuracy. However, recent fluctuation microscopy experiments suggest a possible ordering up until 1 nm reaching the mid-range regime [21], which only seems to be poorly reproduced using these models.

1.3.2 Micro- and Para-Crystalline

Another class of models builds on the assumption that an amorphous semiconductor is characterized by the coexistence of crystalline and disordered (CRN-like) domains. For instance, in the microcrystallite model, which is one of the oldest, amorphous silicon is represented by the aggregation of small nanocrystallites that are packed together. This model has been discarded over the years, because the obtained configurations contain interfaces that are unreasonably highly stressed, which is not compatible with experimental findings.

A related model that resolves this issue is the so-called paracrystalline model, where small crystallites are assumed to be embedded in a larger disordered domain [21]. This combination allows for the emergence of a weak medium range ordering, which is in agreement with the aforementioned fluctuation microscopy experiments. The crystalline domains have been found to be about 10–15 Å in size, and make up approximately 10 % of the whole system [24].

1.3.3 Generation Methods

Finally, various methods and techniques have been proposed for how to generate the amorphous models described above. Those can be classified in three general categories:

1.3.3.1 Building algorithms

The first category concerns building algorithms, which generate a model by following a set of rules. For instance, the earliest CRN models of amorphous silicon were built by hand, with the requirements that all atoms are 4-fold coordinated and, furthermore, that they belong to 6-fold or 5-fold rings [17]. In the later WWW method, the model is obtained by introducing a bond-switching algorithm, which preserves the 4-fold coordination while allowing for the introduction of 5-fold and 7-fold rings. In addition, the introduced stress is relaxed using an interatomic empirical potential and a local minimization algorithm [25]. The topological defect created by the bond switching procedure is also called a IV pair, and it has been shown that a simple random accumulation of these defects is enough to mimic amorphization by implantation [14]. The WWW algorithm has over the years been greatly improved by successive refinements, using both more complex rules and potentials with higher accuracy [1, 16].

1.3.3.2 Simulation of experimental preparation

The second group of approaches for generating an amorphous configuration is intimately related to numerical simulations, especially molecular dynamics. The general idea for these methods is to reproduce experimental processes leading to amorphization. For instance, a widely used technique for preparing models consists of quenching the semiconductor material from the liquid phase by doing molecular dynamics simulations. This tends to mimic the amorphous samples obtained using laser glazing techniques that involve freezing rates in the range 10^{10}–10^{12} K/s. Such cooling rates are indeed accessible from classical molecular dynamics simulations, in which the atomic interactions are modeled with empirical potentials [4, 9, 13]. Using more accurate interaction models, such as tight-binding and density functional theory, has also been shown to be possible, but at the expense of smaller sized models and higher quench rates [2, 12, 19]. Furthermore, the latter models are more prone to be less relaxed, and include more bond defects, or the presence of artifacts caused by significant self-interactions over the periodic boundaries.

In addition to quenching from a melt, it is also possible to design amorphous models using numerical simulations reproducing other experimental techniques. For instance, silicon amorphization by sputtering has also been achieved [20]. We are not aware of simulations where the formation of the disordered silicon relies on growth or indentation procedures, but there is no apparent reason why this should not be possible.

1.3.3.3 Experimental data fit strategy

Finally, it is also possible to build amorphous models using Reverse Monte Carlo (RMC) based techniques [5]. Here, the model is progressively optimized by fitting the model predictions toward experimental structural indicators like the RDF. Such approaches have been demonstrated to be sufficient to achieve models in agreement with experiments, especially when the RMC technique was combined with an energy minimization procedure to relax stresses [3].

1.4 Conclusion

In this chapter, the main structural quantities used for characterizing amorphous semiconductors were introduced. These quantities for, respectively, a crystal and a disordered/amorphous system were compared. The discussion and the listed examples are restricted to the case of amorphous silicon, which is probably the most studied tetrahedrally coordinated amorphous semiconductor. Then, current available models and associated generation methods were described.

At the present state of the art, there are amorphous models of excellent quality, either CRN-like or paracrystalline, which are able to reproduce most experimental findings. On the one hand, this success is may be the consequence of the limited amount of experimental data concerning the detailed structure of amorphous semiconductors. On the other hand, it is not obvious that there is a unique, well-defined, and general amorphous configuration. For instance, it is well known that different preparation methods can yield amorphous systems with different characteristics. Whether these configurations are closely connected states with different degrees of relaxation, or states residing in well-separated regions of configurational phase space, remains an unresolved issue.

On a separate note, it is also noteworthy that the available models aim at representing an idealized and "perfect" amorphous system. As a perfect crystal is a pure abstraction, it is likely that a disordered sample in reality includes a certain quantity of defects. Since short-range order is observed to be present in amorphous semiconductors, possible defects are vacancy-like or interstitial-like defects, which can be viewed as local density inhomogeneities, or bond defects that locally break the perfect coordination.

References

1. Barkema, G., and Mousseau, N. (2000). High-quality continuous random network, *Phys. Rev. B* **62**, 8, p. 4985.
2. Bernstein, N., Feldman, J., and Fornari, M. (2006). Structural model of amorphous silicon annealed with tight binding, *Phys. Rev. B* **74**, p. 205202.
3. Borisenko, K., Haberl, B., Liu, A., Chen, Y., Li, G., Williams, J., Baradby, J., Cockayne, D., and Tracy, M. (2012). Medium-range order in amorphous silicon investigated by constrained structural relaxation of two-body and four-body electron diffraction data, *Acta Mater.* **60**, p. 359.

4. Choudhary, D., and Clancy, P. (2005). Characterizing the nature of virtual amorphous silicon, *J. Chem. Phys.* **122**, p. 174509.
5. Cliffe, M. J., Dove, M. T., Drabold, D. A., and Goodwin, A. L. (2010). Structure determination of disordered materials from diffraction data, *Phys. Rev. Lett.* **104**, p. 125501.
6. Custer, J. S., Thompson, M. O., Jacobson, D. C., Poate, J. M., Roorda, S., Sinke, W. C., and Spaepen, F. (1994). Density of amorphous Si, *Appl. Phys. Lett.* **64**, 4, p. 437.
7. Fortner, J., and Lannin, J. (1989). Radial distribution functions of amorphous silicon, *Phys. Rev. B* **39**, 8, p. 5527.
8. Glover, C., Foran, G., and Ridgway, M. (2003). Structure of amorphous silicon investigated by EXAFS, *Nucl. Instrum. Methods Phys. Res. B* **199**, pp. 195–199.
9. Ishimaru, M. (2001). Molecular-dynamics study on atomistic structures of amorphous silicon, *J. Phys.: Condens. Matter* **13**, p. 4181.
10. Kail, F., Farjas, J., Roura, P., Secouard, C., Nos, O., Bertomeu, J., and Cabarrocas, P. R. (2011). The configurational energy gap between amorphous and crystalline silicon, *Phys. Stat. Sol. RRL* **5**, 10–11, p. 361.
11. Laaziri, K., Kycia, S., Roorda, S., Chicoine, M., Robertson, J. L., Wang, J. and Moss, S. C. (1999). High-energy x-ray diffraction study of pure amorphous silicon, *Phys. Rev. B* **60**, 19, p. 13520.
12. Lewis, L. J., and Mousseau, N. (1998). Tight-binding molecular-dynamics studies of defects and disorder in covalently bonded materials, *Comput. Mater. Sci.* **12**, p. 210.
13. Luedtke, W., and Landman, U. (1989). Preparation, structure, dynamics, and energetics of amorphous silicon: a molecular-dynamics study, *Phys. Rev. B* **40**, 2, p. 1164.
14. Marqués, L. A., Pelaz, L., López, P., Aboy, M., Santos, I., and Barbolla, J. (2005). Atomistic simulations in Si processing: Bridging the gap between atoms and experiments, *Mater. Sci. Eng. B* **124–125**, p. 72.
15. Moss, S. C., and Graczyk, J. F. (1969). Evidence of voids within the as-deposited structure of glassy silicon, *Phys. Rev. Lett.* **23**, pp. 1167–1171.
16. Nakhmanson, S. M., Voyles, P. M., Mousseau, N., Barkema, G. T., and Drabold, D. A. (2001). Realistic models of paracrystalline silicon, *Phys. Rev. B* **63**, p. 235207.
17. Polk, D. (1971). Structural model for amorphous silicon and germanium, *J. Non-Cryst. Solids* **5**, p. 365.

18. Roura, P., Farjas, J., and i Cabarrocas, P. R. (2008). Quantification of the bond-angle dispersion by raman spectroscopy and the strain energy of amorphous silicon, *J. Appl. Phys.* **104**, p. 073521.
19. Štich, I., Car, R., and Parrinello, M. (1991). Amorphous silicon studied by ab initio molecular dynamics: Preparation, structure, and properties, *Phys. Rev. B* **44**, 20, p. 11092.
20. Timonova, M., and Thijsse, B. J. (2011). Molecular dynamics simulations of the formation and crystallization of amorphous Si, *Comput. Mater. Sci.* **50**, p. 2380.
21. Treacy, M. M. J., and Borisenko, K. B. (2012). The local structure of amorphous silicon, *Science* **335**, p. 950.
22. Turnbull, D., and Polk, D. (1972). Structure of amorphous semiconductors, *J. Non-Cryst. Solids* **8–10**, p. 19.
23. Voyles, P. M., Gibson, J. M., and Treacy, M. M. J. (2000). Fluctuation microscopy: a probe of atomic correlations in disordered materials, *J. Electron Microsc.* **49**, 2, pp. 259–266.
24. Voyles, P. M., Zotov, N., Nakhmanson, S. M., Drabold, D. A., Gibson, J. M., Treacy, M. M. J., and Keblinski, P. (2001). Structure and physical properties of paracrystalline atomistic models of amorphous silicon, *J. Appl. Phys.* **90**, 9, p. 4437.
25. Wooten, F., Winer, K., and Weaire, D. (1985). Computer generation of structural models of amorphous Si and Ge, *Phys. Rev. Lett.* **54**, p. 1392.
26. Wright, A., and Thorpe, M. (2013). Eighty years of random networks, *Phys. Stat. Sol. (b)* **250**, 5, p. 931.

Chapter 2

Thermal Conductivity of Disordered Systems

Laurent Chaput
Université de Lorraine, LEMTA, Centre National de la Recherche Scientifique, Unité Mixte de Recherche 7563, BP 70239, 54506 Vandœuvre Cedex, France
laurent.chaput@univ-lorraine.fr

In this chapter a general theory is developed for the lattice thermal conductivity, valid for both crystals and disordered systems.

First, a hydrodynamical description of heat transport is given to define the wavevector and frequency-dependent thermal conductivity tensor $\kappa_{\alpha\gamma}(\boldsymbol{k}, \omega)$. The small wavevector and small frequency limit of $\kappa_{\alpha\gamma}(\boldsymbol{k}, \omega)$ is the uniform static thermal conductivity measured in the experiments, $\kappa_{\alpha\gamma}$. In Section 2.3 $\kappa_{\alpha\gamma}$ is expressed in term of the current–current correlation function, the so called Kubo formula.

To be able to compute the thermal conductivity for a particular system one need to express the correlation function in term of the microscopic quantities which characterize the system: atomic positions, force constants, etc. Therefore the lattice dynamics is derived in Section 2.4 and the general expression for the energy flux operator is obtained in Section 2.4.3. Collecting the results of the previous sections, the thermal conductivity is finally obtained in

Nanostructured Semiconductors: Amorphization and Thermal Properties
Edited by Konstantinos Termentzidis

ISBN 978-981-4745-64-2 (Hardcover), 978-1-315-36445-2 (eBook)
www.panstanford.com

Section 2.5.1 as a sum of the product of heat capacity and diffusivity associated to each vibrational modes.

The rest of the chapter is devoted to applications of this formula. In Section 2.5.2 it is shown to be equivalent to the Peierls–Boltzmann equation used in crystals. In Sections 2.5.3 and 2.5.4 it is applied to silicon systems with structural or configurational disorder. In particular, the mode diffusivity is used to establish a taxonomy of the vibrational modes according to their localized/delocalized and propagating/non-propagating characters.

The writing of this chapter has largely been inspired by the articles of Kadanoff and Martin, and Allen and Feldman,

- Hydrodynamic Equations and Correlation Functions (Kadanoff and Martin, 1963)
- Thermal Conductivity of Glasses: Theory and Application to Amorphous Si (Allen and Feldman, 1989)

Moreover, it is assumed that the reader is familiar with Quantum Mechanics and Statistical Physics at the level of (Landau and Lifshitz, 1977) and (Landau and Lifshitz, 2013), as well as with Solid State Physics (Marder, 2000).

2.1 The Hydrodynamic Limit

In the hydrodynamic limit the physical quantities describing a system are considered to vary slowly in time and space. This way each part of the system can be assumed to be in a local equilibrium which allows to define locally the thermodynamic variables (De Groot and Mazur, 2013). In the following we will denote by $A(\boldsymbol{r}, t)$ the value of the quantity A at position $\boldsymbol{r}$ and time t.

To study the energy and temperature propagations we will need to consider the energy density $h(\boldsymbol{r}, t)$ and the local temperature $T(\boldsymbol{r}, t)$. The definition of the local energy density is, in most cases, unambiguous since it is possible to write the hamiltonian of the system, H, as an integral of the energy density,

$$H = \int d^3r\, h(\boldsymbol{r}, t). \tag{2.1}$$

Any perturbation applied on the system will then produce a change in energy according to

$$\delta h(\mathbf{r}, t) = \langle h(\mathbf{r}, t)\rangle - \langle h(\mathbf{r}, t)\rangle_0, \tag{2.2}$$

where the bracket $\langle . \rangle$ represent the average in the non-equilibrium state, and $\langle . \rangle_0$ at equilibrium.

The local temperature, $T(\mathbf{r}, t)$, and the change with respect to the temperature T at equilibrium,

$$\delta T(\mathbf{r}, t) = T(\mathbf{r}, t) - T, \tag{2.3}$$

is more difficult to define since it is a statistical quantity. It hardly appears in the quantum mechanical dynamic of the system. Therefore we are left to define the local temperature from a generalization of the thermodynamical relation at constant pressure,

$$\delta h(\mathbf{r}, t) = C_p \delta T(\mathbf{r}, t), \tag{2.4}$$

where C_p is the specific heat per unit volume at constant pressure.

Once these two quantities have been defined the problem of temperature propagation may be solved.

First, it is assumed that an initial temperature disturbance

$$\delta T(\mathbf{r}, t = 0) = \delta T(\mathbf{r}) \tag{2.5}$$

is created. If this temperature change is small enough, it will give rise to an energy current which can be expressed as a linear combination of the disturbances at different point in space and time,

$$j_\alpha(\mathbf{r}, t) = -\sum_\gamma \int d^3r' \int dt' \, \kappa_{\alpha\gamma}(\mathbf{r} - \mathbf{r}', t - t') \frac{\partial}{\partial r'_\gamma} \delta T(\mathbf{r}', t'), \tag{2.6}$$

where j_α is the α-Cartesian component of the energy current, and the coefficient of the linear expansion are known as the components of the thermal conductivity tensor. In this formulation it has been assumed that the thermal conductivity only depends on the positions and times through the differences $\mathbf{r} - \mathbf{r}'$ and $t - t'$. It means that, at the scale of our hydrodynamical description, the system is assumed to be translationally invariant. Causality is also requested, which means that $\kappa_{\alpha\gamma}(\mathbf{r} - \mathbf{r}', t - t') = 0$ for $t < t'$.

Thermodynamical quantities being defined locally, it is possible to write the conservation of energy in differential form,

$$\frac{d}{dt} \int d^3r \, h(\mathbf{r}, t) = 0 \Longrightarrow \frac{\partial}{\partial t} h(\mathbf{r}, t) + \nabla \cdot \mathbf{j}(\mathbf{r}, t) = 0. \tag{2.7}$$

It is the equation which governs the relaxation of energy after the disturbance has been applied. Substituting Eq. 2.4 into Eq. 2.7 allow to determine the temperature propagation for the times subsequent the initial perturbation.

A useful representation of these equations is obtained in the Fourier-Laplace space $(\boldsymbol{k}, z)$. The Fourier transform with respect to space variables is defined by

$$A(\boldsymbol{k}, t) = \int d^3r\, e^{-i\boldsymbol{k}\cdot\boldsymbol{r}} A(\boldsymbol{r}, t), \tag{2.8}$$

whereas the Laplace transform with respect to the time variable is

$$A(\boldsymbol{k}, z) = \int_0^{+\infty} dt\, e^{izt} A(\boldsymbol{k}, t) \qquad \mathrm{Im}\, z > 0. \tag{2.9}$$

Therefore in the $(\boldsymbol{k}, z)$ space the thermodynamic description is given by the limits $\boldsymbol{k} \to 0$ and $z \to 0$.

Equations (2.4), (2.6), and (2.7) become

$$\delta h(\boldsymbol{k}, z) = C_p \delta T(\boldsymbol{k}, z) \tag{2.10}$$

$$j_\alpha(\boldsymbol{k}, z) = -i \sum_\gamma \kappa_{\alpha\gamma}(\boldsymbol{k}, z) k_\gamma \delta T(\boldsymbol{k}, z) \tag{2.11}$$

$$-iz\delta h(\boldsymbol{k}, z) - \delta h(\boldsymbol{k}, t = 0) + i \sum_\alpha k_\alpha j_\alpha(\boldsymbol{k}, z) = 0 \tag{2.12}$$

and therefore

$$\delta T(\boldsymbol{k}, z) = \frac{C_p}{\sum_{\alpha\gamma} k_\alpha \kappa_{\alpha\gamma}(\boldsymbol{k}, z) k_\gamma - izC_p} \delta T(\boldsymbol{k}, t = 0). \tag{2.13}$$

The above equation completely determines the temperature evolution for $t > 0$ but neither C_P nor $\kappa_{\alpha\gamma}$ can be obtained from the above hydrodynamical description. They must be obtained from a more microscopic theory which will be presented in the next section. As long as the same hypotheses are fulfilled, the results must coincide in the limits $\boldsymbol{k} \to 0$ and $z \to 0$.

2.2 Correlation Functions

In this section we would like to investigate the response of a system to a temperature gradient from a quantum mechanical point of view

(Luttinger, 1964). The problem is very similar to the study of the electrical conductivity where the system response to an electric field is considered. However, as mentioned previously, the temperature gradient, unlike the electric field, is a statistical quantity whose effects are difficult to include in the dynamics.

Fortunately, as a result of Eq. 2.7 which is first order in time, Eq. 2.13 shows that the temperature evolution is completely determined by the initial energy distribution,

$$\delta h(\boldsymbol{k}, t = 0) = C_p \delta T(\boldsymbol{k}, t = 0),$$

and does not depend on the way it has been created. This lead Luttinger to consider a gravitational potential $\phi(\boldsymbol{r})$, instead of a temperature gradient, as a direct interaction capable of creating the initial energy distribution. This way the dynamics of the system can be studied as the response to the mechanical perturbation

$$\delta H = -\int d^3r\, h(\boldsymbol{r})\phi(\boldsymbol{r}, t). \tag{2.14}$$

Of course, the strength of the gravitational potential corresponding to the initial temperature variation $T(\boldsymbol{k}, t = 0)$ as yet to be determined. This will be done at the end of the section using a very fundamental theorem of physics, the fluctuation dissipation theorem (Callen and Welton, 1951).

The linear response theory (Kubo, 1957) can be used to obtain the change in the average value of any observable $A(\boldsymbol{r}, t)$ after a perturbation is applied. Following (Kadanoff and Martin, 1963), the theory is summarized in Appendix A. In particular, from Eq. A.17 we have

$$\delta A(t) = -\frac{i}{\hbar}\int_{-\infty}^{t} dt' \,\langle [A^I(t), \delta H^I(t')]\rangle_0. \tag{2.15}$$

Equation 2.14 can be seen as a sum of perturbations at different points in space. Therefore, because within linear response the different perturbations can be considered to be independent, we obtain

$$\delta A(\boldsymbol{r}, t) = \frac{i}{\hbar}\int d^3r' \int_{-\infty}^{t} dt' \langle [A^I(\boldsymbol{r}, t), h^I(\boldsymbol{r}', t')]\rangle_0 \phi(\boldsymbol{r}', t'). \tag{2.16}$$

We consider that the gravitational potential is turned on adiabatically from the infinite past to setup the energy distribution at $t = 0$

and then turned off,

$$\phi(\mathbf{r}, t) = e^{\eta t}\phi(\mathbf{r}) \qquad \text{with } \eta \to 0^+ \qquad \text{for } t < 0 \tag{2.17}$$

$$= 0 \qquad \text{for } t > 0. \tag{2.18}$$

This specific form of the interaction guaranties the above integral to converge. Using a Fourier representation for the average value of the commutator,

$$\frac{1}{2\hbar}\langle[A^I(\mathbf{r}, t), h^I(\mathbf{r}', t')]\rangle_0 = \int \frac{d\omega}{2\pi} e^{-i\omega(t-t')}\chi''_{Ah}(\mathbf{r}, \mathbf{r}', \omega), \tag{2.19}$$

we obtain

$$\delta A(\mathbf{r}, t) = i\int d^3r' \int_{-\infty}^{t} dt' \int \frac{d\omega}{\pi}\chi''_{Ah}(\mathbf{r}, \mathbf{r}', \omega)e^{-i\omega(t-t')}e^{\eta t'}\phi(\mathbf{r}') \tag{2.20}$$

for $t < 0$, and

$$\delta A(\mathbf{r}, t) = i\int d^3r' \int_{-\infty}^{0} dt \int \frac{d\omega}{\pi}\chi''_{Ah}(\mathbf{r}, \mathbf{r}', \omega)e^{-i\omega(t-t')}e^{\eta t'}\phi(\mathbf{r}') \tag{2.21}$$

for $t > 0$. It gives

$$\begin{aligned}\delta A(\mathbf{r}, t) &= \int d^3r' \int \frac{d\omega}{\pi}\chi''_{Ah}(\mathbf{r}, \mathbf{r}', \omega)\frac{e^{\eta t}}{\omega - i\eta}\phi(\mathbf{r}') \qquad \text{for } t < 0 \\ &= \int d^3r' \int \frac{d\omega}{\pi}\chi''_{Ah}(\mathbf{r}, \mathbf{r}', \omega)\frac{e^{-i\omega t}}{\omega - i\eta}\phi(\mathbf{r}') \qquad \text{for } t > 0.\end{aligned} \tag{2.22}$$

Using the analytical continuation of $\chi''_{Ah}(\mathbf{r}, \mathbf{r}', \omega)$ in the complex plane,

$$\chi_{Ah}(\mathbf{r}, \mathbf{r}', z) = \int \frac{d\omega}{\pi}\frac{\chi''_{Ah}(\mathbf{r}, \mathbf{r}', \omega)}{\omega - z},$$

the previous equation can be rewritten as

$$\delta A(\mathbf{r}, t) = e^{\eta t}\int d^3r' \chi_{Ah}(\mathbf{r}, \mathbf{r}', i\eta)\phi(\mathbf{r}'), \qquad \text{for } t < 0$$

whereas the Laplace transform of $\delta A(\mathbf{r}, t)$ for $t > 0$ gives

$$\delta A(\mathbf{r}, z) = \int d^3r' \frac{\chi_{Ah}(\mathbf{r}, \mathbf{r}', z) - \chi_{Ah}(\mathbf{r}, \mathbf{r}', i\eta)}{iz}\phi(\mathbf{r}') \quad \text{for Im } z > 0.$$

In Fourier space the result is simply

$$\delta A(\boldsymbol{k}, z) = \frac{\chi_{Ah}(\boldsymbol{k}, z) - \chi_{Ah}(\boldsymbol{k}, i\eta)}{iz}\phi(\boldsymbol{k}) \qquad \text{for Im } z > 0 \tag{2.23}$$

$$\delta A(\boldsymbol{k}, t) = e^{\eta t}\chi_{Ah}(\boldsymbol{k}, i\eta)\phi(\boldsymbol{k}) \qquad \text{for } t < 0. \tag{2.24}$$

The first equation shows that $\chi_{Ah}(\boldsymbol{k}, z)$ control the system response for $t > 0$. The second shows that in the limit $\boldsymbol{k} \to 0$, $\chi_{Ah}(\boldsymbol{k}, i\eta)$ is a thermodynamic derivative. This is proved in the following using the fluctuation-dissipation theorem (see Appendix B and (Kadanoff and Martin, 1963)).

First, from Eq. A.33, we notice that $\chi''_{Ah}(\boldsymbol{k}, 0) = 0$ and therefore

$$\chi_{Ah}(\boldsymbol{k}, i\eta) = P\int \frac{d\omega}{\pi}\frac{\chi''_{Ah}(\boldsymbol{k}, 0)}{\omega}. \tag{2.25}$$

In Eq. 2.19, χ''_{Ah} is defined as the Fourier transform of the average value of the commutator between A and h, at different points in space and time. In the same way, the Fourier transform c_{Ah} of the average of the anticommutator can be defined, and in Appendix B both are shown to be related by the equation

$$c_{Ah}(\boldsymbol{r}, \boldsymbol{r}', \omega) = \chi''_{Ah}(\boldsymbol{r}, \boldsymbol{r}', \omega)\coth\frac{\hbar\omega\beta}{2}. \tag{2.26}$$

This is the fluctuation-dissipation theorem which relates the thermal fluctuations of the system at equilibrium, c_{Ah}, to the response when a perturbation is applied, χ''_{Ah}. A famous example of this relation is the Johnson–Nyquist theorem, which, in a electrical circuit, relates the mean square voltage to the resistance.

The long wavelength value of the fluctuation function is independent of time because $H_0 = \int d^3r h(\boldsymbol{r})$ is a conserved quantity,

$$c_{Ah}(\boldsymbol{k} = 0, t) = \frac{1}{2\hbar}\int d^3r\langle\{\delta A, \delta h(-\boldsymbol{r}, -t)\}\rangle_0 \tag{2.27}$$

$$= \frac{1}{2\hbar}\langle\{\delta A, \int d^3r\delta h(-\boldsymbol{r}, -t)\}\rangle_0. \tag{2.28}$$

This shows that in Fourier space $c_{Ah}(\boldsymbol{k} = 0, t)$ has only a low frequency component $\delta(\omega)$. If we assume this remains true for small but finite $\boldsymbol{k}$ then we can write

$$c_{Ah}(\boldsymbol{k}, \omega) = \chi''_{Ah}(\boldsymbol{k}, \omega)\coth\frac{\hbar\omega\beta}{2} = \frac{2}{\hbar\omega\beta}\chi''_{Ah}(\boldsymbol{k}, \omega) \tag{2.29}$$

and therefore from Eq. 2.25

$$\chi_{Ah}(\boldsymbol{k}, i\eta) = \hbar\beta c_{Ah}(\boldsymbol{k}, t=0). \tag{2.30}$$

The $\boldsymbol{k} \to 0$ limit is then easily written as a thermodynamical derivative,

$$\lim_{\boldsymbol{k}\to 0} \chi_{Ah}(\boldsymbol{k}, i\eta) = \frac{\beta}{2}\int d^3r \langle\{\delta A, \delta h(-\boldsymbol{r})\}\rangle_0 \tag{2.31}$$

$$= \beta\langle\delta A \delta H_0\rangle_0 \tag{2.32}$$

$$= T\frac{\partial\langle A\rangle_0}{\partial T}. \tag{2.33}$$

In the last line we have used the identity $\frac{\partial\langle A\rangle_0}{\partial\beta} = -\langle\delta A\delta H_0\rangle_0$ which can be checked by direct differentiation.

In the special case $A = h$ we obtain

$$\lim_{\boldsymbol{k}\to 0} \chi_{hh}(\boldsymbol{k}, i\eta) = T\frac{\partial\langle h\rangle_0}{\partial T} = TC_p. \tag{2.34}$$

Actually this limit value allows to express the initial temperature distribution in term of the Luttinger gravitational potential. From Eq. 2.24, at small $\boldsymbol{k}$, we obtain

$$\delta h(\boldsymbol{k}, t=0) = TC_p\phi(\boldsymbol{k}), \tag{2.35}$$

which is then compared to the defining equation for the temperature, Eq. 2.4,

$$\delta h(\boldsymbol{k}, t=0) = C_p\delta T(\boldsymbol{k}, t=0), \tag{2.36}$$

to obtain

$$\delta T(\boldsymbol{k}, t=0) = T\phi(\boldsymbol{k}). \tag{2.37}$$

Using Eq. 2.23, the temperature evolution can finally be written in a form analogous to the hydrodynamical equation, Eq. 2.13,

$$\delta T(\boldsymbol{k}, z) = \frac{\chi_{hh}(\boldsymbol{k}, z) - \chi_{hh}(\boldsymbol{k}, i\eta)}{izC_pT}\delta T(\boldsymbol{k}, t=0) \quad \text{Im}\, z > 0. \tag{2.38}$$

2.3 Kubo Formula

A comparison of Eqs. 2.13 and 2.38 gives

$$\sum_{\alpha\gamma} k_\alpha \kappa_{\alpha\gamma}(\boldsymbol{k}, \omega + i\eta)k_\gamma = \frac{i(\omega + i\eta)C_p^2 T}{\chi_{hh}(\boldsymbol{k}, \omega + i\eta) - \chi_{hh}(\boldsymbol{k}, i\eta)} + i(\omega + i\eta)C_p. \tag{2.39}$$

To solve the above equation for the thermal conductivity we can consider the continuity equation, Eq. 2.7, and take its commutator with $h(\boldsymbol{r}', 0)$ and $j_\alpha(\boldsymbol{r}', 0)$ to obtain

$$-i(\omega + i\eta)\chi_{hh}(\boldsymbol{k}, \omega + i\eta) + i\sum_\alpha k_\alpha \chi_{j_\alpha h}(\boldsymbol{k}, \omega + i\eta) = 0 \tag{2.40}$$

$$-i(\omega + i\eta)\chi_{j_\alpha h}(\boldsymbol{k}, \omega + i\eta) + i\sum_\gamma \chi_{j_\alpha j_\gamma}(\boldsymbol{k}, \omega + i\eta)k_\gamma = 0 \tag{2.41}$$

The initial conditions are zero if $\langle[h(\boldsymbol{r}), h(\boldsymbol{r}')]\rangle_0 = 0$. This gives the energy–energy response function in term of the current–current response function,

$$(\omega + i\eta)^2 \chi_{hh}(\boldsymbol{k}, \omega + i\eta) = \sum_{\alpha\gamma} k_\alpha \chi_{j_\alpha j_\gamma}(\boldsymbol{k}, \omega + i\eta)k_\gamma \tag{2.42}$$

which, for small $\boldsymbol{k}$, allow to express the thermal conductivity as

$$\kappa_{\alpha\gamma}(\boldsymbol{k}, \omega + i\eta) = \frac{iC_p}{\omega + i\eta} \frac{\chi_{j_\alpha j_\gamma}(\boldsymbol{k}, \omega + i\eta)}{\chi_{hh}(\boldsymbol{k}, \omega + i\eta) - \chi_{hh}(\boldsymbol{k}, i\eta)}. \tag{2.43}$$

After substituting Eq. 2.42, into Eq. 2.39 we have used Eq. 2.34 to simplify the result (Sham, 1967).

The static and uniform thermal conductivity $\kappa_{\alpha\gamma}$ obtain from the experiments correspond to the $\boldsymbol{k} \to 0$ and $\omega \to 0$ limits. However, one has to specify the order in which those limits are taken since they correspond to very different physical situations. That the order of the limits matters can be seen from the continuity equation or Eq. 2.13 which distinguish two regimes. In the slow regime, $\eta \ll \boldsymbol{k}_\alpha \kappa_{\alpha\gamma} \boldsymbol{k}_\gamma / C_p$, the gravitational potential is turned on so slowly that the system is kept in local equilibrium and no current can flow. This is the *thermodynamic* limit in which we have obtained Eq. 2.34. When this result is used in the above continuity equation it gives

$$\lim_{\boldsymbol{k}\to 0} \chi_{j_\alpha h}(\boldsymbol{k}, i\eta) = 0 \tag{2.44}$$

which by Eq. 2.24 shows that the system is prepared in a state with no initial current.

In the rapid limit, $\eta \gg \boldsymbol{k}_\alpha \kappa_{\alpha\gamma} \boldsymbol{k}_\gamma / C_p$, in response to the fast perturbation, currents are created within the system to bring it back to equilibrium. This is the *transport* limit. Therefore one has

$$\kappa_{\alpha\gamma} = \lim_{\omega\to 0}\lim_{\boldsymbol{k}\to 0} \kappa_{\alpha\gamma}(\boldsymbol{k}, \omega + i\eta) \tag{2.45}$$

This limit is easily computed using Eq. 2.34 and

$$\lim_{\omega\to 0}\lim_{\boldsymbol{k}\to 0} \chi_{hh}(\boldsymbol{k}, \omega + i\eta) = 0 \tag{2.46}$$

which is obtained by a direct computation using that $\int d^3 r h(\boldsymbol{r})$ is conserved. We obtain

$$\kappa_{\alpha\gamma} = \lim_{\omega\to 0}\lim_{\boldsymbol{k}\to 0} \frac{1}{i\omega T} \chi_{j_\alpha j_\gamma}(\boldsymbol{k}, \omega + i\eta). \tag{2.47}$$

This last formula is known as the Kubo formula for the thermal conductivity (Kubo et al., 1957).

It is usually rewritten in a form which allow for a more straightforward evaluation. The first step is to substitute the average current operator,

$$\boldsymbol{J} = \frac{1}{V}\int d^3 r\, \boldsymbol{j}(\boldsymbol{r}) \tag{2.48}$$

for the current density in the previous equation. Then for the frequency-dependent thermal conductivity,

$$\kappa_{\alpha\gamma}(\omega + i\eta) \equiv \kappa_{\alpha\gamma}(\boldsymbol{k} = 0, \omega + i\eta),$$

we obtain

$$\kappa_{\alpha\gamma}(\omega + i\eta) = \frac{V}{\hbar\omega T}\int_0^\infty e^{i(\omega+i\eta)t}\langle [J_\alpha(t) J_\gamma]\rangle_0. \tag{2.49}$$

The time reversal symmetry gives

$$\langle J_\alpha(t) J_\gamma\rangle_0 = \langle (J_\alpha(-t) J_\gamma)^+\rangle_0 = \langle J_\gamma(t) J_\alpha\rangle_0 \tag{2.50}$$

and therefore

$$\kappa_{\alpha\gamma}(\boldsymbol{r}, \omega + i\eta) = \frac{V}{\hbar\omega T}\int_0^\infty e^{i(\omega+i\eta)t}\langle [J_\alpha(-t) j_\gamma]\rangle_0. \tag{2.51}$$

The last step is to removed the commutator. Using the density matrix $\rho_0 = \exp(-\beta H_0)/\mathrm{Tr}\exp(-\beta H_0)$, the average can be rewritten as

$$\langle [J_\alpha(-t) J_\gamma]\rangle_0 = \mathrm{Tr}[\rho_0, J_\alpha(-t)] J_\gamma \tag{2.52}$$

and a direct computation shows that the commutator can be expressed as (Mahan, 2000)

$$i\hbar\rho_0 \int_0^{\beta} d\lambda \frac{\partial}{\partial t} J_\alpha(-t - i\hbar\lambda) = [\rho_0, J_\alpha(-t)]. \tag{2.53}$$

This gives

$$\kappa_{\alpha\gamma}(\omega + i\eta) = \frac{V}{T} \int_0^{\beta} d\lambda \int_0^{+\infty} dt e^{i(\omega+i\eta)t} \langle J_\alpha(-i\hbar\lambda) J_\gamma(t)\rangle_0. \tag{2.54}$$

Avoiding the use of the commutator in the thermal conductivity imply to have one more integration to perform. But, as we will see, it is practically useful. Also, this allows to obtain the classical limit simply by considering $\hbar \to 0$,

$$\kappa^{cl}_{\alpha\gamma}(\omega + i\eta) = \frac{V}{k_B T^2} \int_0^{+\infty} dt e^{i(\omega+i\eta)t} \langle J_\alpha(0) J_\gamma(t)\rangle. \tag{2.55}$$

It is the formula that will be used to obtain the thermal conductivity from force fields in the chapter about classical molecular dynamics simulations.

2.4 Microscopic Model: The Harmonic Solid

The hamiltonian for the atoms in a solid can be written as (Marder, 2000)

$$H = \sum_{I\alpha} \frac{p_{I\alpha}^2}{2m_{I\alpha}} + E(\{R_{I\alpha}\}) \tag{2.56}$$

with $R_{I\alpha}$ and $p_{I\alpha}$ the Cartesian components of the position and momentum operators for atom I, with a mass m_I. $E(\{R_{I\alpha}\})$ is the Born–Oppenheimer potential energy of the system. This function is assumed to be analytic and therefore can be expanded with respect to the atom displacements, $\delta R_{I\alpha}$, around their equilibrium position $R^0_{I\alpha}$,

$$R_{I\alpha} = R^0_{I\alpha} + \delta R_{I\alpha}. \tag{2.57}$$

In the harmonic approximation, the expansion is truncated to the second order. Using for the force constants the notation

$$\Phi_{I\alpha,J\gamma} = \frac{\partial^2 E}{\partial R^0_{I\alpha} \partial R^0_{J\gamma}}, \tag{2.58}$$

we obtain

$$H^0 = \sum_{I\alpha} \frac{p_{I\alpha}^2}{2m_{I\alpha}} + \frac{1}{2} \sum_{\substack{IJ \\ \alpha\gamma}} \delta R_{I\alpha} \Phi_{I\alpha,J\gamma} \delta R_{J\gamma}. \tag{2.59}$$

We have used that the first-order derivatives of the energy are zero at equilibrium, and chosen the value of the energy function at equilibrium, $E(\{R_{I\alpha}^0\})$, to be the origin of energies.

The time evolution of the atomic displacements are obtained from the Heisenberg equations of motion,

$$i\hbar \frac{d}{dt} \delta R_{I\alpha}(t) = [\delta R_{I\alpha}(t), H^0]. \tag{2.60}$$

Using the canonical commutation relations between positions and momentums, $[R_{I\alpha}, p_{J\gamma}] = i\hbar \delta_{IJ} \delta_{\alpha\gamma}$, we obtain

$$\frac{d}{dt} \delta R_{I\alpha}(t) = \frac{p_{I\alpha}}{m_\alpha}. \tag{2.61}$$

Taking the time derivative once more,

$$\frac{d^2}{dt^2} \delta R_{I\alpha}(t) = \frac{1}{i\hbar} \left[\frac{p_{I\alpha}}{m_\alpha}, H^0 \right] \tag{2.62}$$

$$= \frac{1}{i\hbar} \left[\frac{p_{I\alpha}}{m_\alpha}, \frac{1}{2} \sum_{\substack{I'J' \\ \alpha'\gamma'}} \delta R_{I'\alpha'} \Phi_{I'\alpha',J'\gamma'} \delta R_{J'\gamma'} \right] \tag{2.63}$$

it gives finally

$$\frac{d^2}{dt^2} \delta R_{I\alpha}(t) = - \sum_{J\gamma} \Phi_{I\alpha,J\gamma} \delta R_{J\gamma}(t). \tag{2.64}$$

In the harmonic approximation the Heisenberg equation of motion is therefore equivalent to the Newton equation of motion of atoms connected by springs of strength $\Phi_{I\alpha,J\gamma}$. It is not very surprising in regards of the Ehrenfest theorem for quadratic potentials.

2.4.1 Dynamical Matrices

If we look for harmonic solutions to the previous equations (Horton and Maradudin, 1974),

$$\delta R_{I\alpha}(t) = x_{I\alpha} e^{-i\omega t}, \tag{2.65}$$

the frequency ω is found to fulfill the equation,

$$\omega^2 m_I x_{I\alpha} = \sum_{J\gamma} \Phi_{I\alpha,J\gamma} x_{J\gamma}. \tag{2.66}$$

Defining the reduced coordinates and the dynamical matrix by

$$u_{I\alpha} = \sqrt{m_I} x_{I\alpha} \tag{2.67}$$

$$D_{I\alpha,J\gamma} = \frac{\Phi_{I\alpha,J\gamma}}{\sqrt{m_I}\sqrt{m_J}} \tag{2.68}$$

the above equation is rewritten as

$$\sum_{J\gamma} D_{I\alpha,J\gamma} u_{J\gamma} = \omega^2 u_{I\alpha}. \tag{2.69}$$

This is the eigenvalue equation of the dynamical matrix. The solutions are the allowed frequencies, denoted as $\omega^2(m)$, and the associated displacements, $u_{I\alpha}(m)$, or just $|m\rangle$ using the Dirac notation.

If there are $\mathcal{N}$ atoms in the solid the dynamical matrix has dimensions $3\mathcal{N} \times 3\mathcal{N}$. Moreover, it is real and symmetric therefore its $3\mathcal{N}$ eigenvalues are all real,

$$\omega^2(m) \in \mathbb{R} \quad m = 1 \cdots 3\mathcal{N}, \tag{2.70}$$

and its eigenvectors can be chosen to be orthonormal,

$$\langle m|m\rangle \equiv \sum_{I\alpha} \bar{u}_{I\alpha}(m) u_{I\alpha}(m') = \delta_{mm'} \quad m, m' = 1 \cdots 3\mathcal{N}. \tag{2.71}$$

The eigendisplacements being assumed to be complete they fulfill the closure relation,

$$\delta_{IJ}\,\delta_{\alpha\gamma} = \sum_m u_{I\alpha}(m)\bar{u}_{J\gamma}(m), \tag{2.72}$$

or simply

$$1 = \sum_m |m\rangle\langle m|. \tag{2.73}$$

The eigendisplacements can be chosen to be real, but they don't need to. In particular, for a crystal, the translation operator commutes with the dynamical matrix and it is preferred to chose the eigenvectors to be common to both those operators to fulfill the Bloch theorem (Ziman, 1960),

$$u_{l\tau\alpha}(\boldsymbol{k}j) = \frac{1}{\sqrt{N}} e_{\tau\alpha}(\boldsymbol{k}j) e^{i\boldsymbol{k}\cdot\boldsymbol{R}^0_{l\tau}}. \tag{2.74}$$

In the above equation the atoms I have been numbered using a composite index, $I = (l\tau)$, where l is used to label the crystal cells, $l = 1\cdots N$, and τ the atoms within each such a cell, $\tau = 1\cdots N_a$. Clearly $\mathcal{N} = N_a N$. The wavevectors $\boldsymbol{k}$ label the eigenvalues of the translation operators, therefore instead of m they are used to label the eigenvectors of the dynamical matrix together with a band index j. Because there are N wavevectors inside the first Brillouin zone, $j = 1\cdots 3N_a$. The same labeling is used for the eigenvalues, $\omega^2(\boldsymbol{k}j)$.

In Eq. 2.74 $e_{\tau\alpha}(\boldsymbol{k}j)$ are polarization vectors which have yet to be determined. This is done by substituting Eq. 2.74 into Eq. 2.69:

$$\sum_{l'\tau'\alpha'} D_{l\tau\alpha,l'\tau'\alpha'} e^{-i\boldsymbol{k}\cdot(\boldsymbol{R}^0_{l\tau}-\boldsymbol{R}^0_{l'\tau'})} e_{\tau'\alpha'}(\boldsymbol{k}j) = \omega^2(\boldsymbol{k}j) e_{\tau\alpha}(\boldsymbol{k}j) \tag{2.75}$$

The translational invariance of the dynamical matrix is implemented as

$$D_{l\tau\alpha,l'\tau'\alpha'} = D_{l-l'\tau\alpha,0\tau'\alpha'}, \tag{2.76}$$

which allows to define the reciprocal space dynamical matrix

$$C_{\tau\alpha,\tau'\alpha'}(\boldsymbol{k}j) = \sum_{l'} D_{l\tau\alpha,l'\tau'\alpha'} e^{-i\boldsymbol{k}\cdot(\boldsymbol{R}^0_{l\tau}-\boldsymbol{R}^0_{l'\tau'})} \tag{2.77}$$

$$= \sum_{l'} D_{l-l'\tau\alpha,0\tau'\alpha'} e^{-i\boldsymbol{k}\cdot(\boldsymbol{R}^0_{l\tau}-\boldsymbol{R}^0_{l'\tau'})}. \tag{2.78}$$

This gives

$$\sum_{\tau'\alpha'} C_{\tau\alpha,\tau'\alpha'}(\boldsymbol{k}j) e_{\tau'\alpha'}(\boldsymbol{k}j) = \omega^2(\boldsymbol{k}j) e_{\tau\alpha}(\boldsymbol{k}j). \tag{2.79}$$

The polarizations are therefore the eigenvectors of the reciprocal space dynamical matrix. $C(\boldsymbol{k}j)$ being hermitian, as before its eigenvalues are real and the eigenvectors fulfill the orthonormal and closure relations. However, because $C(-\boldsymbol{k}j) = \bar{C}(\boldsymbol{k}j)$ they cannot be chosen to be real. Rather this gives

$$\omega^2(-\boldsymbol{k}j) = \omega^2(\boldsymbol{k}j) \tag{2.80}$$

$$e_{\tau\alpha}(-\boldsymbol{k}j) = \bar{e}_{\tau\alpha}(\boldsymbol{k}j). \tag{2.81}$$

Although the algebra used for a crystal and for an arbitrary solid seems rather different, they can be treated on a common basis by introducing a convenient notation.

First, we will use $m = \boldsymbol{k}j$ and $I = (l\tau)$ in the special case of a crystal. The complex conjugate of the eigenvectors is

$$\bar{u}_{I\alpha}(m) = u_{I\alpha}(m) \qquad \text{in general} \tag{2.82}$$

but

$$\bar{u}_{I\alpha}(m) = u_{I\alpha}(-m) \qquad \text{for a crystal} \tag{2.83}$$

if we use $-m \equiv -\boldsymbol{k}j$.

This can be summarized as

$$\bar{u}_{I\alpha}(m) = u_{I\alpha}(\bar{m}) \tag{2.84}$$

if we define

$$\bar{m} = m \qquad \text{in general} \tag{2.85}$$

$$\bar{m} = -m \qquad \text{for a crystal.} \tag{2.86}$$

Then the orthonormality and closure relations are

$$\sum_{I\alpha} u_{I\alpha}(\bar{m})u_{I\alpha}(m') = \delta_{mm'} \tag{2.87}$$

$$\sum_{m} u_{I\alpha}(m)u_{J\gamma}(\bar{m}) = \delta_{IJ}\,\delta_{\alpha\gamma}. \tag{2.88}$$

2.4.2 Harmonic Oscillators

To the eigenvectors of the previous section is associated a new coordinates system, Q_m, such that

$$Q_m = \sum_{I\alpha} u_{I\alpha}(\bar{m})u_{I\alpha}. \tag{2.89}$$

The orthogonality of the eigendisplacements gives the reciprocal relation

$$u_{I\alpha} = \sum_{m} u_{I\alpha}(m)Q_m \tag{2.90}$$

and it is easy to check that the hermitian conjugate is

$$\left(Q_m\right)^{+} = Q_{\bar{m}}. \tag{2.91}$$

By construction the new coordinates system diagonalize the potential energy,

$$\frac{1}{2}\sum_{\substack{IJ\\ \alpha\gamma}} u_{I\alpha} D_{I\alpha,J\gamma} u_{J\gamma} = \frac{1}{2}\sum_{\substack{IJ\\ \alpha\gamma}} u_{I\alpha}\Big(\sum_m u_{I\alpha}(m)\omega^2(m)u_{J\gamma}(\bar{m})\Big)u_{J\gamma}$$

$$= \frac{1}{2}\sum_m \omega^2(m) Q_{\bar{m}} Q_m, \tag{2.92}$$

but it also diagonalizes the kinetic energy. Indeed, from Eq. 2.89 we have the relations

$$-i\hbar\frac{\partial}{\partial Q_m} = \sum_{I\alpha} u_{I\alpha}(m)\Big(-i\hbar\frac{\partial}{\partial u_{I\alpha}}\Big) \tag{2.93}$$

$$-i\hbar\frac{\partial}{\partial u_{I\alpha}} = \sum_{m} u_{I\alpha}(\bar{m})\Big(-i\hbar\frac{\partial}{\partial Q_m}\Big) \tag{2.94}$$

$$\Big(-i\hbar\frac{\partial}{\partial Q_m}\Big)^+ = -i\hbar\frac{\partial}{\partial Q_{\bar{m}}}, \tag{2.95}$$

which gives

$$\sum_{I\alpha}\frac{p_{I\alpha}^2}{2m_{I\alpha}} = -\frac{\hbar^2}{2}\sum_m \frac{\partial}{\partial Q_{\bar{m}}}\frac{\partial}{\partial Q_m}. \tag{2.96}$$

The harmonic hamiltonian is therefore written as

$$H^0 = \sum_m -\frac{\hbar^2}{2}\frac{\partial}{\partial Q_{\bar{m}}}\frac{\partial}{\partial Q_m} + \frac{1}{2}\omega^2(m) Q_{\bar{m}} Q_m. \tag{2.97}$$

The form we obtained it very similar to the harmonic oscillator problem[a] (Messiah, 1961). This suggests to introduce the Dirac ladder operators,

$$a_m = \sqrt{\frac{\omega_m}{2\hbar}}\left(Q_m + \frac{i}{\omega_m}\left(-i\hbar\frac{\partial}{\partial Q_{\bar{m}}}\right)\right) \tag{2.98}$$

$$a_m^+ = \sqrt{\frac{\omega_m}{2\hbar}}\left(Q_{\bar{m}} - \frac{i}{\omega_m}\left(-i\hbar\frac{\partial}{\partial Q_m}\right)\right), \tag{2.99}$$

which fulfill the canonical commutation relations,

$$[a_m, a_{m'}] = [a_m^+, a_{m'}^+] = 0, \tag{2.100}$$

$$[a_m, a_{m'}^+] = \delta_{mm'}. \tag{2.101}$$

[a] Indeed, combining the Q_n to obtain *real* mode coordinates it can be put to this form exactly.

Substituting

$$Q_m = \sqrt{\frac{\hbar}{2\omega_m}}(a_m + a_{\bar{m}}^+) \tag{2.102}$$

$$-i\hbar\frac{\partial}{\partial Q_m} = -i\sqrt{\frac{\hbar\omega_m}{2}}(a_{\bar{m}} - a_m^+) \tag{2.103}$$

into H^0, we obtain

$$H^0 = \sum_m \hbar\omega_m \left(a_m^+ a_m + \frac{1}{2}\right). \tag{2.104}$$

The last part of the above equation,

$$E_{ZP} = \sum_m \frac{\hbar\omega_m}{2}, \tag{2.105}$$

is known as the zero point energy. By virtue of the Heisenberg uncertainty principle, both position and momentum cannot be determined simultaneously. As a result, even in the ground state of the system, even at zero temperature, they fluctuate. The energy corresponding to this motion is the zero point energy. In some way, it is a measure of the quantumicity of the system.

The ladder operators a_m^+ and a_m creates or destroys a lattice wave of frequency ω_m, the so called *1-particle* excitations of the system (Horton and Maradudin, 1974). Indeed, those excitations, according to the particle-wave duality principle of quantum mechanics, can conveniently be seen as particles, called *vibrons* in general, and *phonon* in the case of a crystal. From Eq. 2.104, the total hamiltonian H^0 is a sum of independent harmonic oscillators hamiltonians, therefore those particles are independent and the total wavefunction can be written as a product of the wavefunctions of the individual harmonic oscillators. It means that the excited vibrational eigenstates of the system are simply obtained by creating an arbitrary number of vibrons in each of the displacement states m. The number of vibrons in one displacement state is called the *occupation*, n_m, and the set of occupations over all displacement states, $\{n_m, m = 1 \cdots 3\mathcal{N}\}$, completely defined the vibrational state of the system.

The vibrons being independent, the vibrational energy is given by the sum of the energy of each vibron,

$$E_{\{n_m\}} = \sum_m n_m \hbar\omega_m + E_{ZP}. \tag{2.106}$$

Once the excitation energies $E_{\{n_m\}}$ are known, the thermodynamical properties can be computed.

The canonical partition function is

$$Z = \sum_{\{n_m\}} e^{-\beta E_{\{n_m\}}} = \sum_{n_1=0}^{\infty} \cdots \sum_{n_{3\mathcal{N}}=0}^{\infty} e^{-\beta \sum_m (n_m + \frac{1}{2})\hbar\omega_m} \tag{2.107}$$

$$= \prod_m \sum_{n=0}^{\infty} e^{-\beta(n+\frac{1}{2})\hbar\omega_m} \tag{2.108}$$

$$= \prod_m \frac{e^{-\frac{\beta\hbar\omega_m}{2}}}{1 - e^{-\beta\hbar\omega_m}}. \tag{2.109}$$

Therefore we obtain the free energy

$$F = -k_B T \ln Z = k_B T \sum_m \ln\left(2\sinh\frac{\hbar\omega_m}{2k_B T}\right) \tag{2.110}$$

the entropy

$$S = -\frac{\partial F}{\partial T}\bigg|_V = k_B \sum_m \frac{\hbar\omega_m}{2k_B T}\coth\frac{\hbar\omega_m}{2k_B T} - k_B \sum_m \ln\left(2\sinh\frac{\hbar\omega_m}{2k_B T}\right) \tag{2.111}$$

the internal energy

$$E = F + TS = k_B T \sum_m \frac{\hbar\omega_m}{2k_B T}\coth\frac{\hbar\omega_m}{2k_B T} \tag{2.112}$$

and the heat capacity at constant volume

$$C_v = \frac{\partial E}{\partial T}\bigg|_V = k_B \sum_m \left(\frac{\hbar\omega_m}{2k_B T}\operatorname{csch}\frac{\hbar\omega_m}{2k_B T}\right)^2. \tag{2.113}$$

At high temperature, $C_v \approx 3\mathcal{N}k_B$, which is the classical or Dulong–Petit limit. The above functions are plotted in Fig. 2.1.

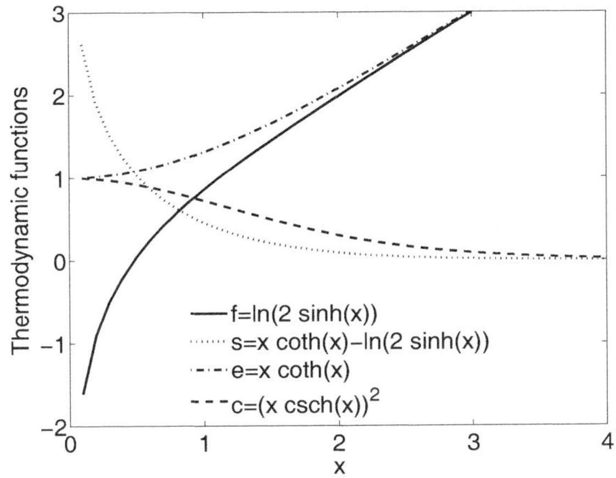

Figure 2.1 Some thermodynamic functions. The factors k_B or $k_B T$ in front of the summations have been omitted. We use the variable $x = \hbar\omega/2k_B T$.

2.4.3 Energy Flux Operator

The goal of this section is to obtain a microscopic expression for the energy flux operator. We will then be able to compute the thermal conductivity according to Eq. 2.54.

In Section 2.1 the energy flux is defined as the operator fulfilling the continuity equation. Therefore, to obtain a microscopic expression our strategy will be to compute the time derivative of the hamiltonian according to the Heisenberg equation of motion, and then rewrite the result as the divergence of a quantity that we will identify with the energy flux. This program was first carried out by (Hardy, 1963).

The harmonic hamiltonian of Eq. 2.59 can be written as

$$H_0 = \sum_I h_I \tag{2.114}$$

$$h_I = \frac{p_{I\alpha}^2}{2m_{I\alpha}} + \frac{1}{2}\sum_{\substack{J\\ \alpha\gamma}} \delta R_{I\alpha} \Phi_{I\alpha,J\gamma} \delta R_{J\gamma} \equiv T_I + V_I. \tag{2.115}$$

This suggests to define the energy density as

$$h(\boldsymbol{r}) = \frac{1}{2}\sum_I \delta(\boldsymbol{r} - \boldsymbol{R}_I^0) h_I + h_I \delta(\boldsymbol{r} - \boldsymbol{R}_I^0), \tag{2.116}$$

which obviously fulfills Eq. 2.1. This equation is identical to the most automatic procedure of (Enz, 1968), and the original papers of (Hardy, 1963) and (Luttinger, 1964).

We must now compute the time derivative,

$$i\hbar\frac{\partial}{\partial t}h(\boldsymbol{r},t) = [h(\boldsymbol{r},t), H^0]. \tag{2.117}$$

Using the rules of commutators and $[T_I, T_J] = [V_I, V_J] = 0$, we obtain

$$\begin{aligned} i\hbar\frac{\partial}{\partial t}h(\boldsymbol{r},t) = \frac{1}{2}\sum_{IJ}&[\delta(\boldsymbol{r}-\boldsymbol{R}_I^0) - \delta(\boldsymbol{r}-\boldsymbol{R}_J^0)][T_I, V_J] \\ &+ [\delta(\boldsymbol{r}-\boldsymbol{R}_I^0), T_J]h_I + \quad \text{h.c.} \end{aligned} \tag{2.118}$$

The difference of two delta functions can be computed by expanding one of them around the position of the other,

$$\delta(\boldsymbol{r}-\boldsymbol{R}_I^0) = \delta(\boldsymbol{r}-\boldsymbol{R}_J^0) + (\boldsymbol{R}_I^0 - \boldsymbol{R}_J^0)\cdot\nabla_{\boldsymbol{R}_I^0}\delta(\boldsymbol{r}-\boldsymbol{R}_I^0)\Big|_{\boldsymbol{R}_I^0=\boldsymbol{R}_J^0} + \cdots \tag{2.119}$$

This gives

$$\delta(\boldsymbol{r}-\boldsymbol{R}_I^0) - \delta(\boldsymbol{r}-\boldsymbol{R}_J^0) = -\nabla_{\boldsymbol{r}}\cdot\Big[(\boldsymbol{R}_I^0 - \boldsymbol{R}_J^0)\delta(\boldsymbol{r}-\boldsymbol{R}_J^0) + \cdots\Big], \tag{2.120}$$

which will allow to rewrite the first term of Eq. 2.118 as a divergence.

Also, using[a]

$$[p_{J\alpha}, \delta(\boldsymbol{r}-\boldsymbol{R}_I^0)] = i\hbar\delta_{IJ}\frac{\partial}{\partial r_\alpha}\delta(\boldsymbol{r}-\boldsymbol{R}_I^0) \tag{2.121}$$

we obtain

$$\begin{aligned} [\delta(\boldsymbol{r}-\boldsymbol{R}_I^0), T_J] &= \frac{1}{2m_J}\sum_\alpha\Big([\delta(\boldsymbol{r}-\boldsymbol{R}_I^0), p_{J\alpha}]p_{J\alpha} \qquad (2.122) \\ &\qquad + p_{J\alpha}[\delta(\boldsymbol{r}-\boldsymbol{R}_I^0), p_{J\alpha}]\Big) \\ &= -\frac{i\hbar}{m_I}\delta_{IJ}\nabla_{\boldsymbol{r}}\cdot\boldsymbol{p}_I\delta(\boldsymbol{r}-\boldsymbol{R}_I^0), \qquad (2.123) \end{aligned}$$

which also expresses the second term of Eq. 2.118 as a divergence.

[a]We have the identity $[p_x, f(x)] = p_x f(x)$ for any function of the variable x.

The average density operator is therefore

$$J = \frac{1}{2i\hbar V}\sum_{IJ}(\boldsymbol{R}_I^0 - \boldsymbol{R}_J^0)\left[\frac{\boldsymbol{p}_I^2}{2m_I}, V_J\right] \quad + \text{h.c.} \tag{2.124}$$

$$+\frac{1}{2V}\sum_I \frac{\boldsymbol{p}_I}{m_I}\left(\frac{\boldsymbol{p}_I^2}{2m_I} + V_I\right) \quad + \text{h.c.} \tag{2.125}$$

In this result the terms beyond the first order in Eq. 2.120 does not appear because in computing the average over the volume to obtain $\boldsymbol{J}$ we integrated derivatives of the delta functions which gives zero.

The commutator in the first line can be easily computed using the explicit expression of V_J in Eq. 2.115,

$$\left[\frac{\boldsymbol{p}_I^2}{2m_I}, V_J\right] = -\frac{i\hbar}{2m_I}\sum_{\alpha\gamma} \boldsymbol{p}_{I\alpha}\Phi_{I\alpha,J\gamma}\delta\boldsymbol{R}_{J\gamma} \quad \text{if } I \neq J \tag{2.126}$$

to obtain

$$J = -\frac{1}{2V}\sum_{\substack{IJ\\ \alpha\gamma}}(\boldsymbol{R}_I^0 - \boldsymbol{R}_J^0)\frac{\boldsymbol{p}_{I\alpha}}{m_I}\Phi_{I\alpha,J\gamma}\delta\boldsymbol{R}_{J\gamma} \tag{2.127}$$

$$+\frac{1}{2V}\sum_I \frac{\boldsymbol{p}_I}{m_I}\left(\frac{\boldsymbol{p}_I^2}{2m_I} + V_I\right) \quad + \text{h.c.}$$

In the first line we recognize the product distance-velocity-force. This term can therefore be interpreted as the rate at which atom J is doing work on atom I times the distance over which the energy is transported. In a solid it is the main contribution to the energy flux. In the second line the energy h_I is transported at velocity $\boldsymbol{v}_I = \frac{\boldsymbol{p}_I}{m_I}$. It is the main contribution in a liquid but it will be neglected in the following because we focus on solid.

The last step is to express the above result for the energy flux operator in terms of the creation and annihilation operators of the vibron states. This will allow to perform actual computations.

Using Eqs. 2.89, 2.94, 2.102, and 2.103, we obtain

$$u_{I\alpha} = \sum_m \sqrt{\frac{\hbar}{2\omega_m}}u_{I\alpha}(m)(a_m + a_{\bar{m}}^+) \tag{2.128}$$

$$-i\hbar\frac{\partial}{\partial u_{I\alpha}} = -i\sum_m \sqrt{\frac{\hbar\omega_m}{2}}u_{I\alpha}(\bar{m})(a_{\bar{m}} - a_m^+). \tag{2.129}$$

Substituting in Eq. 2.127 this gives

$$J = \sum_{m'm} \left(\frac{i\hbar}{4V} \sum_{\substack{IJ \\ \alpha\gamma}} u_{I\alpha}(\bar{m}')(\boldsymbol{R}_I^0 - \boldsymbol{R}_J^0) D_{I\alpha,J\gamma} u_{J\alpha}(m) \right)$$
$$\times \sqrt{\frac{\omega_{m'}}{\omega_m}} (a_{\bar{m}'} - a_{m'}^+)(a_m + a_{\bar{m}}^+). \tag{2.130}$$

Terms like aa and a^+a^+, where two vibrons are created or destroyed, are neglected since they do not contribute when the energy must be conserved. This allows to rearranged the above operator as

$$J = \frac{1}{V} \sum_{m'm} a_{m'}^+ a_m \left(\hbar \frac{\omega_{m'} + \omega_m}{2} \right)$$
$$\times \sum_{\substack{IJ \\ \alpha\gamma}} u_{I\alpha}(\bar{m}') \left(\frac{-i(\boldsymbol{R}_I^0 - \boldsymbol{R}_J^0) D_{I\alpha,J\gamma}}{2\sqrt{\omega_{m'}\omega_m}} \right) u_{J\alpha}(m) \tag{2.131}$$

or simply

$$J = \sum_{m'm} J_{m'm} a_{m'}^+ a_m \tag{2.132}$$

with

$$J_{m'm} = \frac{1}{V} \hbar \frac{\omega_{m'} + \omega_m}{2} v_{m'm} \tag{2.133}$$

$$v_{m'm} = \sum_{\substack{IJ \\ \alpha\gamma}} u_{I\alpha}(\bar{m}') \left(\frac{-i(\boldsymbol{R}_I^0 - \boldsymbol{R}_J^0) D_{I\alpha,J\gamma}}{2\sqrt{\omega_{m'}\omega_m}} \right) u_{J\alpha}(m)$$
$$\equiv \langle m' | \boldsymbol{v}_g | m \rangle. \tag{2.134}$$

It is actually an interesting result since the current matrix elements $J_{m'm}$ appears to be given by the averaged energy of the transition, $\hbar(\omega_{m'} + \omega_m)/2$, times the matrix element of a group velocity operator defined by the brace in Eq. 2.134.

In the case of a crystal, using the eigenvectors of Eq. 2.74, one can check that the diagonal elements of this operator give the usual group velocity,

$$\begin{aligned}
\langle \boldsymbol{k}j|\boldsymbol{v}_g|\boldsymbol{k}j\rangle &= \frac{1}{2\omega(\boldsymbol{k}j)}\sum_{\substack{\tau\tau'\\ \alpha\alpha'}} e^*_{\tau\alpha}(\boldsymbol{k}j)e_{\tau'\alpha'}(\boldsymbol{k}j)\\
&\quad\times\left(\frac{1}{N}\sum_{ll'} -i(\boldsymbol{R}^0_{l\tau}-\boldsymbol{R}^0_{l'\tau'})D_{l\tau\alpha,l'\tau'\alpha'}e^{-i\boldsymbol{k}\cdot(\boldsymbol{R}^0_{l\tau}-\boldsymbol{R}^0_{l'\tau'})}\right)\\
&= \frac{1}{2\omega(\boldsymbol{k}j)}\sum_{\substack{\tau\tau'\\ \alpha\alpha'}} e^*_{\tau\alpha}(\boldsymbol{k}j)\left(\frac{\partial}{\partial\boldsymbol{k}}C_{\tau\alpha,\tau'\alpha'}(\boldsymbol{k})\right)e_{\tau'\alpha'}(\boldsymbol{k}j)\\
&= \frac{\partial\omega(\boldsymbol{k}j)}{\partial\boldsymbol{k}}. \qquad (2.135)
\end{aligned}$$

One also recovers the current used in the Peierls–Boltzmann theory (Peierls, 1955),

$$\boldsymbol{J}_{\boldsymbol{k}j} = \frac{1}{V}\hbar\omega_{\boldsymbol{k}j}\frac{\partial\omega(\boldsymbol{k}j)}{\partial\boldsymbol{k}}, \qquad (2.136)$$

which will be discussed in Section 2.5.2

2.5 Numerical Results

2.5.1 Formula for the Thermal Conductivity

Now that we have an explicit expression for the current operator, we can evaluate the current–current correlation function in Eq. 2.54. Inserting the unit operator we obtain

$$\begin{aligned}
\chi_{\alpha\gamma}(\omega+i\eta) &= \frac{V}{T}\int_0^\beta d\lambda\int_0^{+\infty} dt e^{i(\omega+i\eta)t}\sum_{\{n\}\{n'\}}\frac{e^{-\beta E_{\{n\}}}}{Z}\\
&\quad\times\langle\{n\}|e^{\lambda H^0}\boldsymbol{J}_\alpha e^{-\lambda H^0}|\{n'\}\rangle\langle\{n'\}|e^{\frac{i}{\hbar}Ht}\boldsymbol{J}_\gamma e^{-\frac{i}{\hbar}Ht}|\{n\}\rangle\\
&= \frac{V}{T}\sum_{\{n\}\{n'\}}\frac{e^{-\beta E_{\{n\}}}}{Z}\frac{e^{\beta(E_{\{n\}}-E_{\{n'\}})}-1}{E_{\{n\}}-E_{\{n'\}}}\\
&\quad\times\frac{i}{\omega-\frac{E_{\{n\}}-E_{\{n'\}}}{\hbar}+i\eta}\langle\{n\}|\boldsymbol{J}_\alpha|\{n'\}\rangle\langle\{n'\}|\boldsymbol{J}_\gamma|\{n\}\rangle
\end{aligned}$$

and using 2.132,

$$\chi_{\alpha\gamma}(\omega+i\eta)=\frac{V}{T}\sum_{\{n\}\{n'\}}\sum_{mm'm''m'''}\frac{e^{-\beta E_{\{n\}}}}{Z}\frac{e^{\beta(E_{\{n\}}-E_{\{n'\}})}-1}{E_{\{n\}}-E_{\{n'\}}}$$
$$\times\frac{i}{\omega-\frac{E_{\{n\}}-E_{\{n'\}}}{\hbar}+i\eta}\boldsymbol{J}^{\alpha}_{mm'}\boldsymbol{J}^{\gamma}_{m''m'''}$$
$$\times\langle\{n\}|a^{+}_{m}a_{m'}|\{n'\}\rangle\langle\{n'\}|a^{+}_{m''}a_{m'''}|\{n\}\rangle.$$

To evaluate the matrix elements we must consider separately the two cases, $\{n\}=\{n'\}$ and $\{n\}\neq\{n'\}$.

The former case also implies $m=m'$ and $m''=m'''$, which means that the vibrons are removed and created again in the same state. It is the diagonal contribution to the thermal conductivity, $\chi^{(d)}_{\alpha\gamma}$. In crystals we will see it correspond to the propagative heat transfer given by the Boltzmann equation.

We have,

$$\chi^{(d)}_{\alpha\gamma}(\omega+i\eta)=\frac{V}{T}\sum_{\{n\}}\sum_{mm''}\frac{e^{-\beta E_{\{n\}}}}{Z}\beta\frac{i}{\omega+i\eta}$$
$$\times\boldsymbol{J}^{\alpha}_{mm'}\boldsymbol{J}^{\gamma}_{m''m'''}\left(n_m[\{n\}]\right)\left(n_{m''}[\{n\}]\right)$$

if $n_m[\{n\}]=\langle\{n\}|a^{+}_{m}a_m|\{n\}\rangle$ denotes the number of vibrons in the mode m of the total vibrational state $|\{n\}\rangle$.

The thermodynamic average of the product of vibron occupations is easily computed,

$$\begin{aligned}\langle n_m n_{m''}\rangle&=\sum_{\{n\}}\frac{e^{-\beta E_{\{n\}}}}{Z}\left(n_m[\{n\}]\right)\left(n_{m''}[\{n\}]\right)\\&=\langle n_m\rangle\langle n_{m''}\rangle&&\text{if } m\neq m''\\&=\langle n^2_m\rangle&&\text{if } m=m''.\end{aligned}$$

Using the identity

$$\begin{aligned}\langle(n_m+\frac{1}{2})^r\rangle&=\frac{1}{Z}\sum_{\{n\}}e^{-\beta\sum_{m'}(n_{m'}[\{n\}]+\frac{1}{2})\hbar\omega_{m'}}\left(n_m[\{n\}]+\frac{1}{2}\right)^r\\&=\frac{1}{Z}\frac{\partial^r}{\partial(-\hbar\omega_m\beta)^r}Z\end{aligned}$$

we obtain

$$\langle n_m\rangle=\frac{1}{e^{\beta\hbar\omega_m}-1}\equiv\bar{n}_m,$$

the Bose–Einstein occupation function, and

$$\langle n_m^2 \rangle = \bar{n}_m^2 + \bar{n}_m(\bar{n}_m + 1).$$

This gives

$$\begin{aligned}\chi_{\alpha\gamma}^{(d)}(\omega + i\eta) &= \frac{V}{k_B T^2}\frac{i}{\omega + i\eta}\Big(\sum_m \boldsymbol{J}_{mm}^{\alpha}\bar{n}_m\Big)\Big(\sum_{m''} \boldsymbol{J}_{m''m''}^{\alpha}\bar{n}_{m''}\Big)\\ &+ \frac{V}{k_B T^2}\frac{i}{\omega + i\eta}\sum_m \boldsymbol{J}_{mm}^{\alpha}\boldsymbol{J}_{mm}^{\gamma}\bar{n}_m(\bar{n}_m + 1).\end{aligned}$$

The first line gives the current at equilibrium and must be zero. The second line is rewritten as

$$\begin{aligned}\chi_{\alpha\gamma}^{(d)}(\omega + i\eta) &= \frac{1}{V}\frac{i}{\omega + i\eta}\sum_m k_B\left(\frac{\hbar\omega_m}{2k_B T}\operatorname{csch}\frac{\hbar\omega_m}{2k_B T}\right)^2 \boldsymbol{v}_m^{\alpha}\boldsymbol{v}_m^{\gamma}\\ &= \frac{1}{V}\frac{i}{\omega + i\eta}\sum_m C_m \boldsymbol{v}_m^{\alpha}\boldsymbol{v}_m^{\gamma},\end{aligned} \tag{2.137}$$

where we have defined the mode-dependent heat capacity

$$C_m = k_B\left(\frac{\hbar\omega_m}{2k_B T}\operatorname{csch}\frac{\hbar\omega_m}{2k_B T}\right)^2 \tag{2.138}$$

according to Eq. 2.113.

For a perfect crystal $m \to (\boldsymbol{k}j)$ and the velocity is given by Eq. 2.135. Therefore in the static limit $\omega \to 0$ the thermal conductivity is infinite. It is not surprising since in that case the thermal resistivity comes from the scattering in between the phonons, and such events have been left out of our harmonic model. In the case of a disordered system the thermal conductivity must be finite even within the harmonic approximation. This is only possible if

$$\boldsymbol{J}_{mm} = 0 \qquad \text{or} \qquad \boldsymbol{v}_m = 0 \qquad \text{for every } m, \tag{2.139}$$

which is the mathematical definition of a *non-propagating* state. As we will see later this does not imply that the states are localized.

The non-diagonal part of the thermal conductivity, $\{n\} \neq \{n'\}$, is evaluated much in the same way.

For the matrix elements we have

$$\langle\{n\}|a_m^{+}a_{m'}|\{n'\}\rangle = \sqrt{n_m[\{n\}](n_{m'}[\{n\}] + 1)}\delta_{\{n'\}\{n\}_{\bar{m}m'}}, \tag{2.140}$$

where $\{n\}_{\bar{m}m'}$ is the state $\{n\}$ where one vibron has been removed from mode m, and once created in mode m'. This gives

$$\chi^{(nd)}_{\alpha\gamma}(\omega+i\eta) = \frac{V}{T}\sum_{\{n\}}\sum_{m\neq m'} \frac{e^{-\beta E_{\{n\}}}}{Z}\frac{e^{\beta\hbar\omega_{mm'}}-1}{\hbar\omega_{mm'}}\frac{i}{\omega-\omega_{mm'}+i\eta}$$
$$\times \boldsymbol{J}^{\alpha}_{mm'}\boldsymbol{J}^{\gamma}_{m'm}\left(n_m[\{n\}]\right)\left(n_{m'}[\{n\}]+1\right)$$

with $(E_{\{n\}} - E_{\{n'\}})/\hbar = \omega_m - \omega_{m'} = \omega_{mm'}$.

Performing the thermodynamical average we obtain

$$\chi^{(nd)}_{\alpha\gamma}(\omega+i\eta) = \frac{V}{T}\sum_{m\neq m'}\frac{e^{\beta\hbar\omega_{mm'}}-1}{\hbar\omega_{mm'}}\frac{i}{\omega-\omega_{mm'}+i\eta}$$
$$\times \boldsymbol{J}^{\alpha}_{mm'}\boldsymbol{J}^{\gamma}_{m'm}\bar{n}_m(\bar{n}_{m'}+1)$$
$$= \frac{V}{T}\sum_{m\neq m'}\frac{\bar{n}_{m'}-\bar{n}_m}{\hbar\omega_{mm'}}\frac{i}{\omega-\omega_{mm'}+i\eta}\boldsymbol{J}^{\alpha}_{mm'}\boldsymbol{J}^{\gamma}_{m'm}. \tag{2.141}$$

In the static limit, $\omega \to 0$, $-\frac{\bar{n}_{m'}-\bar{n}_m}{\omega_{m'm}} \to -\frac{\partial\bar{n}_m}{\partial\omega_m}$, and therefore the real part of the above result simplifies to

$$\mathrm{Re}\chi^{(nd)}_{\alpha\gamma} = \sum_{m\neq m'} k_B\left(\frac{\hbar\omega_m}{2k_BT}\mathrm{csch}\frac{\hbar\omega_m}{2k_BT}\right)^2\frac{\pi V}{\hbar^2\omega_m^2}\boldsymbol{J}^{\alpha}_{mm'}\boldsymbol{J}^{\gamma}_{m'm}\delta(\omega_m-\omega_{m'})$$
$$= \frac{1}{V}\sum_m C_m D^{\alpha\gamma}_m \tag{2.142}$$

if one defines the mode diffusivity tensor by (Allen and Feldman, 1993)

$$D^{\alpha\gamma}_m = \frac{\pi V^2}{\hbar^2\omega_m^2}\sum_{m'\neq m}\boldsymbol{J}^{\alpha}_{mm'}\boldsymbol{J}^{\gamma}_{m'm}\delta(\omega_m-\omega_{m'}). \tag{2.143}$$

The physical significance of this result, which was first obtained in (Allen and Feldman, 1989), will be discussed in the following sections.

2.5.2 Perfect Crystal

In a crystal the phonon eigenstates fulfill the Bloch theorem and the wavevector $\boldsymbol{k}$ is therefore a good quantum number. Because the wavevectors are so well defined the picture we can have of a crystal is that of weakly interacting phonons experiencing collisions which creates the thermal resistivity. This picture is correct as long as the collisions are rare enough for the phonon mean free path to be much larger than the phonon wavelength,

$$\lambda_{kj} = v_{kj}\tau_{kj} \gg \frac{2\pi}{|\boldsymbol{k}|}, \tag{2.144}$$

if τ_{kj} the average time in between two collisions. This is the approach used by (Peierls, 1955). In such a case the current is given by the sum of the currents contributed by each mode $\boldsymbol{k}j$,

$$\boldsymbol{J} = \frac{1}{V}\sum_{kj} \hbar\omega_{kj}\boldsymbol{v}_{kj}n_{kj}, \tag{2.145}$$

where n_{kj} is the number of phonons in mode $\boldsymbol{k}j$ when a temperature gradient ∇T is applied on the system.

Because the system is not at equilibrium n_{kj} is not the Bose–Einstein occupation function $\bar{n}_{kj}$, but must be computed. The textbook approach to this problem is to write a Boltzmann equation for n_{kj}, stating that the number of phonons in mode $\boldsymbol{k}j$ is changed because of the diffusion of particles due to the temperature gradient, but also because of the collisions that we did not explicitly include in the hamiltonian H^0. In a stationary regime, these effects must be balanced by each other, therefore we obtain

$$0 = \left.\frac{\partial n_{kj}}{\partial t}\right|_{diff} + \left.\frac{\partial n_{kj}}{\partial t}\right|_{col}. \tag{2.146}$$

Because

$$\left.\frac{\partial n_{kj}}{\partial t}\right|_{diff} = \frac{n_{kj}(\boldsymbol{r}, t+\delta t) - n_{kj}(\boldsymbol{r}, t)}{\delta t} = \frac{n_{kj}(\boldsymbol{r} - \boldsymbol{v}_{kj}\delta t, t) - n_{kj}(\boldsymbol{r}, t)}{\delta t}$$
$$= -\boldsymbol{v}_{kj}\cdot\nabla n_{kj} \tag{2.147}$$

the Boltzmann equation becomes

$$\frac{\partial n_{kj}}{\partial T}\nabla T\cdot\boldsymbol{v}_{kj} = \left.\frac{\partial n_{kj}}{\partial t}\right|_{col}. \tag{2.148}$$

If we assume that the collisions can be described using a relaxation time,

$$\left.\frac{\partial n_{kj}}{\partial t}\right|_{col} \approx \frac{\bar{n}_{kj} - n_{kj}}{\tau_{kj}}, \tag{2.149}$$

this gives, to the first order in ∇T,

$$n_{kj} = \bar{n}_{kj} - \frac{\partial \bar{n}_{kj}}{\partial T}\nabla T\cdot\boldsymbol{v}_{kj}\tau_{kj}. \tag{2.150}$$

Substituting in Eq. 2.145, we obtain for the current,

$$\boldsymbol{J} = -\frac{1}{V}\sum_{\boldsymbol{k}j} \hbar\omega_{\boldsymbol{k}j} \frac{\partial \bar{n}_{\boldsymbol{k}j}}{\partial T} \tau_{\boldsymbol{k}j} \boldsymbol{v}_{\boldsymbol{k}j} \cdot \boldsymbol{v}_{\boldsymbol{k}j} \nabla T, \tag{2.151}$$

and

$$\kappa = \frac{1}{V}\sum_{\boldsymbol{k}j} C_{\boldsymbol{k}j} \tau_{\boldsymbol{k}j} \boldsymbol{v}_{\boldsymbol{k}j} \otimes \boldsymbol{v}_{\boldsymbol{k}j} \tag{2.152}$$

for the thermal conductivity. We have used for the mode-dependent heat capacity $C_{\boldsymbol{k}j} = \hbar\omega_{\boldsymbol{k}j}\partial\bar{n}_{\boldsymbol{k}j}/\partial T$.

This result is equivalent to Eq. 2.137 we obtained from linear response theory for the diagonal part of the thermal conductivity, once the parameter η is replaced by $1/\tau_{\boldsymbol{k}j}$. In doing so we state that the frequency level $\omega_{\boldsymbol{k}j}$ is not infinitely well resolved, $\eta \to 0$, but that it has a finite linewidth $1/\tau_{\boldsymbol{k}j}$ due to the collisions which was left out of our model H^0.

The non-diagonal part we have obtained for the thermal conductivity correspond interband transitions, in a way analogous to the Drude model of optical conductivity for electrons (Marder, 2000).

In the Peierls–Boltzmann approach it is therefore straightforward to compute the thermal conductivity using Eq. 2.152, but the relaxation time $\tau_{\boldsymbol{k}j}$ should be known. To obtain this quantity is actually the most difficult part of the calculation.

At high enough temperature the atoms are vibrating far from their equilibrium positions, and therefore the expansion of the Born–Oppenheimer potential energy cannot be limited to the second order as it was done in Eq. 2.59. Any higher order terms will be regarded as a perturbation to H^0, and will be called *anharmonic* energy terms. The third-order term is

$$\begin{aligned} H' = &\frac{1}{3!}\sum_{ll'l''}\sum_{\tau\tau'\tau''}\sum_{\alpha\alpha'\alpha''} \frac{\partial^3 E}{\partial R^0_{l\tau\alpha}\partial R^0_{l'\tau'\alpha'}\partial R^0_{l''\tau''\alpha''}} \\ &\times \frac{1}{\sqrt{m_{l\tau\alpha}m_{l'\tau'\alpha'}m_{l''\tau''\alpha''''}}} u_{l\tau\alpha}u_{l'\tau'\alpha'}u_{l''\tau''\alpha''} \end{aligned} \tag{2.153}$$

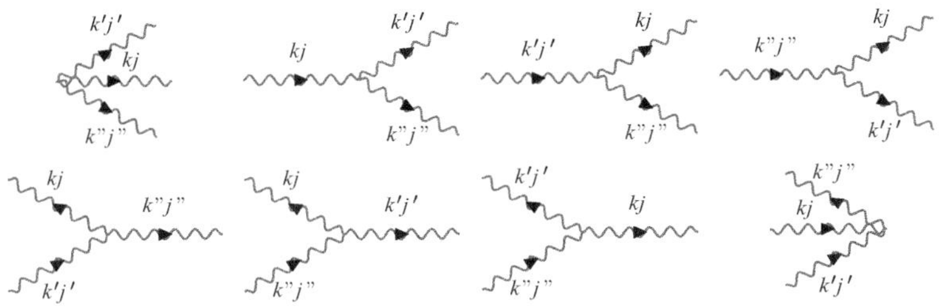

Figure 2.2 Feynman diagrams for the third-order anharmonic interactions.

or, expressing the atoms displacements in term of the creation and annihilation operators,

$$H' = \frac{1}{3!} \sum_{\substack{\boldsymbol{k}\boldsymbol{k}'\boldsymbol{k}'' \\ jj'j''}} \mathcal{F}^{jj'j''}_{\boldsymbol{k}\boldsymbol{k}'\boldsymbol{k}''} (a_{\boldsymbol{k}j} + a^{+}_{-\boldsymbol{k}j})(a_{\boldsymbol{k}'j'} + a^{+}_{-\boldsymbol{k}'j'})(a_{\boldsymbol{k}''j''} + a^{+}_{-\boldsymbol{k}''j''}) \tag{2.154}$$

with

$$\mathcal{F}^{jj'j''}_{\boldsymbol{k}\boldsymbol{k}'\boldsymbol{k}''} = \left(\frac{\hbar}{2N}\right)^{\frac{3}{2}} \sum_{ll'l''} \sum_{\tau\tau'\tau''} \sum_{\alpha\alpha'\alpha''} \frac{\partial^3 E}{\partial R^0_{l\tau\alpha} \partial R^0_{l'\tau'\alpha'} \partial R^0_{l''\tau''\alpha''}} \times \frac{e_{\tau\alpha}(\boldsymbol{k}j) e_{\tau'\alpha'}(\boldsymbol{k}'j') e_{\tau''\alpha''}(\boldsymbol{k}''j'')}{\sqrt{m_{l\tau\alpha} m_{l'\tau'\alpha'} m_{l''\tau''\alpha''}}} \frac{e^{i\boldsymbol{k}\cdot\boldsymbol{R}^0_{l\tau} + i\boldsymbol{k}'\cdot\boldsymbol{R}^0_{l'\tau'} + i\boldsymbol{k}''\cdot\boldsymbol{R}^0_{l''\tau''}}}{\sqrt{\omega_{\boldsymbol{k}j}\omega_{\boldsymbol{k}'j'}\omega_{\boldsymbol{k}''j''}}}.$$

$\mathcal{F}^{jj'j''}_{\boldsymbol{k}\boldsymbol{k}'\boldsymbol{k}''}$ is the strength of the interaction in between the three phonons $\boldsymbol{k}j$, $\boldsymbol{k}'j'$, and $\boldsymbol{k}''j''$ involved in a collision. Therefore going from Eq. 2.153 to Eq. 2.154 the lattice anharmonicity has been traduced has interactions in between the phonons. Expanding the braces containing phonon creation and annihilation operators in Eq. 2.154 those interactions can also be pictured as scattering events. The eight processes are displayed in Fig. 2.2.

To the second order in the strength of the interaction, the probability, per unit time, for a phonon to have such a scattering can be computed using the Fermi–Golden rule. The generic formula is

$$P = \frac{2\pi}{\hbar} |\langle f|H'|i\rangle|^2 \delta(E_f - E_i), \tag{2.155}$$

where E_i and E_f are the energies of the initial and final states $|i\rangle$ and $|f\rangle$.

In our case we must consider that in the initial state there is a phonon of mode $\boldsymbol{k}j$. The perturbation H' contains terms of the form aa^+a^+ and aaa^+ which correspond to scattering for which the energy may be conserved. In Fig. 2.2 they correspond to diagrams 3 to 5 and 6 to 7, respectively. The first and the last diagrams cannot conserve energy since three phonons are created or destroyed (Srivastava, 1990).

The first group of diagrams are called *decay* processes,

$$\boldsymbol{k}j \rightarrow \boldsymbol{k}'j' + \boldsymbol{k}''j'', \tag{2.156}$$

and the second group *collisions* processes,

$$\boldsymbol{k}j + \boldsymbol{k}'j' \rightarrow \boldsymbol{k}''j''. \tag{2.157}$$

For a decay process the energy balance is

$$E_f - E_i = \hbar\omega_{\boldsymbol{k}''j''} + \hbar\omega_{\boldsymbol{k}'j'} - \hbar\omega_{\boldsymbol{k}j} \tag{2.158}$$

and the probability rate of such an event is obtained from the Fermi golden rule,

$$\begin{aligned} P_{\boldsymbol{k}j\rightarrow\boldsymbol{k}'j'+\boldsymbol{k}''j''} &= \frac{2\pi}{\hbar}|\mathcal{F}^{jj'j''}_{\boldsymbol{k}\boldsymbol{k}'\boldsymbol{k}''}|^2 n_{\boldsymbol{k}j}(n_{\boldsymbol{k}'j'}+1)(n_{\boldsymbol{k}''j''}+1) \\ &\times \delta(\hbar\omega_{\boldsymbol{k}''j''} + \hbar\omega_{\boldsymbol{k}'j'} - \hbar\omega_{\boldsymbol{k}j}). \end{aligned} \tag{2.159}$$

The factor $1/3!$ in the perturbation H' is canceled by the $3 \times 2 = 6$ possibilities given to the phonon $\boldsymbol{k}j$ to decay.
For a collision process,

$$E_f - E_i = \hbar\omega_{\boldsymbol{k}''j''} - \hbar\omega_{\boldsymbol{k}'j'} - \hbar\omega_{\boldsymbol{k}j} \tag{2.160}$$

$$\begin{aligned} P_{\boldsymbol{k}j+\boldsymbol{k}'j'\rightarrow\boldsymbol{k}''j''} &= \frac{2\pi}{\hbar}|\mathcal{F}^{jj'j''}_{\boldsymbol{k}\boldsymbol{k}'\boldsymbol{k}''}|^2 n_{\boldsymbol{k}j}n_{\boldsymbol{k}'j'}(n_{\boldsymbol{k}''j''}+1) \\ &\times \delta(\hbar\omega_{\boldsymbol{k}''j''} - \hbar\omega_{\boldsymbol{k}'j'} - \hbar\omega_{\boldsymbol{k}j}). \end{aligned} \tag{2.161}$$

The net change, per unit time, in the number of phonon in mode $\boldsymbol{k}j$ is therefore

$$\begin{aligned} \left.\frac{\partial n_{\boldsymbol{k}j}}{\partial t}\right|_{col} &= \sum_{\substack{\boldsymbol{k}'\boldsymbol{k}'' \\ j'j''}} (P_{\boldsymbol{k}j\leftarrow\boldsymbol{k}'j'+\boldsymbol{k}''j''} - P_{\boldsymbol{k}j\rightarrow\boldsymbol{k}'j'+\boldsymbol{k}''j''}) \\ &+ \frac{1}{2}(P_{\boldsymbol{k}j+\boldsymbol{k}'j'\leftarrow\boldsymbol{k}''j''} - P_{\boldsymbol{k}j+\boldsymbol{k}'j'\rightarrow\boldsymbol{k}''j''}), \end{aligned} \tag{2.162}$$

where $\leftarrow$ is used to denote the time reversed scattering. The factor $1/2$ is inserted not to count twice the same collision process (Ziman, 1960).

Using the previous results for the probabilities the collision kernel is explicitly written as

$$\begin{aligned}\left.\frac{\partial n_{\mathbf{k}j}}{\partial t}\right|_{col} = & \frac{2\pi}{\hbar^2}\sum_{\mathbf{k}'j'\mathbf{k}''j''}|\mathcal{F}_{\mathbf{k}\mathbf{k}'\mathbf{k}''}^{jj'j''}|^2\delta(\hbar\omega_{\mathbf{k}''j''}+\hbar\omega_{\mathbf{k}'j'}-\hbar\omega_{\mathbf{k}j})\\ & \times[(n_{\mathbf{k}j}+1)n_{\mathbf{k}'j'}n_{\mathbf{k}''j''}-n_{\mathbf{k}j}(n_{\mathbf{k}'j'}+1)(n_{\mathbf{k}''j''}+1)]\\ & +\frac{1}{2}\frac{2\pi}{\hbar^2}\sum_{\mathbf{k}'j'\mathbf{k}''j''}|\mathcal{F}_{\mathbf{k}\mathbf{k}'\mathbf{k}''}^{jj'j''}|^2\delta(\hbar\omega_{\mathbf{k}''j''}-\hbar\omega_{\mathbf{k}'j'}-\hbar\omega_{\mathbf{k}j})\\ & \times[(n_{\mathbf{k}j}+1)(n_{\mathbf{k}'j'}+1)n_{\mathbf{k}''j''}-n_{\mathbf{k}j}n_{\mathbf{k}'j'}(n_{\mathbf{k}''j''}+1)].\end{aligned}$$

If the temperature gradient is not too large the system remains close to thermodynamical equilibrium and the occupation function can be linearized around the Bose–Einstein occupation,

$$n_{\mathbf{k}j}=\bar{n}_{\mathbf{k}j}+\delta n_{\mathbf{k}j}. \tag{2.163}$$

The previous equation becomes

$$\begin{aligned}\left.\frac{\partial n_{\mathbf{k}j}}{\partial t}\right|_{col} = & -\frac{2\pi}{\hbar^2}\sum_{\mathbf{k}'j'\mathbf{k}''j''}|\mathcal{F}_{\mathbf{k}\mathbf{k}'\mathbf{k}''}^{jj'j''}|^2\delta(\hbar\omega_{\mathbf{k}''j''}+\hbar\omega_{\mathbf{k}'j'}-\hbar\omega_{\mathbf{k}j})\\ & \times[\delta n_{\mathbf{k}j}(\bar{n}_{\mathbf{k}'j'}+\bar{n}_{\mathbf{k}''j''}+1)\\ & +\delta n_{\mathbf{k}'j'}(\bar{n}_{\mathbf{k}j}-\bar{n}_{\mathbf{k}''j''})+\delta n_{\mathbf{k}''j''}(\bar{n}_{\mathbf{k}j}-\bar{n}_{\mathbf{k}'j'})]\\ & -\frac{1}{2}\frac{2\pi}{\hbar^2}\sum_{\mathbf{k}'j'\mathbf{k}''j''}|\mathcal{F}_{\mathbf{k}\mathbf{k}'\mathbf{k}''}^{jj'j''}|^2\delta(\hbar\omega_{\mathbf{k}''j''}-\hbar\omega_{\mathbf{k}'j'}-\hbar\omega_{\mathbf{k}j})\\ & \times[\delta n_{\mathbf{k}j}(\bar{n}_{\mathbf{k}'j'}-\bar{n}_{\mathbf{k}''j''})\\ & +\delta n_{\mathbf{k}'j'}(\bar{n}_{\mathbf{k}j}-\bar{n}_{\mathbf{k}''j''})+\delta n_{\mathbf{k}''j''}(\bar{n}_{\mathbf{k}j}+\bar{n}_{\mathbf{k}'j'}+1)].\end{aligned}$$

Clearly, in general, the collision kernel cannot assume the form of Eq. 2.149. Therefore the relaxation time is an ill defined quantity. However, if we neglect the change in the number of phonons in states different from $\mathbf{k}j$,

$$\delta n_{\mathbf{k}'j'}=\delta n_{\mathbf{k}''j''}=0 \quad \mathbf{k}'j',\mathbf{k}''j''\neq\mathbf{k}j \tag{2.164}$$

then we obtain

$$\begin{aligned}\left.\frac{\partial n_{\mathbf{k}j}}{\partial t}\right|_{col} = & -\delta n_{\mathbf{k}j}\frac{2\pi}{\hbar^2}\sum_{\mathbf{k}'j'\mathbf{k}''j''}|\mathcal{F}_{\mathbf{k}\mathbf{k}'\mathbf{k}''}^{jj'j''}|^2\\ & \Big[(\bar{n}_{\mathbf{k}'j'}+\bar{n}_{\mathbf{k}''j''}+1)\delta(\hbar\omega_{\mathbf{k}''j''}+\hbar\omega_{\mathbf{k}'j'}-\hbar\omega_{\mathbf{k}j})\\ & +\frac{1}{2}(\bar{n}_{\mathbf{k}'j'}-\bar{n}_{\mathbf{k}''j''})\delta(\hbar\omega_{\mathbf{k}''j''}-\hbar\omega_{\mathbf{k}'j'}-\hbar\omega_{\mathbf{k}j})\Big],\end{aligned} \tag{2.165}$$

which is of the form of Eq. 2.149 if

$$\frac{1}{\tau_{\boldsymbol{k}j}} = \frac{2\pi}{\hbar^2} \sum_{\substack{\boldsymbol{k}'\boldsymbol{k}'' \\ j'j''}} |\mathcal{F}^{jj'j''}_{\boldsymbol{k}\boldsymbol{k}'\boldsymbol{k}''}|^2 \Big[(\bar{n}_{\boldsymbol{k}'j'} + \bar{n}_{\boldsymbol{k}''j''} + 1)\delta(\hbar\omega_{\boldsymbol{k}''j''} + \hbar\omega_{\boldsymbol{k}'j'} - \hbar\omega_{\boldsymbol{k}j})$$

$$+ \frac{1}{2}(\bar{n}_{\boldsymbol{k}'j'} - \bar{n}_{\boldsymbol{k}''j''})\delta(\hbar\omega_{\boldsymbol{k}''j''} - \hbar\omega_{\boldsymbol{k}'j'} - \hbar\omega_{\boldsymbol{k}j}) \Big]. \tag{2.166}$$

Because of Eq. 2.164, the relaxation time approximation, Eq. 2.149 is sometimes called the single-mode relaxation time approximation to pinpoint the origin of the approximation.

When we summarize our results, we see that all what is needed to compute the thermal conductivity are the phonon velocities and the lifetimes. They are completely determined by Eqs. 2.134 and 2.166 once the second- and third-order force constants,

$$\Phi_{l\tau\alpha,l'\tau'\alpha'} = \frac{\partial^2 E}{\partial \boldsymbol{R}^0_{l\tau\alpha} \partial \boldsymbol{R}^0_{l'\tau'\alpha'}}, \tag{2.167}$$

$$\Phi_{l\tau\alpha,l'\tau'\alpha',l''\tau''\alpha''} = \frac{\partial^3 E}{\partial \boldsymbol{R}^0_{l\tau\alpha} \partial \boldsymbol{R}^0_{l'\tau'\alpha'} \partial \boldsymbol{R}^0_{l''\tau''\alpha''}}, \tag{2.168}$$

are known.

The second-order force constants can be computed from ab initio calculations and since a few years it is also possible to compute the third-order derivatives with good accuracy (Chaput et al., 2011). The strategy is very similar for both; they are approximated using finite differences. For the second order,

$$\Phi_{l\tau\alpha,l'\tau'\alpha'} = \frac{\partial}{\partial \boldsymbol{R}^0_{l'\tau'\alpha'}} \frac{\partial E}{\partial \boldsymbol{R}^0_{l\tau\alpha}}$$
$$\approx -\frac{F(l\tau\alpha; \Delta R_{l'\tau'\alpha'})}{\Delta R_{l'\tau'\alpha'}}$$

with $F(l\tau\alpha; \Delta R_{l'\tau'\alpha'})$ the force on atom $l\tau\alpha$ if atom $l'\tau'\alpha'$ is displaced for its equilibrium position by $\Delta R_{l'\tau'\alpha'}$.

For the third-order derivative, the approximation is

$$\Phi_{l\tau\alpha,l'\tau'\alpha',l''\tau''\alpha''} \approx -\frac{F(l\tau\alpha; \Delta R_{l'\tau'\alpha'}, \Delta R_{l''\tau''\alpha''}) - F(l\tau\alpha; \Delta R_{l'\tau'\alpha'})}{\Delta R_{l'\tau'\alpha'} \Delta R_{l''\tau''\alpha''}}.$$

The forces on atoms are output of ab initio computations. Therefore to obtain the above derivatives the strategy is simply to displace atoms, or pair of atoms, for there equilibrium positions in any possible ways to obtain every components of the force constants

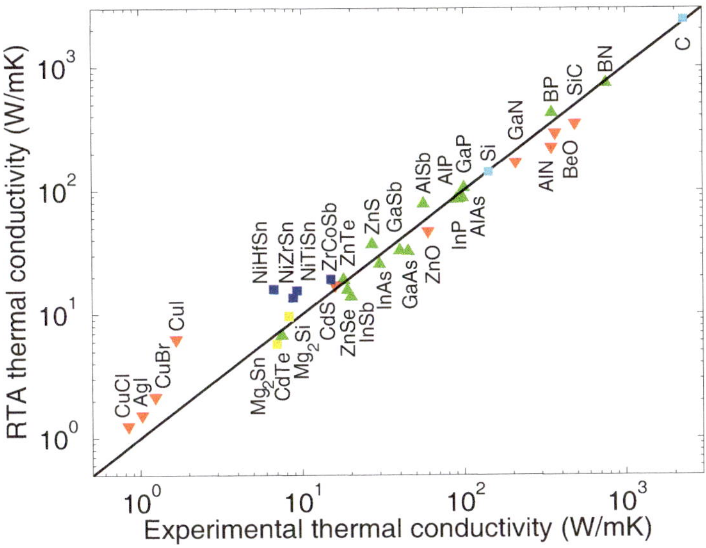

Figure 2.3 Comparison with experiments of thermal conductivities computed with the relaxation time approximation (RTA).

tensors. Eventually the space group symmetry can be used to reduce the number of displaced configurations to be considered.

The results of such calculations are compared to experiments in Fig. 2.3 for zinc-blend, wurtzite, and a few other structures. It is interesting to notice how this approach, conceptually simple, can be quite accurate (Togo et al., 2015).

2.5.3 Structural Disorder

In disorderd systems the translation operator does not commute with the hamiltonian. Therefore the Bloch theorem is not obeyed and wavevectors are no longer good quantum numbers. It means that the vibrational states should be named in a different way. A general classification, useful for transport, has been proposed by Allen and co-workers (Allen et al., 1999). It is reproduced in Fig. 2.4. Vibrons are the general vibrational modes. They can be extended states (*Extendons*) or localized in which case they are called *Locons*. A state is said to be localized if the corresponding eigenvector decay

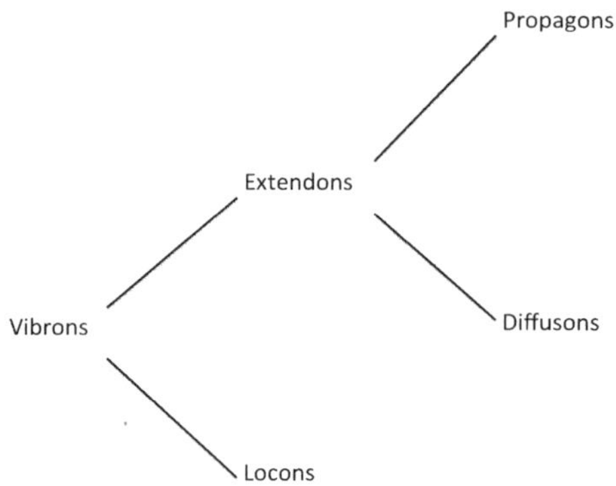

Figure 2.4 Taxonomy of vibrations.

exponentially fast from some center $\boldsymbol{R}_0^0$. For example,

$$|\boldsymbol{u}_l(n)| \propto e^{-\frac{|\boldsymbol{R}_l^0 - \boldsymbol{R}_0^0|}{\zeta_n}}. \tag{2.169}$$

The Extendons themselves can be propagating states, $\langle m|\boldsymbol{J}|m\rangle \neq 0$, or non-propagating. The former are called *Propagons* and the latter *Diffusons*. In crystals, Phonon states are Propagons.

For simulations, structurally disordered systems are often studied using finite-size models. Atomic positions are generated to produce, at least, a radial distribution function compatible with experiments. Amorphous silicon has been extensively studied using the Wooten–Winer–Weaire algorithm (Wooten et al., 1985) to produce the positions and the Stillinger–Weber interatomic potential (Stillinger and Weber, 1985) to obtain the force constants. The dynamical matrix can then be constructed according to Eq. 2.68, and diagonalized to obtain the vibrational states and their frequencies. This was done by Feldman et al. for a cell of 1000 or 4096 atoms periodically repeated in the three directions (Feldman et al., 1993, 1999).

The eigenstates obtained from these calculations can be categorized as locons if they fulfill Eq. 2.169. To identify propagons, the eigenstates can be considered to be wavepackets and then the wavevectors are required to be reasonably well defined for those wavepackets. For example, this can be done by computing the longitudinal and transverse structure factors,

$$S_L(\boldsymbol{k},\omega) = \sum_m \left| \sum_l \boldsymbol{u}_l(m)\cdot \hat{k} e^{i\boldsymbol{k}\cdot \boldsymbol{R}_l^0} \right|^2 \delta(\omega-\omega_m)$$
$$\approx \frac{A(\boldsymbol{k})}{(\omega(\boldsymbol{k})-\omega)^2+(\frac{1}{2\tau(\boldsymbol{k})})^2} \tag{2.170}$$

$$S_T(\boldsymbol{k},\omega) = \sum_m \left| \sum_l \boldsymbol{u}_l(m)\times \hat{k} e^{i\boldsymbol{k}\cdot \boldsymbol{R}_l^0} \right|^2 \delta(\omega-\omega_m)$$
$$\approx \frac{A(\boldsymbol{k})}{(\omega(\boldsymbol{k})-\omega)^2+(\frac{1}{2\tau(\boldsymbol{k})})^2} \tag{2.171}$$

for which a dispersion relation can be extracted from a fit to a Lorentzian if the propagation is not too damped (Larkin and McGaughey, 2014).

Another possibility is to consider the mode diffusivity defined in Eq. 2.143,

$$D_m^{xx} = \frac{\pi V^2}{\hbar^2\omega_m^2} \sum_{m'\neq m} |\boldsymbol{J}_{mm'}^x|^2 \delta(\omega_m-\omega_{m'}). \tag{2.172}$$

The result of (Feldman et al., 1993) is shown in Fig. 2.5. For localized states the diffusivity is zero. This allows to locate the mobility edge in the upper part of the spectrum, around 72 meV. Therefore about 3% of the states are locons. They do not carry heat as long as anharmonicity is not included in the model. For propagons the diffusivity should be large. A strong increase is observed in the diffusivity for states below 12 meV; therefore about 4% of the states are propagons. Of course, 12 meV is only an estimate. The Propagon-Diffuson boundary is not sharp, in contrast to the Extendon–Locon transition. This is summarized in Fig. 2.6.

The solid red line shown in Fig. 2.5 represents the diffusivity of propagons in the Peierls–Boltzmann theory,

$$D_{\boldsymbol{k}}^{xx} = \frac{1}{3}\boldsymbol{v}_{\boldsymbol{k}}^2\tau_{\boldsymbol{k}}, \tag{2.173}$$

where $\tau_{\boldsymbol{k}}$ and $\boldsymbol{v}_{\boldsymbol{k}}$ are the lifetime and the velocity extracted from Eq. 2.171. It shows that both approaches, Eqs. 2.171 and 2.172, agree to evidence propagons in the lower part of the spectrum. It also suggests that at low enough frequency the Peierls–Boltzmann theory could be used to describe energy transport in

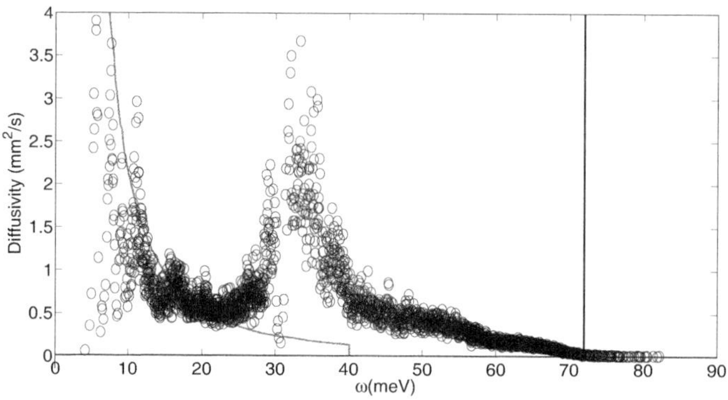

Figure 2.5 Diffusivity of amorphous silicon (Feldman et al., 1993; Allen et al., 1999).

amorphous systems. It is indeed very useful in practice since the low energy states cannot be described by finite-size calculations. Their wavelength could become much larger than the simulation box.

This interpretation allows to rewrite the thermal conductivity as

$$\kappa^{xx} = \frac{1}{V}\frac{1}{3}\int_0^{\omega_{cut}} d\omega N(\omega) C\left(\frac{\hbar\omega}{2k_B T}\right)\tau(\omega)\mathbf{v}^2(\omega) + \frac{1}{V}\sum_{\omega_n > \omega_{cut}} C_n D_n, \tag{2.174}$$

where $N(\omega) = \sum_n \delta(\omega - \omega_n)$ is the density of states and $C(x) = k_B x^2 \text{csch}^2 x$ is the heat capacity defined in Eq. 2.113. ω_{cut} is a cut off frequency which can be chosen to separate the propagons from the diffusons. $\tau(\omega)$ is a lifetime which can be obtained from Eq. 2.171. However, when the thermal conductivity is written this way, it is straightforward to also include the additional inelastic scattering experienced by the propagons. As shown in Fig. 2.7, this was done by the authors of (Feldman et al., 1993) who succeeded in reproducing the behavior of the thermal conductivity as a function of temperature. It is clear from this result that the plateau around 20 K comes from the superposition of contributions of both propagons and diffusons. At low temperature the propagons dominate, but at higher temperature they are more strongly scattered leading to a decrease of the conductivity from which the plateau emerge. On the contrary, the contribution of diffusons to the thermal conductivity is very smooth. It increases monotonically with temperature and

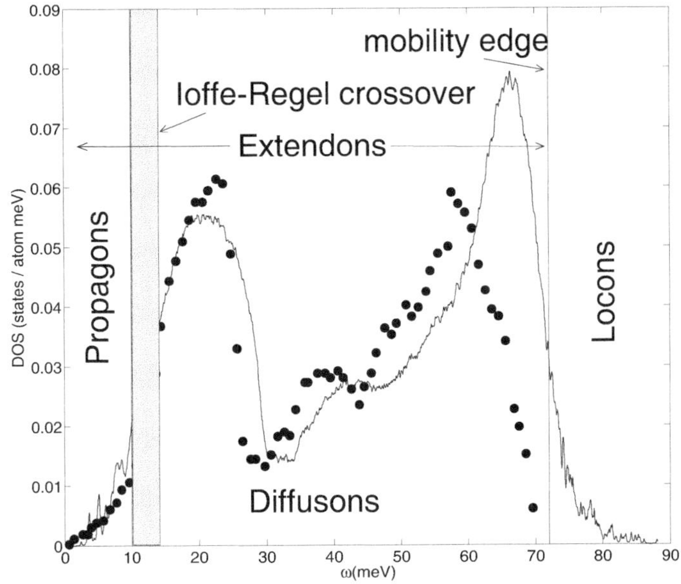

Figure 2.6 Density of states of amorphous silicon (Allen et al., 1999).

saturates to a finite value very much like the heat capacity. It is interesting to notice that even if only 4% of the states are propagons, around room temperature they carry 30–40% of the heat. Such a large contribution has been confirmed by other theoretical studies (He et al., 2011; Larkin and McGaughey, 2014) as well as by experiments (Regner et al., 2013). The long wavelength modes being very sensitive to geometrical modulations, porosification and nanostructuration should reduce the thermal conductivity of amorphous silicon (Newby et al., 2013).

Similar studies were reported for amorphous SiO_2 (Larkin and McGaughey, 2014) and it was shown that in this case the propagons contribute to the thermal conductivity for less than 6%. Diffusons can therefore contribute for a very large part. The underlying mechanism is, however, subtle enough.

Let us consider a system where a temperature gradient is established. If the vibrons were localized, it would be easy to construct the wavefunction representing such a state just by

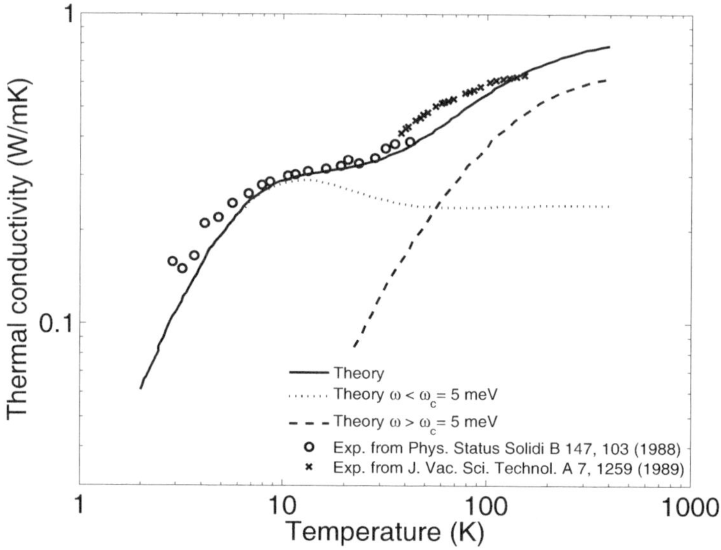

Figure 2.7 Thermal conductivity of amorphous silicon. Diamond and crossed are experimental data. The solid line represent the total thermal conductivity while the dashed and dotted lines are the contributions above and below $\omega_c = 5$ meV (Feldman et al., 1993).

populating more vibrons on the hot side of the system. But if we assume the vibrons to be delocalized then the wavefunction cannot be a stationary state. Instead, it must be a linear combination of such a states, which may eventually be regarded as a wavepacket picked around some typical occupation configuration. The expectation value of an observable A in such a state will always assume the form

$$\sum_{\{n\}\{n'\}} \bar{c}_{\{n\}} c_{\{n'\}} \langle \{n\} | A | \{n'\} \rangle. \tag{2.175}$$

This shows that non-diagonal elements ($\{n\} \neq \{n'\}$) contribute to the average like in Eq. 2.141. It means that energy is transferred from state $\{n\}$ to state $\{n'\}$ through the coupling to the energy current operator.

The mechanism is analogous to the electrical conductivity problem where electrons couple through the current operator to transfer energy. In this case the hamiltonian is

$$\delta H = -\int d^3r\, \boldsymbol{j}(\boldsymbol{r}) \cdot \boldsymbol{A}(\boldsymbol{r}, t) \tag{2.176}$$

$$= \int d^3r \int_{-\infty}^{t} dt' \boldsymbol{j}(\boldsymbol{r}) \cdot \boldsymbol{E}(\boldsymbol{r}, t') \tag{2.177}$$

$$= T \int d^3r \int_{-\infty}^{t} dt' \sigma_S \tag{2.178}$$

because in the transverse gauge $\boldsymbol{E} = -\partial A/\partial t$ and the entropy production is $\sigma_S = \boldsymbol{E} \cdot \boldsymbol{J}/T$.

In our case, $\nabla(1/T)$ being the thermodynamic force creating the energy flux, we obtain

$$\sigma_S = \boldsymbol{j} \cdot \nabla \frac{1}{T} \tag{2.179}$$

and therefore

$$\delta H = -\frac{1}{T} \int d^3r \int_{-\infty}^{t} dt' \boldsymbol{j}(\boldsymbol{r}) \cdot \nabla T. \tag{2.180}$$

This way an effective perturbation including the temperature gradient is exposed. Indeed, it was the basic assumption made by the authors of (Allen and Feldman, 1989, 1993) who first derived Eq. 2.141.

2.5.4 Configurational Disorder

In the previous section, the computational cell must be chosen large enough for the low energy vibrons to be accurately described, but also for any configuration of the local environment to be statistically well represented.

In the case of configurational disorder the topology do not change from site to site, but instead each site is randomly occupied by one of several atomic species. The prototypical example of such model is a binary compound A_xB_{1-x} with the atoms A and B randomly distributed on a lattice (Ducastelle, 1991).

In such disordered systems measurable quantities like the thermal conductivity are assumed to be given by their average over all possible distributions of the atoms A and B compatible with the chemical composition, the so-called atomic configurations. The average over atomic configuration is usually denoted using double brackets. For example, $<< \kappa >>$ for the thermal conductivity.

To obtain this average by brute force calculations is practically impossible due to the prohibitive cost the computations. Instead this configuration averaging can be done statistically, by considering a single crystal together with a probability occupation function for the sites. This way, once the averaging is performed, an effective medium is defined. The advantage of this approach is that approximations can be obtained, and some of them can even be carried out analytically.

The general idea of such approximations is to say that the averaging need not be done on the thermal conductivity itself, but on more elementary quantities which still contain enough relevant physical information to obtain a good approximation for the thermal conductivity. For example, one could average the masses and the force constants and then compute the thermal conductivity using such average quantities. This approximation is called a *virtual crystal approximation* because an average crystal is actually *defined* by the average masses and force constants.

The authors of (Garg et al., 2011) uses this approximation to study the $Si_{1-x}Ge_x$ crystal alloys. They assume a Boltzmann-Peierls form for the thermal conductivity, as given by Eq. 2.152, together with a lifetime due to anharmonicity, Eq. 2.166, and another one due to mass fluctuations (Tamura, 1983),

$$\frac{1}{\tau_{\mathbf{k}j}} = \frac{\pi}{2} g \omega_{\mathbf{k}j}^2 \frac{N(\omega_{\mathbf{k}j})}{3\mathcal{N}}. \tag{2.181}$$

g is the relative mass fluctuation,

$$g = \frac{\Delta m^2}{\bar{m}^2}, \tag{2.182}$$

and $\bar{m}$ and Δm^2 the average mass and variance

$$\bar{m} = x m_{Si} + (1-x) m_{Ge} \tag{2.183}$$

$$\Delta m^2 = x(m_{Si} - \bar{m})^2 + (1-x)(m_{Ge} - \bar{m})^2. \tag{2.184}$$

In other words the heat is assumed to be carried by long wavelength propagons, and the thermal resistivity created through their scattering, like in crystals. Within such an approximation the details of the bonding around each atom in any particular

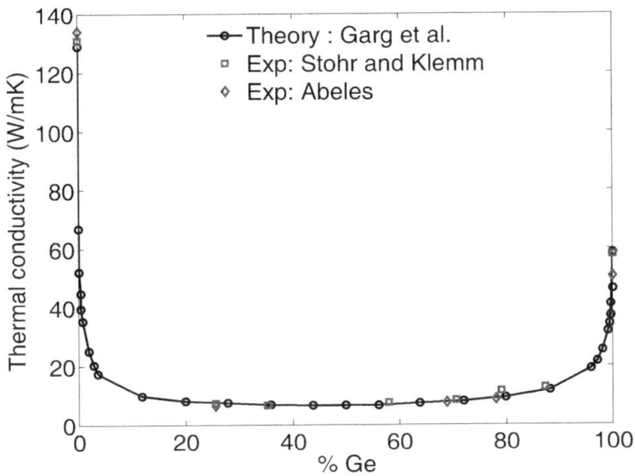

Figure 2.8 Thermal conductivity of $Si_{1-x}Ge_x$ at 300 K (Garg et al., 2011).

configuration must not be too important. The force constants are computed from an interpolation between the parents compounds, Si and Ge, and, say, the ordered alloy $Si_{0.5}Ge_{0.5}$. Again, this define a virtual crystal.

When the thermal conductivity is computed this way, at room temperature, 82% of the heat is carried by the phonons with a frequency lower than 1 THz. It seems compatible with the approximations which were made. This result is reproduced in Fig. 2.8 as function of the germanium concentration. At low doping the thermal conductivity drop rapidly because of the strong scattering of the phonon by the mass fluctuations. It reaches an almost constant value in the range $0.2 < x < 0.8$, with a minimum around $x = 0.5$ where the mass fluctuations are the largest.

Feldman and co-workers (Feldman et al., 1993) studied the amorphous $Si_{1-x}Ge_x$ alloys. They included configurational disorder through mass fluctuations, but also consider the structural disorder of the previous section. To do that a large supercell with randomly distributed Si and Ge atoms is considered and the thermal conductivity computed according to Eq. 2.174. They obtained the results shown in Fig. 2.9, which clearly shows similarity with the crystalline alloys of Fig. 2.9, but with much lower values.

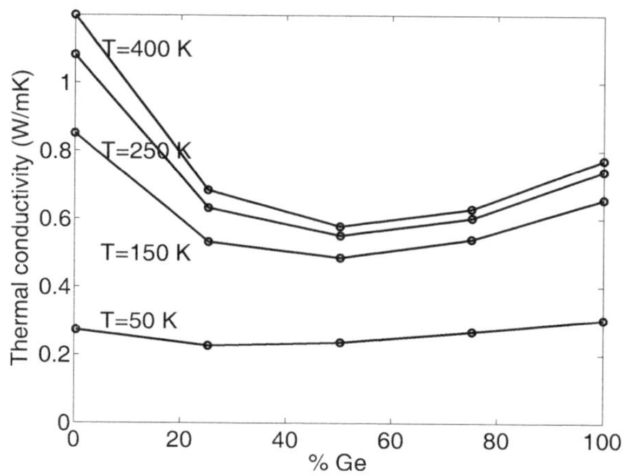

Figure 2.9 Thermal conductivity of amorphous $Si_{1-x}Ge_x$ at different temperatures (Feldman et al., 1993).

To end this section, it should be said that the virtual crystal, as presented here, is an uncontrolled approximation, even if it can lead to very good results. In principle, what should be computed is $<< \kappa >>$. A step in this direction as been taken in (Alam and Mookerjee, 2005; Alam et al., 2011).

2.6 Summary

In this chapter we have reviewed the theory of heat transport in a setting adequate to describe ordered and disordered solids.

The hydrodynamical description of heat conduction is given first, after which a microscopic theory is derived. In particular, the Kubo formula for the thermal conductivity is obtained from the linear response theory and the expression of the energy current operator.

The practical implementation of those formulas is presented in the last section. In the case of ordered bulk solids, the equations can be cast into the form of the Boltzmann theory of transport, and, when the parameters are obtained within the framework of density functional theory, this approach has been proved to be accurate in many cases (see Fig. 2.3).

Even if the theory seems to be based on solid grounds, in the case of disordered systems not enough examples have been investigated from first-principle calculations to be able to draw a very unified view of transport. It is clear that both propagons and diffusons should be considered, but their relative contribution is unclear and would need to be discussed, for example, in term of the range of disorder and the chemical elements which constitute the solid. In this context the finite-size models that are used to study disordered systems may be a severe limitation. Also, in these studies the effects of anharmonicity have not been included from first principles. This may alter the relative contribution of propagons and diffusons, which then compete with locons. Therefore, much work remains to be done on disordered phases for the picture of the solid to become as clear and detailed as it is for the ordered phases.

Appendix A: Linear Response Theory

Let us consider the system response to a perturbation

$$\delta H(t) = -BF(t) \qquad t > t_0 \tag{A.1}$$

applied at a time t_0. $F(t)$ can seem as a driving force, and B an operator to which it couples.

The hamiltonian of the system is

$$H = H_0 + \delta H(t) \qquad t > t_0 \tag{A.2}$$

if H_0 is the hamiltonian prior to the perturbation.

Our goal is to compute the expectation value of an observable $A(t)$ for $t > t_0$, i.e. $\langle \Psi(t)|A(t)|\Psi(t)\rangle$. $\Psi(t)$ is a state which can be obtained from a solution of the Schrödinger equation,

$$i\hbar\frac{\partial}{\partial t}\Psi(t) = (H_0 + \delta H(t))\Psi(t), \tag{A.3}$$

or

$$\Psi(t) = \Psi(t_0) - \frac{i}{\hbar}\int_{t_0}^{t} dt'\,(H_0 + \delta H(t'))\Psi(t') \tag{A.4}$$

$$\equiv T\exp\left[-\frac{i}{\hbar}\int_{t_0}^{t} dt'\,(H_0 + \delta H(t'))\right]\Psi(t_0). \tag{A.5}$$

In the last equation the time ordered exponential, T exp, is defined by the power series expansion obtained from successive iterations of the previous line. We obtain

$$\langle\Psi(t)|A(t)|\Psi(t)\rangle = \langle\Psi(t_0)|T e^{\frac{i}{\hbar}\int_{t_0}^{t} dt' \,(H_0+\delta H(t'))} A(t) T e^{\frac{-i}{\hbar}\int_{t_0}^{t} dt' \,(H_0+\delta H(t'))}|\Psi(t_0)\rangle \quad \text{(A.6)}$$

if the perturbation is hermitian.

Expanding the exponentials allow to obtain the expectation value of $A(t)$ as a polynomial in $H_0 + \delta H(t')$. This is very inconvenient because it is the perturbation $\delta H(t')$ which is expected to be small. A polynomial in $\delta H(t')$ instead would allow to truncate the series with good accuracy. To rewrite the previous equation in such a form the trick is to look for a solution written as

$$\Psi(t) = e^{-\frac{i}{\hbar}H_0(t-t_0)}\Phi(t). \quad \text{(A.7)}$$

Substituting in Eq. A.3 gives

$$i\hbar\frac{\partial}{\partial t}\Phi(t) = \left[e^{\frac{i}{\hbar}H_0(t-t_0)}\delta H(t)e^{-\frac{i}{\hbar}H_0(t-t_0)}\right]\Phi(t) \quad \text{(A.8)}$$

$$\equiv \delta H^I(t)\Phi(t) \quad \text{(A.9)}$$

with $\delta H^I(t)$ the interaction representation of $\delta H(t)$. In this chapter, all operators have a representation in the interaction picture defined in a similar way.

Solving the Schrödinger equation as before we obtain

$$\Phi(t) = T\,\exp\left[-\frac{i}{\hbar}\int_{t_0}^{t} dt'\,\delta H^I(t')\right]\Phi(t_0) \quad \text{(A.10)}$$

$$= T\,\exp\left[-\frac{i}{\hbar}\int_{t_0}^{t} dt'\,\delta H^I(t')\right]\Psi(t_0) \quad \text{(A.11)}$$

and therefore

$$\langle\Psi(t)|A(t)|\Psi(t)\rangle = \langle\Psi(t_0)|T e^{\frac{i}{\hbar}\int_{t_0}^{t} dt'\,\delta H^I(t')} A^I(t) T e^{\frac{-i}{\hbar}\int_{t_0}^{t} dt'\,\delta H^I(t')}|\Psi(t_0)\rangle. \quad \text{(A.12)}$$

Our program has therefore been accomplished, but at the price of having the observable in the interaction picture, $A^I(t)$.

In many applications it is necessary to average the pure states over a thermal ensemble,

$$\langle A(t)\rangle = \sum_i p_i\langle\Psi_i(t)|A(t)|\Psi_i(t)\rangle \quad \text{(A.13)}$$

with $p_i = \exp(-\beta E_i)/Z$ if one considers a canonical ensemble. E_i are the eigenvalues of the hamiltonian and Z is the partition function.

We obtain

$$\langle A(t)\rangle = \langle T e^{\frac{i}{\hbar}\int_{t_0}^{t} dt'\,\delta H^I(t')} A^I(t) T e^{\frac{-i}{\hbar}\int_{t_0}^{t} dt'\,\delta H^I(t')}\rangle_0, \tag{A.14}$$

where $\langle . \rangle_0$ means that we average over the initial states, $\Psi_i(t_0)$, which can be taken to be the eigenvectors $|i\rangle$ of H_0.

To linear order in $\delta H(t)$,

$$\begin{aligned}\langle A(t)\rangle &\approx \langle (1 + \frac{i}{\hbar}\int_{t_0}^{t} dt'\,\delta H^I(t'))A^I(t)(1 - \frac{i}{\hbar}\int_{t_0}^{t} dt'\,\delta H^I(t'))\rangle_0 \\ &\approx \langle A(t)\rangle_0 - \frac{i}{\hbar}\int_{t_0}^{t} dt'\,\langle [A^I(t), \delta H^I(t')]\rangle_0 \end{aligned} \tag{A.15}$$

and therefore, for the first variation, $\delta A(t) = \langle A(t)\rangle - \langle A(t)\rangle_0$, we obtain

$$\delta A(t) = -\frac{i}{\hbar}\int_{t_0}^{t} dt'\,\langle [A^I(t), \delta H^I(t')]\rangle_0. \tag{A.16}$$

If the perturbation is vanishingly small in the past, the integral will converge when we choose $t_0 = -\infty$. In such a case

$$\delta A(t) = -\frac{i}{\hbar}\int_{-\infty}^{t} dt'\,\langle [A^I(t), \delta H^I(t')]\rangle_0 \tag{A.17}$$

$$= \frac{i}{\hbar}\int_{-\infty}^{t} dt'\,\langle [A^I(t), B^I(t')]\rangle_0 F(t') \tag{A.18}$$

$$= \int_{-\infty}^{+\infty} dt'\,\chi_{AB}(t, t')F(t'). \tag{A.19}$$

In the second line, we have used the explicit form of the perturbation, as given by Eq. A.1, and in the third line, we have defined

$$\chi_{AB}(t, t') = \frac{i}{\hbar}\theta(t - t')\langle [A^I(t), B^I(t')]\rangle_0 \tag{A.20}$$

where $\theta(t)$ is the Heaviside step function.

$\chi_{AB}(t, t')$ is known as the response function of the observables A and B. It has the important properties to be translationally invariant in time,

$$\chi_{AB}(t, t') = \chi_{AB}(t - t'). \tag{A.21}$$

It is easily proved by inserting a complete set of eigenvectors in the expectation value and assuming that A and B does not have an explicit time dependence,

$$
\begin{aligned}
\langle [A^I(t), B^I(t')] \rangle_0 &= \langle A^I(t) B^I(t') - B^I(t') A^I(t) \rangle_0 \\
&= \sum_{ij} p_i \langle i | e^{\frac{i}{\hbar} H_0 (t-t_0)} A e^{-\frac{i}{\hbar} H_0 (t-t_0)} | j \rangle \\
&\quad \langle j | e^{\frac{i}{\hbar} H_0 (t'-t_0)} B e^{-\frac{i}{\hbar} H_0 (t'-t_0)} | i \rangle \\
&\quad - \sum_{ij} p_i \langle i | e^{\frac{i}{\hbar} H_0 (t'-t_0)} B e^{-\frac{i}{\hbar} H_0 (t'-t_0)} | j \rangle \\
&\quad \langle j | e^{\frac{i}{\hbar} H_0 (t-t_0)} A e^{-\frac{i}{\hbar} H_0 (t-t_0)} | i \rangle \\
&= \sum_{ij} p_i \Big(A_{ij} B_{ji} e^{\frac{i}{\hbar}(t-t')(E_i - E_j)} \\
&\quad - B_{ij} A_{ji} e^{-\frac{i}{\hbar}(t-t')(E_i - E_j)} \Big) \\
&= \sum_{ij} (p_i - p_j) A_{ij} B_{ji} e^{i\omega_{ij}(t-t')}
\end{aligned}
\tag{A.22}
$$

with $A_{ij} = \langle i | A | j \rangle$ and $\omega_{ij} = \frac{E_i - E_j}{\hbar}$.

Therefore we obtain that the response of observable A is simply given by the convolution of the response function χ_{AB} with the force which couple to observable B,

$$
\delta A(t) = \int_{-\infty}^{+\infty} dt' \, \chi_{AB}(t - t') F(t'). \tag{A.23}
$$

At this point it is useful to introduce the Fourier representations,

$$
\frac{1}{2\hbar} \langle [A^I(t), B^I(t')] \rangle_0 = \int \frac{d\omega}{2\pi} e^{-i\omega(t-t')} \chi''_{AB}(\omega) \tag{A.24}
$$

$$
\theta(t - t') = \int \frac{d\omega}{2\pi} e^{-i\omega(t-t')} \frac{i}{\omega + i\eta} \qquad \eta \to 0^+ \tag{A.25}
$$

$$
F(t) = \int \frac{d\omega}{2\pi} e^{-i\omega t} F(\omega) \tag{A.26}
$$

$$
\delta A(t) = \int \frac{d\omega}{2\pi} e^{-i\omega t} \delta A(\omega). \tag{A.27}
$$

We obtain

$$\delta A(t) = -2\int dt' \int \frac{d\omega}{2\pi}\int \frac{d\bar{\omega}}{2\pi}\int \frac{d\bar{\bar{\omega}}}{2\pi} e^{i(\omega+\bar{\omega}-\bar{\bar{\omega}})t'} e^{-i(\omega+\bar{\omega})t} \frac{\chi''_{AB}(\omega)}{\bar{\omega}+i\eta} F(\bar{\bar{\omega}}) \tag{A.28}$$

$$= \int \frac{d\bar{\bar{\omega}}}{2\pi} e^{-i\bar{\bar{\omega}}t}\left(\int \frac{d\omega}{\pi}\frac{\chi''_{AB}(\omega)}{\omega-\bar{\bar{\omega}}-i\eta}\right) F(\bar{\bar{\omega}}) \tag{A.29}$$

$$\tag{A.30}$$

and therefore

$$\delta A(\omega) = \chi_{AB}(\omega + i\eta)F(\omega) \tag{A.31}$$

if we define the analytic continuation of χ''_{AB} in the complex plane by

$$\chi_{AB}(z) = \int \frac{d\omega}{\pi}\frac{\chi''_{AB}(\omega)}{\omega - z}. \tag{A.32}$$

This function contains all the information about the system response, and moreover it is also the one which can be computed from many-body perturbation theory.

It possesses several additional important properties. First, we compute the Fourier transform of Eq. A.22 to obtain

$$\chi''_{AB}(\omega) = \frac{\pi}{\hbar}\sum_{ij}(p_i - p_j)A_{ij}B_{ji}\delta(\omega - \omega_{ji}). \tag{A.33}$$

Substituting in Eq. A.32 it gives

$$\chi_{AB}(z) = -\frac{1}{\hbar}\sum_{ij}(p_i - p_j)\frac{A_{ij}B_{ji}}{z - \omega_{ji}}. \tag{A.34}$$

This relation is known as the spectral representation of $\chi_{AB}(z)$. It shows that the excitation energies of the system ω_{ji} are given by the poles of the analytical continuation of the response function. Using

$$\int_0^\infty dt\, e^{it(z-\omega_{ji})} = \frac{i}{z-\omega_{ji}} \qquad \text{Im}\, z > 0 \tag{A.35}$$

we can rewrite $\chi_{AB}(z)$ as

$$\chi_{AB}(z) = \int_0^\infty dt\, e^{izt}\left(\frac{i}{\hbar}\sum_{ij}(p_i - p_j)A_{ij}B_{ji}e^{it\omega_{ij}}\right)$$

$$= \int_{-\infty}^\infty dt\, e^{izt}\chi_{AB}(t). \tag{A.36}$$

where Eq. A.22 has been used to recognize the commutator in the brace. Therefore the analytical continuation of Eq. A.32 can be seen as a Laplace transform in the upper part of the complex plane.

Finally, using the identity

$$\frac{1}{\omega + i\eta} = P\frac{1}{\omega} - i\pi\delta(\omega) \tag{A.37}$$

in Eq. A.32 we obtain

$$\chi_{AB}(\omega + i\eta) = P\int \frac{d\bar{\omega}}{\pi}\frac{\chi''_{AB}(\bar{\omega})}{\bar{\omega} - \omega} + i\chi''_{AB}(\omega) \tag{A.38}$$

$$\equiv \chi'_{AB}(\omega) + i\chi''_{AB}(\omega). \tag{A.39}$$

If $B = A^{+}$, Eq. A.33 shows that $\chi''_{AA^{+}}(\omega)$ is real. Therefore in this case the previous equation shows that the real part of the function $\chi_{AA^{+}}(\omega + i\eta)$, $\chi'_{AA+}(\omega)$, can be computed as the Hilbert transform of its imaginary part $\chi''_{AA^{+}}(\omega)$. This property is known as a Kramers–Kroning relation.

More properties of the χ_{AB} functions, including the Onsager reciprocity relations, can be found in (Kadanoff and Martin, 1963).

Appendix B: The Fluctuation-Dissipation Theorem

The fluctuation-dissipation theorem relates the Fourier transform of the average values of the commutator and anti-commutator of two observables. Those observables should not have an explicit time dependence. They should follow the Heisenberg equation of motion with respect to the hamiltonian at equilibrium, H_0.

To prove the theorem one first needs an identity which follows from the cyclic invariance of the trace. From Eq. A.13 we obtain

$$\begin{aligned}\langle A(\mathbf{r}, t)B(\mathbf{r}', t')\rangle_0 &= \frac{1}{Z}\mathrm{Tr}e^{-\beta H_0}A(\mathbf{r}, t)B(\mathbf{r}', t')\\ &= \frac{1}{Z}\mathrm{Tr}e^{\frac{i}{\hbar}(i\hbar\beta)H_0}A(\mathbf{r}, t)e^{-\frac{i}{\hbar}(i\hbar\beta)H_0}e^{-\beta H_0}B(\mathbf{r}', t')\\ &= \frac{1}{Z}\mathrm{Tr}e^{-\beta H_0}B(\mathbf{r}', t')A(\mathbf{r}, t + i\hbar\beta)\\ &= \langle B(\mathbf{r}', t')A(\mathbf{r}, +i\hbar\beta')\rangle_0.\end{aligned} \tag{B.1}$$

Now we define the Fourier transform, $f_{AB}(\boldsymbol{r}, \boldsymbol{r}', \omega)$, of the average of the product of two fluctuations.

$$\int \frac{d\omega}{2\pi} e^{-i\omega(t-t')} f_{AB}(\boldsymbol{r}, \boldsymbol{r}', \omega) = \langle \delta A(\boldsymbol{r}, t) \delta B(\boldsymbol{r}', t') \rangle_0.$$

It gives

$$\begin{aligned}\int \frac{d\omega}{2\pi} e^{-i\omega(t-t')} f_{AB}(\boldsymbol{r}, \boldsymbol{r}', \omega) &= \langle A(\boldsymbol{r}, t) B(\boldsymbol{r}', t') \rangle_0 - \langle A(\boldsymbol{r}) \rangle_0 \langle B(\boldsymbol{r}') \rangle_0 \\ &= \langle B(\boldsymbol{r}', t') A(\boldsymbol{r}, t + i\hbar\beta) \rangle_0 - \langle A(\boldsymbol{r}) \rangle_0 \langle B(\boldsymbol{r}') \rangle_0 \\ &= \int \frac{d\omega}{2\pi} e^{-i\omega(t'-t-i\hbar\beta)} f_{BA}(\boldsymbol{r}, \boldsymbol{r}', \omega)\end{aligned}$$

and therefore

$$f_{AB}(\boldsymbol{r}, \boldsymbol{r}', \omega) = f_{BA}(\boldsymbol{r}, \boldsymbol{r}', -\omega) e^{\hbar\omega\beta}. \tag{B.2}$$

In the second line, we have used the average value with respect to H_0 of an observable is time independent, and changed ω to $-\omega$ to obtain the result.

The anti-correlation of two fluctuations,

$$\frac{1}{2\hbar} \langle \{ \delta A(\boldsymbol{r}, t), \delta B(\boldsymbol{r}', t') \} \rangle_0 = \int \frac{d\omega}{2\pi} e^{-i\omega(t-t')} c_{AB}(\boldsymbol{r}, \boldsymbol{r}', \omega) \tag{B.3}$$

is therefore given by

$$c_{AB}(\boldsymbol{r}, \boldsymbol{r}', \omega) = \frac{1}{2\hbar} (f_{AB}(\boldsymbol{r}, \boldsymbol{r}', \omega) + f_{BA}(\boldsymbol{r}, \boldsymbol{r}', -\omega)) \tag{B.4}$$

$$= \frac{1}{2\hbar} f_{AB}(\boldsymbol{r}, \boldsymbol{r}', \omega)(1 + e^{-\hbar\omega\beta}). \tag{B.5}$$

In the same way the correlation of the observables,

$$\frac{1}{2\hbar} \langle [A(\boldsymbol{r}, t), B(\boldsymbol{r}', t')] \rangle_0 = \int \frac{d\omega}{2\pi} e^{-i\omega(t-t')} \chi''_{AB}(\boldsymbol{r}, \boldsymbol{r}', \omega) \tag{B.6}$$

is given by

$$\chi''_{AB}(\boldsymbol{r}, \boldsymbol{r}', \omega) = \frac{1}{2\hbar} (f_{AB}(\boldsymbol{r}, \boldsymbol{r}', \omega) - f_{BA}(\boldsymbol{r}, \boldsymbol{r}', -\omega)) \tag{B.7}$$

$$= \frac{1}{2\hbar} f_{AB}(\boldsymbol{r}, \boldsymbol{r}', \omega)(1 - e^{-\hbar\omega\beta}). \tag{B.8}$$

Therefore we obtain

$$c_{AB}(\boldsymbol{r}, \boldsymbol{r}', \omega) = \chi''_{AB}(\boldsymbol{r}, \boldsymbol{r}', \omega) \coth \frac{\hbar\omega\beta}{2}. \tag{B.9}$$

This is the fluctuation-dissipation theorem. It relates the fluctuation of two observables for a system in thermodynamical equilibriums, c_{AB}, to the dissipation of the system when it is coupled to an external perturbation. We have seen that χ''_{AB} is the response function to such perturbation, which therefore control the way energy is dissipated.

References

1. Alam, A., Chouhan, R. K. and Mookerjee, A. (2011). Thermal conductivity and diffusion-mediated localization in $fe_{1-x}cr_x$ alloys from first principles, *Phys. Rev. B* **84**, p. 224309, doi:10.1103/PhysRevB.84.224309.
2. Alam, A. and Mookerjee, A. (2005). Lattice thermal conductivity of disordered binary alloys, *Phys. Rev. B* **72**, p. 214207, doi:10.1103/PhysRevB.72.214207.
3. Allen, P. B. and Feldman, J. L. (1989). Thermal conductivity of glasses: Theory and application to amorphous si, *Phys. Rev. Lett.* **62**, pp. 645–648, doi:10.1103/PhysRevLett.62.645.
4. Allen, P. B. and Feldman, J. L. (1993). Thermal conductivity of disordered harmonic solids, *Phys. Rev. B* **48**, pp. 12581–12588, doi:10.1103/PhysRevB.48.12581.
5. Allen, P. B., Feldman, J. L., Fabian, J. and Wooten, F. (1999). Diffusons, locons and propagons: character of atomic vibrations in amorphous Si, *Philosophical Magazine, Part B* **79**, pp. 1715–1731, doi:10.1080/014186399255836.
6. Callen, H. B. and Welton, T. A. (1951). Irreversibility and generalized noise, *Phys. Rev.* **83**, pp. 34–40, doi:10.1103/PhysRev.83.34.
7. Chaput, L., Togo, A., Tanaka, I. and Hug, G. (2011). Phonon-phonon interactions in transition metals, *Phys. Rev. B* **84**, p. 094302, doi:10.1103/PhysRevB.84.094302, URL http://link.aps.org/doi/10.1103/PhysRevB.84.094302.
8. De Groot, S. and Mazur, P. (2013). *Non-Equilibrium Thermodynamics*, Dover Books on Physics (Dover Publications), ISBN 9780486153506.
9. Ducastelle, F. (1991). *Order and Phase Stability in Alloys*, Cohesion and structure (North-Holland), ISBN 9780444869739.
10. Enz, C. P. (1968). One-particle densities, thermal propagation, and second sound in dielectric crystals, *Ann. Phys.* **46**, pp. 114–173, doi:10.1016/0003-4916(68)90305-9.
11. Feldman, J. L., Allen, P. B. and Bickham, S. R. (1999). Numerical study of low-frequency vibrations in amorphous silicon, *Phys. Rev. B* **59**, pp. 3551–3559, doi:10.1103/PhysRevB.59.3551.
12. Feldman, J. L., Kluge, M. D., Allen, P. B. and Wooten, F. (1993). Thermal conductivity and localization in glasses: Numerical study of a model of amorphous silicon, *Phys. Rev. B* **48**, pp. 12589–12602, doi:10.1103/PhysRevB.48.12589.

13. Garg, J., Bonini, N., Kozinsky, B. and Marzari, N. (2011). Role of disorder and anharmonicity in the thermal conductivity of silicon-germanium alloys: A first-principles study, *Phys. Rev. Lett.* **106**, p. 045901, doi: 10.1103/PhysRevLett.106.045901.
14. Hardy, R. J. (1963). Energy-flux operator for a lattice, *Phys. Rev.* **132**, pp. 168–177, doi:10.1103/PhysRev.132.168.
15. He, Y., Donadio, D. and Galli, G. (2011). Heat transport in amorphous silicon: Interplay between morphology and disorder, *Applied Physics Letters* **98**, 14, 144101, doi:http://dx.doi.org/10.1063/1.3574366.
16. Horton, G. and Maradudin, A. (1974). *Dynamical Properties of Solids*, no. vol. 1 in Dynamical Properties of Solids (North-Holland), ISBN 9780720402780.
17. Kadanoff, L. P. and Martin, P. C. (1963). Hydrodynamic equations and correlation functions, *Ann. Phys.* **24**, pp. 419–469, doi:10.1016/0003-4916(63)90078-2.
18. Kubo, R. (1957). Statistical-mechanical theory of irreversible processes. i. general theory and simple applications to magnetic and conduction problems, *J. Phys. Soc. Jpn.* **12**, 6, pp. 570–586, doi:10.1143/JPSJ.12.570.
19. Kubo, R., Yokota, M. and Nakajima, S. (1957). Statistical-mechanical theory of irreversible processes. ii. response to thermal disturbance, *J. Phys. Soc. Jpn.* **12**, 11, pp. 1203–1211, doi:10.1143/JPSJ.12.1203.
20. Landau, L. and Lifshitz, E. (1977). *Quantum Mechanics: Non-Relativistic Theory*, Butterworth Heinemann (Butterworth-Heinemann), ISBN 9780750635394.
21. Landau, L. and Lifshitz, E. (2013). *Statistical Physics*, vol. 5 (Elsevier Science), ISBN 9780080570464.
22. Larkin, J. M. and McGaughey, A. J. H. (2014). Thermal conductivity accumulation in amorphous silica and amorphous silicon, *Phys. Rev. B* **89**, p. 144303, doi:10.1103/PhysRevB.89.144303.
23. Luttinger, J. M. (1964). Theory of thermal transport coefficients, *Phys. Rev.* **135**, pp. A1505–A1514, doi:10.1103/PhysRev.135.A1505.
24. Mahan, G. (2000). *Many-Particle Physics*, Physics of Solids and Liquids (Springer), ISBN 9780306463389.
25. Marder, M. (2000). *Condensed Matter Physics*, Wiley-Interscience (John Wiley), ISBN 9780471177791.
26. Messiah, A. (1961). *Quantum Mechanics*, Dover books on physics (Dover Publications), ISBN 9780486409245.

27. Newby, P. J., Canut, B., Bluet, J.-M., Gomès, S., Isaiev, M., Burbelo, R., Termentzidis, K., Chantrenne, P., Fréchette, L. G. and Lysenko, V. (2013). Amorphization and reduction of thermal conductivity in porous silicon by irradiation with swift heavy ions, *J. Appl. Phys.* **114**, 1, p. 014903, doi: 10.1063/1.4812280.
28. Peierls, R. (1955). *Quantum Theory of Solids*, International series of monographs on physics (Clarendon Press), ISBN 9780198507819.
29. Regner, K. T., Sellan, D. P., Su, Z., Amon, C. H., McGaughey, A. J. H. and Malen, J. A. (2013). Broadband phonon mean free path contributions to thermal conductivity measured using frequency domain thermoreflectance, *Nat. Commun.* **4**, 1640, doi:10.1038/ncomms2630.
30. Sham, L. J. (1967). Temperature propagation in anharmonic solids, *Phys. Rev.* **163**, pp. 401–407, doi:10.1103/PhysRev.163.401.
31. Srivastava, G. (1990). *The Physics of Phonons* (Taylor & Francis), ISBN 9780852741535.
32. Stillinger, F. H. and Weber, T. A. (1985). Computer simulation of local order in condensed phases of silicon, *Phys. Rev. B* **31**, pp. 5262–5271, doi:10.1103/PhysRevB.31.5262.
33. Tamura, S.-i. (1983). Isotope scattering of dispersive phonons in ge, *Phys. Rev. B* **27**, pp. 858–866, doi:10.1103/PhysRevB.27.858.
34. Togo, A., Chaput, L. and Tanaka, I. (2015). Distributions of phonon lifetimes in brillouin zones, *Phys. Rev. B* **91**, p. 094306, doi:10.1103/PhysRevB.91.094306, URL http://link.aps.org/doi/10.1103/PhysRevB.91.094306.
35. Wooten, F., Winer, K. and Weaire, D. (1985). Computer generation of structural models of amorphous si and ge, *Phys. Rev. Lett.* **54**, pp. 1392–1395, doi:10.1103/PhysRevLett.54.1392.
36. Ziman, J. (1960). *Electrons and Phonons: The Theory of Transport Phenomena in Solids*, International series of monographs on physics (OUP Oxford), ISBN 9780198507796.

Chapter 3

Nanostructures and Heat Transport

Samy Merabia[a] **and Konstantinos Termentzidis**[b]
[a] *ILM, Université de Lyon; UMR 5306 Université Lyon 1 et CNRS, F-69622 Villeurbanne, France*
[b] *Université de Lorraine, LEMTA, Centre National de la Recherche Scientifique, UMR 7563, BP 70239, 54506 Vandoeuvre Cedex, France*
samy.merabia@univ-lyon1.fr, konstantinos.termentzidis@univ-lorraine.fr

3.1 Introduction

In this chapter, we develop the main ideas of the heat transport in the nanoscale. As we will explain above, the physical phenomena change considerably when we pass from the bulk materials to the nanostructured materials or nanostructures. The ratio interfaces-surfaces/total volume increases and the phonon mean free path (PMFP) becomes at the same order as the characteristic sizes of the nanostructures (width of the nanofilms, diameter for the nanowires, periodicity for the superlattices). When this happens, we pass from the diffusive to so-called ballistic regime. In this regime, the thermal conductivity is no longer a material property, but it depends on the geometrical parameters of the nanostructures, their stoichiometries, the doping, the point or line defects etc. Another

Nanostructured Semiconductors: Amorphization and Thermal Properties
Edited by Konstantinos Termentzidis

ISBN 978-981-4745-64-2 (Hardcover), 978-1-315-36445-2 (eBook)
www.panstanford.com

important difference between the macro and the nanoscale heat transfer is that in the macroscale we usually treat the carriers as particles, while in the nanoscale the wave effects can become dominant and we shall treat the carriers as waves to include wave phenomena as tunneling or interference.

With the advent of nanotechnologies during the early 1990s, there is a growing interest in understanding and tailoring thermal transport at the nanoscale. Nanostructuration may be viewed as a way to tune the thermal transport properties of a material, as compared with the thermal conductivity characterizing the bulk state. Typically, nanostructuration is a way to reduce the phononic thermal conductivity of a bulk material. This reduction is realized because of the large density of interfaces present in the nanostructured materials, which strongly scatter the energy carriers thus contributing to deteriorate energy transport at least in the direction perpendicular to the interfaces. This short discussion anticipates the relation between the thermal conductivity which measures the ability of a material to transport energy and the scattering mechanisms. This section aims at providing a quantitative link.

The temperature dependence of the thermal conductivity for insulating solids and semiconductors follows in general three distinct regimes. At low temperatures, there are very few phonons as the atoms do not vibrate and the thermal conductivity is low. Increasing the temperature, the long-wavelength phonons are limited only by the size of the sample and from the Debye model the overall, thermal conductivity depends on the third power of the temperature ($\lambda \propto T^3$). The maximum is reached around $1/10\,\theta_D$ and it is governed by the boundary or imperfections scattering. In the third regime at higher temperatures, the $\lambda \propto T^{-1}$ and the thermal resistivity is dominated by the Umklapp scattering. We will explain in the following paragraphs the different scattering mechanisms, which provide thermal resistance over the different temperature regimes. In low temperatures phonons scatter mainly on the grain boundaries, while at high temperatures the resistive processes are the phonon–phonon anharmonic scattering. At intermediate temperatures isotopic mass defects are the dominant resistive processes. We first remind the reader with the classical Fourier's law:

$$\vec{J} = -\lambda\vec{\nabla}T, \tag{3.1}$$

which relates the heat flux vector $\vec{J}$ to the local temperature gradient $\vec{\nabla}T$. The latter equation is strictly speaking valid in an isotropic medium, where the relation between the heat flux and the temperature gradient is independent of the direction of the temperature gradient. Nanostructured materials are generally not isotropic. Take the example of a solid superlattice made of alternating solid layers. Clearly it could be anticipated that the conductivity in the direction parallel to the interface is different from the conductivity in the direction perpendicular to the interfaces. For such anisotropic materials, Fourier's law takes the tensorial form:

$$J_\alpha = -\lambda_{\alpha,\beta}\nabla_\beta T, \tag{3.2}$$

where α and β denote two spatial directions, and we have introduced the conductivity tensor $\lambda_{\alpha,\beta}$. Commonly, only the diagonal elements of the conductivity tensor take nonvanishing values. For instance, for a superlattice grown in the z direction, the conductivity tensor is

$$\begin{pmatrix} \lambda_{\mathrm{IP}} & 0 & 0 \\ 0 & \lambda_{\mathrm{IP}} & 0 \\ 0 & 0 & \lambda_{\mathrm{CP}} \end{pmatrix} = \lambda, \tag{3.3}$$

where λ_{IP} and λ_{CP} are, respectively, the in-plane and cross-plane thermal conductivities of the superlattice [51]. We shall conclude this paragraph by mentioning that strictly speaking Fourier's law means the thermal conductivity is a constant at a given temperature, and as such is independent of the size of the system. However, in some situations it is rather possible to define an effective thermal conductivity through

$$\vec{J} = -\lambda(L)\vec{\nabla}T, \tag{3.4}$$

which depends on the size L of the system, and we have assumed isotropy of the material for simplicity. We insist that although Eq. (3.4) is similar to Eq. (3.1), it does not always encompass the Fourier assumption of diffusive regime. Such a situation may be encountered when a fraction of the energy carriers has a mean free path comparable with the system size L. We will briefly discuss this issue in the section concerning boundary scattering.

Having defined the thermal conductivity, we now derive an expression of the thermal conductivity. The thermal conductivity has different contributions, each one stemming from a different excitation. The most common excitations in solids are electrons and phonons. Other excitations such as spin waves may contribute in specific cases but we shall ignore them in the rest of this chapter. Phonons are collective excitations of the lattice in a solid. The thermal conductivity may then be decomposed in

$$\lambda = \lambda_{\rm e} + \lambda_{\rm ph}, \tag{3.5}$$

where the first term is the electronic thermal conductivity, while the latter term is the phononic thermal conductivity. The first term is typically dominant in noble metals or in some transition metals where the phononic contribution may be discarded. For metal alloys or impure metals, the electronic and the phononic conductivities have equal importance. In semiconductors, heat transport is mostly carried by phonons and the electronic contribution may be ignored. The electronic thermal conductivity is generally well described by the Wiedemann–Franz law [56]:

$$\frac{\lambda_{\rm e}}{\sigma_{\rm e}} = LT, \tag{3.6}$$

where $\sigma_{\rm e}$ is the electronic conductivity , and L is the metals' Lorenz number. The Wiedemann–Franz law states that the electronic thermal conductivity is directly proportional to the electronic conductivity. This may be used for instance to estimate the value of the electronic thermal conductivity provided we are able to measure the electronic transport properties of a metal. The Lorenz number L depends on the metal and may take values close to the Sommerfeld value of the Lorenz number:

$$L \equiv L_0 = \frac{\pi^2}{3}\left(\frac{k_B}{e}\right)^2 \simeq 2.44\ 10^{-8} V^2 K^{-2}, \tag{3.7}$$

which is obviously independent of the metal. For a broad range of metals, the value of the Lorenz number measured experimentally does not differ significantly from its Sommerfeld value [9].

The Wiedemann–Franz law is valid at high temperatures $T > \theta_D$, where θ_D is the metal's Debye temperature or at very low temperatures, where electron scattering is elastic. At intermediate temperatures, one observes large deviations of the Wiedemann's law due to the inelastic nature of electron scattering.

In the following, we will focus our attention on the phononic contribution to the conductivity, and consider semiconductors as a paradigm example. As said before, the thermal conductivity is then controlled by collective lattice dynamics, the latter being described by phonons. Phonons are quanta of vibrations, in the same manner as photons are quanta of electromagnetic radiation. Consider a simple solid lattice, and denote $\vec{r}_i^0$ the equilibrium positions of the ions, which are the average atom positions occupied by the ions at the considered temperature. We can define an instantaneous displacement vector $\vec{u}_i(t) = \vec{r}_i(t) - \vec{r}_i^0$, which measures the displacement of the atom labeled i from its equilibrium position. The Hamiltonian of the crystal may be written:

$$\mathcal{H} = \mathcal{H}_0(T) + \sum_{i,j} \frac{1}{2} \Phi_{\alpha,\beta}(i, j) u_{i,\alpha} u_{j,\beta} + \sum_{i,j,k} \frac{1}{6} \Psi_{\alpha,\beta,\gamma}(i, j, k) u_{i,\alpha} u_{j,\beta} u_{k,\gamma}, \tag{3.8}$$

where we have introduced the reference Hamiltonian $\mathcal{H}_0(T)$ which might depend on the temperature, and the harmonic and anharmonic force constant tensors $\Phi_{\alpha,\beta}(i, j)$ and $\Psi_{\alpha,\beta,\gamma}(i, j, k)$. The indexes α, β denote spatial directions, and we have used Einstein convention for the summation in Eq. (3.8). The force constants may be calculated using ab initio calculations or they may be derived from an empirical potential. Of course, we have truncated the development of the Hamiltonian to the third order terms and highest order anharmonic terms may be defined as well, but for the prediction of the thermal conductivity of most of the semiconductors this truncation is sufficient. At the leading order, the displacement obeys the equation of motion:

$$m_i \frac{d^2 u_{i,\alpha}}{dt^2} = -\sum_{j,\beta} \Phi_{\alpha,\beta}(i, j) u_{j,\beta} \tag{3.9}$$

The solution may be decomposed in normal modes:

$$u_{i,\alpha}(t) = \sum_{\vec{k}} \frac{Q(\vec{k})}{m_i} e_\alpha(\vec{k}) \exp(i(\vec{k} \cdot \vec{r}_i^0 - \omega(\vec{k})t)), \tag{3.10}$$

where the sum runs over the wave vectors $\vec{k}$ of the reciprocal lattice and we have introduced the normal coordinates $Q(\vec{k})$, the polarization vector $\vec{e}(\vec{k})$ and the angular frequency $\omega(\vec{k})$. It is a rather simple exercise to show that the normal coordinates obey the equipartition theorem [55]:

$$\omega^2(\vec{k})\langle|Q(\vec{k})|^2\rangle = \hbar\omega(\vec{k})\left(n_{\mathrm{eq}}(\omega(\vec{k}), T) + \frac{1}{2}\right), \tag{3.11}$$

where $\hbar$ is Planck's constant, brackets indicate a thermal average and we have introduced the Bose–Einstein factor:

$$n_{\mathrm{eq}}(\omega, T) = \frac{1}{\exp(\hbar\omega/k_B T) - 1} \tag{3.12}$$

The latter relation justifies the term phonons, i.e., quanta of vibrations. The polarizations vectors are solutions of the secular equation:

$$D_{\alpha,\beta}(\vec{k})e_\alpha(\vec{k}) = \omega^2(\vec{k})e_\alpha(\vec{k}), \tag{3.13}$$

where D is the D-type dynamical matrix [8]:

$$D_{\alpha,\beta}(\vec{k}) = \sum_{j \text{ neighb i}} \frac{\Phi_{\alpha,\beta}(i, j)}{\sqrt{m_i m_j}} \exp(i(\vec{k}\cdot(\vec{r}_j^{\,0} - \vec{r}_i^{\,0})), \tag{3.14}$$

where the sum runs over the atoms j neighbors of i and the result is independent on the reference atom i because of the translational invariance of the crystal. Indeed, here we have considered simple crystal structures with one atom per primitive cell. For these structures, the dynamical matrix has dimensions 3×3. For more complex crystal structures containing p atoms per primitive cell, the dynamical matrix has dimensions $3p \times 3p$.

So far, the collective excitations that we described have infinite lifetime. This is understandable since for the harmonic theory that we consider, the Hamiltonian may be decomposed in independent normal modes. This means that the initial energy present in each of the normal mode is constant, and two different mode do not interact at all. This conclusion is of course contrary to our intuition, since it would lead to an infinite thermal conductivity. The element that is missing in our analysis is the anharmonic terms in the crystal Hamiltonian Eq. (3.8). The effect of the anharmonic terms is to induce scattering between two different normal modes, so that the lifetime of a mode becomes finite. To put this on quantitative grounds, we consider the normal mode decomposition Eq. (3.10), where we replace the constant amplitude $Q(\vec{k})$ by a time dependent amplitude $Q(\vec{k}, t)$. This way, we implicitly assume that the characteristic time scale of the variation of the amplitude $Q(\vec{k}, t)$ is larger than the phonon mode period $T = 2\pi/\omega(\vec{k})$. This assumption is partially justified, since the previous harmonic theory describes accurately the phonon dispersion curves as measured experimentally. We will come to this assumption later on.

To determine the thermal conductivity, we shall write an evolution equation for the phonon occupation number $n(\vec{k}, t)$. In numerous of situations, this equation may take the simple form:

$$\frac{\partial n}{\partial t} + \frac{\partial n_{\mathrm{eq}}}{\partial T}\vec{v}_g \cdot \vec{\nabla} T = -\frac{(n - n_{\mathrm{eq}})}{\tau(\vec{k})} \tag{3.15}$$

This form is called the Boltzmann equation under the relaxation time approximation. Its interpretation is quite simple: the first term on the left hand side describes the local change in the phonon occupation number, the second describes the advection in the local temperature gradient $\vec{\nabla} T$, while the term on the right hand side accounts for phonon relaxation with a mode dependent relaxation time $\tau(\vec{k})$. This relaxation time may have different physical origins, as discussed in the next section.

It is this latter term that is approximated as a term proportional to the difference between the instantaneous phonon number and its equilibrium value n_{eq}. From the Boltzmann equation, it is a rather simple exercise to show that the thermal conductivity writes:

$$\lambda_{\alpha,\beta} = \frac{1}{V}\sum_{\vec{k}} v_{g,\alpha} v_{g,\alpha} \tau(\vec{k}) \hbar\omega(\vec{k}) \frac{\partial n_{\mathrm{eq}}}{\partial T}, \tag{3.16}$$

where V is the volume of the system, and the sums runs over the first Brillouin zone of the solid. In case of isotropic material, the thermal conductivity takes the form:

$$\lambda = \frac{1}{3}\int_0^{\omega_{\max}} g(\omega) v_g^2(\omega) \tau(\omega) \hbar\omega \frac{\partial n_{\mathrm{eq}}}{\partial T} d\omega, \tag{3.17}$$

where we have introduced the vibrational density of states $g(\omega) = \frac{1}{V}\sum_{\vec{k}}\delta(\omega - \omega(\vec{k}))$. Now the challenge is to find the frequency dependence of the mode dependent relaxation time, and to do this we must first say some words about the physical mechanisms behind the relaxation processes. Note also to conclude that the form Eq. (3.17) may take the familiar form:

$$\lambda = \frac{1}{3}C_v v_g \Lambda, \tag{3.18}$$

where C_v is the solid heat capacity, and Λ is a phonon mean free path.

3.2 Scattering Mechanisms from Bulk to Nano

The harmonic vibration is rather a theoretical approach and it is valid only in very low temperatures. In a perfect infinite crystal, if one considers only harmonic scattering events, where the momentum is conserved, then the phonon mean free path would be infinite. This is not the case as even the materials with the highest thermal conductivity have a certain thermal resistance. The introduction of the anharmonic scattering events are then necessary to describe the heat transport in real materials.

Peierls first showed [99] that there are two kinds of phonon–phonon scattering, the normal (N) and the Umklapp (U) processes, depending if the process is harmonic or not. At the nanoscale the most important scattering mechanisms are the phonon-boundaries, phonon-impurities, and phonon-defects. Each scattering mechanism is related with a characteristic relaxation time τ. The total relaxation time considering that the scattering mechanisms are not related with each other can be given using Matthiessen's rule:

$$\frac{1}{\tau_{total}} = \sum_i \frac{1}{\tau_i}, \tag{3.19}$$

where i can be all scattering mechanisms.

Three phonons, N processes The normal processes vary linearly with ω. In this process, usually two acoustic phonons merge in a third one or a single acoustic phonon decay in two distinct phonons. In both cases, energy and momentum are preserved. In the specific case when two acoustic phonons create an optical one, again the process does not reduce the thermal conductivity, but this time affects the phonon populations.

Three phonon Umklapp, U processes Umklapp scattering dominates at high frequencies, as U processes vary with ω^2,

$$\frac{1}{\tau_U} = 2\gamma^2 \frac{k_B T}{\mu V_0} \frac{\omega^2}{\omega_D} \tag{3.20}$$

where γ is the Gruneisen anharmonic parameter, μ is the shear modulus, V_0 is the volume per atom and ω_D is the Debye frequency. These are the only phonon–phonon interactions (with the four-phonon scattering) who can affect the thermal conductivity. In these processes, the energy is always conserved but not the momentum. As Umklapp scattering requires a large population of phonons to interact with each other, it becomes the most dominant scattering event at room temperature and above.

Four-phonon scattering Four-phonon scattering processes are only important at high temperatures. The temperature dependence of the thermal conductivity related to the four phonon scattering processes is proportional to T^{-2}. The four-phonon relaxation rate varies as the square of the frequency and the square of the temperature [100]. It has been estimated that the four-phonon scattering processes are at least two to three orders of magnitude weaker than the three-phonon scattering. Four-phonon processes were first discussed by Pomeranchuk (J. Phys. USSR 4, 259 (1941); 7, 197 (1943)), who invoked them not as a direct source of thermal resistance, but to eliminate a divergence in the thermal-conductivity integral due to low-frequency longitudinal waves.

Phonon-boundaries (grain boundaries, interfaces, free surfaces) Boundary scattering is due to the interfaces such as grain boundaries in polycrystalline materials, or planar/rough interfaces in superlattices, or free surfaces in several nanostructures or nanostructured systems like nanowires, nanofilms, and nanoporous materials. This scattering mechanisms becomes very important in nanostructures as the interface/volume ratio is very high in comparison to bulk materials.

Phonon-impurities, point defects, dislocations Impurity scattering occurs when the material contains point defects as vacancies, interstitial or substitutional atoms, Frenkel pairs or line defects as screw, edge, or mixed dislocations.

Phonon-isotopes Phonon-isotope scattering occurs in compound materials with a variety of atoms with different masses. This process affects mainly the peak of the thermal conductivity, and it has been measured that the thermal conductivity of pure Ge is three times greater than the sample of Ge with mixed isotope.

Phonon-magnon There is evidence for strong anharmonic coupling between the ferroelectric transverse optic modes and the longitudinal acoustic modes in several magnetic materials (e.g., PbTe). This interaction directly affects the heat-carrying longitudinal acoustic phonons. The longitudinal acoustic-transverse optic anharmonic coupling might play a central role in explaining the very low thermal conductivity of PbTe [101].

Phonon-electrons, polaritons, photon Phonon-electron scattering occurs either inside bulk materials or at the interfaces of a metal with a semiconductor. We will not discuss further other scattering mechanisms of phonons with polaritons or photons, which are important and they need a detailed analysis, which is beyond the scope of the book. For the motivated readers, there are several books and review articles about this issue. For example, *Nanoscale Energy Transport and Conversion*, by G. Chen; *Microscale and Nanoscale Heat Transfer* and *Thermal Nanosystems and Nanomaterials*, edited by S. Volz; and *Nano/Microscale Heat Transfer*, by Z. M. Zhang, are some good references for further reading. Among numerous articles, we advice the readers to read the following: A. Balandin, "Phonon engineering," *J. Nanosci. Nanotech.* 5, 1015 (2005); D. Cahill et al., "Nanoscale thermal transport," *J. Appl. Phys.* 93, 793 (2003); A. Henry and G. Chen, *J. Comp. Theo. Nanosci.* 5, 1 (2008).

3.3 Interfacial Scattering: Models for the Thermal Boundary Conductance

Heat transfer at the nanoscale is controlled, as mentioned above, by the large density of interfaces. When the distance between interfaces becomes comparable with a few nanometers, heat transfer becomes no longer dominated by the intrinsic conduction of the materials between consecutive interfaces, but it is limited by the intrinsic thermal resistance displayed by the interface (see Fig. 3.1). Generally, interfacial thermal resistance may be understood in terms of a "thermal slip length" also called Kapitza length, namely the existence of a temperature discontinuity when a heat flux flows across the interface:

$$R = \frac{1}{G} = \frac{(T_2 - T_1)}{q}, \tag{3.21}$$

where R and G are the thermal interfacial resistance and conductance, and q is the heat flux per unit area flowing normal to the interface, and ΔT is the temperature discontinuity at the interface, as schematically represented in Fig. 3.1. The thermal boundary conductance is the inverse of the thermal boundary resistance. Alternately, it may be useful to define the Kapitza length l_K

$$l_K = \lambda / G \tag{3.22}$$

which depends on the conductivity of the conductor medium. This latter Kapitza length share some similarities with the hydrodynamic slip length introduced in nanofluidics to quantify the discontinuity of the fluid velocity parallel to the interface [20]. The Kapitza length provides an estimation of the characteristic length scale where the effect of a finite boundary resistance need to be accounted for. As we will see in the following, l_K is on the nanometer range. Hence, thermal boundary effects will be important when at least one of the dimensions is nanometric. On the contrary, when the characteristic dimensions of a device are larger than the micron, then one can assume the temperature field to be continuous at the interfaces, and the thermal boundary resistance does not play any role. As we anticipate, thermal boundary resistance is a prominent factor in the heat transport in nanostructured materials.

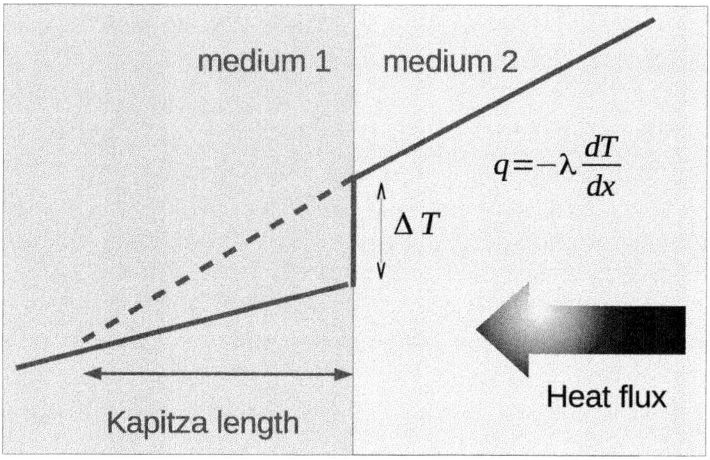

Figure 3.1 (Color online) Temperature jump across the interface between two media crossed by an interfacial flux q. The temperature profile in each medium is schematically represented by red solid lines.

The first observation of the existence of a finite thermal resistance at the interface between two media can be traced back to the experiments performed by Kapitza in the early forties on Helium II [21]. At the interface between Helium and metals, the Kapitza length is micronic owing to the large thermal conductivity of Helium II, which makes the effects observable.

The first measurements had been hence realized at cryogenic temperatures, smaller than 1 K. It is only during the years 1990 that the first experimental measurements of the interfacial thermal resistance between two solids at room temperature have been published [5]. More recently, thermoreflectance experiments have opened the way to measure the values of the interfacial resistance of a number of solids on a broad range of temperatures [6, 15]. Experimental data have shown that the existing theoretical models could highly underestimate the interfacial conductance. Here, we first consider interfacial thermal transport mediated by phonon–phonon scattering, and the role of free electrons will be discussed later on. Most of the theoretical models generally rely on Landauer's formalism which relates the interfacial conductance to the phonon transmission coefficient α_{12}:

$$G_{\text{eq}} = \frac{1}{2}\int_0^{\omega_c}\int_0^{\pi/2} C_1(\omega)v_1(\omega)\cos\theta_1\alpha_{12}(\omega)d\omega d\cos\theta_1 \qquad (3.23)$$

where C_1 is the heat capacity characterizing medium 1, v_1 is the phonon group velocity in medium 1 and ω_c is the maximal angular frequency transmitted at the interface, which is the maximal frequency above which the energy transmission vanishes. In Eq. (3.23), the integration variable θ_1 corresponds to the incidence angle, while ω denotes the angular frequency. Detailed balance ensures besides that the expression of the conductance Eq. (3.23) does not depend on the choice of medium 1. Let us mention also that in the previous expression, the heat capacities and group velocities may be easily calculated using the harmonic crystal model [56]. On the contrary, it is more difficult to predict the transmission coefficient α_{12}. Overall, we can distinguish two categories of theoretical models that may predict analytically the transmission coefficient α_{12}: DMM (diffuse mismatch models) [2] which assume totally diffuse phonon interfacial scattering; The phonon transmission coefficient does not depend on the angle of incidence θ_1 and writes then

$$\alpha_{12}(\omega) = \frac{g_2(\omega)/v_2}{g_1(\omega)/v_1 + g_2(\omega)/v_2}, \qquad (3.24)$$

where $g_i(\omega)$ is the vibrational density of states characterizing medium i, and v_i is the group velocity at the angular frequency ω. In general, it is equally assumed that the transmission vanishes beyond the cut-off frequency of the softer medium, which has the lowest Debye frequency. Here, the Debye frequency is equal to $\omega_D = (6\pi^2\rho)^{1/3}v_{\text{ac}}$ where ρ and v_{ac} are the solid number density and average acoustic velocity, respectively. AMM (acoustic mismatch models) type models [4, 44] assume specular phonon scattering at interfaces, and as a result the transmission is supposed to be governed by the acoustic impedances of the two media:

$$\alpha_{12}(\theta_1, \theta_2) = \frac{4Z_1Z_2\cos\theta_1\cos\theta_2}{(Z_1\cos\theta_1 + Z_2\cos\theta_2)^2}, \qquad (3.25)$$

where $Z_i = \rho_i v_i$ are the acoustic impedances of the two media, and $\cos\theta_1$, $\cos\theta_2$ are the cosines of the incidence and refraction angles. These angles are assumed to obey the equivalent of Snell–Descartes laws which rule the refraction of electromagnetic waves at the interface between two dielectric media:

$$\frac{\sin\theta_1}{v_1} = \frac{\sin\theta_2}{v_2} \tag{3.26}$$

if the angle is smaller than the critical angle $\theta_c = \sin^{-1}(v_1/v_2)$, in the case where medium 1 is softer than medium 2. At this point, it is important to bear in mind that the two above-mentioned models depend only on the *bulk* vibrational properties of the two media, and any information characterizing the interface structure is lost. In particular the effect of the strength of the interface is absent from the description, while it is known experimentally to affect thermal transport [27, 28]. To remedy this pitfall, Prasher developed recently a generalization of the AMM model which incorporates the finite energy of adhesion of the interface [4]. The phonon transmission coefficient writes then

$$\alpha_{12}(\theta_1, \theta_2) = \frac{4Z_1Z_2\cos\theta_1\cos\theta_2}{(Z_1\cos\theta_1 + Z_2\cos\theta_2)^2 + \frac{\omega^2}{K_{12}^2}(Z_1Z_2\cos\theta_1\cos\theta_2)^2}, \tag{3.27}$$

which depends on the interaction strength K_{12} (force constant per interfacial area) between two media. If the interaction between the solids is supposed to be of van der Waals type, the interaction strength writes $K_{12} = 72/2^{1/3}n_s\epsilon_{12}/\sigma_{12}^2$ where n_s is the number of solid atoms per interfacial area and ϵ_{12}, σ_{12} are the Lennard–Jones interaction energy and diameter. Additionally, one can show that K_{12} is related to the work of adhesion W_{12} between the solids according to $\epsilon_{12} = W_{12}/2\pi\rho_1\rho_2\sigma_{12}^4$. These relations allow to calculate the thermal boundary conductance as a function of the energy adhesion. Finally, we should note that we recover the original AMM model in the limit of strong bonding $K_{12} \to \infty$.

Another limitation of the previous models is the use of Landauer formula for the calculation of the thermal conductance, which describes the phonon population in each medium as a Bose–Einstein equilibrium distribution. The Landauer formalism is generally at the basis of transport calculations in mesoscopic physics, in particular to model electronic transport across a junction. The equilibrium hypothesis can be reasonable to describe electronic transport, in the sense that charge carriers are emitted by reservoirs large enough so that they are not perturbed by the existence of an electric current. The hypothesis is more problematic for the description of heat flux at the interface between two solids which cannot be described as a junction. In particular, the expression 3.23 shows that for an ideal interface which transmits all phonon modes, $\forall\omega, \alpha_{12}(\omega) = 1$, the thermal conductance is finite $G_\infty = \rho_1 v_1 k_B/4$. This is of course contrary to the intuition since an ideal interface should be characterized by a vanishing temperature discontinuity and thus an infinite conductance. Since the work of Simon in the seventies [22], several authors have proposed a correction to Landauer formalism which takes into account the heat flux in a self consistent manner [14, 22, 23]. The correction writes

$$G_{\text{neq}} = G_{\text{eq}}/(1 + \beta_{12} + \beta_{21}), \quad (3.28)$$

where the expression of the two coefficients β_{12} and β_{21} are given in [23], and are such that the conductance G_{neq} diverges when the two solids display a low acoustic contrast.

If we compare the predictions of both models AMM and DMM, it is curious to conclude that these two classes of models yield predictions generally close one from each other, but which may be far from the experimental values [10]. Different reasons have been proposed to explain these discrepancies: interfacial roughness, interfacial dislocations, interfacial stress generated by lattice mismatch between the two solids [16]. In this context, it is rather instructive to run molecular dynamics simulations, which allow to consider model interfaces which do not display imperfections thus opening the way to a fundamental understanding of interfacial phonon scattering [16]. Simulations of the thermal transport at perfect interfaces show that the DMM model fails to describe thermal transport at perfect interfaces, while the AMM model gives a good description of the energy flow over a broad range of acoustic contrasts. In this situation, phonon scattering is almost specular for all the phonon modes, except for high frequency modes that have a wavelength comparable with the atomic roughness [39].

The effect of interfacial roughness has been addressed from both experimental and theoretical sides. Indeed, it has been long recognized that roughness plays a key role in interfacial energy transfer [30–32]. In the context of liquid helium, the dominant phonon wavelength is comparable with the root mean square (rms) roughness $\sim$1 nm, and it is expected in this situation that the interface strongly scatters phonons beyond the critical angle. Because the critical angle between liquid Helium and metallic solids is rather small, roughness induced interfacial scattering may enhance the thermal conductance by orders of magnitude, as shown by Adamenko and Fuks [32]. At room temperature, and for common solids the effect of the interfacial roughness is far less pronounced because the acoustic contrast between the solids is smaller and the dominant phonon wavelength is also smaller. Experimentally, a way to systematically control the roughness is through chemical etching [33–35]. In these latter experimental investigations a decrease of the interfacial conductance with the rms roughness is demonstrated, and also interestingly they have shown that the thermal conductance of partially rough interfaces is less temperature dependent [33]. On the theory side, the effect of roughness on interfacial heat transport at solid interface has been addressed in detail since the work of Kekrachos in the early 1990s [36, 37], who used tools of field theory. More recently, molecular dynamics simulations have been used to probe the effect of roughness in heat transport mostly in the context of superlattices [50, 51, 59, 92]. Small thermal conductivity reductions have been also reported for Si/Ge superlattices with a one layer of interfacial mixing [42]. If it is reasonable to rationalize such variations in terms of an atomic interfacial roughness, it is less clear for superlattices having thicker mixed layer. Large interfacial conductance enhancements have been observed in this latter case, using MD [16]. Enhanced heat transfer is also observed for very rough interfaces, as phonon scattering in this limit becomes proportional to the true interfacial area [39–41]. MD results seem to conclude that these patterned surfaces are good candidates to enhance the intrinsic boundary conductance between two semiconductors [39]. On the other hand, randomly rough surfaces should be considered if one prefers to reduce the Kapitza conductance between two solids [33]. In particular, in the context of solid superlattices, randomly rough interfaces should be designed if one aims at designing materials with the lowest cross-plane thermal conductivity.

Up to now, we have essentially discussed thermal interfacial transport mediated by phonon scattering. There are, however, some situations where electron scattering may play a role in the thermal boundary conductance. Maybe the simplest case to consider is the interface between two metals [52]. The thermal conductance between a series of metal bilayers has been measured recently [53, 54], with reported values of the thermal boundary conductance commonly above 1000 MW m^{-2} K^{-1}. This indicates negligible effect of phonon–phonon interfacial scattering and heat transport dominated by free electron transport. This conclusion is confirmed by comparison between the experimental results and the interfacial form of the Wiedemann–Franz Law [54].

The case of metal-semiconductor interfaces is by far the less understood from a fundamental point of view. Note, however, that it is a very important issue for thermal management of microelectronic devices. Also, it is one of the most experimentally documented situation, because of the ability to heat up free electron metals, and the need to work with a nonmetal substrate. The boundary resistance of metal-diamond interfaces at room temperature has been first experimentally determined by Stoner and Maris in 1993 [5], using time domain thermoreflectance (TDTR). This work has been followed by the characterization of various interfaces, extending the early measurements to higher temperatures [6, 15, 66, 67]. More recently, TDTR has been applied to investigate heat transfer across metal–graphite [62] and metal–carbon nanotube interfaces as well [63]. In parallel, theoretical progress have been achieved through molecular dynamics [51, 68] which ignores electronic degrees of freedom, or treat them in the spirit of the two-temperature model [79]. A recurrent conclusion of this body of works is that the Kapitza conductances measured between two solids depart significantly from the classical AMM and DMM models, which often yield comparable values. For instance, the thermal conductance of one of the most studied system, the Pb diamond interface, is found to be typically sevenfold higher than the DMM prediction.

After the early room temperatures measurements of Stoner and Maris, several authors have proposed to relate this excess of conductance to electron–phonon scattering taking place across

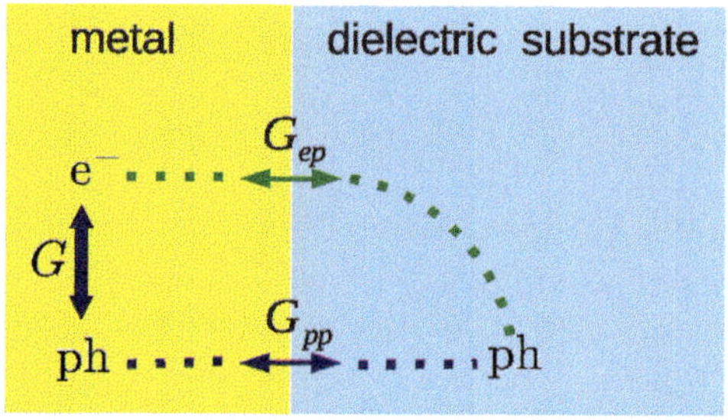

Figure 3.2 (Color online) Illustration of the different energy channels involved in thermal transport at metal/dielectric interfaces. The indirect coupling (blue) involves electron–phonon coupling in the bulk metal, characterized by g followed by transfer via the phonon–phonon conductance G_{pp}. This channel has been studied in [80]. The direct route is represented in green, involves the electron–phonon conductance G_{ep}, for which theoretical predictions exist [69, 70, 83]. Diffusive transport represented by dots plays a role in both cases.

the interface [69, 70, 80, 83]. Indeed, there are at least two mechanisms according to which electrons may heat up the substrate, as represented in Fig. 3.2: The first one has been extensively studied and involves an indirect coupling between the free electrons and the substrate through phonon-phonon processes [65, 74–76, 79–81]. This mechanism sets in when the electrons are not in equilibrium with the lattice. In this situation, the effect of the electron–phonon process is to introduce an additional resistance $1/\sqrt{gk_p}$ where g is the electron–phonon coupling factor, and k_p the phononic thermal conductivity of the metal. Hence, it is found that electron–phonon processes deteriorate the interfacial energy transfer, as compared to the case where electronic degrees of freedom are neglected.

There is another energy channel that has been recently unveiled from the experimental side [11, 17, 61, 73], and which has been previously predicted theoretically [69, 70, 83]. This mechanism is a direct coupling of the electron–phonon coupling through the interface, characterized by an interfacial heat flux of the form $G_{ep}(T_e - T_s)$, which depends on the temperature difference between the electrons and the phonons substrate, through the electron–phonon interfacial conductance G_{ep}. If we account for this latter direct coupling, Lombard et al. [60] have shown that the effective thermal conductance takes the form

$$1/G_{\text{eff}} = 1/gh + 1/(G_{\text{pp}} + G_{\text{ep}}), \tag{3.29}$$

where h is the metal film thickness. The resistance corresponding to electron–phonon coupling in the metal $1/gh$ decreases with the metal film thickness h, and with the metal electron–phonon coupling constant g. For a gold thin film having a thickness $h = 10$ nm, one obtains $gh = 250$ MW/K/m using the value $g = 2.5\ 10^{16}$ W m^{-3}/K as reported by DFT calculations [72]. For an aluminum film with the same thickness, $gh = 2500$ MW/K/m. Hence, electron–phonon coupling in the metal is important for thin metal films that have a small g, while it is negligible for thick metal films that have a large constant g.

Throughout the decade 2000 and until recently, the direct electron–phonon transfer mechanism has been disregarded essentially because experiments where the metal is replaced with bismuth, a semi-metal, and which conclude that the values of the interfacial conductance seems to be independent on the nature of the electronic transport properties of the film [6]. There are, however, actually two hints that the direct electron–phonon channel could play a significant role. First, the above-mentioned models for the conductance σ_e predict values that are comparable or even largely exceed the typical values for the phonon–phonon conductance σ_p. Second, recent transient thermoreflectance experiments on a Au film [11, 17, 61] concluded on the existence of an electron-substrate energy transfer, with estimate of σ_e in the range 100 to 1000 MW m^{-2} K^{-1} , compatible with the theoretical predictions. The consequences of this direct channel of energy have been scarcely discussed in the literature. There are three analytical models predicting the electron–phonon interfacial conductance at a metal/dielectric interface. Huberman and Overhauser [69] first proposed that metal electrons couple with joint vibrational modes at the interface leading to a significant enhancement of the energy transfer, as compared with direct phonon–phonon coupling. Later, Sergeev developed a field theory to predict the effect of the boundaries on the inelastic scattering of conduction electrons [70]. Mahan derived a theory where metal electrons may transfer heat to the substrate through the image charges created by the vibrating ions of the nonmetal [83]. This model should mainly apply to ionic crystals or polar semiconductors, rather than to apolar nonmetals. In both Huberman and Mahan's models, the conductance is predicted to be constant at high temperatures, while in Sergeev's model, it is found to increase linearly with temperature [70], in agreement with experimental observations [11]. The expression of the electron–phonon thermal conductance predicted by Sergeev writes

$$\sigma_e = \frac{3\pi\hbar}{35\zeta(3)k_B}\frac{\gamma u_L}{\tau_{\text{e-p}}}\left[1+2\left(\frac{u_L}{u_T}\right)^3\right], \quad (3.30)$$

where $\hbar$ is the reduced Planck constant, k_B is Boltzmann constant, ζ is Riemann zeta function, γ is the Sommerfeld constant of the metal, u_L and u_T are, respectively, the metal longitudinal and transverse sound velocities and $\tau_{\text{e-p}}$ is the electron–phonon energy relaxation time of the bulk metal. Because at high temperature $1/\tau_{\text{e-p}} \sim T_e$ [72] while other factors have weak or no temperature dependence, one immediately obtains $G_{ep} \sim T$. We can estimate the electron–phonon energy relaxation time using its relation with the electron–phonon coupling factor [71, 72]:

$$g = \frac{\pi^2 m_e c_s^2 n_e}{6\,\tau_{\text{e-p}} T_e}, \quad (3.31)$$

where m_e is the effective mass of the electron, c_s is the speed of sound and n_e is the free electrons number density. Besides, one has

$$\gamma_T \equiv \frac{g}{C_e} = \frac{3\hbar\lambda\langle\omega^2\rangle}{\pi k_B T_e}, \quad (3.32)$$

where λ is the electron–phonon coupling constant, so that the electron–phonon conductance Eq. (3.30) may be cast as

$$G_{ep} = \frac{54\gamma^2\hbar^2\lambda\langle\omega^2\rangle}{35\pi^2 k_B^2 m_e c_s n_e} T_e\left[1+2\left(\frac{u_L}{u_T}\right)^3\right]. \quad (3.33)$$

To calculate G_{ep}, we may use the values of $\lambda\langle\omega^2\rangle$ from the DFT calculations of Lin et al. [72] for noble metals. In [60], it is shown that Sergeev's model yields acceptable agreement for the electron–phonon interfacial conductance measured at low temperatures, $T <$ 100 K, while it tends to overestimate the conductance at room temperature [60]. Further work is clearly needed to build models that incorporate electron–phonon processes both in the bulk and at interface, to predict the thermal conductance at metal/dielectric interfaces.

3.4 Coherence Effects

So far, we have treated phonons as if they had no phase. This is the *particle* approach. However, phonons are displacement waves, and as such they have a well-defined phase, and the question is when is it important and when can it be ignored. When the phase information is important, we say that phonons should be treated as *waves*. The quantity of prime importance here is the phonon coherence length l_c, which is a measure of the spatial extension of phonon wavepackets. The phonon coherence length is inversely proportional to the spectral width $\Delta\nu$ of the thermally excited phonons. Typically, one has $l_c \simeq 1 - 2$ nm at room temperature [86].

Coherence effects may arise in devices which have at least one characteristic length comparable with the phonon coherence length. Needless to say, coherence effects are only possible for nanostructured materials due to the smallness of the phonon coherence length. Maybe, the most pedagogical example to understand the transition between the wave-like picture and the particle picture is to consider phonon transport in a superlattice. Superlattices are composite solids obtained by periodically repeating motif. The superlattice structure is characterized by its period, which usually takes values between 1 and 100 nm. Now, when we discuss possible coherence effects in superlattices we need to compare the superlattice period with the phonon coherence length. Therefore, two extreme cases may be envisioned: when the coherence length is much larger than the superlattice period, thermal transport is said to be *coherent*. In this situation, phonons see the superlattice as a homogeneous medium, and as such experience mini band folding and also feel the band gaps opened by the superlattice periodicity. On the contrary, if the coherence length is much smaller than the periodicity, thermal transport is said to be *incoherent* and phonons may be treated as particles. In this regime, phonons experience scattering at the interfaces which destroy phase coherence. Mini band and the existence of band gaps are not seen by phonons in this regime. Strictly speaking, our previous discussion of the different regimes concerns superlattice with perfect interfaces, i.e., superlattices with atomically sharp interfaces. Interfacial roughness or intermixing may destroy phonon coherence, and we will come back to this below. Finally, it is worth mentioning that the first regime-coherent transport is hard to observe because of the smallness of the phonon coherence length l_c. For superlattices with small periodicity, the phonon coherence length is commonly comparable with the superlattice period and thermal transport is said to be *partially coherent*. In this situ action, phonons do not see the superlattice as a homogeneous medium and also interfacial scattering does not destroy phase relation. Rather, the partial phonon transmission in this regime should be computed using wave physics. In particular, at the superlattice interfaces, a wave packet is partially transmitted and reflected. In the partially coherent regime, the total transmission coefficient arises from the interference of all the transmitted waves.

Coherence effects have important consequences regarding the thermal conductivity and again we refer to superlattices to discuss coherent effects. Indeed, when the superlattice period is large, i.e., larger than a few nanometers thermal transport is incoherent and the thermal conductivity—in the cross plane direction perpendicular to the layers—is expected to increase approximately with the period. Such an increase is controlled by interface scattering that destroy coherence and provide a thermal resistance which is proportional to the interface density. Phonon–phonon scattering in the superlattice sublayers induce also a thermal resistance that increases with the interface density. In the coherent regime, the superlattice thermal conductivity is expected to decrease with the period. The interpretation of this decrease is related to the fact that on increasing the period, the Brillouin zone of the superlattice is folded leading to decrease the group velocity [9, 84, 85]. As a consequence, a minimum in the thermal conductivity should be observed for a superlattice period close to the phonon coherence length. Such a minimum has been predicted a long time ago and evidenced by molecular dynamics simulations [42, 91, 94]. However, it has been hard to observe experimentally the predicted minimum of the thermal conductivity [87–90]. Interface roughness and stresses are the two main factors that have been proposed to explain the difficulty to probe coherent thermal transport [1, 42, 89, 92, 93]. A direct evidence of coherent transport and a minimum of thermal conductivity has been clearly shown for epitaxial-grown perovskite superlattices [95]. The thermal conductivity minimum is possibly explored in these superlattices, thanks to the precise surface treatment procedures such as laser molecular beam epitaxy (MBE) which realize superlattices with atomically sharp interfaces. We should also mention another experimental investigation where it is claimed that phonon coherent thermal transport is confirmed [96]. However, in this latter study the thermal conductivity is measured as function of the superlattice length, and not as a function of the period. We have mostly discussed coherent effects in superlattices, but recent experimental work has also probed coherence in two dimensional nanostructured phononic crystal membranes [97, 98].

3.5 Phonon Rectification

In the conventional diodes and transistors, there are two states, on and off, which can be alternated by controlling the electrical conduction. The idea of thermal rectifier, which is the analogue of the electrical diode, is to realize thermal circuits by the use of heat valves. We have to admit that to control the flow of electrons is much easier procedure than this of the phonons, as phonons are quasiparticles without mass or charge [102]. Phonon or thermal rectification refers to a phenomenon in materials in which the thermal conductivity along a specific direction is dependent on the sign of the heat current or the temperature gradient. A perfect thermal rectifier would have high thermal conductivity in one direction (on state), while in the opposite direction it would be thermal insulator (off state) [103]. The thermal rectification has been predicted theoretically in several systems and in very few cases has been also observed experimentally.

First observation of the thermal rectification was done in 1930s by Starr [104] in a copper/cuprous oxide interface, where he observed electrical rectification characteristics in such interfaces and he also noticed thermal rectification. The measured thermal conductance was greater in the direction of the greater electrical conduction (from the metal to oxide). He explained this effect to the asymmetric electron transport in the system resulting in the asymmetric thermal transport. In the specific case, the reason for the rectification is the electron–phonon coupling (it is not authors' intention to analyze in this book). This coupling occurs at the interfaces and it is not in equilibrium. Starr claimed that the transfer from electrons to phonons (from metal to insulator) can be different than that from phonons to electrons (from insulator to metal).

They have been reported several bulk or nanostructured systems the last decade which demonstrate thermal rectification. Most of the reported systems are heterojunctions or graded systems [103]. Phonon localization, phonon confinement or different temperature dependence in the case of two materials were suggested to play an important role. For the moment, there are two main strategies to obtain thermal rectification: the use of lattice nonlinearity and the structural asymmetry. In the first class of materials, there are nonlinear lattices with strong anharmonicity [105] or asymmetric mass or defect distributions [106]. The second type of materials that can be used as thermal rectifier are asymmetric nanostructures, where mainly the acoustic phonon scattering from the boundaries is important [107]. The idea within this strategy is to tune geometrically the rectification, or by introducing asymmetric defects or edges, even amorphous phase or even phase transitions to obtain different boundary scattering in opposite directions.

3.6 Amorphous Limit to the Thermal Conductivity

The so-called thermal insulating limit refers sometimes to the alloy limit or to the amorphous limit. In alloys due to the heavily scattered phonons at the atomic substitutions, one can obtain a very low thermal conductivity, which often referred as the "alloy limit." The atomic substitutions in alloys scatter phonons because of the bond stiffness and the mass differences between the two or more species. They have thermal conductivities typically between 5 and 10 W/mK [108]. The important reduction of the thermal conductivity is mainly due to the low wavelength phonons, which interact with the point defects. During the past decade, there have been a lot of theoretical and experimental studies of nanostructures and nanostructured materials that overpass this alloy limit of thermal conductivity. The reason is the influence of the middle and higher wavelength phonons which are affected by the nanostructuration. On the other hand, there is the so-called amorphous limit of the thermal conductivity. As it has been explained in chapter 1, amorphous solids do not exhibit long-range crystalline order. This lack of order destroys coherence of lattice vibrations and limits the mean free path of phonons to a few nanometers. The thermal conductivities of amorphous oxides is in the range between 1.3 and 3.0 W/mK at room temperature [108]. To overpass the amorphous limit of the thermal conductivity is a more difficult task than to beat the alloy limit, as the phonon mean free path in amorphous solids is extremely small.

The thermal transport in amorphous materials is also driven by lattice vibrations as in crystalline materials, but in a different way. The temperature dependencies of their thermal conductivities are different, as in crystalline systems, the thermal conductivity has a maximum at low temperatures as explained in the introduction of this chapter and then the thermal conductivity decreases in increasing the temperature, in contrast to the amorphous materials where the thermal conductivity above the Debye temperature increases weakly with increasing the temperature. Furthermore, in amorphous materials, most lattice vibrations are strongly localized or they are nonpropagating modes (zero velocity) and they do not contribute to the thermal conductivity. The energy transport in amorphous materials occurs through a hopping mechanism between localized modes or by "diffusion" of nonpropagating modes. In the first case, anharmonicity leads to a coupling between different localized modes and an increase in the thermal transport, opposite to its effect in crystalline solids.

Recent research pushed further the alloy and amorphous limits in thermal transport. Generally electrically insulating amorphous solids and glasses exhibit usually the lowest thermal conductivity [109]. In such materials, the majority of the vibrational modes are the diffusions which are localized and therefore unable to contribute to heat conduction, and only a very small fraction of the vibrational modes propagate as waves. The lowest thermal conductivity have been achieved by multilayer thin films, which are constituted by metals and oxides this lowest thermal conductivity limit has been overpassed [110, 111]. In such materials, the Kapitza resistance (thermal boundary resistance) of the interfaces between metal and oxide reduces the thermal conductivity far below the thermal conductivity of the homogeneous amorphous oxide, in this case when these interfaces are very close to each other (few nanometers). The lowest thermal conductivity for a dense solid (disordered WSe_2) ever reported till now is 0.05 W/mK [109, 112]. In this material the thermal conductivity is dominated by the interfaces.

There are several examples in the literature in which it has been shown that with the nanostructuration, we can beat the amorphous limit of the thermal conductivity. For example, as mentioned above, adding interfaces increases the scattering processes in superlattices, or adding roughness in nanofilms or nanowires, the thermal conductivity decreases rapidly. Interestingly the combination of the crystalline and amorphous phases in nanostructures is another very effective strategy to reduce the thermal conductivity. In the last couple of years, several structures have been proposed with this approach: silicon nanowires with amorphous silicon [113, 114] or silicon oxide [113, 115], amorphous/crystalline superlattices [116] or nanoporous materials with or without amorphous shell around the porous [117–119].

References

1. G. Chen, *Phys. Rev. B* **57**, 14958 (1998).
2. E. T. Swartz and R. O. Pohl, *Rev. Mod. Phys.* **61**, 605 (1989).
3. D. G. Cahill, W. K. Ford, K. E. Goodson, G. D. Mahan, A. Majumdar, H. J. Maris, R. Merlin and S. R. Philpot, *J. App. Phys.* **93**, 793 (2003).
4. R. Prahser, *Ap. Phys. Lett.* **94**, 041905 (2009).
5. R. J. Stoner and H. J. Maris, *Phys. Rev. B* **48**, 16373, (1993).
6. Lyeo H. K. and D. G. Cahill, *Phys. Rev. B* **73**, 144301 (2006).
7. V. Juve, M. Scardamaglia, P. Maioli, A. Crut, S. Merabia, L. Joly, N. Del Fatti and F. Vallée, *Phys. Rev. B* **80**, 195406 (2009).
8. G. P. Srivastava, *The Physics of Phonons*, CRC Press, New-York, 1990.

9. T. M. Tritt, *Thermal Conductivity, Theory, Properties, and Applications*, Kluwer Academic/Plenum Publishers, ISBN 0-306-48327-0, New York, pp. 1–17 (2004).
10. P. E. Hopkins, P. M. Norris and R. J. Stevens, *ASME J. Heat Transfer* **130**, 022401 (2008).
11. P. E. Hopkins and P. M. Norris, *ASME J. Heat Transfer* **131**, 022402 (2009).
12. D. A. Young and H. J. Maris, *Phys. Rev. B* **40**, 3685 (1989).
13. W. Zhang, T. S. Fisher and N. Mingo, *AMSE J. Heat Transfer* **129**, 483 (2007).
14. E. S. Landry and A. J. H. McGaughey, *Phys. Rev. B* **80**, 165304 (2009).
15. R. J. Stevens, A. N. Smith and P. M. Norris, *J. Heat Transf.* **127** 315 (2005).
16. R. J. Stevens, L. V. Zhigilei and P. M. Morris, *Int. J. Heat Mass Transfer* **50**, 3977 (2007).
17. P. E. Hopkins and P. M. Norris, *Appl. Surf. Sci.* **253**, 6289 (2007).
18. P. K. Schelling, S. R. Phillpot and P. Keblinski, *Phys. Rev. B* **65**, 144306 (2002).
19. A. Rajabpour and S. Volz, *J. Appl. Phys.* **108**, 094324 (2010).
20. J.-L. Barrat and F. Chiaruttini, *Mol. Phys.* **101**, 1605 (2003).
21. P. L. Kapitza, *J. Phys.* (USSR) **4**, 181 (1941).
22. S. Simons, *J. Phys. C.* **7**, 4048 (1974).
23. S. Merabia and K. Termentzidis, *Phys. Rev. B* **86**, 094303 (2012).
24. E. T. Swartz and R. O. Pohl, *Appl. Phys. Lett.* **51**, 2200 (1987).
25. S. Pettersson and G. D. Mahan, *Phys. Rev. B* **42**, 7386 (1990).
26. G. Chen, *App. Phys. Lett.* **82**, 991 (2003).
27. M. D. Losego, M. E. Grady, N. R. Sottos, D. G. Cahill and P. V. Braun, *Nat. Mater.* **11**, 502 (2012).
28. P. J. O'Brien,S. Shenogin, J. X. Liu, P. K. Chow, D. Laurencin, P. H. Mutin, M. Yamaguchi, P. Keblinski, G. Ramanath, *Nat. Mater.* **12**, 118 (2013).
29. W. A. Little, *Can. J. Phys.* **37**, 334 (1959).
30. J. Amrit, *Phys. Rev. B* **81**, 054303 (2010).
31. N. S. Shiren, *Phys. Rev. Lett.* **47**, 1466 (1981).
32. I. N. Adamenko and I. M. Fuks, *Sov. Phys. JETP* **32**, 1123 (1971).
33. P. E. Hopkins, L. M. Phinney, J. R. Serrano and T. E. Beechem, *Phys. Rev. B* **82**, 085307 (2010).
34. P. E. Hopkins, J. C. Duda, C. W. Petz and J. A. Floro, *Phys. Rev. B* **84**, 035438 (2011).

35. J. C. Duda and P. E. Hopkins, *App. Phys. Lett.* **100**, 111602 (2012).
36. D. Kekrachos, *J. Phys. Condens. Matter* **2**, 2637 (1990).
37. D. Kekrachos, *J. Phys. Condens. Matter* **3**, 1443 (1991).
38. B. C. Daly, H. J. Maris, K. Imamura and S. Tamura, *Phys. Rev. B* **66**(2), 024301 (2002).
39. S. Merabia and K. Termentzidis, *Phys. Rev. B* **89**, 054309 (2014).
40. X. W. Zhou, R. E. Jones, C. J. Kimmer, J. C. Duda, and P. E. Hopkins, *Phys. Rev. B* **87**, 094303 (2013).
41. Z. Tian, K. Esfarjani and G. Chen, *Phys. Rev. B* **86**, 235304 (2012).
42. E. S. Landry and A. J. H. McGaughey, *Phys. Rev. B* **79**, 075316 (2009).
43. J. Callaway, *Phys. Rev.* **113**, 1046 (1959).
44. W. A. Little, *Can. J. Phys.* **37**, 334 (1959).
45. R. Landauer, *Phil. Mag.* **21**, 863 (1970).
46. J. M. Ziman, *Electrons and Phonons*, Oxford University Press, New York, 2001.
47. J. A. Katerberg, C. L. Reynolds and A. C. Anderson, *Phys. Rev. B* **16**, 673 (1977).
48. S. Aubry, C. J. Kimmer, A. Skye and P. K. Schelling, *Phys. Rev. B* **78**, 064112 (2008).
49. Y. Chalopin, K. Esfarjani, A. Henry, S. Volz and G. Chen, *Phys. Rev. B* **85**, 195302 (2012).
50. K. Termentzidis, P. Chantrenne, J-Y. Duquesne and A. Saci, *J. Phys. Condens. Matt.* **22**, 475001 (2010).
51. K. Termentzidis, J. Parasuraman, C. A. Da Cruz, S. Merabia, D. Angelescu, F. Marty, T. Bourouina, X. Kleber, P. Chantrenne and P. Basset, *Nanoscale Res. Lett.* **6**, 288 (2011).
52. B. M. Clemens, G. L. Eesley, and C. A. Paddock, *Phys. Rev. B* **37**(3), 1085–1096 (1988).
53. B. C. Gundrum, D. G. Cahill, and R. S. Averback, *Phys. Rev. B* **72**(24) 5 (2005).
54. R. B. Wilson and D. G. Cahill, *Phys. Rev. Lett.* **108**, 255901 (2012).
55. M. T. Dove, *Introduction to Lattice Dynamics* (Cambridge University Press, Cambridge England, 1993).
56. N. W. Ashcroft and N. D. Mermin, *Physique des Solides* (EDP Sciences, Les Ulis, 2002).
57. K. Termentzidis, S. Merabia, P. Chantrenne and P. Keblinski, *Int. J. Heat Mass Transf.* **54**, 2014 (2011).

58. A. J. H. McGaughey and M. Kaviany, *Phys. Rev. B* **69**, 094303 (2004).
59. K. Termentzidis, P. Chantrenne and P. Keblinski, *Phys. Rev. B* **79**, 214307 (2009).
60. J. Lombard, F. Detecheverry and S. Merabia, *J. Phys.: Condens. Matter* **27**, 015007 (2015).
61. L. Guo, S. L. Hodson, T. S. Fischer and X. Xu, *J. Heat Transfer* **134**, 042402 (2012).
62. A. J. Schmidt, K. C. Collins, A. J. Minnich and G. Chen, *J. App. Phys.* **107**, 104907 (2010).
63. P. Hopkins, M. Baraket, E. V. Barnat, B. E. S. P. Beechem, J. C. Duda, J. T. Robinson, and S. G. Walton, *Nano Lett.* **12**, 590 (2012).
64. P. E. Hopkins, J. C. Duda, B. Kaehr, X. W. Zhou, C. Y. Peter Yang, and R. E. Jones, *App. Phys. Lett.* **103**, 211910 (2013).
65. R. E. Jones, J. C. Duda, X. W. Zhou, C. J. Kimmer, and P. E. Hopkins, *App. Phys. Lett.* **102**, 183119 (2013).
66. C. Monachon, G. Schusteritsch, E. Kaxiras, and L. Weber, *J. App. Phys.* **115**, 123509 (2014).
67. P. Komarov, M. G. Burzo, G. Kaytaz, and P. E. Raad, *Microelectronic J.* 34, 1115 (2003).
68. C. Abs da Cruz, P. Chantrenne and X. Kleber, *J. Heat. Trans.* **134**, 062402 (2012).
69. M. L. Huberman and A. W. Overhauser, *Phys. Rev. B* **50**, 2865 (1994).
70. Sergeev, A. *Phys. Rev. B* **58**, R10199–R10202 (1998).
71. Kaganov, M. I. and Lifshitz, I. M. and Tanatarov, L. V., *Sov. Phys. JETP* **4**, 173 (1957). (2014).
72. Z. Lin, L. Zhigilei, and V. Celli, *Phys. Rev. B* **77**, 075133 (2008).
73. Z. Li, S. Tan, E. Bozorg-Grayeli, T. Kodama, M. Asheghi, G. Delgado, M. Panzer, A. Pokrovsky, D. Wack, and K. E. Goodson, *Nano Lett.* **12**, 3121 (2012).
74. E. Dechaumphai, D. Lu, J. J. Kan, J. Moon, E. E. Fullerton, Z. Liu, R. Chen, *Nano Lett.* **14**, 2448 (2014).
75. P. E. Hopkins, J. C. Duda, B. Kaehr, X. W. Zhou, C. Y. Peter Yang, and R. E. Jones, *App. Phys. Lett.* **103**, 211910 (2013).
76. R. B. Wilson, J. P. Feser, G. T. Hohensee, and D. G. Cahill, *Phys. Rev. B* **88**, 144305 (2013).
77. A. Giri, B. M. Foley and E. E. Hopkins, *J. Heat. Trans.* **136**, 092401 (2014).

78. P. B. Allen, *Phys. Rev. Lett.* **59**, 1460 (1987).
79. Y. Wang, X. Ruan and A. K. Roy, *Phys. Rev. B* **85**, 205311 (2012).
80. A. Majumdar and R. Reddy, *App. Phys. Lett.* **84**, 4768 (2004).
81. P. Singh, M. Seong and S. Sinha, *App. Phys. Lett.* **102**, 181906 (2013).
82. P. Zhangh, J. T. Lu, J. S. Wang and B. Li, *J. Phys. Condens. Matter* **25**, 445801 (2013).
83. G. Mahan, *Phys. Rev. B* **79**, 075408 (2009).
84. S.-I. Tamura, Y. Tanaka and H. J. Maris, Phonon *Phys. Rev. B* **60**, 2627 (1999).
85. P. Hyldgaard and G. D. Mahan, *Phys. Rev. B* **56**, 10754 (1997).
86. G. Chen, *Nanoscale Energy Transport and Conversion: A Parallel Treatment of Electrons, Molecules, Phonons, and Photons*, MIT-Pappalardo Series in Mechanical Engineering, 1st ed.
87. W. Capinski and H. J. Maris, *Physica B* **219–220**, 699 (1996).
88. R. Venkatasubramanian, *Phys. Rev. B* **61**, 3091 (2000).
89. S. M. Lee, D. G. Cahill and R. Venkatasubramanian, *App. Phys. Lett.* **70**, 2957 (1997).
90. S. T. Huxtable, A. R. Abramson, C.-L. Tien, A. Majumdar, C. LaBounty, X. Fan, G. Zeng, J. E. Bowers, A. Shakouri, J. Baskin and E. T. Croke, *App. Phys. Lett.* **80**, 1737 (2002).
91. M. V. Simkin and G. D. Mahan, *Phys. Rev. Lett.* **84**, 927 (2000).
92. B. C. Daly, H. J. Maris, K. Imamura and S. Tamura, *Phys. Rev. B* **66**, 024301 (2002).
93. Y. K. Koh, Y. Cao, D. G. Cahill, D. Jena, *Adv. Funct. Mater.* **19**, 610 (2009).
94. Y. Chen, D. Li, J. R. Lukes, Z. Ni and M. Chen, *Phys. Rev. B* **72**, 174302 (2005).
95. J. Ravichandran, A. K. Yadav, R. Cheaito, P. B. Rossen, A. Soukiassian, S. J. Suresha, J. C. Duda, B. M. Foley, C.-H. Lee, Y. Zhu, A. W. Lichtenberger, J. E. Moore, D. A. Muller, D. G. Schlom, P. E. Hopkins, A. Majumdar, R. Ramesh and M. A. Zurbuchen, *Nat. Mater.* **13**, 168 (2014).
96. M. N. Luckyanova, J. Garg, K. Esfarjani, A. Jandl, M. T. Bulsara, A. J. Schmidt, A. J. Minnich, S. Chen, M. S. Dresselhaus, Z. Ren, E. A. Fitzgerald, G. Chen, *Science* **338**, 936 (2012).
97. N. Zen, T. A. Puurtinen, T. J. Isotalo, S. Chaudhuri and I. J. Maasilta, *Nat. Commun.* **5**, 3435 (2014).
98. M. Maldovan, *Phys. Rev. Lett.* **110**, 025902 (2013).
99. R. Peierls, Zur kinetischen Theorie der Wärmeleitung in Kristallen, *Annalen der Physik* **395**, 1055 (1929).

100. D. J. Ecsedy and P. G. Klemens, *Phys. Rev. B.* **15**, 5957 (1977).
101. O. Delaire, J. Ma, K. Marty, A. F. May, M. A. McGuire, M.-H. Du, D. J. Singh, A. Podlesnyak, G. Ehlers, M. D. Lumsden, and B. C. Sales, *Nat. Mater.* **10**, 614 (2011).
102. N. Li, J. Ren, L. Wang, G. Zhang, P. Hänggi, and B. Li, *Rev. Mod. Phys.* **84**, 1045 (2012).
103. Y. Wang, A. Vallabhaneni, H. Jiuning, B. Qiu, Y. P. Chen and X. Ruan, *Nano Lett.* **14**, 592 (2014).
104. C. Starr, *J. Appl. Phys.* **7**, 15 (1936).
105. M. Terraneo, M. Peyrard and G. Casati, *Phys. Rev. Lett.* **88**, 094302 (2002).
106. C. W. Chang , D. Okawa, A. Majumdar and A. Zettl, *Science.* **314** 1121 (2006).
107. T. Ouyang, Y. Chen, Y. Xie, X. L. Wei, K. Yang, P. Yang, and J. Zhong, *Phys. Rev. B.* **82**, 245403 (2010).
108. W. Kim, R. Wang, and A. Majumdar, *Nano Today.* **2** 40 (2007).
109. C. Chiritescu, D. G. Cahill, N. Nguyen, D. Johnson, A. Bodapati, P. Keblinski, and P. Zschack, *Science.* **315**, 351 (2007).
110. R. M. Costescu, D. G. Cahill, F. H. Fabreguette, , Z. A. Sechrist and S. M. George, *Science.* **303**, 989 (2004).
111. Y. S. Ju, M.-T. Hung, M. J. Carey, M.-C. Cyrille and J. R. Childress, *Appl. Phys. Lett.* **86**, 203113 (2005).
112. A. Mavrokefalos, N. T. Nguyen, M. T. Pettes, D. C. Johnson and L. Shi, *Appl. Phys. Lett.* **91**, 171912 (2007).
113. Y. He and G. Galli, *Phys. Rev. Lett.* **108**, 215901 (2012).
114. E. Blandre, L. Chaput, S. Merabia, D. Lacroix, and K. Termentzidis, *Phys. Rev. B.* **91**, 115404 (2015).
115. T. Zushi, K. Ohmori, K. Yamada, and T. Watanabe, *Phys. Rev. B.* **91**, 115308 (2015).
116. A. France-Lanord, S. Merabia, T. Albaret, D. Lacroix, and K. Termentzidis, *J. Phys. Condensed Matter.* **26**, 355801 (2014).
117. M. Isaiev, S. Tutashkonko, V. Jean, K. Termentzidis, T. Nychyporuk, D. Andrusenko, and O'Marty, R. M. Burbelo, D. Lacroix, and V. Lysenko, *Appl. Phys. Lett.* **105**, 031912 (2014).
118. R. Dettori, C. Melis, X. Cartoixà, R. Rurali and L. Colombo, *Phys. Rev. B.* **91**, 054305 (2015).
119. J. Lee, J. Lim and P. Yang, *Nano Lett.* **15**, 3273 (2015).

Chapter 4

Applications

Maxime Verdier,[a] Rodolphe Vaillon,[b] Samy Merabia,[c] David Lacroix,[a] and Konstantinos Termentzidis[a]

[a]*Université de Lorraine, LEMTA, Centre National de la Recherche Scientifique, UMR 7563, BP 70239, 54506 Vandoeuvre Cedex, France*
[b]*Univ Lyon, CNRS, INSA-Lyon, Université Claude Bernard Lyon 1, CETHIL UMR5008, F-69621 Villeurbanne, France*
[c]*ILM, Université de Lyon, UMR 5306 Université Lyon 1 et CNRS, F-69622 Villeurbanne, France*
konstantinos.termentzidis@univ-lorraine.fr

Heat transport at the micro/nanoscale is very important for several nanotechnology applications, ranging from micro/nanoelectronics to bioegineering and energy harvesting. Management of heat transfer is crucial for reliability, lifetime and maintenance of functionality of nanoscale devices. Semiconductor lasers, integrated circuits, thermoelectric devices, thermovoltaic devices, data storage are few examples as it will be shown hereafter of importance of manipulation of the heat through them.

For several applications, such as thermoelectric devices and thermal barrier coatings, thermal management itself is the key parameter. There are many other non-thermal technologies for which the ability to manipulate heat is extremely important, like phase change memory devices, heat assisted magnetic recording, thermal management of high power and nanoscale electronic

Nanostructured Semiconductors: Amorphization and Thermal Properties
Edited by Konstantinos Termentzidis

ISBN 978-981-4745-64-2 (Hardcover), 978-1-315-36445-2 (eBook)
www.panstanford.com

and optoelectronic devices, and heated nanoparticles in medical therapies [6].

In general, there are two tendencies concerning thermal management: one is to increase the thermal conductivity (dielectric heat spreaders are ideal for laser diodes, and all semiconductor integrated circuits) and the other one is to decrease it (thermoelectric energy conversion, thermal barriers). Electronic and optoelectronic devices need packaging materials with extremely high thermal conductivity comparable to diamond's conductivity and at the same time less expensive than diamond itself. Very good thermal conductors for industrial use are considered materials with thermal conductivity more than 1000 W/mK. On the other hand, thermoelectric devices need materials with ultra-low thermal conductivity, lower than 1 W/mK as it will be explained in this chapter.

4.1 Electronics, Information Technologies

The revolution in nanofabrication techniques during the last decades opens fascinating opportunities but also creates several challenges. The decrease in device's sizes is followed by increase in their operating speeds and frequencies [6], with a negative secondary effect which is the drastically increased heat generation.

Thermal management is one of these challenges as the characteristic size of devices becomes comparable to the bulk phonon mean free path. With the device size reduction, this becomes more sensitive to temporal fluctuations and spatial inhomogeneities. These fluctuations and inhomogeneities cause decoherence, noise, device irreproducibility, as well interruption of electron flow [17]. In integrated circuits, the manipulation of information is operated by electron transport. With miniaturization of devices, the excited electrons will generate more heat in smaller volumes, which can lead to the device mechanical failure. The so called "power densities" generated by electronic devices is increasing rapidly causing huge reliability and device lifetime issues. Engineers should avoid the existence of hot spots, and the increasing temperature rise in on-chip interconnects.

To overcome the issue of destructive heat generation, several cooling technologies are proposed (internal and external module cooling, thermal interface materials, etc). In general, the strategy is first to spread heat out, and then sink heat away. Concerning the heat spreader, the material should not retain heat, which means a combination of high thermal diffusivity and low heat capacity. A heat sink, on the other hand, should store heat so it needs high thermal diffusivity and high heat capacity. Tailoring the thermal properties of materials is an important way to achieve efficient devices.

4.2 Optoelectronics

For obtaining reliable devices with improved performances, the optoelectronic industry has to solve an important drawback which is the junction temperature management. The performance of optoelectronic devices decreases drastically with increasing chip temperature. High power laser diodes or laser devices usually convert electric energy into light energy with efficiency between 10% to 50%. The remaining energy, which is the majority, is lost as a waste heat and it can create thermal stresses with irreversible damage to the laser if it is not removed rapidly. The temperature of the device has apart the reduction of the lifetime a very crucial influence on the output power, the threshold current, the output wavelength and the band gap (every 3 K change can shift the wavelength of the diode laser of 1 nm). To remove the excessive heat which is generated in the laser diodes as fast and uniformly as possible, effective thermal packaging solutions are required.

The two main types of semiconductor diode lasers are the vertical-cavity surface emitting lasers (VCSEL) and the edge-emitting lasers (EEL). In both there are two crucial parameters involved in their optimal design: reliable optical performance and high thermal conductivity. While the first design parameter is nowadays well controlled, the second one is more difficult to be achieved, as heating is the limiting factor to obtain high power output. The problematic issue in diode lasers is the high number of interfaces and the initial low thermal conductivities of the layers. Both limit the efficient removal of heat, leading to a short device lifetime.

4.3 Thermoelectrics

Thermoelectricity offers attractive perspectives in the field of renewable energies. As it is a technology with no moving parts it can be used in a broad variety of mobile devices as well as in remote or autonomous systems (e.g., in space applications). Furthermore, the continuous increase of energy demand in smart systems and the miniaturization of components used in everyday life apparatuses require new sources of energy that can provide new technological solutions. In this frame, conversion of heat-wasted energy into electricity is a major challenge and this field may benefit from nanoscale energy transport science. Harvesting of the so-called "low-temperature waste heat," which is produced by automobile exhaust or by various industrial plants/processes, is a very important energetic target. Knowing that an important part of global energy production in electronic apparatuses is lost as heat, there is here a source of cheap energy. In this frame, improving the efficiency of thermoelectric materials for power generation becomes a very attractive technology.

Thermoelectric cooling and power generation based on the Seebeck and the Peltier effects were discovered a century ago. With nanostructuration, one can achieve phonon engineering and tuning the transport properties of materials to obtain higher thermoelectric conversion efficiency. A good thermoelectric material has a very low thermal conductivity, while it maintains high electrical conductivity. The concept of a "phonon-glass, electron-single-crystal" is the main research challenge for engineering thermoelectric efficient materials. It is not our aim to make an exhaustive description of the thermoelectric effect here. There are several reviews and books in the literature dealing with it. What it is important to discuss in the frame of the book is the incorporation of the amorphous phase and its impact on the thermal conductivity even if the amorphous phase fraction is very small.

Performance of a thermoelectric material is rated by the so-called Figure of Merit, written as

$$ZT = \frac{S^2 \sigma T}{\kappa_{\text{electrons}} + \kappa_{\text{phonons}}}, \tag{4.1}$$

where S is the Seebeck coefficient, σ is the electrical conductivity, and κ is the thermal conductivity of the material which includes both the lattice and the electronic contributions.

To achieve high figure of merit, one has to optimize the electrical transport while minimizing the thermal conductivity. From a macroscopic thermodynamic point of view, the efficiency of the thermoelectric energy conversion depends on both the Carnot efficiency of the system and the thermoelectric figure of merit. The power factor of the electronic component, $S^2 T$, is typically optimized as a function of carrier concentration ($\sim 10^{19}$ carriers/cm^3) through doping to give the largest ZT. High-mobility carriers are desirable in order to reach a high electrical conductivity for a given carrier concentration. The clever use of nanocomposites/nanoinclusions could lead to improve the thermoelectric figure of merit, while high power factor can be retained simultaneously.

Several strategies were developed to increase ZT: reducing the dimensions of materials such as superlattices, nanofilms, and nanowires, or introducing interfaces in bulk materials with different nanocrystallite sizes, or adding nanoinclusions of a second phase. In general, nanostructured materials have a much lower thermal conductivity than bulk materials, due to phonon confinement and boundary scattering. All successful strategies to reduce lattice thermal conductivity are related to some physical phenomena occurring between phonons and the material itself: (i) scattering of phonons within the nearest neighbors with point defects (interstitials, vacancies or alloying) (ii) use of complex crystal structures to separate the transport behavior of electrons from these of phonons, with the main idea being to tune the mean free path of the two carriers in a separate way (iii) scattering of phonons at interfaces, as, for example, in the case of multiphase composites at the nanoscale.

It will be shown in Chapters 7 and 8 that a very efficient way to increase the figure of merit is to incorporate small fractions of amorphous phase. This is quite efficient for reducing the thermal conductivity.

4.4 Photovoltaics, Thermophotovoltaics

Photovoltaic (PV) devices convert electromagnetic radiation power into electrical power. There are currently different types of converters, i.e., solar cells (also called solar photovoltaic devices), solar concentrated photovoltaics (CPV) where the incoming radiation is optically concentrated to yield high illumination power and in turn high electrical current densities, organic photovoltaics (OPV) where the photoactive elements are made of organic semiconducting materials, thermophotovoltaics (TPV) where the source of photons is not the sun, solar thermophotovoltaics (STPV) where the source of thermal radiation is an element heated by the sun, and near-field thermophotovoltaics (NF-TPV) where the thermal source transfers much more radiative power to the PV device thanks to the tunneling of evanescent waves. Whatever the technology, most of the radiation power which is not converted into electrical power is turned into heat. As a result, the operating temperature of PV devices is usually above the ambient, at a level which depends on the illumination power density and on the conditions of thermal exchange with the environment. Moreover, the efficiency of PV devices generally decreases with temperature. This temperature dependence is characterized by the so-called temperature coefficients. As an example, the most common solar cell technology uses crystalline silicon and the order of magnitude of the temperature coefficient of efficiency is $-0.4\%\ K^{-1}$ [1]. There are several ways to overcome this detrimental effect: tuning the temperature coefficient to small values, lowering the internal heat source or cooling down the device (see, for example, [7]).

For the third strategy to operate, heat generated within the device should be conducted efficiently towards the envelope surface which should in turn exchange thermal power with a cooler environment. The volume-to-surface area ratio should then be small. Also, thermal conductivity of photovoltaic devices should not be too small. Most solar cells are composed of homo- or hetero-

(sometimes sandwiched multi-) junction layers. In the case of thin-film photovoltaics (TFPV), the thickness of the semiconducting layers ranges from tens of nanometers to tens of micrometers. This means that the corresponding thermal conductivity could be lower than the bulk value because of both ballistic transport and interface effects. Because of layering, interface thermal resistance may play a significant role. Since miniaturization and the use of nanostructures are entering the field of photovoltaics, the thermal properties of semiconductor materials, crystalline or amorphous, are becoming critical design parameters.

4.5 Phononics, Thermal Diodes

Possible applications for thermal rectification are thermal diodes, thermal transistors which will switch or amplify heat current, thermal memories and thermal logic gates [10, 16]. The rapid emergence of phononics makes possible the use of phonons to carry and process information, besides electrons and photons. As it has been mentioned in Chapter 3, to enhance the rectification, asymmetric nanostructures, and incorporation of the amorphous phase have been proposed. The main concept of the diodes for either electrons or phonons is the presence of a symmetry-breaking mechanism. This might be obtained either by an asymmetric Kapitza resistance or by a negative differential thermal resistance.

Nanophononic crystals are new interesting structures for thermoelectric applications as their thermal conductivity can be very low. They are the elastic analogues of photonic crystals. Such systems are generally obtained by patterning cylindrical pores into a bulk crystalline semiconductor or more often into membrane [21–24], and less frequently by nano-inclusions [25, 26]. The shape of the pores or inclusions is not essential, one can obtain similar results with spherical or cubic pores. The crucial parameters are the nanopores or nanoinclusions size, the distance between two neighboring pores/inclusions and mostly the periodicity of their distribution, which leads to uncommon thermal conductivity lower than expected. This periodicity distinguishes phononic crystals from simple nanoporous materials.

The current phononic crystal elaboration techniques with nanometer characteristic dimensions are still challenging for experimentalists because of the necessary fabrication precision. The most usual fabrication method used is the electron-beam lithography [22], which enables to obtain a periodic mesh of cylindrical pores with a minimum diameter of the order of tenths of nanomaterials [21]. It is well known from nanoporous material studies that in such systems the majority of the phonons are scattered at the pore's surfaces. The reduction in thermal conductivity in usual nanoporous materials depends on the scattering surface to volume ratio, like in other nanostructured systems (nanowires or nanofilms).

Measurements of the thermal conductivity or thermal conductance have proved that the introduction of organized pores is indeed an efficient way to limit thermal transport [22, 23]. Simulations with equilibrium molecular dynamics were also performed and showed the same trend ([26–28], but numerical and experimental results are not always in agreement; the measured thermal conductivity is lower than the simulated one. This discrepancy cannot be attributed only to the experimental defects related to the fabrication of the samples or the errors of the measurement methods. Moreover, Yu et al. [22] noticed that, for a same ratio S/V, a phononic crystal is less heat conductor than a nanowire. There is still a need of a consequent theory to explain the extremely low thermal conductivity of nanophononic crystals compared to other nanostructures or nanostructured materials.

Researchers proposed several scenarios to explain the difference between what is expected and the real thermal conductivity of phononic crystals. The old Eucken model, just taking into account the effect of voids presence on thermal transport, is valid for macroscopic systems but fails to describe the thermal behavior when the material is nanostructured. Indeed, periodic modification of the structure at the nanoscale seems to change a lot the phonon properties. According to several works, the periodicity length can modify the Brillouin zone boundaries: it usually extends from $-\pi/a$ to π/a with a the lattice constant; but in nanophononic crystals, a will be replaced by the periodicity length (distance between the centers of two neighboring pores). Thus, the Brillouin zone is smaller than in bulk materials, and this modifies the phonon band structure [22], dispersion relations [25–27] and the density of states [23, 24].

The reduction of the Brillouin zone generates a band folding that leads to multiple dispersion relation branches, band flattening and modifications of the band gap [23]. The flattening of different branches causes a reduction of the group velocity, as the latter is defined as the derivative of the dispersion relation. The energy transport is then less efficient. Furthermore, the band gap implies that some phonon frequencies are forbiden in the material. This can result in a great reduction of the thermal conductivity [24]. The density of states can also be reduced over all the frequencies, just by adding periodic cylindrical holes. Lastly, some studies reached the conclusion that the phononic crystals could destroy the coherence of phonons. The sum of all these effects could explain their low thermal conductivity, but one does not know yet in which proportion each process affects the thermal transport. Further investigations are necessary to control thermal properties of nanostructured systems.

4.6 Biomedical Applications

Medical device designers and developers increase their attention to heat-management concerns as there are two crucial issues related to the heat, the biocompatibility and even the use of the heat for medical treatment. Concerning biocompatibility, the knowledge of the thermal properties, including the melting point, phase changes, specific heat, thermal conductivity, thermal expansion coefficient, etc is crucial. Furthermore, managing heat in medical device applications is important engineering issue. Taking into account the miniaturization of the medical devices, the designers are often challenged with limits the size of the device, performance, operating temperature, noise, and budget requirements. Heat-management could be used to remove, spread, and dissipate heat efficiently. While this helps improve system reliability, speed, precision, and service life, it can also help designers reduce their devices' packaging size and weight. The second very promising use of heat management in nanomaterials is related to the medical treatment (metallic nanostructures as photothermal agents for cancer, cancer diagnosis and therapy).

Gold nanoparticles illuminated at a wavelength corresponding to their plasmon resonance, offer the unique possibility to realize nano-sources of heat [2] in a liquid environment. When the nanoparticle is irradiated by a laser of sufficient energy, nanoscale boiling occurs leading to the formation of a nanobubble. Vapor nanobubbles around plasmonic nanoparticles enable numerous applications in nanomedicine [4, 8, 9, 18] and in solar energy conversion [13, 14]. The nanobubbles are generated in extreme conditions by intense and short duration laser pulses (femtosecond or nanosecond pulses), that may induce very localized hot spots and very large temperature gradients amounting to hundredths of Kelvins per nanometer. The nanobubbles produced are generally transient, growing and collapsing in a few tenths of nanoseconds. In addition, the pressure wave generated during their production is strong enough to enhance the efficiency of photoacoustic imaging [12].

Of prime importance is the maximal nanobubble size that can be attained during the bubble lifetime. Large nanobubbles are typically desired in photothermal biomedical therapy, so as to maximize mechanical injuries within the cell, to efficiently disrupt the cell membrane [3, 15, 20], and for liposomal and endosomal release [5, 19]. Conversely, small photothermal nanobubbles serve as image contrast agents while keeping the integrity of the cell [12].

A control of the mechanisms that drive nanobubble production is thus an important issue for all the applications. It is also an important issue from a fundamental point of view since nanoscale boiling is an example of phase transition induced in extreme out-of-equilibrium conditions where the usual nucleation-growth scenario has no time to occur, as already discussed in [11].

4.7 Synthesis of Nanostructures, Nanofabrication and Reliability/Calibration of Experimental Measurements

In several nanofabrication techniques, the heat generation and its impact on the structural properties of the elaborating nanostructures has not received enough attention. For example, when the fabrication technique is related with a laser, then huge amounts of energy is dissipated into the material and its impact on the fabrication performance and accuracy should be investigated. In general, there are several open heat transfer issues during the synthesis of nanostructures and nanostructured materials. In order to achieve repeatable and well-controlled nanofabrication techniques, we should address all these open questions.

References

1. Virtuani, A., Friesen, G., and Pavanello, D. (2010). Overview of temperature coefficients of different thin film photovoltaic technologies, *Overview of Temperature Coefficients of Different Thin Film Photovoltaic Technologies*.
2. Baffou, G., and Quidant, R. (2012). Thermo-plasmonics: using metallic nanostructures as nano-sources of heat, *Laser Photonics Rev.* **7**, 2, p. 171.
3. Baumgart, J., Humbert, L., Boulais, E., Lachaine, R., Lebrun, J.-J., and Meunier, M. (2012). Off-resonance plasmonic enhanced femtosecond laser optoporation and transfection of cancer cells, *Biomaterials* **33**, p. 2345.
4. Boriskina, S., Ghasemi, H., and Chen, G. (2013). Plasmonic materials for energy: from physics to applications, *Mater. Today* **16**, p. 375386.
5. Braun, G. B., Pallaoro, A., Wu, G., Missirlis, D., Zasadzinski, J. A., Tirrell, M., and Reich, N. O. (2009). Laser-activated gene silencing via gold nanoshell-sirna conjugates, *ACS Nano* **3**, 7, p. 2007, doi:10.1021/nn900469q.
6. Cahill, D. G., Braun, P. V., Chen, G., Clarke, D. R., Fan, S., Goodson, K. E., Keblinski, P., King, W. P., Mahan, G. D., Majumdar, A., Maris, H. J., Phillpot, S. R., Pop, E., and Shi, L. (2014). Nanoscale thermal transport. ii. 2003–2012, *Appl. Phys. Rev.* **1**, 1, 011305, doi:http://dx.doi.org/10.1063/1.4832615, http://scitation.aip.org/content/aip/journal/apr2/1/1/10.1063/1.4832615.
7. Dupré, O., Vaillon, R., Green, M. A. (2017). Thermal behavior of photovoltaic devices. Physics and Engineering. Springer. doi:10.1007/978-3-319-49457-9, http://www.springer.com/gp/book/9783319494562
8. Lapotko, D. (2009). Therapy with gold nanoparticles and lasers: what really kills the cells? *Nanomedicine* **4**, p. 253.

9. Lapotko, D. (2011). Plasmonic nanobubbles as tunable cellular probes for cancer theranostics, *Cancers* **3**, 1, p. 802, doi:10.3390/cancers3010802, http://www.mdpi.com/2072-6694/3/1/802.
10. Li, N., Ren, J., Wang, L., Zhang, G., Hänggi, P., and Li, B. (2012). *Colloquium*: Phononics: manipulating heat flow with electronic analogs and beyond, *Rev. Mod. Phys.* **84**, pp. 1045–1066, doi:10.1103/RevModPhys.84.1045, http://link.aps.org/doi/10.1103/RevModPhys.84.1045.
11. Löwen, H., and Madden, P. A. (1992). A microscopic mechanism for shock-wave generation in pulsed-laser-heated colloidal suspensions, *J. Chem. Phys.* **97**, 11, p. 8760, doi:10.1063/1.463345, http://dx.doi.org/10.1063/1.463345.
12. Lukianova-Hleb, E., Hanna, E., Hafner, J., and Lapotko, D. O. (2010). Tunable plasmonic nanobubbles for cell theranostics, *Nanotechnology* **21**, p. 085102.
13. Neumann, O., Feronti, C., Neumann, A. D., Dong, A., Schell, K., Lu, B., Kim, E., Quinn, M., Thompson, S., Grady, N., Nordlander, P., Oden, M., and Halas, N. J. (2013). Compact solar autoclave based on steam generation using broadband light-harvesting nanoparticles, *PNAS* **110**, p. 11677.
14. Neumann, O., Urban, A. S., Day, J., Lal, S., Norlander, P., and Halas, N. J. (2013). Solar vapor generation enabled by nanoparticles, *ACS Nano* **7**, p. 42.
15. Pitsillides, C. M., and Joe, E. K. (2003). Intracellular hyperthermia: nanobubbles and their biomedical applications, *Int. J. Hyperthermia* **25**, p. 533.
16. Sawaki, D., Kobayashi, W., Moritomo, Y., and Terasaki, I. (2011). Thermal rectification in bulk materials with asymmetric shape, *Appl. Phys. Lett.* **98**, 8, 081915, doi:http://dx.doi.org/10.1063/1.3559615, http://scitation.aip.org/content/aip/journal/apl/98/8/10.1063/1.3559615.
17. See, A. M., Hamilton, A. R., Micolich, A. P., Aagesen, M., and Lindelof, P. E. (2015). Using light and heat to controllably switch and reset disorder configuration in nanoscale devices, *Phys. Rev. B* **91**, p. 085417, doi:10.1103/PhysRevB.91.085417, http://link.aps.org/doi/10.1103/PhysRevB.91.085417.
18. Wen, D. (2009). Intracellular hyperthermia: nanobubbles and their biomedical applications, *Int. J. Hyperthermia* **25**, p. 533.
19. Wu, G., Mikhailovsky, A., Khant, H. A., Fu, C., Chiu, W., and Zasadzinski, J. A. (2008). Remotely triggered liposome release by near-infrared light absorption via hollow gold nanoshells, *J. Am. Chem. Soc.* **130**, 7, p. 8175.

20. Yao, C., Rahmanzadeh, R., Endl, E., Zhang, Z., Gerdes, J., and Hüttmann, G. (2005). Elevation of plasma membrane permeability by laser irradiation of selectively bound nanoparticles, *J. Biomed. Opt.* **10**, p. 064012.
21. V. Lacatena, M. Haras, J.-F. Robillard, S. Monfray, T. Skotnicki, and E. Dubois. (2014). Phononic engineering of silicon using dots on the γ e-beam lithography and plasma etching. *Microelectronic Engineering*, **121**, 131–134.
22. Jen-Kan Yu, Slobodan Mitrovic, Douglas Tham, Joseph Varghese, and James R. Heath. (2010). Reduction of thermal conductivity in phononic nanomesh structures. *Nature Nanotechnology*, **5**(10), 718–721.
23. Nobuyuki Zen, Tuomas A. Puurtinen, Tero J. Isotalo, Saumyadip Chaudhuri, and Ilari J. Maasilta. (2014). Engineering thermal conductance using a two-dimensional phononic crystal. *Nature Communications*, **5**.
24. P. E. Hopkins, L. M. Phinney, P. T. Rakich, R. H. Olsson, and I. El-Kady. (2011). Phonon considerations in the reduction of thermal conductivity in phononic crystals. *Applied Physics A*, **103**(3), 575–579.
25. J.N. Gillet, Y. Chalopin, S. Volz, (2009). Atomic-scale three-dimensional phononic crystals witrh a very low thermal conductivity to design crystalline thermoelectric devices. *J. Heat Transfer* **131**, 043206.
26. Lina Yang, Nuo Yang, and Baowen Li. (2013). Reduction of Thermal Conductivity by Nanoscale 3d Phononic Crystal. *Scientific Reports*, 3.
27. Sanghamitra Neogi, J. Sebastian Reparaz, Luiz Felipe C. Pereira, Bartlomiej Graczykowski, Markus R. Wagner, (2015). Marianna Sledzinska, Andrey Shchepetov, Mika Prunnila, Jouni Ahopelto, Clivia M. Sotomayor-Torres, and Davide Donadio. Tuning Thermal Transport in Ultrathin Silicon Mem- branes by Surface Nanoscale Engineering. *ACS Nano*, **9**(4), 3820–3828.
28. V. Lacatena, M. Haras, J.-F. Robillard, S. Monfray, T. Skotnicki, and E. Dubois. (2015). Toward quantitative modeling of silicon phononic thermocrystals. *Applied Physics Letters*, **106**(11), 114104.

Chapter 5

Monte Carlo Resolution of the Boltzmann Transport Equation

David Lacroix
University of Lorraine—LEMTA, CNRS-UMR 7563, 2 Av. de la Forêt de Haye, 54518 Vandœuvre lès Nancy, France
david.lacroix@univ-lorraine.fr

5.1 Introduction

Heat transport in solid materials can be understood in the framework of lattice dynamic. Collective motion of atoms generates acoustic waves that carry heat. According to the wave-corpuscle duality the latter ones can be considered as pseudo-particles, the "phonons." The motion (advection) and the interactions (collisions) of phonons rule heat transport at the microscale and these mechanisms can be described by the Boltzmann transport equation (BTE) for structures length in the range of tens of nanometer till tens of micron. For smaller length scales, atomistic approaches such as molecular dynamic shall be considered and for larger ones heat transport equation applies with the Fourier formalism. BTE can be solved by different techniques as it will be discussed hereafter.

Nanostructured Semiconductors: Amorphization and Thermal Properties
Edited by Konstantinos Termentzidis

ISBN 978-981-4745-64-2 (Hardcover), 978-1-315-36445-2 (eBook)
www.panstanford.com

5.2 Kinetic Theory of Gases Applied for Phonons

5.2.1 Basic Formulation

The kinetic theory (KT) of gases was first proposed by Peierls [23] in his work on thermal transport in solids. In the 1950s and 1960s, a broad attention has been paid to heat transport in semiconductors and a phonon theory of thermal conductivity has become more popular [15, 16]. The KT applied to phonons lies on the following consideration. Phonons are pseudo-particles that carry energy in a material and as long as their wavelength is small in comparison to the size of the studied media their behavior can be depicted as the ones of molecules of a gas that move with given velocities and might scatter with other molecules or material boundaries. In respect of this assumption the wave nature of phonons is neglected and consequently the coherence effects too.

The basic formulation of the KT allows to calculate the thermal conductivity k of a material (Eq. 5.1). It depends on the heat capacity C per unit of mass, the density ρ, the sound velocity v_s which is constant ($v_s^{-1} = (v_{LA}^{-1} + 2v_{TA}^{-1})/3$) and the phonon mean free path Λ, it reads

$$k = \frac{1}{3}\rho C v_s \Lambda. \tag{5.1}$$

In this framework, thermal conductivity dependence in bulk crystalline material can be simply understood and its variation according to the temperature too. When temperature is below 100 K, the phonon's mean free path is large (ballistic transport) and the thermal conductivity increases like the heat capacity as a function of the third power of temperature. On the contrary, when the temperature is high enough, in the order of the Debye temperature θ_D, heat capacity becomes constant according to the Dulong and Petit law while mean free path decreases linearly with T. In the latter case phonon–phonon interactions are dominant and the thermal conductivity is proportional to $1/T$. In between, the thermal conductivity exhibits a maximum and the mechanisms that rule phonon scattering are impurity and lattice defect scattering. Figure 5.1 illustrates this behavior for several crystalline materials. For amorphous materials, things are partly different as explained

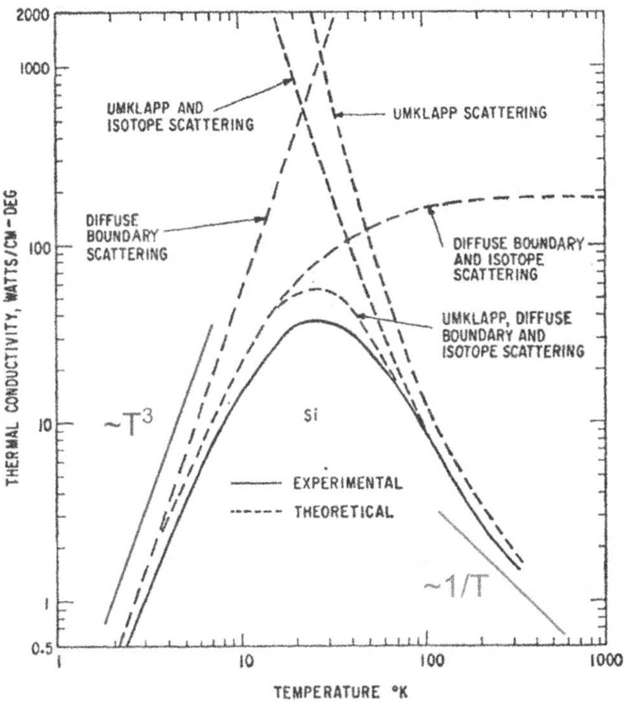

Figure 5.1 Thermal conductivity of Silicon, impact of scattering mechanisms [9].

in Chapter 2. In the low-temperature regime, below 10 K, thermal conductivity is proportional to the square of the temperature, while it very slowly increases with T at higher temperature. Such a behavior is complex to address as several theories have been proposed to explain TC variations in amorphous materials. The current theories that predict the thermal conductivity of amorphous structures are not accurate in the whole range of temperatures. There are distinct modes of vibration in amorphous materials with different mean free paths which contribute differently to the thermal conductivity [1]. In amorphous materials, the thermal conductivity increases weakly above the Debye temperature in increasing the temperature. In amorphous materials most lattice vibrations are localized or have negligible velocities. They are the so-called non-propagating modes. The energy transport in amorphous materials takes place mainly by two mechanisms: (i) by the diffusion of the non-propagating modes or (ii) by a hopping mechanism between localized modes. The latter one explains why the thermal conductivity increases with temperature in contrast to the thermal conductivity of crystalline materials, as the increasing of the anharmonicity in increasing the temperature multiplies the coupling between different localized modes [10].

5.2.2 KT Accounting for Material Dispersion

On the basis of the previously described formulation (Eq. 5.1), the thermal conductivity of bulk materials can be expressed as a function of their intrinsic dispersion properties. In such a case, one has to consider each phonon mode vibrations and sum them over the first Brillouin zone. Following the work by Chantrenne et al. [3], at given wave vector **K** and polarization p, thermal conductivity component in direction z reads

$$k_z(\mathbf{K}, p) = \rho C(\mathbf{K}, p) v_g(\mathbf{K}, p) \Lambda(\mathbf{K}, p) \cos\theta_z(\mathbf{K}), \tag{5.2}$$

where $\theta_z(\mathbf{K})$ is the angle between the wave vector **K** and the direction z and $v_g(\mathbf{K}, p) = d\omega(\mathbf{K}, p)/d\mathbf{K}$ the corresponding group velocity. The later equation can be also written as a function of the scattering collision time $\tau(\mathbf{K}, p)$ of the considered mode, it leads to

$$k_z(\mathbf{K}, p) = \rho C(\mathbf{K}, p) v_g^2(\mathbf{K}, p) \tau(\mathbf{K}, p) \cos^2\theta_z(\mathbf{K}). \tag{5.3}$$

The specific heat $C(\mathbf{K}, p)$ is given as the temperature derivative of the internal energy $U(\mathbf{K}, p)$ [2, 4, 8]. It depends on the average thermal occupation distribution given by the Bose–Einstein statistic $f[\omega(\mathbf{K}, p), T] = [\exp(\hbar\omega(\mathbf{K}, p)/k_B T) - 1]^{-1}$ times the energy of each single modes:

$$C(\mathbf{K}, p) = \frac{dU(\mathbf{K}, p)}{dT} = \frac{d\{\hbar\omega(\mathbf{K}, p) f[\omega(\mathbf{K}, p), T]\}}{dT} \tag{5.4}$$

Specific heat C and thermal conductivity k are obtained summing Eqs. 5.4 and 5.3 for all the polarizations (acoustic and optic branches) over all the wave-vector of the first Brillouin zone of the primitive cell in the case of bulk material:

$$C = \sum_{p} \sum_{\mathbf{K}} C(\mathbf{K}, p) \tag{5.5}$$

$$k_z = \sum_{p} \sum_{\mathbf{K}} k_z(\mathbf{K}, p) \tag{5.6}$$

For this model, the big issue is to find the appropriate values of the scattering collision term $\tau(\mathbf{K}, p)$. In most of the studies achieved until now, calculations are carried out assuming analytic expressions like the ones proposed by Holland for silicon and germanium [11]. The latter technique will be introduced in the next section. However, progress done in the field of ab initio modeling of material properties now allows to obtain phonon–phonon lifetimes from DFT calculations for a broad range of compounds.

5.2.3 KT as a Function of Frequency

The thermal conductivity through the KT formalism can be derived as a function of the phonon mode frequencies. Assuming isotropic dispersion properties (see Fig. 5.2), Eqs. 5.2 and 5.3 can be simplified and the summation over wave vectors is replaced by an integration on all the possible frequencies.

$$\begin{aligned} k &= \frac{\rho}{3} \sum_{p} \int_{0}^{\omega_{max}} C(\omega, p) v_g(\omega, p) \Lambda(\omega, p) d\omega \\ &= \frac{\rho}{3} \sum_{p} \int_{0}^{\omega_{max}} C(\omega, p) v_g^2(\omega, p) \tau(\omega, p) d\omega \end{aligned} \tag{5.7}$$

With the latter formulation, wave vector dependence no longer exists and the number of modes involved in thermal transport is to be reconsidered according to the density of states (DOS) at a given frequency. Assuming isotropic properties $U = \sum_p \sum_{\mathbf{K}} U(\mathbf{K}, p)$, the average internal energy in the K-space becomes

$$U = \frac{1}{\Delta V_K} \sum_{p} \int \hbar\omega f(\omega, T) d^3K = \sum_{p} \int \hbar\omega f(\omega, T) D_p(\omega) d\omega \tag{5.8}$$

with $\Delta V_K = (2\pi^3)/V$. The relation between mode counting in K-space and DOS in the frequency range implies to integrate on shell modes that are in the frequency range $[\omega, \omega + d\omega]$ as illustrated in Fig. 5.3.

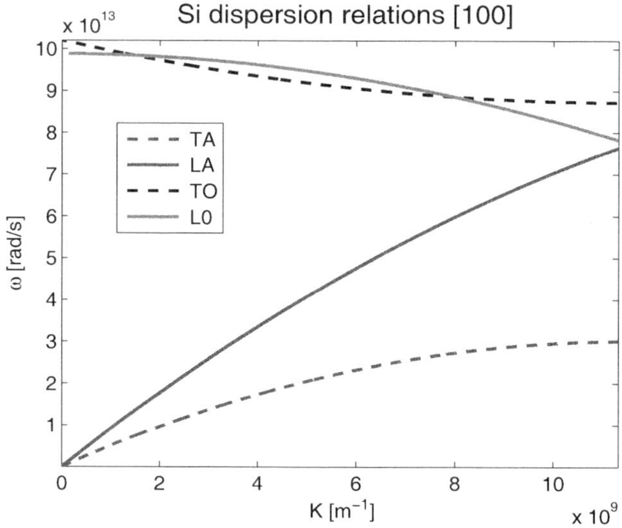

Figure 5.2 Silicon dispersion relations in the direction [100], quadratic fit [26].

$$
\begin{aligned}
D_p(\omega)d\omega &= \int_{Shell} \frac{d^3K}{\Delta V_K} = \int \frac{dS_\omega dK}{\Delta V_K} \\
&= \frac{4\pi}{\Delta V_K} K^2 dK \\
&= \frac{V}{2\pi^2} \frac{K^2}{d\omega/dK} d\omega \\
&= \frac{V}{2\pi^2} \frac{K^2}{v_g(\omega,\, p)} d\omega \\
&= \frac{V}{2\pi^2} \frac{\omega^2}{v_g(\omega,\, p) v_\varphi^2(\omega,\, p)} d\omega, \qquad (5.9)
\end{aligned}
$$

where $v_\varphi = \omega/K$ stands for the phase velocity. Using Eq. (5.4) and Eq. (5.8), the derivation in the frequency domain of heat capacity is straightforward:

$$
C = \sum_p \int_0^{\omega_{p,max}} k_B \left(\frac{\hbar\omega}{k_B T}\right)^2 D_p(\omega) \frac{\exp\left(\frac{\hbar\omega}{k_B T}\right)}{\left[\exp\left(\frac{\hbar\omega}{k_B T}\right) - 1\right]^2} d\omega \qquad (5.10)
$$

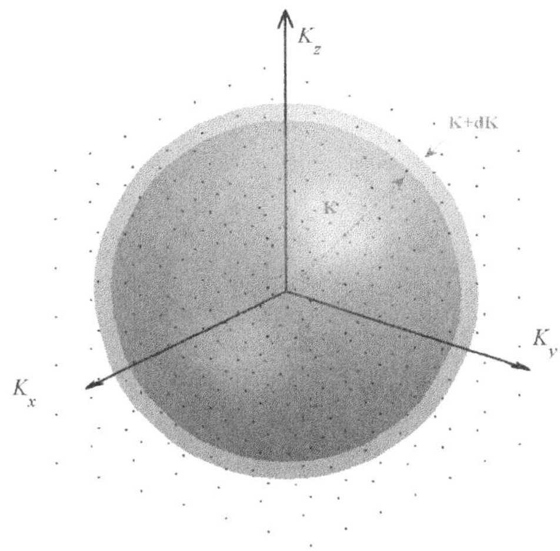

Figure 5.3 Density of States, mode counting in 3-D isotropic case.

The latter expression can be rewritten using the DOS Eq. (5.9) with $x = \hbar\omega / k_B T$:

$$C = \frac{k_B V}{2\pi^2} \left(\frac{k_B T}{\hbar} \right)^3 \sum_p \int_0^{x_{p,max}} \frac{1}{v_g v_\varphi^2} \frac{x^4 \exp(x)}{[\exp(x) - 1]^2} dx \qquad (5.11)$$

The thermal conductivity is thus derived considering Eq. (5.7):

$$k = \frac{k_B}{6\pi^2} \left(\frac{k_B T}{\hbar} \right)^3 \sum_p \int_0^{x_{p,max}} \frac{v_g \tau(x)}{v_\varphi^2} \frac{x^4 \exp(x)}{[\exp(x) - 1]^2} dx \qquad (5.12)$$

In the latter expression, the unknown is $\tau(x)$, the relaxation collision term that describes all the phonon scattering processes (see Table 5.1). In the case of crystalline materials such as Si and Ge, they are given according to analytic expressions depending on the collision mechanism [11]. Currently, three kinds or processes are usually considered:

- boundary collision
- mass difference scattering (point defect or impurity)
- three phonon interactions (Normal (N) or Umklapp (U))

Here L is a characteristic length of the system (required for low-temperature recovering of the thermal conductivity variation when phonon MFP is long), F a geometric factor and A, B_L, B_T, B_{TU})

Table 5.1 Relaxation time expressions in crystalline lattice

Scattering process	Inverse relaxation time (s^{-1})
Boundary	$\tau_B^{-1} = v_s/LF$
Impurity	$\tau_I^{-1} = A\omega^4$
3-phonon N processes	
LA branch	$\tau_{N_{LA}}^{-1} = B_L\omega^2 T^3$
TA branch ($\omega < \omega 12$)	$\tau_{N_{TA}}^{-1} = B_T \omega T^4$
TA branch ($\omega > \omega 12$)	$\tau_{N_{TA}}^{-1} = 0$
3-phonon U processes	
LA branch	$\tau_{U_{LA}}^{-1} = B_L\omega^2 T^3$
TA branch ($\omega < \omega 12$)	$\tau_{U_{TA}}^{-1} = 0$
TA branch ($\omega > \omega 12$)	$\tau_{U_{TA}}^{-1} = B_{TU}\omega^2/\sinh(\hbar\omega/k_B T)$

Source: [11].

adjustable parameters found by matching Eq. (5.12) with thermal conductivity experimental data. In the case of the TA branch, Umklapp processes are likely to occur only if the phonon frequency is large enough, i.e., if it is greater than ω_{12} which is equal to $\omega(K_{max}/2)$. As all the scattering processes are supposed to be independent, the inverse relaxation times can be added into a total one (Matthiesen's rule):

$$\tau(\omega, T, p)^{-1} = \tau_B^{-1}(\omega, p) + \tau_I^{-1}(\omega) + \tau_N(\omega, T, p)^{-1} + \tau_U(\omega, T, p)^{-1} \tag{5.13}$$

The last input to achieve calculation is the knowledge of dispersion properties. Their choice has a direct impact on the determination A, B_L, B_T, B_{TU} constants. In Holland's work, linear dispersion curves were assumed. Results presented in the following sections have been obtained considering quartic polynomial fit [26]. In the case of silicon, bulk properties like specific heat and thermal conductivity are plotted in Figs. 5.4 and 5.5. In what concerns heat capacity, frequency integration must be carried out on all the branches (acoustic and optic). For the thermal conductivity appraisal, the optic polarization can be neglected in several materials as the phonon group velocity is often weak. It means that the latter phonon are not energy carriers. Yet, while temperature increases sufficiently, the k dependence to the optical process is no longer negligible.

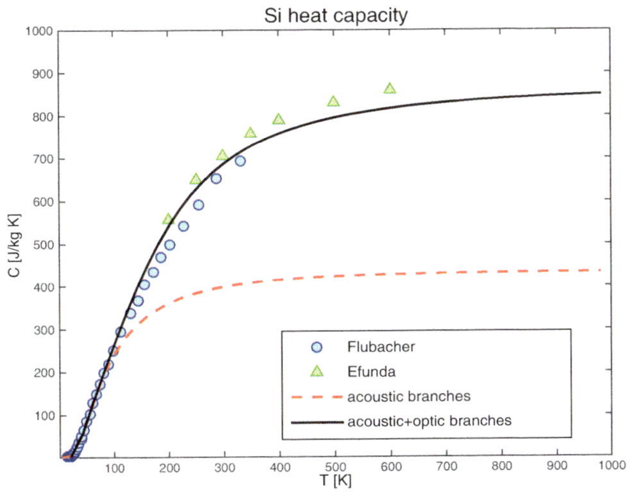

Figure 5.4 Specific heat of crystalline silicon, KT calculations assuming isotropic dispersion relations, comparison to experimental data by Flubacher [7] and Efunda [6].

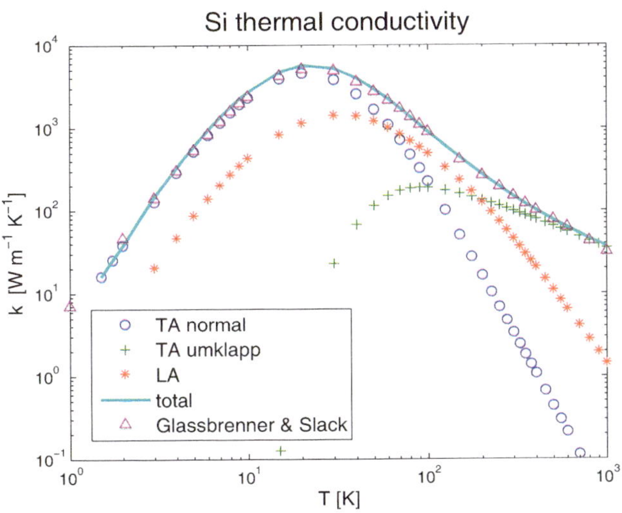

Figure 5.5 Thermal conductivity of crystalline silicon, KT calculations assuming isotropic dispersion relations, comparison to experimental data by Glassbrenner and Slack [9].

5.3 Boltzmann Transport Equation for Phonons

5.3.1 Introduction

The Boltzmann transport equation (BTE) is used in physics to describe the behavior of a system which is not in thermodynamic equilibrium from a statistical standpoint. Typically, BTE can be used to model the transport and the interactions of moving particles. Firstly applied to ideal gases by Boltzmann, this equation has been applied to fluids and its derivation leads to Navier–Stokes equations. During the second half of the XXth century, the BTE was successfully applied to energy carriers transport modeling. The latter ones can be electrons, photons and neutrons assuming that they behave like individual particles. This lead for instance to the radiative transfer equation (RTE), in the case of photons, which is commonly used to model radiative heat exchanges. At the same time, the BTE has been proposed to describe phonon transport in materials [28] but its resolution by mean of different numerical techniques is much recent. In the next subsections the BTE formalism and connections with the relaxation time approximation are briefly recalled. Then, a deterministic technique such as the discrete ordinate method applied to the BTE is discussed before dealing with the MC statistical approach.

5.3.2 BTE Formulation for Phonons

The BTE is an integro-differential equation that provides the variations of a particle distribution function f in the phase space. This distribution function is simply related to the number N of particles through Eq. (5.14):

$$dN = f(\mathbf{r}, \mathbf{p}, t)d^3\mathbf{r}d^3\mathbf{p}, \tag{5.14}$$

where dN is the number of particles in a small volume $d^3\mathbf{r}$ at the location $\mathbf{r}$ having a momentum within $d^3\mathbf{p}$ around $\mathbf{p}$ at time t. The general formulation of the BTE is thus

$$\frac{\partial f}{\partial t} = \left.\frac{\partial f}{\partial t}\right|_{\text{force}} + \left.\frac{\partial f}{\partial t}\right|_{\text{adv}} + \left.\frac{\partial f}{\partial t}\right|_{\text{coll}}, \tag{5.15}$$

where "force" term corresponds to external forces applied on particles, "adv" is the advection term that characterizes particles displacement and "coll" is the collision term accounting for interaction between particles. For phonons, Eq. (5.15) can be simplified, in practice the "force" term is neglected as calculations are usually achieved assuming that there is no stress applied on the material. Finally it comes

$$\frac{\partial f}{\partial t} + \nabla_{\mathbf{K}}\omega \cdot \nabla_{\mathbf{r}} f = \left.\frac{\partial f}{\partial t}\right|_{\text{coll}} \tag{5.16}$$

with $\nabla_{\mathbf{K}}\omega \cdot \nabla_{\mathbf{r}}$ the advection term where $v_g = \nabla_{\mathbf{K}}\omega$. In this equation the left hand side term is related to the displacement of phonons (material derivative) while the right hand side term characterize phonon scattering that restore equilibrium. The latter term cannot be expressed in a general way. In the case of phonon transport it is usually treated under the "relaxation time approximation."

Relaxation time approximation In the framework of the relaxation time approximation we assume that the phonon population N for a given mode is close to the equilibrium one N^0. In such a case the collision term of Eq. (5.15) is given as a function of a relaxation τ which depends on the considered mode.

$$\left.\frac{\partial f}{\partial t}\right|_{\text{coll}} \simeq \frac{f - f^0}{\tau(\mathbf{K})}, \tag{5.17}$$

where f^0 is the Bose–Einstein distribution function.

$$f^0(\omega, T) = \frac{1}{\exp\left(\frac{\hbar\omega}{k_B T}\right) - 1} \tag{5.18}$$

The general theory underlying the relaxation time approximation has been detailed by P. G. Klemens [15, 16]. It lies on the application of the perturbation theory to the Hamiltonian that characterizes the lattice vibrations. According to that, the rate of change of a phonon in mode $\mathbf{K}$ with a population N into a another phonon characterized by $(\mathbf{K}', N')$ (destruction & creation processes) during an interval of time t can be appraised and give rise to the identification of a relaxation time for impurity (atom substitution) or defect (dislocation). Similarly, when three phonons are involved in a scattering mechanism (modes $\mathbf{K}$ and $\mathbf{K}'$ which combine to form mode $\mathbf{K}''$) the perturbation theory can help to find semi-analytical expressions of relaxation times for normal (N) and Umklapp (U) processes. The latter ones usually depend on frequency, polarization and temperature as it has been seen in Section 5.2.3 and involve adjustable parameters which need to be fitted through comparison with measurements. This approach has been applied for decades successfully for several material thermal properties assessment. However, the improvement of numerical modeling combined to the fast development of computation resources make now possible to achieve direct calculation of relaxation time on dense K-grid with DFT approaches (Density Functional Theory) as it is explained in the chapter by L. Chaput.

5.3.3 Phonon Radiative Transfer Equation

As explained above the Boltzmann transport equation can be applied to several kinds of energy carriers. In this context the radiative transfer equation is a BTE where particles of interest are photons. There is an extensive literature dealing with the RTE, most of useful derivations of this equation and its numerical treatment can be found in the reference book by M. Modest [22]. Due to the numerous similarities existing between phonons and photon transposition of well known numerical techniques developed for photons was suggested in the 1990s by A. Majumdar [20] and G. Chen [5]. Using this analogy it is possible to define a spectral intensity of phonons I_ω t location z and time t:

$$I_\omega(\theta, \phi, z, t) = \sum_p v_g(\theta, \phi) f_\omega(z, t)\hbar\omega D_p(\omega) \tag{5.19}$$

with $v_g(\theta, \phi)$ the group velocity vector of the phonon bundle in a direction Ω defined by polar θ and azimuth ϕ angles. Using this formulation, multiplying Eq. (5.16) by $v_{\omega,p} f_\omega \hbar\omega D_p(\omega)$ one can derived the equation of phonon radiative transfer (EPRT) in the one dimensional case (wall).

$$\frac{1}{v_g}\frac{\partial I_\omega}{\partial t} + \mu\frac{\partial I_\omega}{\partial z} = \frac{I_\omega^0(T(z)) - I_\omega}{v_g \tau(\omega, T)}, \tag{5.20}$$

where μ is the direction cosine (i.e., the cosine of the angle between the phonon propagation direction and the z direction. $I_\omega^0(T(z))$ is the equilibrium intensity corresponding to the blackbody intensity for photons.

$$I^0_\omega(T) = \frac{1}{4\pi}\sum_p \frac{v_g \hbar\omega D_p(\omega)}{\exp\left(\frac{\hbar\omega}{k_B T}\right) - 1} \quad (5.21)$$

The EPRT can be rewritten in a more "radiative way" with the phonon scattering coefficient $\sigma_{ph}(\omega, p, T) = 1/v_g\tau(\omega, T)$. Eq. (5.20) can be solved by different numerical techniques including statistical ones. Yet, in order to achieve fast resolution of the EPRT, techniques based on the use of Discrete Ordinates Method (DOM) are more suitable. With this approach finite differences are used to discretize spatial and temporal differential terms, while the phonon propagation direction are arbitrarily assigned with respect to angular quadratures. The most often used are the S_N quadratures [22]. In steady state regime the discretized problem is described by a set of $N_z \times N_d \times N_\omega$ linear equations (N_z is the number of cell of the mesh, N_d the number of directions and N_ω the number of spectral bands). Boundary conditions usually considered are blackbody emission or reflection (diffuse or specular). Both are easily treated using methods developed for RTE. Once the intensities are known at each position z of the layer and in every phonon propagation direction Ω, temperature field and heat flux can be appraised. The integration (discrete summation) of the intensities $I^i_{\omega,p}$ on all the directions, spectral bands and polarization branches allows the calculation of the temperature with the energy balance (Eq. (5.22)), while the heat flux q_z along z direction is given by the integration over the same quantities of the intensities times the direction cosine (Eq. (5.39)).

$$\sum_p \sum_\omega \sigma_{ph}(\omega, p, T)\left[4\pi I^0_{\omega,p}(T(z)) - 2\pi \sum_{i=1}^{Nd} I^i_{\omega,p}(z) w_i\right] = 0 \quad (5.22)$$

and

$$q_z(z) = 2\pi \sum_p \sum_\omega \sum_{i=1}^{Nd} \mu_i I^i_{\omega,p}(z) w_i \quad (5.23)$$

with w_i the angular weight corresponding to i direction of the angular quadrature associated to the direction cosine μ_i. In transient state, EPRT numerical handling is similar, the time derivative

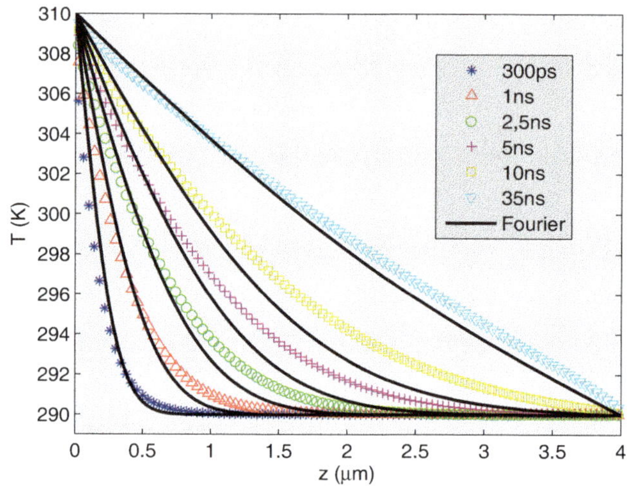

Figure 5.6 Temperature profiles in a $L_z = 4$ μm thin film of silicon at several time steps - EPRT solution by the DOM; comparison with the Fourier solution.

can be discretized with an explicit scheme that leads to a similar formulation to the one of the steady state. In order to illustrate the ability of EPRT to model phonon heat transport, the case of a thin film is addressed hereafter. In Fig. 5.6, the transient variations of the temperature within a thin silicon film are plotted until the steady state. Here, Holland [11] relaxation times formalism combined to quartic dispersion fits are used to determine the phonon scattering coefficients. As the heat flux is also obtained by this method, from the steady state one can extract the thermal conductivity of the material. In Fig. 5.7 the variations of the thermal conductivity as a function of the temperature (up) and the film thickness (down) are plotted.

The DOM method can be straightforwardly applied to 2D and 3D geometries as long as they remain simple (parallelepiped, cylinder). In such a case, this method is usually faster than the statistical ones. When geometry is more complex or in the case of material with interfaces, inclusions, porosities, etc. the DOM method applied to EPRT becomes worthless.

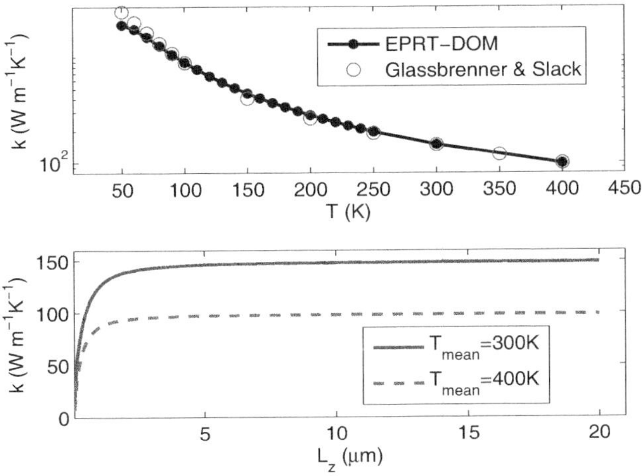

Figure 5.7 (up) Thermal conductivity as a function of T for Si, comparison to [9]; (down) Thermal conductivity as a function of film thickness L_z for Si at 300 K and 400 K.

5.4 Resolution of the BTE with the Monte Carlo Technique

5.4.1 The Monte Carlo Method

The Monte Carlo (MC) method is a numerical technique based on the use of random numbers sampling according to probability laws to model and solve physical or mathematical problems. It finds its origin in the 1940s with contributions of several scientists of the "Manhattan project" as John Von Neumann and Stanislaw Ulam. This method is particularly adapted to problems with several degrees of freedom (radiation propagation, fluids dynamics, molecule transport, etc.) and many other problems as long as a probabilistic model can be established to describe their physics. In that sense, the MC method is well adapted to BTE, which is inherently an equation that models particle transport and interactions characterized by probability laws. In the case of phonon transport, the MC method was first applied by Peterson [25] for an ideal Debye crystal and later by Mazumder and Majumdar [21] for silicon. Those first approaches were improved, first to ensure efficiently energy balance [17] in the collision procedure and secondly to reduce simulation time through variance-reduction formulation [24]. Several other studies have been conducted with this technique to model different nanostructures and materials. In the following section, the derivation of the MC method to the BTE is addressed, then some examples are provided.

5.4.2 BTE and MC Method

The main steps of the BTE numerical handling are detailed in the following: domain discretization, phonon sampling and tracking, scattering, temperature and heat flux evaluation, etc.

Simulation domain and boundary conditions As it has been mentioned before, the geometry of the studied material does not matter with MC modeling. However, in order to give a clear description of the methodology, presently, a simple cubic cells stack (Fig. 5.8) is considered. It can be readily related to the plane wall geometry commonly used in thermal problems. Cylindrical cells or multidimensional stacks can be also considered in order to model nanowires or real semi-conductor devices.

Concerning boundary conditions, we assume that the lateral walls of the cells (in x and y directions) are specularly reflecting in most of the simulation cases. This means that walls are adiabatic and perfectly smooth. Note also that in that case, the dimension in the x and y directions should not affect the simulation's results. Indeed, when reflection is specular on the lateral cells boundaries,

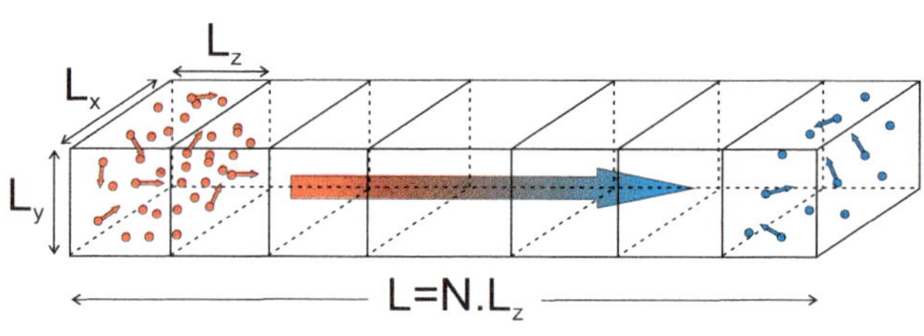

Figure 5.8 Studied model. Phonons location, energy and velocity are randomly chosen in each cell according to dispersion curves and local temperature.

the momentum is preserved in the z direction. Thus this geometry is a film. If diffuse boundaries are considered in x and y directions, the modeled structure is a square nanowire. At both end of the medium, temperature is assumed to be constant. Therefore, energy in the first and the last cell is calculated from equilibrium distribution functions. Incoming phonons in these cells are thermalized at each time step. Consequently, these cells act as blackbodies.

At this stage, an important point is the choice of the three discretizations: temporal, spatial, and spectral. Spatial discretization is directly related to the material geometry: usually cell lengths are about $L_z \sim 100$ nm for micrometric objects and can be smaller in the case of thin films or nanowires, for instance. The time step choice depends on two parameters: the cell size and the group velocity at a given frequency. In order to consider all scattering events and to avoid ballistic jump over several cells, we state that the time step must be lower than $\Delta t < L_z/v_{gmax}$. The spectral discretization is uniform, we used at least $N_b = 1000$ spectral bins in the range $[0, \omega_{LAmax}]$.

Phonon sampling The first step of the simulation procedure, once medium, geometry and mesh have been chosen, is to initialize the state of phonons within each cell describing the material. Hence, the number of phonons present in each cell is required. It will be obtained considering the local temperature within the cell and using a modified expression of Eq. (5.8). In this equation, energy is given for all the quanta $\hbar\omega$ associated to a spectral bin. Therefore, it can be rewritten to give the total number of phonons in a cell as

$$N = V \sum_{p=TA,LA} \sum_{b=0}^{N_b} \left[\frac{1}{\exp\left(\frac{\hbar\omega_{b,p}}{k_B T}\right) - 1} \right] \frac{K_{b,p}^2}{2\pi^2 v_{g\,b,p}} g_p \Delta\omega \tag{5.24}$$

The number of phonons obtained with Eq. (5.24) is usually very large; for instance in a 10 nm silicon cube at 300 K, N can be estimated around 5.5×10^5. In the case of nanoscale structures, direct simulations can be achieved if the temperature is relatively low. In the case of microscale samples or multidimensional cell stacking, a weighting factor shall be used to achieve Monte Carlo simulations. Hence, Peterson's [25] technique has been used. The actual number of phonons N is divided by a constant weight W in order to obtain the number of simulated phonons $N^\star$:

$$N^\star = \frac{N}{W} \tag{5.25}$$

In simulations W maximal values are usually around $W \sim 10^4$ for micrometric structures. During the initialization process, a temperature step is prescribed in the medium. The first cell is raised to the hot temperature T_h and the last to the cold one T_c. All the phonons in the intermediate boxes are also initially at T_c. Associated theoretical energy in the whole structure is obtained from Eq. (5.8). This energy should match the calculated energy $U^\star$ within all the cells, written into the following form:

$$U^\star = \sum_{c=1}^{N_{cell}} \sum_{n=1}^{N^\star} W \times \hbar\omega_{n,c} \tag{5.26}$$

As a consequence, during the initialization, phonons should be added by packs of W at a given frequency, sampled from a normalized number density function F_N. According to Mazumder and Majumdar [21], this function is constructed doing the cumulative summation of the number of phonons in the ith spectral bin over the total number of phonons Eq. (5.24):

$$F_N(j, T) = \frac{\sum_p \sum_{i=1}^{j} g_p D(p, \omega_i) f^0(p, \omega_i, T)}{\sum_p \sum_{i=1}^{N_b} g_p D(p, \omega_i) f^0(p, \omega_i, T)}, \tag{5.27}$$

where g_p is the polarization degeneracy and N_b is the number of spectral bins used for the phonon frequency discretization. In this process, a random number R is drawn (all the random numbers discussed here check $0 \leq R \leq 1$) and the corresponding value $F_{N,i}$ gives the frequency ω_i, knowing that $F_{N,i-1} \leq R \leq F_{N,i}$ location is achieved with bisection algorithm. The actual frequency of the phonon is randomly chosen in the spectral interval prescribing

$$\omega_i = \omega_{0,i} + (2R - 1)\frac{\Delta\omega}{2}, \tag{5.28}$$

where $\omega_{0,i}$ is the central frequency of the ith interval. At this stage, a first observation can be done. When phonons are randomly sampled the process is stopped when their accumulated energy in a cell becomes larger than the theoretical one. This mean that a little overtaking of the material energy occurs, the error decreases while W is increase. Besides, with this approach, the energy carried by each bundle differs as it depends on its frequency. Among the numerous simulations carried out with this approach, sometimes energy balance is not respected and the numerical resolution of the BTE diverges.

Recently a second method for phonon sampling, based on energy consideration was proposed by Péraud et al. [24]. In the latter one, the energy of a sampled phonon bundle is prescribed and remains constant $u_{i,\mathrm{MC}} = u_{\mathrm{Bundle}}$ whatever the value of the ith sampled frequency ω_i is, whereas in the previous case it was equal to $u_{i,\mathrm{MC}} = W\hbar\omega_i$. This also means that each bundle does not carry the same number of phonons. With this method the sampling is done on the basis of a normalized cumulative distribution function based on phonon energy F_U until the total sampled energy $U^\star = N^\star \times u_{\mathrm{Bundle}}$ is equal to U like in the previous case. Distribution functions are thus given by

$$F_U\left(j, T\right) = \frac{\sum_p \sum_{i=1}^{j} g_p D(p, \omega_i) f^0(p, \omega_i, T)\hbar\omega_i}{\sum_p \sum_{i=1}^{N_b} g_p D(p, \omega_i) f^0(p, \omega_i, T)\hbar\omega_i}, \tag{5.29}$$

This second formulation, based on energy scheme for sampling, provides accurate results for the same number of drawn phonons.

The third point which is important to consider when dealing with a MC technique is the accuracy of the achieved results. In the case of BTE modeling, uncertainties decrease as the number of sampled bundles increases. This dilemma often works against the use of MC methods because consequent numerical resources are required to provide reliable results. However, recent studies [24, 27] have suggested to solely sample "useful phonons," namely those which are above a reference temperature T_{ref} chosen at the beginning of the simulation. This reference temperature is usually set just below the lowest temperature prescribed by boundary conditions. Thus, the total number of phonons beyond T_{ref} is $\widetilde{N}(T) = N(T) - N(T_{ref})$; it reads

$$\widetilde{N}(T) = \int_V \int_\omega \sum_p g_p D(p,\omega)\left[f^0(p,\omega,T) - f^0(p,\omega,T_{ref})\right] d\omega dV. \tag{5.30}$$

This technique, also called "variance-reduction," can be easily implemented in both normalized cumulative distribution functions relying on Eq. (5.29) to achieve phonon sampling. It is also implemented in the numerical procedures that calculate the polarization, the energy, and the derived pseudo-temperature of the phonon bundle ensembles. The latter points are not recalled here but can be found in the previously cited papers [17, 24, 27].

Once the frequency is known, the polarization of the phonon has to be determined. It can belong to the TA or LA branch with respect to the Bose–Einstein distribution and the density of states. For a given frequency ω_i, the number of phonon on each branch is: $N_{LA}(\omega_i) = f^0_{LA}(\omega_i)D_{LA}(\omega_i)$ and $N_{TA}(\omega_i) = 2 \times f^0_{TA}(\omega_i) f^0 D_{TA}(\omega_i)$. The associated probability to find a LA phonon is expressed as

$$P_{LA}(\omega_i) = \frac{N_{LA}(\omega_i)}{N_{LA}(\omega_i) + N_{TA}(\omega_i)}. \tag{5.31}$$

A new random number R is drawn: if $R < P_{LA}(\omega_i)$ the phonon belongs to the LA branch otherwise it is a transverse one.

The knowledge of the frequency and the polarization leads to the estimation of the phonon group velocity and the phonon wave vector merely using the dispersion curves and their derivatives. Assuming isotropy within the crystal, the direction Ω of the wave vector and the group velocity are obtained from two random numbers R and R' in the Cartesian coordinates $(\mathbf{i}, \mathbf{j}, \mathbf{k})$ associated to lengths L_x, L_y and L_z. Hence Ω is written as

$$\Omega = \begin{cases} \sqrt{1-(2R-1)^2}\cos(2\pi R') \\ \sqrt{1-(2R-1)^2}\sin(2\pi R') \\ (2R-1) \end{cases}. \tag{5.32}$$

The last operation of the initialization procedure is to give a random position to the phonon within the cell. In the grid previously considered, location of the nth phonon in the cell c is

$$\mathbf{r}_{\mathrm{n,c}} = \mathbf{r}_{\mathrm{c}} + L_x R\,\mathbf{i} + L_y R'\,\mathbf{j} + L_z R''\,\mathbf{k}, \tag{5.33}$$

where $\mathbf{r}_{\mathrm{c}}$ are the coordinates of the cell and R, R' and R'' three random numbers.

Phonon drift and scattering Once the initialization stage is achieved, phonons are allowed to drift inside the nanostructure. Considering the time step Δt and their velocities, each phonon position is updated: $\mathbf{r}_{\mathrm{drift}} = \mathbf{r}_{\mathrm{old}} + \mathbf{v}_{\mathrm{g}}\Delta \mathrm{t}$. In the case of shifting outside of the lateral boundaries (in **i** and **j** directions) the phonon is specularly reflected at the wall. In the case of diffuse reflection with a particular degree d ($0 \leq d \leq 1$, $d = 0$ purely specular, $d = 1$ purely diffuse), a random number R is drawn. When R is lower than d a new phonon propagation direction is calculated using Eq. (5.32).

When a phonon reaches the bottom (z_{min}) or the top (z_{max}) of a cell, it is allowed to carry on its way in the previous or next cell respectively. As a result, it is going to modify the cell energy and by extension its local temperature. At the end of the drift phase, the actual energy $\widetilde{U^{\star}}$ is computed in all the cells using Eq. (5.26). Then, the actual temperature $\widetilde{T}$ is obtained with Eq. (5.8) doing a Newton–Raphson inversion. Phonons drifting in the first and last cells are thermalized to the cold or hot temperature in order to keep boundary cells acting as blackbody sources.

In the Monte Carlo simulation, the scattering process has been treated independently from the drift. The phonon–phonon scattering aims at restoring local thermal equilibrium in the crystal since it changes phonons frequency. Collisions with impurities or crystal defects as well as boundary scattering do not change frequency but solely the direction Ω. These last phenomena are significant when low temperatures are reached and the phonon mean free path becomes large.

Direct simulation of phonons scattering is an awkward challenge. With Monte Carlo simulations, it is possible to model phonons collisions with neighbors as in the gas kinetic theory calculating a three-particle interaction cross section. Thus the collisional process is treated in the relaxation time approximation. Several studies on that topic have been carried out since the early work of Klemens [15]. Relaxation times τ have been proposed for several crystals. They depend on the scattering processes, the temperature and the frequency. Holland's work on silicon and germanium [11] provides various τ values. The independence of the scattering processes is used to consider a global three phonons inverse relaxation times accounting for N and U processes τ_{NU}. It has been obtained using the Matthiessen rule ($\tau_{NU}^{-1} = \tau_N^{-1} + \tau_U^{-1}$).

In order to be implemented in the Monte Carlo simulation, the scattering routine requires an associated collision probability P_{scat}. This one is derived saying that the probability for a phonon to be scattered between t and $t + dt$ is dt/τ. Thus,

$$P_{scat} = 1 - \exp\left(\frac{-\Delta t}{\tau_{NU}}\right). \tag{5.34}$$

A random number R is drawn, if $R < P_{scat}$ the phonon is scattered. As a result, new frequency, polarization, wave vector, group velocity and direction have to be re-sampled with respect of energy and momentum conservation. In order to discriminate between N and U processes a probability $P_U(\omega, j)$ is calculated and a random number is drawn to decide if a N $(R > P_U)$ or U $(R < P_U)$ process shall be considered.

$$P_U = \frac{\tau_U^{-1}}{\tau_N^{-1} + \tau_U^{-1}} \tag{5.35}$$

In former studies on that topic [21, 25], the frequency sampling after collision was achieved from the normalized number density function F at the actual temperature $\widetilde{T}$ of the cell obtained at the end of the drift procedure. In this approach, the actual energy after the scattering stage is usually different from the "target" one obtained with temperature $\widetilde{T}$. Hence a subsequent "creation/destruction" scheme is necessary to ensure the energy conservation. In fact, in the preceding procedure at thermal equilibrium, the probability of destroying a phonon of frequency ω and polarization p is different from the probability of creating this phonon. This means that the Kirchhoff law (creation balances destruction) is not respected. In order to create phonons at the same rate, they are destroyed at thermal equilibrium, the distribution function used to sample the frequencies of the phonons after scattering has to be modulated by the probability of scattering. So we define a new distribution function (here described in terms of energy sampling without

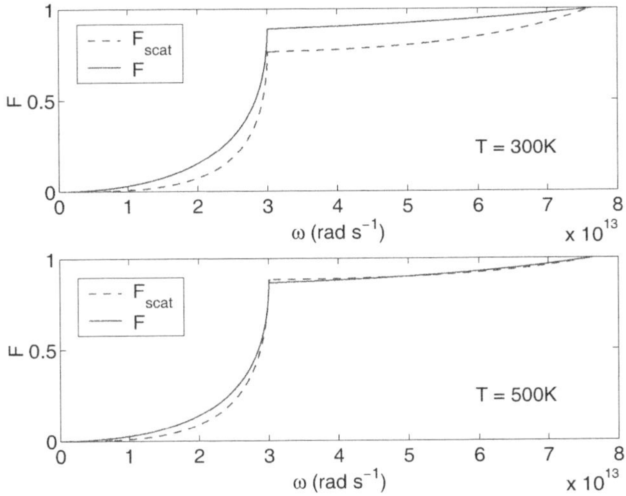

Figure 5.9 Normalized number density function in silicon with and without P_{scat} correction.

variance-reduction for the sake of simplicity):

$$F_{U,scat}(\widetilde{T}) = \frac{\sum_{j=1}^{i} N_j \hbar\omega_j \left(\widetilde{T}\right) \times P_{scat\,j}}{\sum_{j=1}^{N_b} N_j \hbar\omega_j \left(\widetilde{T}\right) \times P_{scat\,j}} \tag{5.36}$$

Taking into account the scattering probability in the distribution function, F_{scat} ensures that a destructed phonon on both transverse and longitudinal branches can be re-sampled with a not too weak energy as it can be seen in Fig. 5.9.

According to the described simulation procedure after initialization step, phonons in cell c are described by $[T_c,\ F_U(T_c),\ N^{\star}(T_c),\ U^{\star}(T_c)]$. They are allowed to drift and the state of cell c before scattering is $[\widetilde{T}_c,\ F(T_c),\ N'^{\star}(\widetilde{T}_c),\ U'^{\star}(\widetilde{T}_c)]$. Then three-phonon collisions occur and change energy by frequency resetting of the colliding phonons (using the distribution function $F_{U,scat}$, leading to the final state $[\widetilde{T}_c,\ F_{U,scat}(\widetilde{T}_c),\ N'^{\star}(\widetilde{T}_c),\ U''^{\star}(\widetilde{T}_c)]$. Hence energy can be expressed as

$$U''^{\star} = \underbrace{\sum_{i=1}^{N'^{\star}_{scat}} \hbar\widetilde{\omega}_i}_{\Rightarrow F_{U,scat}(\widetilde{T}_c)} + \underbrace{\sum_{i=1}^{N'^{\star}-N'^{\star}_{scat}} \hbar\omega_i}_{\Rightarrow F(T_c)} \tag{5.37}$$

Furthermore, the number of colliding phonons can simply be expressed as $N'^{\star}_{scat}(\widetilde{T}_c) = P_{scat} \times N'^{\star}(\widetilde{T}_c)$. One then sees that if we want to preserve energy at thermal equilibrium, the new normalized number density function $F_{U,scat}$ must take into account this collisional probability. Energy conservation during the Monte Carlo simulation might be noticed from frequency distribution (Fig. 5.10) that matches the theoretical distribution given by Eq. (5.24).

Concerning momentum conservation, the task is harder to address since the Monte Carlo process considers phonons one by one. Consequently triadic N or U interactions cannot be rigorously treated. In a first approach, we propose the following procedure to take into account the fact that U processes contribute to the thermal resistance, whereas the N processes do not. When the phonons scatter through a U process, their directions after scattering are randomly chosen as in the initialization procedure. Therefore, these phonons are randomly scattered and contribute to the diffusion of heat. On the contrary, it is assumed that scattering phonons experiencing a N process do not change their propagation direction

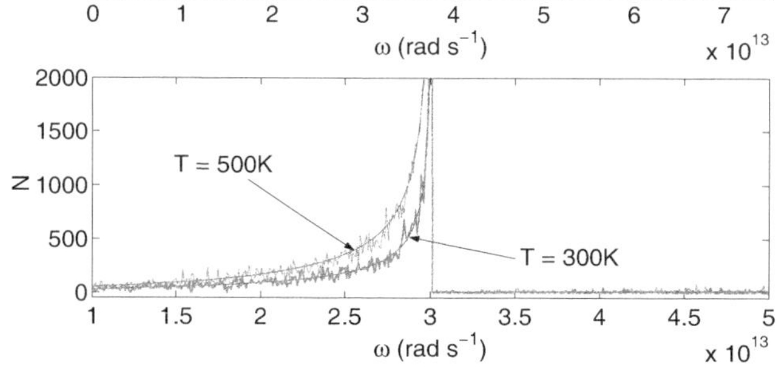

Figure 5.10 Frequency spectra at 300 K and 500 K for silicon, comparison of theoretical and sampled phonon populations; above full spectra, below enlargement of the spectra.

Ω. Statistically, for a given temperature and frequency, the phonons are destroyed by scattering at the same rate that they appear. A phonon which scatters has a great chance to be replaced in the computation by a phonon of a near frequency. Therefore, by this treatment, the N processes "approximately" preserve momentum. Nevertheless, a more accurate treatment should be done in order to respect exactly the momentum in the N processes. For a plan-parallel geometry, it seems possible to guarantee the momentum conservation in a single direction.

In the MC procedure, the three-phonon processes are not the only one to be considered. For impurities and lattice defect scattering, a collision probability is also drawn using the same expression as Eq. (5.34) with the associated relaxation time $\tau_I(\omega)$. Assuming the Matthiessen rule, a total relaxation time that includes N, U and I process is constructed $\tau_{NU\,I}^{-1}$ and a selection rule is set as for Eq. (5.35).

In case of boundary scattering (edge of the structure, porosity), with the MC method, the phonon is reflected at the surface assuming specular or diffuse behavior according to the considered physic. For a porous material, an effective approach based on a scattering mean free path that takes into account the porosity of the media can be used. In such a case, a new relaxation time related to phonon scattering with pore boundaries must be added to the other ones. It can be derived from geometric considerations as

$$\frac{1}{\tau_{pores}} = \frac{v_g(p,\omega)}{\Lambda_{pores}}, \tag{5.38}$$

where Λ_{pores} is the characteristic length related to the collision between phonons and pores. The MFP Λ_{pores} can be appraised on the basis of the medium pore organization through analytical expression or using a computational ray tracing technique derived from our MC calculations [12].

Temperature, heat flux and thermal conductivity derivations In the previous paragraphs, all the stages that occur in a single simulation time step have been described (initialization, drift and scattering). They are repeated until the steady state is reached within the sample. To check the convergence of the procedure, the temperature is obviously calculated (inversion from energy in each cell of the structure). Furthermore, the heat flux is also calculated; considering heat transport along the z direction, it reads

$$q_z(z) = \frac{1}{V_{\text{Cell}}} \sum_{i}^{N_{\text{tot}}} W\hbar\omega_i v_{gz,i}, \tag{5.39}$$

where N_{tot} is the total number of phonons sampled in the considered cell and v_{gz} the group velocity projected along the z direction. Knowing temperature and heat flux profiles, the thermal conductivity can be extracted using Fourier's law formalism.

5.4.3 Examples of the MC Method Applied to BTE

First, let consider Si and Ge thin films. Using the previously discussed methodology, one can calculate the transient evolution of the temperature within the nanofilm. Two examples, taken from [17], illustrate heat transport at room temperature (Fig. 5.11) and low temperature (Fig. 5.12). Room temperature calculations are in good agreement with the Fourier theory, which was expected

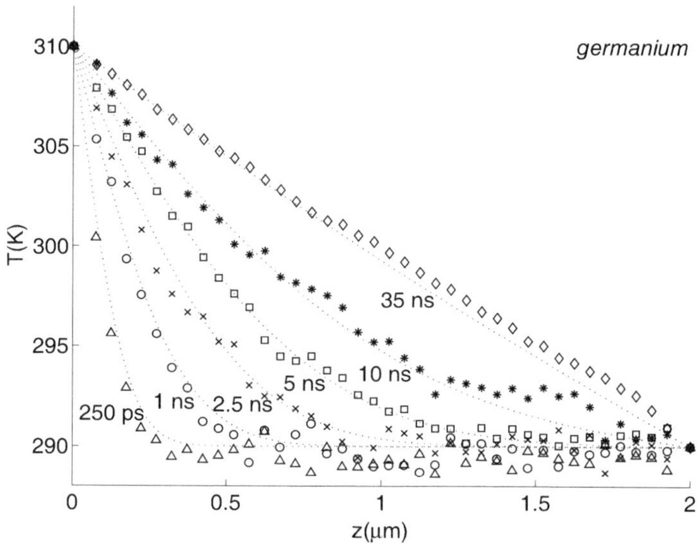

Figure 5.11 Transient temperature in Fourier's regime for germanium and comparison with the analytical solution of heat conduction equation with a constant thermal diffusivity ($\alpha_{Ge} = 0.36 \cdot 10^{-4}\ \text{m}^2\text{s}^{-1}$) (dotted curves).

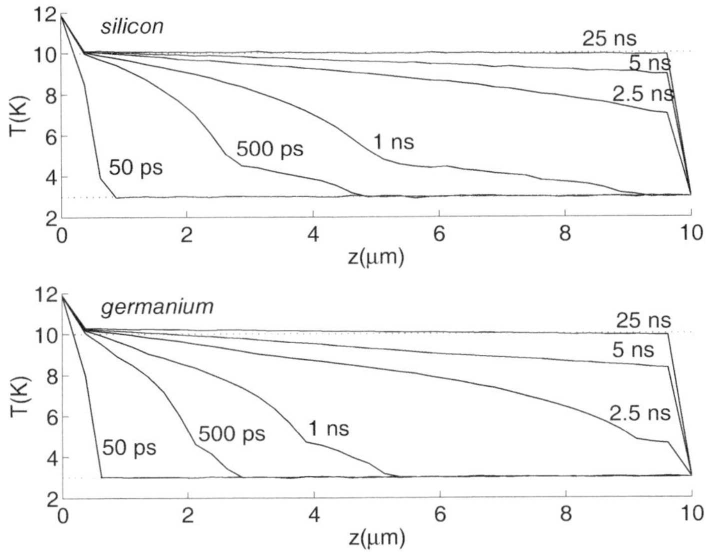

Figure 5.12 Transient temperature in the ballistic regime for silicon and germanium.

as thermal conductivity of 4 μm-thick film is almost the bulk one. For very low temperatures, here close to 10 K, heat transport is ballistic and phonon mean free path is very large. In such case, the only resistive processes are the one related to the scattering with the boundaries of the sample. In such case, Fourier's law is no longer valid. In this ballistic limit, there is temperature jump at boundaries. Furthermore, it can be seen that hot phonons do not fly straight toward the cold limit. More than 1 ns is necessary to heat the last cell in the case of silicon. This is in agreement with velocities prescribed by dispersion curves. Heat propagation in germanium is slower since phonons group speed is also lower. Results at low temperatures obtained with our method have been already be predicted by Joshi and Majumdar [14] in similar cases, who applied successfully the Equation of Phonon Radiative Transfer (EPRT) in ballistic regime ($T_{mean} = [(T_H^4 + T_C^4)/2]^{1/4}$).

Second, the MC method is able to handle complex geometries; to illustrate that point, let us consider nanowires with constrictions having different shapes (Fig. 5.13). Those results are taken from

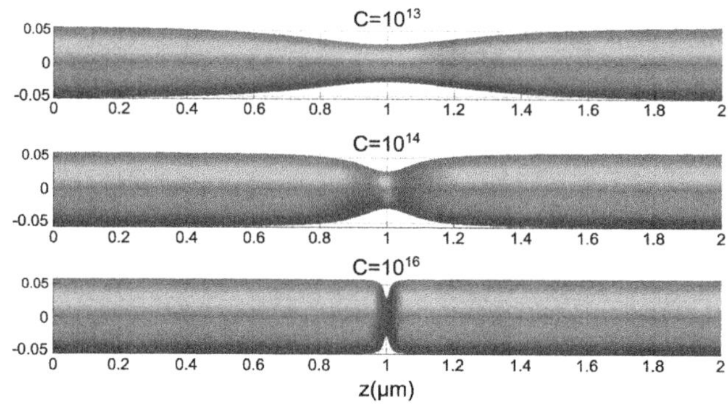

Figure 5.13 Different shapes of nanowire constriction; $L_z = 2$ μm, $d_{\max} =$ 115 nm, $d_{\min} = 57.5$ nm; top $C = 10^{13}$ m^{-2}, middle $C = 10^{14}$ m^{-2} and bottom $C = 10^{16}$ m^{-2}; the dimensionless parameter C characterizes the steepness of the constriction.

[13]. As it can be observed in Fig. 5.14, temperature and heat flux profiles exhibit characteristic features. Heat flux density increases in the constriction as the cross section of the nanowire is reduced while temperature presents some kind of temperature jump at the same location as the constriction. The latter one corresponds to the local variation of the thermal conductivity of the nanowire. It is well known that this parameter decreases as the nanowire diameter is reduced. Besides, if one calculates the equivalent thermal conductivity of such nanostructures, it appears that it is reduced as a constriction resistance limits the phonon flow. Such calculations are hardly achievable with classical methods like the KT or the DOM-EPRT.

Third, using the MC method for BTE solution can help to determine the thermal properties of nanostructures. On the next two plots are given the thermal conductivity of silicon thin films and silicon nanowires, respectively, as a function of the thickness (Fig. 5.15) and the radius (Fig. 5.16). In both cases, the MC method captures well the fast decrease of the thermal conductivity as the characteristic length of the sample is getting smaller and smaller. On the contrary, the diffuse behavior (large thickness or radius) is also observed when the film thermal conductivities reach Si and Ge bulk values. In the case of nanowires, the MC model has also been

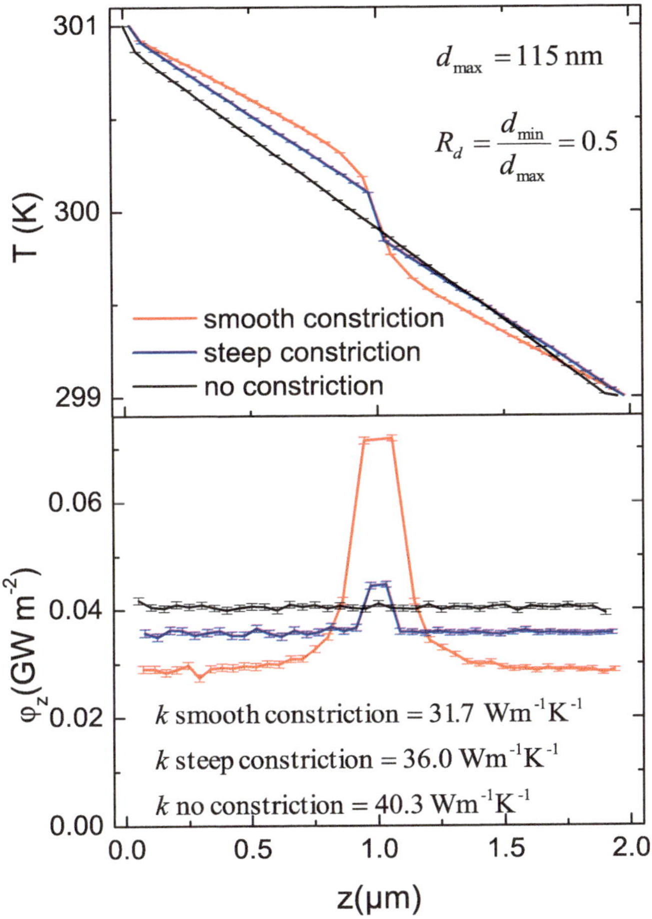

Figure 5.14 Temperature profiles and heat flux density, $d_{max} = 115$ nm, $R_d = 0.5$; no constriction (black), smooth constriction $C = 10^{14}$ m^{-2} (red), steep constriction $C = 10^{16}$ m^{-2} (blue).

tested as a function of the temperature [18] in order to reproduce numerical measurements achieved by Li et al. [19].

To conclude, it can be stated that the Monte Carlo method applied to the resolution of the BTE for phonons is a powerful tool that can be used for a broad range of materials and nanostructures. As long as the physics underlying phonon transport (dispersion and relaxation times) is known, the main limitations of this approach

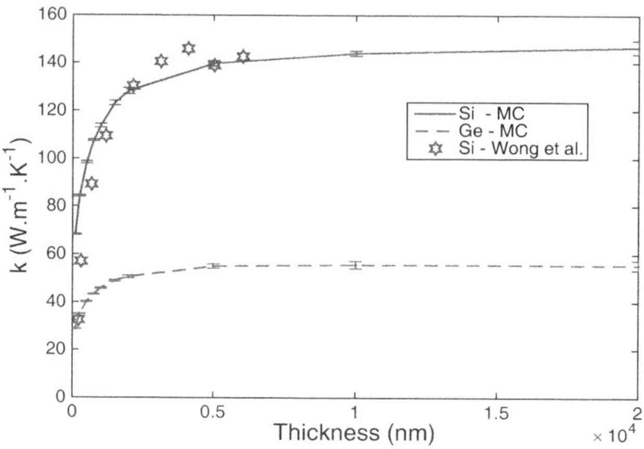

Figure 5.15 Nanofilm thermal conductivity as a function of film thickness at 300 K; comparison with Wong et al. calculations [27].

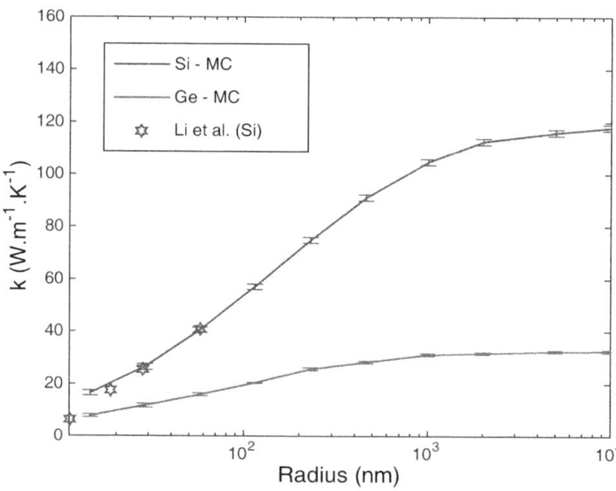

Figure 5.16 Nanowire thermal conductivity as a function of wire radius at 300 K; comparison with Li et al. measurements [19].

are related to the simulation time to achieve reliable calculations. As explained in this chapter, several improvements were brought to this technique during the last 10 years (collision energy balance, variance reduction, etc.). New improvements are expected in order to address more complex materials or models (nanoinclusion, electron–phonon coupling, non-isotropic dispersion, etc).

References

1. Allen, P. B., Feldman, J. L., Fabian, J., and Wootend, F. (1999). Diffusons, locons and propagons: Character of atomic vibrations in amorphous Si, *Phil. Mag. B* **79**, 11/12, pp. 1715–1731.
2. Ashcroft, N. W., and Mermin, D. N. (1976). *Solid State Physics* (Saunders College, Philadelphia).
3. Chantrenne, P., Barrat, J. L., Blase, X., and Gale, J. D. (2005). An analytical model for the thermal conductivity of silicon nanostructures, *J. Appl. Phys.* **97**, 10, p. 104318.
4. Kittel, C. (1983). *Physique de l'état solide*, 5th ed. (Bordas-Dunod Université).
5. Chen, G. (1998). Thermal conductivity and ballistic-phonon transport in the cross-plane direction of superlattices, *Phys. Rev. B* **57**, 23, pp. 14958–14973.
6. Efunda (2015). http://www.efunda.com/materials/elements/element_info.cfm?Element_ID=Si%.
7. Flubacher, P., Leadbetter, A. J., and Morrison, J. A. (1959). The heat capacity of pure silicon and germanium and properties of their vibrational frequency spectra, *Phil. Mag.* **4**, 39, pp. 273–294.
8. Gerald Burns (1985). *Solid State Physics* (Academic Press, Inc.), ISBN 0-12-146070-3.
9. Glassbrenner, C., and Slack, G. (1964). Thermal conductivity of silicon and germanium from 3K to the melting point, *Phys. Rev.* **134**, 4A, pp. A1058–A1069.
10. Goddard, W. A., Brenner, D., Lyshervski, S. E., and Lafrate, G. J. (2012). *Handbook of Nanosciences, Engineering and Technologies* (CRC Press).
11. Holland, M. (1963). Analysis of lattice thermal conductivity, *Phys. Rev.* **132**, 6, pp. 2461–2471.
12. Jean, V., Fumeron, S., Termentzidis, K., Tutashkonko, S., and Lacroix, D. (2014). Monte Carlo simulations of phonon transport in nanoporous silicon and germanium, *J. Appl. Phys.* **115**, 2, p. 024304.
13. Jean, V., Fumeron, S., Termentzidis, K., Zianni, X., and Lacroix, D. (2015). Monte Carlo simulations of phonon transport in Si nanowires with constrictions, *Int. J. Heat Mass Trans.* **86**, pp. 648–655.

14. Joshi, A., and Majumdar, A. (1993). Transient ballistic and diffusive phonon heat transport in thin films, *J. Appl. Phys.* **74**, 1, pp. 31–39.
15. Klemens, P. G. (1958). *Thermal Conductivity and Lattice Vibrational Modes* (Academic Press).
16. Klemens, P. G. (1969). *Theory of the Thermal Conductivity of Solids* (Academic Press).
17. Lacroix, D., Joulain, K., and Lemonnier, D. (2005). Monte Carlo transient phonon transport in silicon and germanium at nanoscales, *Phys. Rev. B* **72**, 6.
18. Lacroix, D., Joulain, K., Terris, D., and Lemonnier, D. (2006). Monte Carlo simulation of phonon confinement in silicon nanostructures: Application to the determination of the thermal conductivity of silicon nanowires, *Appl. Phys. Lett.* **89**, 10, p. 103104.
19. Li, D., Wu, Y., Kim, P., Yang, P., and Majumdar, A. (2003). Thermal conductivity of individual silicon nanowires, *Appl. Phys. Lett.* **83**, 14, pp. 2934–2936.
20. Majumdar, A. (1993). Microscale heat conduction in dielectric thin films, *J. Heat Transf.* **115**, 1, pp. 7–16.
21. Mazumder, S., and Majumdar, A. (2001). Monte Carlo study of phonon transport in solid thin films including dispersion and polarization, *J. Heat Transf.* **123**, 4, p. 749.
22. Modest, M. F. (2013). *Radiative Heat Transfer* (Academic Press).
23. Peierls, R. E. (1955). *Quantum Theory of Solids* (Clarendon Press).
24. Péraud, J. M., and Hadjiconstantinou, N. G. (2011). Efficient simulation of multidimensional phonon transport using energy-based variance-reduced Monte Carlo formulations, *Phys. Rev. B* **84**, 20, p. 205331.
25. Peterson, R. B. (1994). Direct simulation of phonon-mediated heat transfer in a Debye crystal, *J. Heat Transf.* **116**, 4, p. 815.
26. Pop, E., Dutton, R. W., and Goodson, K. E. (2004). Analytic band Monte Carlo model for electron transport in Si including acoustic and optical phonon dispersion, *J. Appl. Phys.* **96**, 9, p. 4998.
27. Wong, B. T., Francoeur, M., and Menguc, P. (2011). A Monte Carlo simulation for phonon transport within silicon structures at nanoscales with heat generation, *Int. J. Heat Mass Tras.* **54**, 9–10, pp. 1825–1838.
28. Ziman (1972). *Electrons and Phonons: The Theory of Transport Phenomena in Solids* (Clarendon Press).

Chapter 6

Equilibrium Molecular Dynamics

Samy Merabia
ILM, Université de Lyon, UMR 5306, Université Lyon 1 et CNRS, F-69622 Villeurbanne, France
samy.merabia@univ-lyon1.fr

6.1 Introduction

This chapter presents an overview of the equilibrium Molecular Dynamics (EMD) simulation technique to predict thermal transport properties of nanostructured materials. This covers systems having characteristic lengths of the order of a few nanometres like carbon nanotubes, nanowires and also superlattices, i.e. composite materials made of submicronic thickness of solid layers. The common features of these systems is the small ratio between their characteristic system size and the phonon mean free path, which leads to ballistic heat transport and deviations from the classical Fourier's law. Also when the density of interfaces gets large, the energy transport properties of the materials can no longer be described solely by the thermal conductivities of the constituents of the material, but depend also on the thermal boundary resistance

Nanostructured Semiconductors: Amorphization and Thermal Properties
Edited by Konstantinos Termentzidis

ISBN 978-981-4745-64-2 (Hardcover), 978-1-315-36445-2 (eBook)
www.panstanford.com

which measures the transmission of phonons across an interface. In this context, molecular dynamics was proven to be a very useful technique to study heat transport in nanostructured materials. The main reasons are; the length scale probed by the method is in the nanometre range, and it does not make any assumption on the phonons dynamics except their classical nature. In this contribution, we present the equilibrium method, which is now commonly used to determine both the thermal conductivity and the thermal boundary resistance of nanostructured materials. We explain also how to extract information regarding phonon lifetimes using equilibrium simulations. These methods are illustrated here using simple example of silicon or argon-like systems.

6.2 Molecular Dynamics

The development of molecular simulations began in the early fifties after the considerable development of computer facilities in the United States during World War II. A few years after the first Monte-Carlo simulation, Alder and Wainwright, introduced in the late 1950s the Molecular Dynamics (MD) method [1, 2]. The aim of the first simulations both Monte-Carlo or MD was to probe the different phases of model hard spheres. The necessity to model liquids motivated the development of realistic potentials. Rahman was the first to model Argon using Lennard–Jones potential, which is still considered as a standard potential for MD [3]. This opened the way to consider a broad range of condensed matter systems, ranging from liquid water first modelled by Rahman and Stillinger to silicon [4]. The efforts towards realistic modelling have permitted to applicate MD to characterize the collective excitations in solids [5]. The last step came with the implementation of MD thermostats opening the way to probe the phonon dynamics in rare-gas solids [6]. More recently, there have been an increasing number of MD studies on nano-scale heat transport motivated by fine measurements of energy transport in nano-materials [7]. We refer the reader to the classical textbooks on MD ([8] and [9]) for a thorough introduction. MD is a simulation method based on the numerical integration of Newton's equation of motion:

$$m\frac{d^2\vec{r}_i}{dt^2} = -\frac{\partial V}{\partial \vec{r}_i} \tag{6.1}$$

where m, $\vec{r}_i$ and V denote respectively the mass of the particle i, its position vector and the inter-atomic potential. The integration scheme commonly used is the Verlet algorithm [10, 11] which predicts the positions at time $t+dt$ given the positions at the earlier times t and $t-dt$:

$$\vec{r}_i(t+dt) = 2\vec{r}_i(t) - \vec{r}_i(t-dt) + \frac{d^2\vec{r}_i}{dt^2}dt^2 + O(dt^4), \tag{6.2}$$

where the acceleration is calculated using the inter-atomic forces. The velocities can be computed using

$$\vec{v}_i(t) = \frac{(\vec{r}_i(t+dt) - \vec{r}_i(t-dt))}{2dt} + O(dt^2) \tag{6.3}$$

and the error is larger than the error on the positions, but as soon as the velocities are only used to compute the instantaneous kinetic energy the consequences are minor. Of course there are other integration schemes, but the Verlet algorithm has the advantage to be simple, easy to implement, stable and time-reversible. The typical time step used in the integration algorithms is a fraction of the characteristic atomic period $\tau = m\sigma^2/k_BT \simeq 1$ ps where m is the mass of the particles and σ denotes a typical atomic diameter. Given the power of modern computers, it takes a few hours on a mono-processor machine to follow the trajectory of a set of 10000 particles over a time comparable to 1 ns. Using parallel MD codes or GPUs (graphics processing units) opens the way to model larger systems or reach orders of magnitude longer times.

The systems represented in a MD simulation have microscopic dimensions, and usually one is not so much interested in simulating a nano system with free surfaces, but rather a part of a bulk system. To embed the small system simulated in a bulk-like system, one can use periodic boundary conditions. These conditions assume that the system is repeated periodically in all space directions and thus any particle close to the boundaries of the simulation box interacts with an image particle, and when a particle crosses one of the face of the central simulation box, one of its images will enter the central box through the opposite face. One of the consequences of the use of periodic boundary conditions is the cut-off of long wavelengths fluctuations, i.e. those having a wavelength larger than the simulation box length. In the context of heat transport simulations this will imply that phonons with long wavelengths are not present.

The strength of MD lies on the versatility and the flexibility in the choice of the inter-atomic potential. The inter-atomic potential can generally be decomposed following:

$$V(\vec{r}_i) = \sum_i V_1(\vec{r}_i) + \sum_{i<j} V_2(\vec{r}_i, \vec{r}_j) + \sum_{i<j<k} V_3(\vec{r}_i, \vec{r}_j, \vec{r}_k) + \dots \quad (6.4)$$

where the first term represents the one body contribution to the potential and is often associated to an external field like e.g. an electric field. The second and third terms represent respectively two-body and three-body potentials of the inter-atomic interaction. The choice of the inter-atomic potential depends on the problem at hand (the type of atomic structure and the properties to be studied), and on the necessity to model a real system or an experiment.

One of the simplest potential used in condensed matter physics is the Lennard–Jones (LJ) potential, where the total potential decomposes in a sum of pair potentials:

$$V = \sum_{i<j} V_2(\vec{r}_{ij}) \quad (6.5)$$

with

$$V_2(\vec{r}_{ij}) = 4\epsilon \left((\sigma/r_{ij})^{12} - (\sigma/r_{ij})^6 \right) \quad (6.6)$$

This is the LJ pair potential, which depends only on the distance between neighbouring particles $r_{ij} = |\vec{r}_{ij}|$. The interaction energy ϵ identifies with the depth well of the potential, and σ is the atomic diameter. Although very simple, the LJ potential is often thought to be a good potential to model rare gas and in particular Argon, using the following parameters: $\epsilon = 1,67 \times 10^{-21}$ J and $\sigma = 3,40 \times 10^{-10}$ m. Due to the algebraic decay of the LJ potential, all the atoms of the simulation box interact. From a practical point of view however, it would be computationally costly to estimate $N^2/2$ inter-atomic forces at each time step, with $N \simeq 10000$. To bring back the computational cost of the force calculation to $O(N)$, the inter-atomic pair potential is often truncated at a cut-off radius, usually

$r_c = 2.5\sigma$, which represents typically a neighbourhood of 50 atoms per particle. It should be mentioned at this point that even if the potential at the radius of truncation is small compared to the well depth: $V(r_c) \simeq 0.016\epsilon$, truncating the potential may have a non negligible effect on the thermodynamics of the system modelled. As an example, let us estimate the effect of truncating the inter-atomic potential respectively on the internal energy per particle and on the pressure. The respective errors are given by

$$\Delta U = \rho \int_{r_c}^{+\infty} 4\pi r^2 V_2(r) dr \simeq 1.07 \rho \sigma^3 \epsilon \tag{6.7}$$

and

$$\Delta P = -\frac{\rho^2}{6} \int_{r_c}^{+\infty} r \frac{\partial V_2}{\partial r} 4\pi r^2 dr \simeq -1.07 \rho^2 \sigma^3 \epsilon, \tag{6.8}$$

where we have used the previous value of the cut-off radius and we have assumed that beyond r_c, the medium is structureless, i.e. the pair correlation function may be considered close to unity. The previous errors are found to be typically 10% the values of the internal energy and pressure for a condensed phase. This is not worrying since the calculation of thermodynamic quantities can be corrected using the previous estimations, but this has to be kept in mind when calculating a phase diagram for instance. Although the LJ potential is quite simple, some situations require to use more sophisticated semi-empirical potentials whose parameters are chosen to reproduce either microscopic or macroscopic properties of a model system. These semi-empirical potentials are usually no longer pair potentials like the LJ potential but many-body, where the many-body terms describes how does the potential energy of an atom depends on its coordination. As an illustration, let us mention the embedded atom model (EAM) used in the context of metals [12], where the potential

$$\sum_{i<j} \phi(r_{ij}) + \sum_{i} u(\rho(r_i)) \tag{6.9}$$

consists of a pair term $\phi(r)$ accounting for repulsion at short distances and a many body term $u(\rho(r))$ which accounts for the cohesion of the metallic bond and which depends on the local density measured using the positions of neighbouring particles

$\rho(r_i) = \sum_j w(r_{ij})$. The choice of the functional $u(\rho)$ and the weight function $w(r)$ depends on the microscopic and macroscopic properties to be reproduced, usually the lattice constant, the elastic constants and the sublimation energy of the metal [12].

Another limitation of the LJ potential is the description of solids crystallizing in non-compact structures. Indeed, Lennard Jones atoms form close-packed structures at low temperatures (usually fcc at low pressure), and thus LJ potentials are also not adapted to model materials which crystallize in opened structures such as diamond. Semiconductors like silicon or germanium are usually modelled using many-body semi-empirical potentials, in which the many body terms account for the local preferential bond ordering of the semiconductor. Among the most famous potentials for silicon, let us quote the Tersoff potential [13–15] and the Stillinger-Weber potential [16]. The parameters of the former are chosen so as to reproduce the elastic properties of bulk silicon, while the latter describes satisfactorily the structural properties of liquid silicon. Of course the list of potentials is non exhaustive and there exists plethora of empirical potentials to model as disparate systems as ionic crystals, two-dimensional solids, polymers ... and interfaces between them.

As we have already briefly mentioned, the output of MD simulations can be compared to experiments if we know how to relate the microscopic state of the material under study to the macroscopic observables typically measured in an experiment. Generally speaking, one can distinguish three classes of information that can be extracted from MD. On the one hand, MD allows to compute global quantities which correspond usually to the thermodynamics of the system modelled. For instance, the internal energy can be expressed as

$$U = \left\langle \sum_i V(\vec{r}_i) \right\rangle + KE \tag{6.10}$$

and the total kinetic energy

$$KE = \left\langle \sum_i \frac{1}{2} m \vec{v}_i^2 \right\rangle \tag{6.11}$$

where the brackets denote an ensemble average. If sufficient time has been left to the system to reach equilibrium, the ensemble average may be performed by averaging over a sufficient long time thanks to ergodicity. On the other hand, in non-equilibrium situations, different initial conditions (usually the positions and the velocities of the atoms) may be used to generate independent trajectories in phase space. The latter definition of the kinetic energy is intimately related to the definition of the local temperature:

$$T(\vec{r}) = \sum_{i \in \mathrm{Vol}(\vec{r})} \left\langle \frac{m\vec{v}_i^2}{3n(\vec{r})k_B} \right\rangle \tag{6.12}$$

where the sum runs over particles i in a small volume $\mathrm{Vol}(\vec{r})$ centred around $\vec{r}$ and which contains $n(r)$ particles. Another important quantity in heat transport simulations is the energy flux, defined as

$$\vec{J} = \frac{d}{dt}\left(\sum_i E_i \vec{r}_i\right) \tag{6.13}$$

where E_i is the total energy of particle i, yielding for pair potentials:

$$\vec{J} = \sum_i E_i \vec{v}_i + \frac{1}{2}\sum_{i<j} \vec{F}_{ij} \cdot (\vec{v}_i + \vec{v}_j)\vec{r}_{ij} \tag{6.14}$$

For simulations in crystals at low temperatures, the previous energy flux displays large oscillations due to optical phonons which carry a negligible amount of heat, and sometimes it can be more suited to work with the equivalent definition of the flux:

$$\vec{J} = \frac{1}{2}\sum_{i<j} \vec{F}_{ij} \cdot (\vec{v}_i + \vec{v}_j)\, \vec{r}^{\,0}_{ij} \tag{6.15}$$

where the superscript '0' denotes the equilibrium positions of the atoms. From the knowledge of the local temperature and the energy flux, it is possible to measure the thermal conductivity and we would come back to the measurement of transport coefficients in MD in the next section, with a particular emphasis on heat transport. The other type of information that one can extract from MD is structural. Since the output of a MD simulation is the set of positions $\vec{r}_{ij}$, it is possible to compute the pair distribution function

$$\rho g(r) = \left\langle \frac{1}{N}\sum_{i \neq j} \delta(\vec{r} - \vec{r}_{ij}) \right\rangle \tag{6.16}$$

where ρ is the mean number density of the system. Physically, $g(r)$ represents the probability to find two particles separated by a distance r, and $4\pi r^2 \rho g(r)dr$ is the average number of particles located at a distance between r and $r + dr$ from a given particle. Note that sometimes, it may be more useful to compute the structure factor:

$$S(\vec{q}) = \left\langle \frac{1}{N} \sum_{i,j} \exp\left(i\vec{q} \cdot (\vec{r}_i - \vec{r}_j)\right) \right\rangle \tag{6.17}$$

which is simply related to the pair distribution function through a Fourier transform: $S(q) = 1 + \rho \int g(r) \exp(-iq \cdot r) d\vec{r}$. Finally, MD may be used to compute the vibrational properties of a model system. In particular, the vibrational density of states (DOS)

$$g(\omega) = \frac{1}{V} \sum_{\vec{k}} \delta(\omega - \omega_k) \tag{6.18}$$

where the sum runs over the eigenmodes of the system can be estimated using the Fourier transform of the velocity autocorrelation function: $g(\omega) = \int \langle v(t) \cdot v(0) \rangle \exp(-i\omega t) dt$ where here $v(t)$ is the velocity of an atom. One has to keep in mind that the previous equation is only approximate, as it assumes small deviations from harmonicity, and thus is only valid strictly speaking at low temperatures. Note also that the resulting DOS mixes information about the different polarizations of the atomic vibrations, e.g. transverse and longitudinal, and to determine the DOS relative to a particular polarization, a more refined analysis of the atomic displacement should be undertaken.

As we have already mentioned, the conversion of the microscopic information given by a MD simulation to macroscopic observables requires to evaluate averages over phase space. The phase space is a multidimensional space generated by the positions and the momenta of a classical system, and has dimension of 6 N for a system of N particles in three dimensions. A molecular dynamics simulation generates a sequence of points in phase space as a function of time, whose points belong to the same thermodynamic ensemble. The first MD simulations were performed in the microcanonical ensemble or NVE ensemble, which corresponds to a fixed number of atoms N, a fixed volume V, and a fixed energy E. Although these simulations are very easy to implement, they do not represent a realistic system as the set of molecules studied is completely isolated. The need to model real situations led to the development of thermostats for MD, which allow to simulate the dynamics of a system in the canonical or the grand-canonical ensemble for instance [9]. Although MD can be used to calculate the thermodynamics, structural and vibrational properties of systems at equilibrium, it is more designed to study the non-equilibrium situations where the system under study is driven by an external force; the latter can be of thermodynamic nature. In the language of out of equilibrium statistical physics, MD can help in determining the relation between the forces and the fluxes. Statistical physics predicts that there is proportionality between the forces and the fluxes if the system under study is driven weakly out of equilibrium. The coefficients of proportionality are called transport coefficients and measure the susceptibility of the system to respond to a given thermodynamic force. For instance, the thermal conductivity quantifies the amount of energy flowing in a material submitted to a temperature gradient.

MD is a very powerful tool to calculate the transport coefficients of a model system. To this end, two routes can be traditionally followed: Either the model system is submitted to an external force or the thermally induced fluctuations of an internal variable are probed at equilibrium. The Onsager regression principle expresses the idea that the typical time decay of the spontaneous thermal fluctuations is proportional to the appropriate transport coefficient. This will be illustrated in the context of heat transfer in the next section.

Before focusing on thermal transport applications of MD, let us mention that MD can be used also beyond the linear response regime. This includes systems for which intrinsically the linear relationship between forces and fluxes is violated. As an example consider a nano-structured system at low temperatures for which the phonon mean free path is comparable with the characteristic system size. Heat transfer in such systems is no longer described by Fourier's law, but is rather described by an effective conductivity which depends on the strength of the thermal flux flowing across the system. Molecular dynamics has been shown to predict the effective conductivity under conditions where the heat carriers travel ballistically in the system and are only scattered by the nanostructure boundaries.

6.2.1 Limitations of Molecular Dynamics

As any simulation technique, molecular dynamics suffers from some intrinsic limitations. The most obvious is the limitation in system size. This may be critical in the modelling of real scale devices with disparate length scales ranging from 1 nm to microns. To address such situations, MD should be coupled with a more mesoscopic method, such as the Boltzmann Transport Equations, where the microscopic information on the phonon lifetime is used as input in a Boltzmann equation. We have also mentioned that a finite system size cuts the long wavelength phonon modes. As we will see, this affects only mildly the measurement of the thermal conductivity unless we couple the system with heat reservoirs. The second limitation of MD is its classical nature. Each phonon mode is equally populated, and the heat capacity is given to a good approximation by the Dulong–Petit law. This is a good approximation if one ever considers solids above or just below their Debye temperature. However, many materials do not obey this condition at ambient temperature. Quantum effects in MD may be accounted for in different ways. The phonon lifetimes computed from MD may be used in a Boltzmann transport equation which includes quantum statistics in the phonon occupation number [17]. Quantum effects may be also directly incorporated in the course of a MD simulation, using a Langevin thermostat with a coloured noise consistent with a Bose-Einstein distribution for the phonon modes [18]. Finally, the electronic degrees of freedom are not explicitly simulated in MD. Hence, it is not possible to probe thermal transport in electrical conductors, and in its basic version the contribution of electron-phonon scattering to the transport in semiconductors is not accounted for. In general the heat energy is transmitted through a solid by electrons (mainly in metals) or phonons (mainly in insulators) or other excitations such as spin waves. In this chapter, we focus on the lattice thermal conductivity, which is the dominant mechanism in semiconductors [19].

There are three principal techniques used to evaluate the thermal conductivity with molecular dynamic simulations [8, 20]; a. the equilibrium approach based on the Green–Kubo formulae [9], b. the non-equilibrium MD, also called the direct method, which uses a heat source and a heat sink, or the so-called direct method, based on the creation of a temperature gradient, and c. the homogeneous non-equilibrium MD, where a heat flux is induced [21, 22]. The comparison between the two methods has been undertaken by many researchers, who have concluded that the two methods give consistent results [23–25].

6.2.2 Non-Equilibrium Molecular Dynamics Method (NEMD)

In the early nineties, Kotake and Wakuri proposed the direct method [26] which is similar in spirit to the hot-plate experiment setup. A temperature gradient is imposed across the structure under study by allowing thermal power exchange between the heat source and sink and measure the resulting heat flux [27–29]. The thermal conductivity is then obtained as the ratio of the heat flux and the temperature gradient. An alternative, but equivalent way consists in inducing a heat flux and to measure the resulting temperature gradient [30]. In both cases the system is first allowed to reach a steady state, after which long simulations are conducted allowing to obtain correct statistical measurements [31]. The NEMD method is often the method of choice for studies of nanomaterials [32] while for bulk thermal conductivity, particularly of high-conductivity materials, the equilibrium method is typically preferred due to less severe size effects.

Finite size effects in the determination of the thermal conductivity using NEMD can be understood using the following simple line of thought [23]. Imagine a phonon travelling in the crystal between the two heat reservoirs. This phonon can experience at least two kinds of scattering events: Either it can be scattered by other phonons travelling in the material, or it can be scattered by the reservoirs which are seen by the considered phonon as a different material with an almost infinite thermal conductivity. The mean free path associated with this former mechanism can be roughly approximated by $\Lambda_{\text{reservoir}} = L/2$, where L is the distance between the reservoirs, because on average a phonon will travel ballistically a distance $L/2$ before being scattered by the heat source/sink. Now invoking Matthiessen rule to predict the effect of reservoir scattering on the overall thermal conductivity, the effective mean free path is given by

$$1/\Lambda(L) = 2/L + 1/\Lambda(L \to \infty), \tag{6.19}$$

where the last term $\Lambda(L \to \infty)$ describes the phonon-phonon interaction in a bulk- like medium. The length dependence of the thermal conductivity can now be estimated using elementary kinetic theory of gas: $\lambda(L) = cv\Lambda(L)/3$ where c is the heat capacity per unit of volume, v is the mean group velocity. Of course, this analysis is simplified because we have assumed that the effect of phonon-phonon scattering can be described in terms of a single mean free path, and we should have rather done the analysis mode by mode [33]. However, the previous simple kinetic model predicts

$$1/\lambda(L) = 1/\lambda(L \to \infty) + 6/(cvL) \tag{6.20}$$

and the size dependence is found to be consistent with MD NEMD simulations [23, 29]. Because of the algebraic decay, finite size effects in NEMD are severe, and sometimes it is faster to run equilibrium simulations to determine the thermal conductivity.

6.2.3 Equilibrium Molecular Dynamics Method (EMD)

Alternatively, the thermal conductivity of a model system can be determined using equilibrium simulations [23]. The principle relies on the fact that the regression of the thermal fluctuations of an internal variable, in our case the thermal flux, obeys macroscopic laws. Hence, the time decay of the fluctuations of the flux is proportional to the thermal conductivity. This is mathematically expressed by the Green–Kubo formulae which state that the time integral of the heat flux autocorrelation function is proportional to the thermal conductivity tensor [34]:

$$\lambda_{\alpha,\beta} = \frac{1}{Vk_BT^2}\int_0^\infty \langle J_\alpha(t)J_\beta(0)\rangle dt, \tag{6.21}$$

where V is the volume of the system, and J_α denotes the component of the heat flux vector along the direction α. The expression of the energy flux vector is given in Eq. (6.13), and for pairwise potentials, it writes

$$\vec{J} = \sum_i E_i \vec{v}_i + \frac{1}{2} \sum_{i<j} \vec{F}_{ij} \cdot (\vec{v}_i + \vec{v}_j) \vec{r}_{ij} \equiv \vec{J}_{\mathrm{KE}} + \vec{J}_{\mathrm{pot}}. \qquad (6.22)$$

For simulations of complex crystals or superlattices, the previous energy flux displays large oscillations due to optical phonons which carry a negligible amount of heat, and sometimes it can be more suited to work with the equivalent definition of the flux:

$$\vec{J}_{\mathrm{pot}} = \frac{1}{2} \sum_{i<j} \vec{F}_{ij} \cdot \left(\vec{v}_i + \vec{v}_j\right) \vec{r}^{\,0}_{ij} \qquad (6.23)$$

where the superscript '0' denotes the equilibrium positions of the atoms. For simulations of systems described by many-body potentials, the above formula should not strictly speaking be applied. For many empirical potentials including Tersoff, Stillinger-Weber or Brenner potentials, the total potential of the system may be decomposed in the form

$$U = \sum_i V_i, \qquad (6.24)$$

then, the potential part of the heat flux vector is [35]

$$\vec{J}_{\mathrm{pot}} = \sum_i \sum_{i \neq j} \vec{r}_{ij} \left(\frac{\partial V_j}{\partial \vec{r}_{ij}} \cdot \vec{v}_i \right) \qquad (6.25)$$

and it differs from the expression used in the open source code LAMMPS:

$$\vec{J}_{\mathrm{pot}} = -\sum_i \vec{S}_i \bigotimes \vec{v}_i, \qquad (6.26)$$

where $\vec{S}_i$ is the stress tensor of atom i:

$$\vec{S}_i = -\frac{1}{2} \sum_{j \neq i} \vec{r}_{ij} \bigotimes \vec{F}_{ij} \qquad (6.27)$$

and $\vec{F}_{ij}$ is the pairwise force $\vec{F}_{ij} = -\frac{\partial V_i}{\partial \vec{r}_{ij}}$. However, as shown in Fan et al. [35], the difference in the prediction of the thermal conductivity is small for bulk systems, but significant differences may appear for the study of two-dimensional systems such as graphene layers.

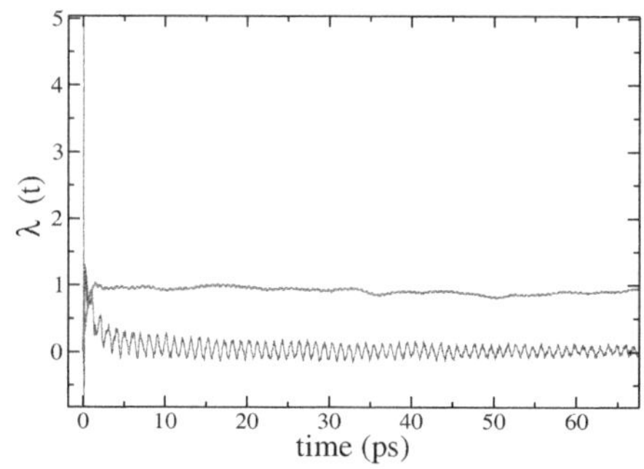

Figure 6.1 Heat flux autocorrelation function (blue curve) and the running integral $\lambda(t)$ (red curve) for a bulk amorphous silicon at 300 K modelled by the Stillinger-Weber potential.

We now come back to the determination of the thermal conductivity from the Green–Kubo formula Eq. (6.21). The determination of the conductivity requires the computation of the running integral:

$$\lambda(t) = \frac{1}{3Vk_BT^2}\int_0^t \langle \vec{J}(t') \cdot \vec{J}(0)\rangle dt', \tag{6.28}$$

where we have considered here an isotropic system for simplicity. The equilibrium method consists then in computing the heat flux autocorrelation function, which requires following the dynamics of a system over time scales a few times larger than the longest relaxation time present in the system. In the case of heat transfer in solid materials, the longest relaxation times correspond to long wavelengths phonons (a few nm) with a lifetime on the order of 100 picoseconds. An example of the heat flux autocorrelation function and the running integral Eq. (6.28) is displayed in Fig. 6.1. Here, an example of bulk silicon represented by the Stillinger-Weber potential is considered. In practice, one needs to record the instantaneous values of the energy flux vector during a time which is roughly ten times the typical decay time of the autocorrelation function. Once the autocorrelation function is constructed, the running integral $\lambda(t)$ maybe calculated and the Green Kubo conductivity is identified with the plateau of the running integral, see Fig. 6.1. To improve the statistics on the determination of the conductivity, one needs to run several independent simulations to obtain different running integrals and averaging the corresponding integrals to obtain a converged value for the conductivity. In some cases, it may be difficult to unambiguously define a plateau, and different recipes have been proposed. One of the simplest is to fit the heat flux autocorrelation function by a finite sum of exponentials [36, 37]:

$$\langle J(t)J(0)\rangle = \sum_k C_k \exp(-t/\tau_k) \tag{6.29}$$

so that the thermal conductivity is

$$\lambda = \frac{1}{3Vk_B T^2}\sum_k C_k\tau_k \tag{6.30}$$

This is the case typically for simple crystals such as LJ Argon. For more complex crystals, an exponential fit may no longer fit the autocorrelation function [23, 38]. This is also the case for the silicon system considered in Fig. 6.1. An alternative is to use the first maximum of the running integral $\lambda(t)$ [39]. In some situations however, it is even difficult to define the first maximum of $\lambda(t)$. Such a situation may be encountered for complex crystals or superlattices as illustrated in Fig. 6.2. It is shown that $\lambda(t)$ displays large high frequency oscillations, that are due to optical modes which do not carry significant energy [38]. A first operation to filter out these oscillations is to compute the heat current vector with the equilibrium positions, as in Eq. (6.23). Even after this operation, it may be difficult to identify clearly a plateau in the running integral. A practical solution consists in calculating the time Fourier transform of the running integral $\lambda(t)$, and fitting the result with

$$\tilde{\lambda}(\omega) = \frac{\lambda}{\sqrt{1+(\omega\tau)^2}} \tag{6.31}$$

and the thermal conductivity is obtained by taking the zero-frequency limit of the latter equation [40].

We now comment on some advantages of the equilibrium method. The advantage of the equilibrium method is that it allows to compute the full conductivity tensor from one simulation, which may be appreciated in superlattice simulations for instance, which display large thermal anisotropies [40, 41]. Another advantage of the equilibrium method is that it does not suffer from severe finite size effects as NEMD, mainly because in EMD the phonons are not strongly scattered by the boundaries of the simulation

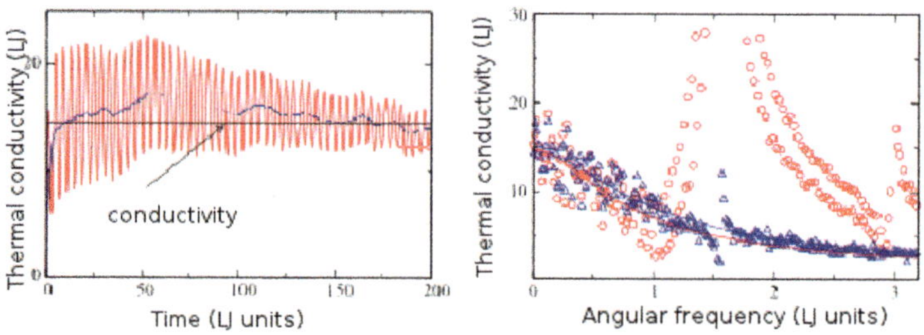

Figure 6.2 Left: Running integral $\lambda(t)$ for a Argon superlattice. Here, the cross-plane conductivity is considered. The red curve corresponds to the usual definition of the heat current vector, and the blue curve corresponds to the heat current vector calculated with the equilibrium positions. Right: Fourier transform of the heat current autocorrelation function. Same colour code as the left figure.

box. We can estimate the finite size effects in EMD using the following assumptions: The thermal conductivity measured in a MD simulation is given by

$$\lambda = \int_{\omega_c(L)}^{\omega_D} g(\omega)c(\omega)v^2(\omega)\tau(\omega)d\omega, \tag{6.32}$$

where

$$\omega_c(L) = \omega(k = \pi/L) \tag{6.33}$$

is the low pulsation cut-off introduced by the periodic boundary conditions, and the upper bound is

$$\omega_D = \omega(k = \pi/a) \tag{6.34}$$

where a is the inter-atomic distance. In the latter expression of the thermal conductivity, $c(\omega) = k_B$ is the heat capacity which obeys Dulong and Petit law, $v(\omega)$ is the phonon group velocity, and $\tau(\omega)$ is the relaxation time of phonon having an angular frequency ω. In the following, we have ignored the possible different polarization states of the phonons to simplify the discussion. The error in the determination of the thermal conductivity is thus

$$\delta\lambda = \int_0^{\omega_c(L)} g(\omega)c(\omega)v(\omega)\tau(\omega)d\omega \tag{6.35}$$

and if we assume that for sufficient low wavevectors that the medium behaves like a Debye solid, $g(\omega) \sim \omega^2$, the group velocity is constant and the phonon relaxation time is supposed to obey Callaway's law: $\tau(\omega) \sim \omega^{-2}$, then the error in the conductivity scales as $\delta\lambda \sim \omega_c(L) \sim a/L \ll 1$ and thus decreases faster as the error in NEMD $\delta\lambda_{NEMD} \sim \Lambda/L$. This allows obtaining good estimates of the thermal conductivity with rather small systems. The main drawback is intrinsic to the method, as we need to probe small thermal fluctuations around equilibrium over long time scales, which requires also performing several statistical averages over different initial conditions.

In conclusion, there are several advantages of the Green–Kubo technique, as compared to the non-equilibrium method for the calculation of the thermal conductivity: -a single simulation may be used to compute the full thermal conductivity tensor -finite size effects are less severe for equilibrium simulations, as the boundaries do not introduce a cut-off in the phonon mean free paths distribution.

6.2.4 Phonon Lifetimes

In addition to the calculation of the thermal conductivity, equilibrium simulations may be very efficient in providing information on phonon dynamics. Here again, we are mostly interested in heat transport in semiconductors, for which the energy carriers are phonons, i.e. lattice dynamics collective excitations. The thermal conductivity may be related to the phonon mean free paths according to

$$\lambda = \sum_{\vec{k}} c_v(\vec{k}) v_g(\vec{k}) \Lambda(\vec{k}), \tag{6.36}$$

where the index k here denotes a phonon mode characterized by its frequency $\omega(k)$ and its polarization. The phonon mean free path is simply related to the phonon lifetime $\tau(\vec{k})$:

$$\Lambda(\vec{k}) = v_g(\vec{k})\tau(\vec{k}) \tag{6.37}$$

Hence, a knowledge of the mode dependent phonon lifetime is important for understanding its relative contribution to the thermal conductivity λ. In the following, we briefly illustrate how to extract such an information from equilibrium simulations.

In the quasi-harmonic picture, the equations of motion obeyed by the atoms write, at the leading order,

$$m_i \frac{d^2 u_{i,\alpha}}{dt^2} = -\sum_{j,\beta} \Phi_{\alpha,\beta}(i, j) u_{j,\beta}, \tag{6.38}$$

where m_i is the mass of atom i, $\vec{u}_i$ is the vector displacement of atom i, and $\Phi_{\alpha,\beta}(i, j)$ is the second order force constant corresponding to the interaction between atoms i and j, and the indexes α, β refer to spatial dimensions. Here, we have considered the simple case where there is only one atom per unit cell. Generalizations to more complex crystals are discussed in [42]. The solution may be decomposed in normal modes:

$$u_{i,\alpha}(t) = \sum_{\vec{k}} \frac{Q(\vec{k})}{m_i} e_\alpha(\vec{k}) \exp(i(\vec{k} \cdot \vec{r}_i^{\,0} - \omega(\vec{k})t)), \tag{6.39}$$

where the sum runs over the wave vectors $\vec{k}$ of the reciprocal lattice and we have introduced the normal coordinates $Q(\vec{k})$, the polarization vector $\vec{e}(\vec{k})$ and the angular frequency $\omega(\vec{k})$. The indexes 0 indicate again equilibrium atomic positions. Here, we have considered a simple crystal structure with one atom per primitive cell, but generalizations to more complex structures are straightforward [42]. At equilibrium, the normal coordinates obey the equipartition theorem [42]:

$$\omega^2(\vec{k})\langle |Q(\vec{k})|^2\rangle = k_B T, \tag{6.40}$$

where the brackets indicate a thermal average, and we have used the fact that the statistics relevant to MD is fully classical.

The polarizations vectors are solutions of the secular equation:

$$D_{\alpha\beta}(\vec{k}) e_\beta(\vec{k}) = \omega^2(\vec{k}) e_\alpha(\vec{k}), \tag{6.41}$$

where D is the D-type dynamical matrix [43]:

$$D_{\alpha\beta}(\vec{k}) = \sum_{j \text{ neighb i}} \frac{\Phi_{\alpha\beta}(i, j)}{\sqrt{m_i m_j}} \exp(i(\vec{k} \cdot (\vec{r}_j^{\,0} - \vec{r}_i^{\,0})), \tag{6.42}$$

where the sum runs over the atoms j neighbours of i and the result is independent on the reference atom i because of the translational invariance of the crystal. Indeed, here we have considered simple crystal structures with one atom per primitive cell. For the simple crystal structure considered here, the dynamical matrix has dimensions 3×3. For more complex crystal structures containing p atoms per primitive cell, the dynamical matrix has dimensions $3p \times 3p$. It is also customary to introduce the vibrational density of states:

$$g(\omega) = \frac{1}{V} \sum_{\vec{k}} \delta(\omega - \omega(\vec{k})) \tag{6.43}$$

which measures the density of vibrational modes, and should be normalized so that

$$\int_0^{\omega_{\max}} g(\omega) d\omega = 3\frac{N}{V}, \tag{6.44}$$

where $\omega_{\max}$ is the maximal frequency of the solid, that is supposed to be three-dimensional. The vibrational DOS is a quantity easily accessible with molecular dynamics, as one can show that it is simply related to the velocity autocorrelation function:

$$g(\omega) = \frac{m\rho}{2\pi k_B T} \int_{-\infty}^{\infty} \langle v(t) \cdot v(0) \rangle \exp(-i\omega t) dt, \tag{6.45}$$

where the prefactor ensures the normalization Eq. (6.44). One has to keep in mind that the previous equation is only approximate, as it assumes small deviations from harmonicity, and thus is only valid strictly speaking at low temperatures. Note also that the resulting DOS mixes information about the different polarizations of the atomic vibrations, e.g. transversal or longitudinal, and to determine the DOS relative to a particular polarization, a more refined analysis of the atomic displacement should be undertaken.

Coming back to the normal mode decomposition, we can rewrite

$$Q_{k,\nu}(t) = \frac{1}{N^{1/2}} \sum_i m_i^{1/2} \exp(-ik \cdot \vec{r}_i^{\,0}) \vec{e}(\vec{k}, \nu) \cdot \vec{u}_i(t) \tag{6.46}$$

Hence, the normal modes may be obtained through the course of a MD simulation by projecting the instantaneous displacement on the polarization vectors $\vec{e}(\vec{k}, \nu)$. The phonon lifetimes may be inferred from the time dependence of the normal modes, as explained below. Indeed, we have from the normal mode projection that

$$\dot{Q}_{k,\nu}(t) = \frac{1}{N^{1/2}} \sum_i m_i^{1/2} \exp(-ik \cdot \vec{r}_i^{\,0}) \vec{e}(\vec{k}, \nu) \cdot \dot{\vec{u}}_i(t) \tag{6.47}$$

From the normal modes and its time derivatives, one can compute the total energy of the mode as

$$E_{\vec{k},\nu}(t) = U_{\vec{k},\nu}(t) + T_{\vec{k},\nu}(t), \tag{6.48}$$

where U and T are respectively the potential energy and the kinetic energy given by

$$U_{\vec{k},\nu}(t) = \frac{1}{2}\omega^2(\vec{k},\nu)|Q_{\vec{k},\nu}(t)|^2 \tag{6.49}$$

and

$$T_{\vec{k},\nu}(t) = \frac{1}{2}|\dot{Q}_{\vec{k},\nu}(t)|^2. \tag{6.50}$$

The phonon lifetime $\tau_{\vec{k},\nu}$ may be obtained from the energy autocorrelation function:

$$\frac{\langle E_{\vec{k},\nu}(t)E_{\vec{k},\nu}(0)\rangle}{\langle E^2_{\vec{k},\nu}(t)\rangle} = \exp(-t/\tau_{\vec{k},\nu}) \tag{6.51}$$

or alternately the phonon lifetime may be calculated using the following running integral:

$$\tau_{\vec{k},\nu} = \int_0^{+\infty} \frac{\langle E_{\vec{k},\nu}(t)E_{\vec{k},\nu}(0)\rangle}{\langle E^2_{\vec{k},\nu}(t)\rangle} dt. \tag{6.52}$$

In practice, one needs to compute the normal modes $Q_{\vec{k},\nu}(t)$ from the atomic positions, and calculate the time dependent energy to extract the phonon lifetime.

Alternatively, the phonon lifetimes may be obtained using normal mode decomposition in the frequency domain. Indeed, from the Wiener–Khinchin theorem, the energy spectrum is equal to the Fourier transform of the energy autocorrelation function Eq. (6.51). Hence, the energy spectrum has a Lorentzian shape:

$$\tilde{E}_{\vec{k},\nu}(\omega) = C_{\vec{k},\nu}\frac{1/\tau_{\vec{k},\nu}}{(\omega-\omega_a(\vec{k},\nu))^2 + 1/(4\tau^2_{\vec{k},\nu})}, \tag{6.53}$$

where $\tilde{E}_{\vec{k},\nu}$ is the Fourier transform of the energy $E_{\vec{k},\nu}$:

$$\tilde{E}_{\vec{k},\nu} = \lim_{T\to\infty}\int_0^T E_{\vec{k},\nu}(t)\exp(-i\omega t)dt. \tag{6.54}$$

The Lorentzian in Eq. (6.53) involves the anharmonic frequency $\omega_a(\vec{k},\nu)$, and a mode-dependent constant $C_{\vec{k},\nu}$. The anharmonic frequency $\omega_a(\vec{k},\nu)$ differs from the harmonic solution of the secular Eq. (6.41) as it includes anharmonicities of the inter-atomic potential. However, in most of the cases the anharmonic frequency is found not to differ too much from the harmonic frequency, which tends to conclude that the harmonic crystal is a good model at least to describe phonon dispersion in solids.

The analysis of phonon lifetimes with MD has been performed in different contexts. The first to run out such a framework was Ladd et al. in the middle of the eighties [6]. The systems considered was Argon solid modelled by Lennard–Jones interactions, and at that time only small systems have been analysed. They calculated phonon lifetimes and showed agreement with the thermal conductivity found with Green–Kubo simulations. During twenty years, there was basically no further effort in the predictions of phonon lifetimes. In the mid 2000s, McGaughey and Kaviany [17, 44] took up the pioneering work of Ladd et al. to analyze larger systems. In particular, phonon lifetimes calculated from MD have been shown to be in good agreement with anharmonic calculations with the Lennard–Jones potential [44]. Henry and Chen performed normal mode decomposition to predict phonon mean free paths in bulk silicon [45]. This type of information opens the way to construct a thermal conductivity accumulation function, which quantifies the relative contribution of long mean free paths phonons. The same authors applied the normal mode decomposition to the heat conduction in single polymer chains of polyethylene [46]. Of interest here is the relatively large thermal conductivity displayed by a single chain, as compared with bulk polyethylene. Apart from 1 D chain, the normal mode decomposition has been also applied to the conductivity of silicon nanowires [47] or carbon nanotubes [48]. Finally, we shall mention that normal mode lifetimes may be predicted for amorphous materials, and this has been done for amorphous silicon [49] and silica [50]. Due to the lack of periodicity of the amorphous solid, the whole analysis is performed at the gamma point. It is found that in amorphous silicon, some energy carriers may have a mean free path longer than their wavelength, hence behaving like phonons. On the other hand, in amorphous silica almost all the energy carriers have a mean free path smaller than their wavelength and are localized. A microscopic interpretation for this difference of behaviour is still lacking.

6.2.5 Thermal Boundary Resistance

Apart from the thermal conductivity, both NEMD and EMD simulation techniques may be used to calculate the thermal boundary resistance characterizing heat transport across the interface between two media [51]. The thermal boundary resistance (TBR), also known as the Kapitza resistance is defined as the ratio of the temperature jump at the interface ΔT over the heat flux J crossing the interface:

$$R_K = \Delta T / J. \tag{6.55}$$

The Kapitza resistance is thus a measure of the ability of phonons to be transmitted by the interface, with common values falling in the range between 0.01 and 0.1 MW·m^2·K [52] depending on the dissimilarity between the two media. There have been several models to predict the TBR between two solids [53, 54]. Among the most popular, let us mention the acoustic mismatch model (AMM) [55] in which the transmission of phonons is assumed to be specular and depends only the acoustic impedance mismatch between the two media, in a way similar to Snell–Descartes law ruling the transmission of an electromagnetic wave across the interface between two dielectrics with different optical indexes. The other popular model is the so-called diffuse mismatch model (DMM) [54], in which it is assumed that the phonons are diffusively scattered by the interface, and the transmission coefficient depends in this limit in the mismatch of the density of states characterizing the two media. Generally speaking, both models fail to predict the thermal boundary resistance of real interfaces [52], and a deep physical understanding of interfacial heat transport between two solids is still missing. Indeed, it is not surprising that the two above-mentioned models fail in describing energy transfer at interfaces, since they depend only on the *bulk* properties of the two media, and any information characterizing the interface is lost. In particular, it is clear that the two models cannot predict energy transfer at imperfect interfaces. Also, it is known from the experimental side that the TBR strongly depends on the presence of interfacial defects including dislocations, vacancies or intermixing [56]. In this context, MD may be a convenient tool to probe the transmission of phonons across ideal surfaces, and has also the flexibility to introduce defects at the interface and study their effect on interfacial heat transport. Again, two routes may be followed to calculate the interfacial resistance: In NEMD simulations, the interface to be characterized is placed between two heat reservoirs fixed at a distance a few atomic layers away from the interface. A net heat flux in the direction perpendicular to the interface is created, and the resulting temperature jump at the interface is measured, allowing to derive the interfacial resistance. With NEMD method only the TBR of smooth interfaces can be calculated, as for rough interfaces there are geometrical issues about the cross-section/effective surface of the interfaces. Equilibrium simulations rely again on the Green–Kubo formula for the interfacial conductance $G = 1/R$. Puech et al. were the first to derive a Green–Kubo formula for TBR [57]:

$$G = \frac{1}{\mathcal{A}k_B T^2} \int_0^{+\infty} \langle q(t)q(0)\rangle dt, \tag{6.56}$$

where A is the interfacial area and $q(t)$ is the instantaneous value of the flux flowing across the interface. The latter quantity may be computed in a MD simulation using the power of interfacial forces [51]:

$$q_1(t) = \sum_{i\in 1,\, j\in 2} \vec{F}_{ij} \cdot \vec{v}_i \tag{6.57}$$

where the indexes 1 and 2 denote the two media separating the interface to be studied, and we have assumed pair potentials to simplify. More recently, a symmetrical formulation of the energy flux has been proposed [58, 59]:

$$q(t) = \sum_{i\in 1,\, j\in 2} \vec{F}_{ij} \cdot \frac{(\vec{v}_i + \vec{v}_j)}{2} \tag{6.58}$$

and it is also found that for many-body potentials, the contributions of three, fourth body terms maybe neglected in the expression of the flux.

An example of Green–Kubo calculations for the interface between Argon and heavy Argon is shown in Fig. 6.3. Here the two curves correspond to the interfacial fluxes q_1 and q_2, estimated from Eq. (6.57). A plateau may be identified at long times, which value corresponds to the conductance G predicted from Eq. (6.56). It is important to mention that a plateau is clearly visible only after averaging the running integrals over 30 independent simulations, as the calculations of the flux Eq. (6.57) involves only atoms close to the interface and thus displays large fluctuations. Some oscillations

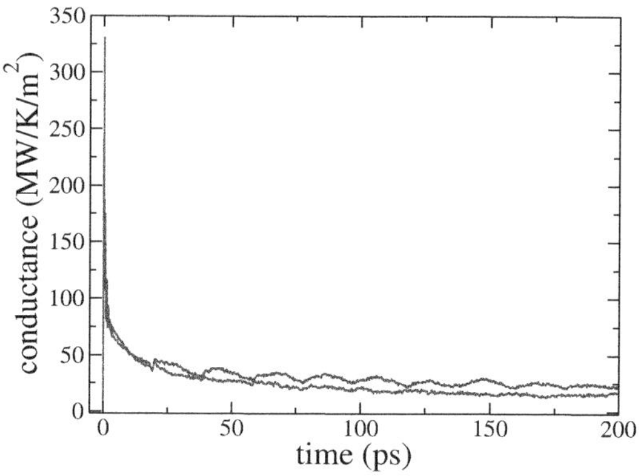

Figure 6.3 (Colour online) Interfacial conductance of the interface between Argon/heavy Argon calculated using the Puech Green–Kubo formula Eq. (6.56). The two curves correspond to the two interfaces of the system. Each curve is an average over 30 independent trajectories. The mass ratio between heavy Argon and Argon is here $m_r = 2$.

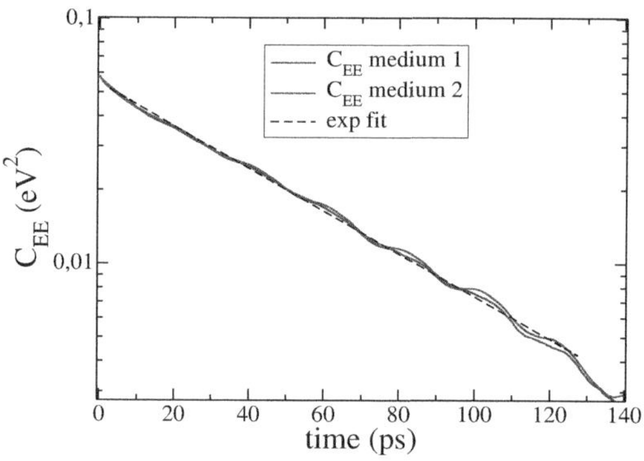

Figure 6.4 (Colour online) Energy autocorrelation functions for the Argon/heavy Argon system considered in Fig. 6.3. Dashed lines show the exponential fit.

are visible at long times, which correspond to acoustic echoes of long wavelength phonons that are reflected by the interfaces (remember that the system studied is a superlattice due to the periodic boundary conditions).

Some other Green–Kubo have been also formulated for the prediction of the TBR. Rajabpour and Volz proposed to calculate the

TBR from the temperature autocorrelation function [60]:

$$\frac{1}{G} = \frac{1}{\mathcal{A}k_B} \int_0^{+\infty} \frac{\langle \delta T(t) \delta T(0) \rangle}{\langle \delta T(0)^2 \rangle} dt \left(\frac{1}{N_1} + \frac{1}{N_2} \right), \tag{6.59}$$

where N_1 and N_2 are the number of degrees of freedom characterizing each medium. Practically, the interfacial resistance is obtained by fitting the kinetic energy autocorrelation function with a single exponential having a decay time τ. The conductance is then given by $G = \mathcal{A}k_B \rho V/(2\tau)$.

Also, Merabia and Termentzidis showed an alternative formula equivalent to Puech's formula [61]:

$$G = -\frac{1}{\mathcal{A}k_B T^2} \left(\frac{dC_{EE}}{dt} \right)_{t=0^+}, \tag{6.60}$$

where $C_{EE}(t) = \langle \delta E_1(t) \delta E_1(0) \rangle$ is the energy auto-correlation characterizing solid 1 and E denotes the total energy of one medium:

$$E_i = \sum_{j \in i} \frac{1}{2} m \vec{v}_j^2 + \sum_{j,k \in i} V(\vec{r}_j - \vec{r}_k), \tag{6.61}$$

where we have considered pair potentials, and again one can slightly improve the statistics by considering separately the energy autocorrelations of the two media. An example of the energy correlation function calculated for an Argon system is depicted in Fig. 6.4. This formula has some advantages as compared with the usual Green–Kubo formula. In particular, it avoids estimating a plateau in the calculation of a running integral. Instead, one has to calculate the slope of the derivative of the energy correlation function, for which we have better statistics.

Equilibrium simulations have been performed for interface between simple Lennard-Jones solids, where the dissimilarity between the two solids is introduced by changing the acoustic impedance ratio between the two solids. The results were found to disagree with NEMD determinations of the Kapitza conductance [61–63], especially for solids with a weak acoustic mismatch. These discrepancies are not surprising and can be traced back to the formulation of the equations themselves [64], as the Green–Kubo formula above predicts a finite conductance for the interface between two identical media, while of course in NEMD one measures an infinite conductance in this situation. Maybe for this

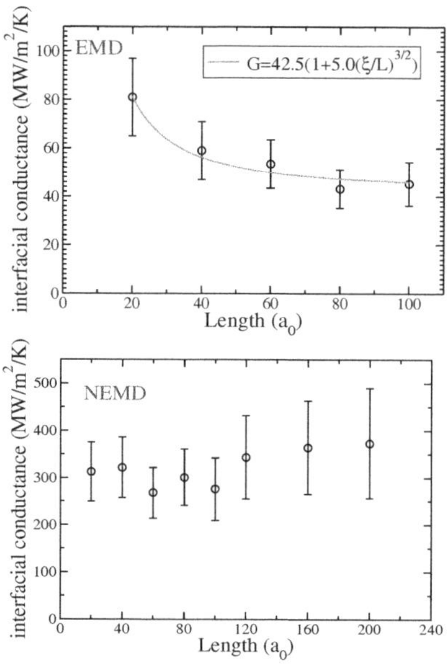

Figure 6.5 Interfacial conductance obtained by EMD (top figure) and NEMD (bottom figure) as a function of the length of the system in units of the crystal monolayer $a_0 \simeq 0.54$ nm. The solid red line on the top figure displays the theoretical formula Eq. (6.62). Same other parameters as in Fig. 6.4.

reason, most of the MD works on Kapitza resistance have considered the direct method. We end up this section by discussing finite size effects for the determination of the interfacial thermal conductance.

Figure 6.5 compares finite size effects in the determination of the interfacial conductance for EMD and NEMD simulations. It is found that EMD is more sensitive to the size of the simulation box than NEMD. Again, this is related to the fact that the systems used in EMD are superlattices, for which thermal correlations may be present between consecutive interfaces. These correlations have been quantified in [61], with the resulting effective conductance:

$$G(L) = G_\infty \left(1 + \alpha \left(\frac{\xi}{L}\right)^{3/2}\right), \tag{6.62}$$

where G_∞ is the conductance characterizing the interface between semi-infinite media, α is a numerical constant and $\xi = \lambda/G$ is the Kapitza length. In particular, if the system size is not large enough there may be appreciable correlations in the thermal flux Eq. (6.57) between two neighbouring interfaces, and the Puech formula does not give information on a single interface. To extract the thermal conductance of an isolated interfaces, one then needs to work with superlattices having large enough periodicities. On the other hand, NEMD simulations show typically less severe finite size effects as illustrated in Fig. 6.5. The absence of appreciable size effects may be related again to the different boundary conditions.

6.3 Conclusion

In conclusion, we illustrated the principle of equilibrium molecular dynamics in the calculation of thermal transport properties of nanostructured materials. The general idea is to extract information on the bulk and interfacial transport coefficients from an educated observation of the dynamics of thermally induced fluctuations. For conductive materials, equilibrium molecular dynamics is the method of choice as the distribution of phonon mean free paths is significantly less affected than in non-equilibrium situations. In addition, equilibrium simulations may provide valuable information to reconstruct the mode-dependent mean free path. This knowledge helps in a fundamental understanding of phonon-phonon scattering processes, especially in complex crystals and disordered systems for which DFT ab initio calculations cannot be considered. The true advantage of molecular dynamics is the possibility to study nanostructures, in particular to probe energy transmission at interfaces. Several Green–Kubo formulations have been proposed recently, and they have been applied mostly for superlattices. Further effort should be pursued to use equilibrium simulations to the calculation of the energy transmission coefficient [65, 66].

References

1. B. J. Alder and T. E. Wainwright. Phase transition for hard sphere system. *J. Chem. Phys.*, 27(5):1208–1209, 1957.
2. B. J. Alder and T. E. Wainwright. Studies in molecular dynamics. i. general method. *J. Chem. Phys.*, 31(2):459–466, 1959.

3. A. Rahman. Correlations in the motion of atoms in liquid argon. *Phys. Rev.*, 136(No. 2A):A405–A411, 1964.
4. A. Rahman and F. Stillinger. Propagation of sound in water. a molecular-dynamics study. *Phys. Rev. A*, 10(1):368–378, 1974.
5. J. P. Hansen and M. L. Klein. Dynamical structure factor of rare-gas solids. *Phys. Rev. B*, 13:878–887, 1976.
6. A. J. C. Ladd, B. Moran, and W. G. Hoover. Lattice thermal conductivity: a comparison of molecular dynamics and anharmonic lattice dynamics. *Phys. Rev. B*, 34(8):5058–5064, 1986.
7. S. Volz and G. Chen. Molecular dynamics simulation of thermal conductivity of silicon nanowires. *Appl. Phys. Lett.*, 75(14):2056–2058, 1999.
8. M. P. Allan and D. J. Tildesley. *Computer Simulation of Liquids*. Oxford University Press, New York, USA, 1987.
9. D. Frenkel and B. Smit. *Understanding Molecular Simulation: From Algorithms to Applications*. San Diego, 1996.
10. L. Verlet. Computer "experiments" on classical fluids. i. thermodynamical properties of Lennard–Jones molecules. *Phys. Rev.*, 159, (July 1967):98–103, 1967.
11. L. Verlet. Computer "experiments" on classical fluids. ii. equilibrium correlation functions. *Phys. Rev.*, 165:201–214, 1968.
12. M. S. Daw and M. I. Baskes. Embedded-atom method: derivation and application to impurities, surfaces and other defects in metals. *Phys. Rev. B*, 29:6443–6453, 1984.
13. J. Tersoff. New empirical model for the structural properties of silicon. *Phys. Rev. Lett.*, 56:632–635, 1986.
14. J. Tersoff. New empirical approach for the structure and energy of covalent systems. *Phys. Rev. B*, 37:6991–7000, 1988.
15. J. Tersoff. Empirical interatomic potential for silicon with improved elastic properties. *Phys. Rev. B*, 38:9902– 9905, 1988b.
16. F. H. Stillinger and T. A. Weber. Computer simulations in condensed phases of silicon. *Phys. Rev. B*, 31:5262–5271, 1985.
17. A. J. H. McGaughey and M. Kaviany. Quantitative validation of the Boltzmann transport equation phonon thermal conductivity model under the single-mode relaxation time approximation. *Phys. Rev. B*, 69:094303–1, 094303–12, 2004.
18. Dammak H., Chalopin Y., M. Laroche, M. Hayoun, and J-J. Greffet. Quantum thermal bath for molecular dynamics simulation. *Phys. Rev. Lett.*, 103:190601, 2009.

19. J. Yang. Theory of thermal conductivity, in: Thermal conductivity, theory, properties, and applications. pages 1–17, 2004.
20. P. Chantrenne. Molecular dynamics, in: *Microscale and Nanoscale Heat Transfer*. pages 155–179, 2007.
21. D. J. Evans. Homogeneous nemd algorithm for thermal conductivity-application of non-canonical linear response theory. *Phys. Lett. A*, 91(9):457–460, 1982.
22. D. J. Evans. Thermal conductivity of the Lennard-Jones fluid. *Phys. Rev. A*, 34(2):1449–1453, 1986.
23. P. K. Schelling, S. R. Phillpot, and P. Keblinski. Comparison of atomic-level simulation methods for computing thermal conductivity. *Phys. Rev. B*, 65(14):144306, 2002.
24. S. Mahajan, G. Subbarayan, and B. G. Sammakia. Estimating thermal conductivity of amorphous silica nanoparticles and nanowires using molecular dynamics simulations. *Phys. Rev. E*, 76(5):056701, 2007.
25. K. Termentzidis, J. Parasuraman, C. Abs Da Cruz, S. Merabia, D. Angelescu, F. Marty, T. Bourouina, X. Kleber, P. Chantrenne, and Basset P. Thermal conductivity and thermal boundary resistance of nanostructures. *Nanoscale Res. Lett.*, 6(1):288, 2011.
26. S. Kotake and S. Wakuri. Molecular dynamics study of heat conduction in solid materials. *JSME Int. J. Ser. B Fluids Thermal Eng.*, 37(1):103–108, 1994.
27. P. Chantrenne and J. L. Barrat. Analytical model for the thermal conductivity of nanostructures. *Superlatt. Microstruct.*, 35:173–186, 2004.
28. P. Chantrenne and J. L. Barrat. Finite size effects in determination of thermal conductivities: comparing molecular dynamics results with simple models. *J. Heat Transfer*, 126:577–585, 2004.
29. K. Termentzidis, P. Chantrenne, and P. Keblinski. Nonequilibrium molecular dynamics simulations of the in-plane thermal conductivity of superlattices with rough interfaces. *Phys. Rev. B*, 79(21):214307, 2009.
30. F. Mueller-Plathe. A simple nonequilibrium molecular dynamics method for calculating the thermal conductivity. *J. Chem. Phys.*, 106(14):6082–6085, 1997.
31. R. J. Stevens, L. V. Zhigilei, and P. M. Norris. Effects of temperature and disorder on thermal boundary conductance at solid-solid interfaces: nonequilibrium molecular dynamics simulations. *Int. J. Heat Mass Transfer*, 50(19–20):3977–3989, 2007.

32. R. H. Poetzsch and H. Bottger. Interplay of disorder and anharmonicity in heat conduction: molecular dynamics study. *Phys. Rev. B*, 50(21):15757–15763, 1994.
33. Sellan D. P., Landry E. S., Turney J. E., McGaughey A. J. H., and Amon C. H. Size effects in molecular dynamics thermal conductivity prediction. *Phys. Rev. B*, 81:214305, 2010.
34. D. J. Evans and G. P. Morris. *Statistical Mechanics of Nonequilibrium Liquids, ISBN-13 978-0-521-85731-8*. New York, 1990.
35. Fan Z., Pereira L. F. C., Wang H.-Q., Zheng J.-C., Donadio D., and Harju A. Force and heat current formulas for many-body potentials in molecular dynamics simulations with applications to thermal conductivity calculations. *Phys. Rev. B*, 92(094301), 2015.
36. A. J. H. McGaughey and M. Kaviany. Thermal conductivity decomposition and analysis using molecular dynamics simulations. part i. Lennard–Jones argon. *Int. J. Heat Mass Transfer*, 47:1783–1798, 2004.
37. A. J. H. McGaughey and M. Kaviany. Thermal conductivity decomposition and analysis using molecular dynamics simulations. part ii. complex silica structures. *Int. J. Heat Mass Transfer*, 47:1799–1816, 2004.
38. M. I. Landry, E. S. Hussein and A. J. H. McGaughey. Complex superlattice unit cell designs for reduced thermal conductivity. *Phys. Rev. B*, 77(18):18184302, 2008.
39. Chen J., Zhang G., and Li B. How to improve the accuracy of equilibrium molecular dynamics for computation of thermal conductivity? *Phys. Lett. A*, 374:2392–2396, 2010.
40. K. Termentzidis, S. Merabia, P. Chantrenne, and P. Keblinski. Cross-plane thermal conductivity of superlattices with rough interfaces using equilibrium and non- equilibrium molecular dynamics. *Int. J. Heat Mass Transfer*, 54(9):2014–2020, 2011.
41. France-Lanord A., Merabia S., Albaret T., Lacroix D., and Termentzidis K. Thermal properties of amorphous/crystalline silicon superlattices. *J. Phys. Condensed Matter*, 26(35):355801, 2014.
42. M. T. Dove. *Introduction to Lattice Dynamics, ISBN 0521398940*. Cambridge United Kingdom, 2005.
43. G. P. Srivastava. *The Physics of Phonons*. 1990.
44. J. E. Turney, E. S. Landry, A. J. H. McGaughey, and C. H. Amon. Predicting phonon properties and thermal conductivity from anharmonic lattice dynamics calculations and molecular dynamics simulations. *Phys. Rev. B*, 79:064301, 2009.

45. A. S. Henry and G. Chen. High thermal conductivity of single polyethylene chains using molecular dynamics simulations. *Phys. Rev. Lett.*, 101:235502, 2008.
46. A. S. Henry and G. Chen. Anomalous heat conduction in polyethylene chains: theory and molecular dynamics simulations. *Phys. Rev. B*, 79:144305, 2009.
47. D. Donadio and G. Galli. Atomistic simulations of heat transport in silicon nanowires. *Phys. Rev. Lett.*, 102:195901, 2009.
48. D. Donadio and G. Galli. Thermal conductivity of isolated and interacting carbon nanotubes: comparing results from molecular dynamics and the Boltzmann transport equation. *Phys. Rev. Lett.*, 99:255502, 2007.
49. Y. He, D. Donadio, and G. Galli. Heat transport in amorphous silicon: interplay between morphology and disorder. *App. Phys. Lett.*, 98(144101), 2011.
50. J. M. Larkin and A. J. H. McGaughey. Thermal conductivity accumulation in amorphous silica and amorphous silicon. *Phys. Rev. B*, 89(144303), 2014.
51. J. L. Barrat and F. Chiaruttini. Kapitza resistance at the liquid-solid interface. *Mol. Phys.*, 101(11):1605–1610, 2003.
52. D. G. Cahill, W. K. Ford, K. E. Goodson, G. D. Mahan, A. Majumdar, H. J. Maris, R. Merlin, and S. R. Phillpot. Nanoscale thermal transport. *J. Appl. Phys.*, 93(2):793–818, 2003.
53. E. T. Swartz and R. O. Pohl. Thermal resistance at interfaces. *Appl. Phys. Lett.*, 51(26):2200–2202, 1987.
54. E. T. Swartz and R. O. Pohl. Thermal boundary resistance. *Rev. Modern Phys.*, 61(3):605–668, 1989.
55. I. M. Khalatnikov. *Zh. Eksp. Teor. Fiz.*, 22:687, 1952.
56. Hopkins P. E. *ISRN Mechanical Engineering*, page 682586, 2013.
57. L. Puech, G. Bonfait, and B. Castaing. Mobility of the he-3 solid liquid interface-experience and theory. *J. Low Temp. Phys.*, 62:315–327, 1986.
58. Chalopin Y., Esfarjani K., Henry A., Volz S., and Chen G. Thermal interface conductance in Si/Ge superlattices by equilibrium molecular dynamics. *Phys. Rev. B*, 85:195302, 2012.
59. Liang Z., Evans W., and Keblinski P. Equilibrium and nonequilibrium molecular dynamics simulations of thermal conductance at solid-gas interfaces. *Phys. Rev. E*, 87:022119, 2013.
60. Rajabpour A. and Volz S. *J. Appl. Phys.*, 108:094324, 2010.

61. Merabia S. and Termentzidis K. Thermal conductance at the interface between crystals using equilibrium and nonequilibrium molecular dynamics. *Phys. Rev. B*, 86:094303, 2012.
62. A. J. H. Mcgaughey and J. Li. Molecular dynamics prediction of the thermal resistance of solid-solid interfaces in superlattices. *Proceedings of IMECE 2006, ASME International Mechanical Engineering Conference and Exhibition, November 5–10, 2006, Chicago USA*, 2006.
63. E. S. Landry and A. J. H. McGaughey. Thermal boundary resistance predictions from molecular dynamics simulations and theoretical calculations. *Phys. Rev. B*, 80(16):16165304, 2009.
64. S. Pettersson and G. D. Mahan. Theory of the thermal boundary resistance between dissimilar lattices. *Phys. Rev. B*, 42(12):7386–7390, 1990.
65. S. R. Schelling, P. K. Phillpot and P. Keblinski. Phonon wave-packet dynamics at semiconductor interfaces by molecular-dynamics simulation. *Appl. Phys. Lett.*, 80(14):2484–2486, 2002.
66. K. Saaskilahti, J. Oksanen, J. Tulkki, and S. Volz. Role of anharmonic phonon scattering in the spectrally decomposed thermal conductance at planar interfaces. *Phys. Rev. B*, 90(13):134312, 2014.

Chapter 7

Non-Equilibrium Molecular Dynamics

Arthur France-Lanord,[a,b] **Etienne Blandre,**[c]
and Konstantinos Termentzidis[d]

[a]*Materials Design SARL, 92120 Montrouge, France*
[b]*SPEC, CEA, CNRS, Université Paris-Saclay, CEA Saclay 91191, Gif-sur-Yvette, France*
[c]*Université de Lyon, CNRS, INSA-Lyon, UCBL, CETHIL, UMR5008, F-69621 Villeurbanne, France*
[d]*Université de Lorraine, LEMTA, UMR 7563, CNRS F-54506 Vandoeuvre-lès-Nancy, France*
arthur.flanord@gmail.com, etienne.blandre@gmail.com,
konstantinos.termentzidis@univ-lorraine.fr

7.1 Introduction

The non-equilibrium state is at the core of transport phenomena. For instance, applying a temperature gradient to a system leads to a flow of energy until equilibration is achieved. This is the basis of non-equilibrium molecular dynamics (NEMD) simulations: in a way similar to experiments, one applies a known cause and measures the outcome (its effect). For thermal conductivity evaluation, the cause is usually a temperature gradient, and the effect a heat flux: this is known as the *direct method.* However, it is also possible to consider the opposite problem. Applying a heat flux to the system and

Nanostructured Semiconductors: Amorphization and Thermal Properties
Edited by Konstantinos Termentzidis

ISBN 978-981-4745-64-2 (Hardcover), 978-1-315-36445-2 (eBook)
www.panstanford.com

measuring the induced temperature gradient is known as reverse-NEMD [14]. This method presents the advantage of sampling a rapidly converging quantity instead of a slowly converging one. Both direct and reverse techniques have been applied to many different systems: bulks [3], thin films [1], superlattices [5–7], nanowires [2, 4], two-dimensional materials [8–10], fluids [11–13], etc.

The classical nature of molecular dynamics (MD) restricts the scope of NEMD to the high temperature limit. Quantum effects such as zero-point vibrations are not accounted for, since in MD nuclear degrees of freedom are treated classically. Sometimes, quantum corrections to the thermal conductivity by rescaling the heat capacity are applied [15]. However, these corrective treatments are not generally applicable, and usually oversimplified [16]. Quantum effects are generally assumed to be negligible if, in the quantum regime, the classical description of the thermal conductivity as obtained from MD calculations is reliable. One strong asset of NEMD methods is that, as opposed to Boltzmann transport calculations, they are "classically exact" and do not rely on theoretical approximations.

When the temperature gradient and the heat flux present in the system are known, whether they are imposed or measured, it is possible to evaluate the lattice thermal conductivity using Fourier's law:

$$\boldsymbol{J} = -\boldsymbol{\kappa} \cdot \nabla T, \tag{7.1}$$

where $\boldsymbol{J}$ is the heat flux vector, $\boldsymbol{\kappa}$ is the thermal conductivity tensor, and ∇T the temperature gradient. It is not possible to obtain the full thermal conductivity tensor at once, as with fluctuation-based methods such as EMD, we can only obtain one component of $\boldsymbol{\kappa}$, which depends on the direction of the temperature gradient we choose to use. Hence, one needs to define *a priori* the thermal transport direction of the system that will be evaluated during the simulation. In any case, only the diagonal components of the thermal conductivity tensor can be evaluated with NEMD.

In NEMD simulations, a spatial decomposition of the system in slabs is first performed. The temperature gradient is obtained by coupling different regions of the system with thermostats at different temperatures. Time integration in the microcanonical

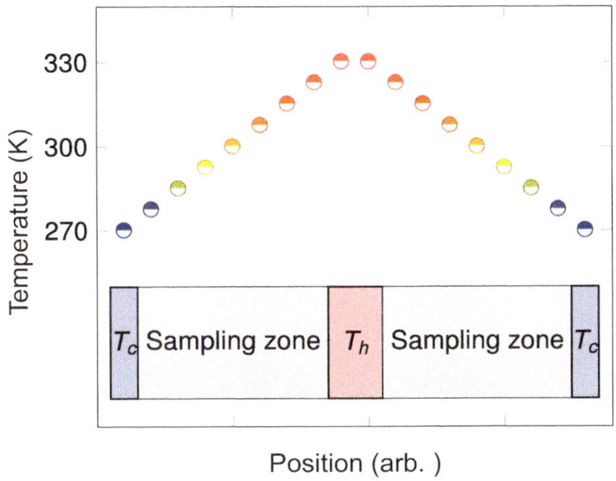

Figure 7.1 A setup for non-equilibrium molecular dynamics simulations. The sample is subdivided in several zones, both for thermostatting and for sampling.

(NVE) ensemble is performed on the rest of the system. Once the steady state is reached, when the energy creation rate is equal to the energy removal rate at the thermostatted regions, one can sample the temperature profile, using the spatial binning of the sample. In the reverse method, swapping highest and lowest atomic kinetic energies in different regions, at a certain rate, generates the heat flux. Figure 7.1 shows a typical setup for both NEMD and rNEMD methods, and an idealized temperature gradient through the structure.

In practice, the temperature gradient is often nonlinear near the thermostats. The validity of Fourier's law dictates to make linear fits using only parts of the system, at a certain distance of the thermostatted regions. This is achieved by making use of buffer zones near the heat source and sink. Nonlinearities in the temperature profile are due to strong scattering at the heat baths, where the energy transport becomes diffusive.

When the sample considered is smaller than the mean free path of heat carriers in an infinite system, the lattice thermal conductivity is limited by the system size. This regime is known as the Casimir limit. Finite size effects, arising from scattering

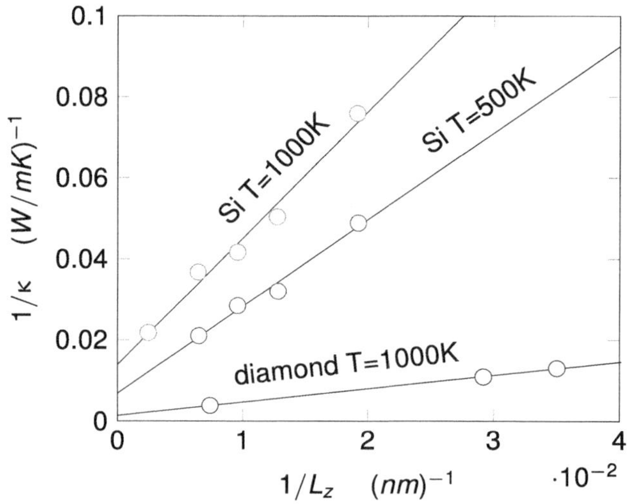

Figure 7.2 System size dependence of the thermal conductivity for two systems: crystalline Si at 500 K and 1000 K, and diamond at 1000 K. We note that the slope depends on the nature of the system, and slightly on its temperature. All data come from [3].

events at the interfaces of thermostats, are almost unavoidable in systems exhibiting good thermal transport properties. However, one can extrapolate the thermal conductivity at infinite system size through linear regression by evaluating the inverse of the thermal conductivity as a function of the inverse of the system size, in the transport direction [17]. Examples of infinite size extrapolation on different systems are given in Fig. 7.2.

In the next two subsections, we will review two examples of how the non-equilibrium approach can be used in systems presenting complex geometries, to evaluate thermal conductivity. Both examples present amorphous-crystalline interfaces of silicon: planar in the first place, and radial in the second one. These systems are promising candidates for future thermoelectric devices, with very high figure of merit *ZT*. We will show how to use NEMD in order to tailor the properties of the studied system, in this case to maximize *ZT* by minimizing the lattice thermal conductivity.

All simulations were performed using a modified version of the Stillinger–Weber potential by Vink et al. [18], which is specially parametrized to reproduce correctly the vibrational density of states of amorphous silicon. A time step value of 4 fs was found to be sufficient to ensure the absolute conservation of the total energy, during the full simulation time. The structures are built using an initial random distribution of Si atoms in the amorphous areas, and an annealing process [19]. Every system has been equilibrated in the isoenthalpic-isobaric ensemble (constant enthalpy and pressure, NPH), prior to the NEMD simulation.

7.2 Example I: Amorphous/Crystalline Silicon Superlattices

The first geometry we will study is a superlattice, formed with successive amorphous-crystalline silicon interfaces, which are defect free and atomically sharp. Amorphous/crystalline superlattices generally present high thermal conductivity anisotropy, which could make them appropriate for a plethora of potential applications: for thermal barrier coatings, heat spreading in optoelectronics or for heat shields for example. NEMD is the appropriate tool to study the dependence of the system's thermal conductivity on the thickness of the amorphous layer. The full study can be found in [20].

7.2.1 NEMD Simulation Settings

Special care must be given to the positioning of the thermostats. We are making use of periodic boundary conditions in the thermal transport direction (z), hence thermostats can be placed anywhere in the system if (a) the distance between hot and cold zones is the same and (b) the thermostats cover both amorphous and crystalline phases, in equal parts. Following these conditions, hot and cold thermostats have to be placed at amorphous-crystalline interfaces. Constraint (b) is related to the fact that the sampling regions should be identical and therefore should also contain equal parts of amorphous and crystalline phases. In this study, it has revealed to be necessary in order to obtain consistent values of the thermal

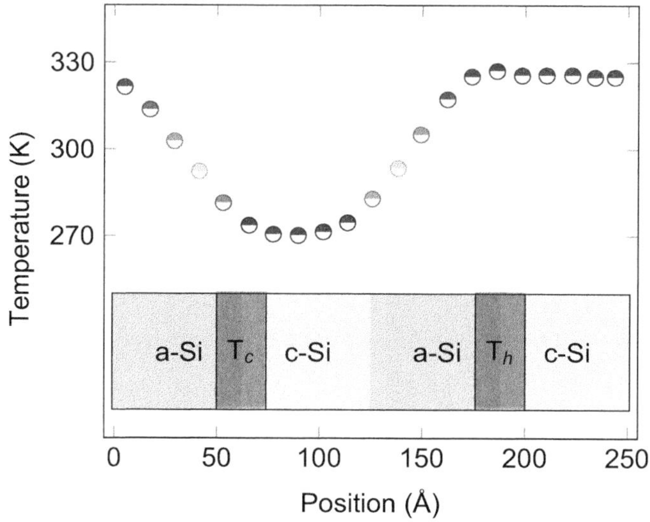

Figure 7.3 Temperature profile and thermostatting in a a-Si/c-Si superlattice.

conductivity. However, this is not a general rule and depends on the nature of system [21]. Once the steady state is reached, sampling of the temperature profile is performed, over a period of several ns. The setup and a typical temperature gradient are presented in Fig. 7.3. The obtained temperature gradients show two different slopes, corresponding to the amorphous and to the crystalline phases. Since linearity is obtained in both phases, the temperature gradient is computed over several amorphous-crystalline repetitions, at a reasonable distance of thermostats.

7.2.2 Results

We have considered two periodicities (the thickness of both amorphous and crystalline layers): $9a_0$, and $18a_0$. The lattice parameter a_0 of Si is equal to 5.43 Å. The amorphous proportion has been varied from 6.25% to 50%. Figure 7.4 presents the dependence of the lattice thermal conductivity on the thickness of the amorphous layer.

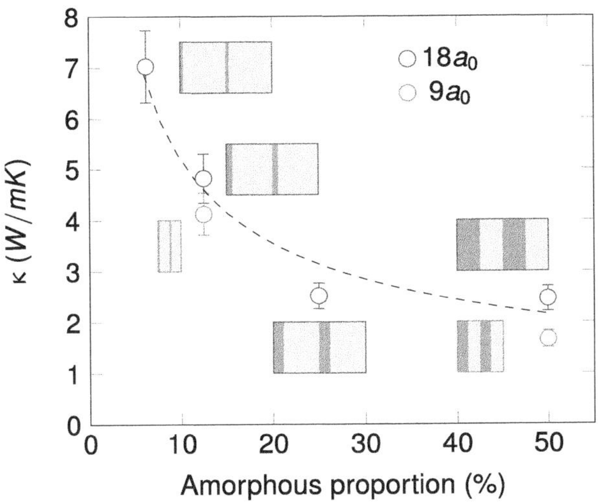

Figure 7.4 Lattice thermal conductivity as a function of the amorphous proportion. Two different periodicities are considered: 9a_0, and 18a_0.

As we can see, the conductivity decreases with the amorphous proportion, and converges to a value of 2 W/m/K for superlattices containing more than 25% of amorphous Si. This value is 75 times smaller than the room temperature thermal conductivity of bulk crystalline silicon, 148.0 W/m/K, and close to the thermal conductivity of bulk amorphous silicon 0.68 W/m/K, as computed using the modified Stillinger–Weber potential. Also, it seems that decreasing the superlattice period leads to a slight reduction (30%) in the converged value of the conductivity. Classical crystalline superlattices usually show a stronger dependence on the periodicity; this dissimilarity can be related to the diffusive nature of energy transport in amorphous Si, which leads to the system having an effective thermal conductivity independent of the thickness of the amorphized regions, when it is less than 4 nm. Finally, the critical impact of disorder on energy transport is obvious when looking at the thermal conductivity of superlattices with very small amorphous proportion; it only takes 6.25% of amorphous phase to reduce the SL's crossplane thermal conductivity by a factor of 20, when compared to the one of bulk crystalline. This is related to the existence of a finite Kapitza resistance between crystalline and amorphous silicon.

7.3 Example II: Amorphous/Crystalline Core/Shell Nanowires

We will now see how NEMD methods can help explore how the combination of nanostructuration and partial amorphization may be used to reduce the thermal conductivity of silicon nanowires. Our system of interest is a partially amorphized diameter-modulated silicon nanowire. The amorphization can occur at the core (ac/cs NW), through swift and spatially resolved heavy-ion bombardment [23] or at the free surface, the shell, by ion irradiation or oxidation [23] (cc/as NW). The full study can be found at [24].

7.3.1 NEMD Simulation Settings

We study three different morphologies: diameter-modulated pristine crystalline nanowires (labeled DM), core/shell type nanowires (CS) with constant diameter, and a combination of the two effects, labeled DMCS nanowires. We consider both crystalline core/amorphous shell (CC/AS) and the inverse configuration (AC/CS) for CS wires, with an external diameter of 6 nm, and different phase proportions. Finally, we examine three different DMCS wires: (i) the C1 configuration, in which the diameter of the crystalline core remains constant and where only the amorphous shell diameter is modulated; (ii) the C2 configuration, in which both the core and the shell diameters are modulated; and (iii) the C3 configuration, in which a diameter-modulated amorphous shell contains crystalline nanoinclusions.

We set periodic boundary conditions in the direction of the axis of the wire, and free boundary conditions in the two others. To avoid any drifting or large distortions of the entire structure, we freeze atomic positions in regions that are very close to the simulation box. Directly after these fixed regions, we set the hot and the cold thermostats, which are $2a_0$ wide. The entire setup for the C3 configuration is described in Fig. 7.5.

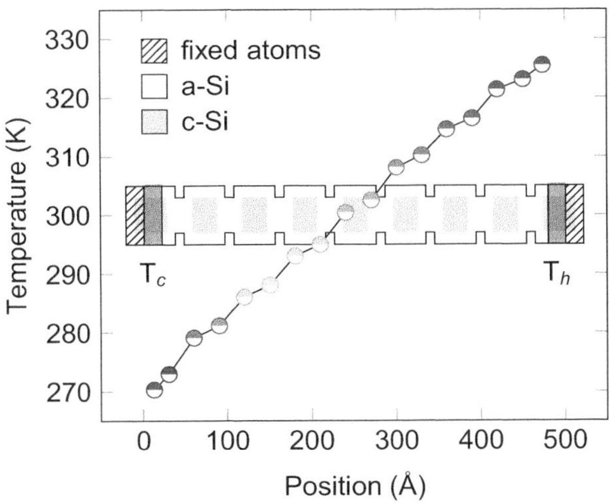

Figure 7.5 Temperature profile of a diameter-modulated nanowire with crystalline nanoinclusions (C3). The different slopes in the temperature profile are due to the modulated diameter.

7.3.2 Results

The full set of results is presented in Fig. 7.6. First, we consider the influence of the periodicity (P) in simple pristine DM wires on the thermal conductivity. P is varied from 60 to 120 Å, and for these nanowires, the modulations are 2 nm high. For comparison, we also report in Fig. 7.6a the limiting cases of nanowires with constant radii equal to the small section (3 nm), and the high section (5 nm). Diameter modulation induces new scattering mechanisms, which in turn, as we can see in Fig. 7.6a, reduce the thermal conductivity of the system. Reducing P is equivalent to increasing the number of modulations, and the scattering rate. Here, we can see that a shorter periodicity can lead to at least a twice smaller thermal conductivity.

In Fig. 7.6b,c, we present results on CS nanowires. The amorphous proportion is varied, while the external diameter (6 nm) is kept constant. In other words, we only vary the core diameter of both AC/CS and CC/AS configurations. Two limiting cases are also included in Fig. 7.6b: purely crystalline and purely amorphous nanowires. As a general result, and as one could intuitively expect, an increase in the amorphous proportion leads to a decrease in the

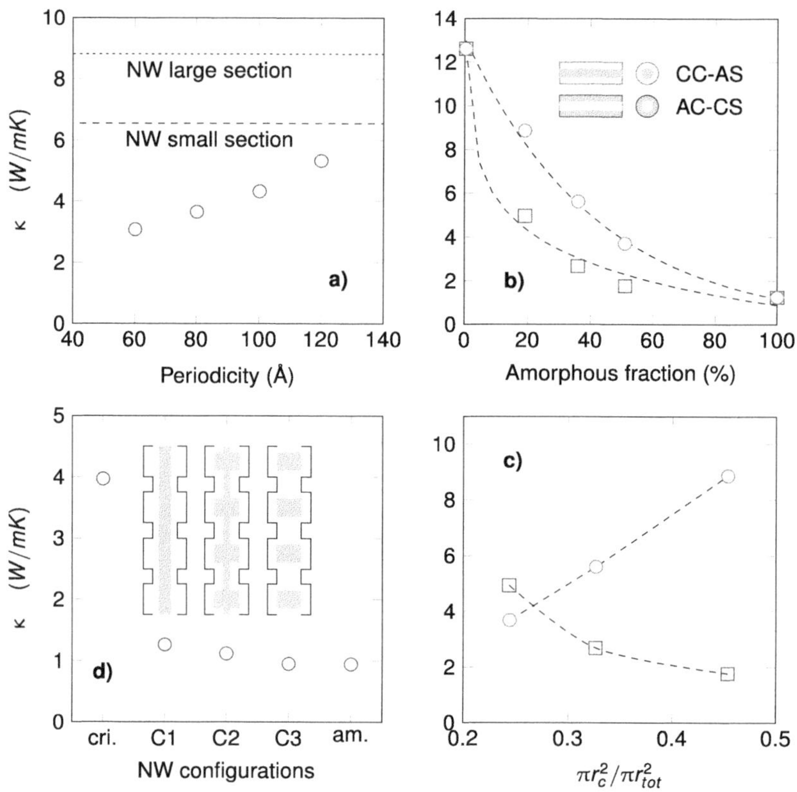

Figure 7.6 Thermal transport properties of nanowires. a) modulated pristine wire, b) and c) core–shell systems, and d) DMCS nanowires.

thermal conductivity. The thermal conductivity of AC/CS nanowires is smaller and behaves less linearly with increasing amorphous proportion than the one of CC/AS wires. Assuming that most of the transport occurs in the crystalline part of the system, the difference in transport between the two configurations can be understood as a direct consequence of the much higher surface area in the crystalline shell case, leading to additional scattering of modes which in the crystalline core case, would contribute to thermal transport. This is best presented in Fig. 7.6c, where the linear relation between the thermal conductivity for both configurations and the ratio of the cross-section areas of the core part (πr_c^2) and the whole nanowire (πr_{tot}^2) is shown.

We now move to DMCS wires. The three different cases are presented in Fig. 7.6d. For comparison, we also included the thermal conductivity of fully crystalline and amorphous diameter-modulated nanowires, sharing the same characteristics with the C(1, 2, 3) configurations. In these three cases, the central crystalline conducting channel is more and more affected (by modulations, and then by nanoinclusions), which is why the thermal conductivity slightly decreases from C1 to C3. Interestingly, cases C2 and C3 conduct heat in a very similar fashion than the modulated amorphous nanowire. The C2 configuration is probably the most interesting geometry for thermoelectrical applications, since a good charge transport channel is preserved at the center of the nanowire. In a previous work, He and Galli [25] studied similar DMCS wires, with a larger diameter of 15 nm—three times the diameter of our systems—with quite similar results.

7.4 Conclusions

We showed that non-equilibrium molecular dynamics can be a very efficient method for probing thermal transport properties of both crystalline and amorphized semiconductors. Keeping in mind the inherent limitations of molecular dynamics due to its classical nature, one can easily access complex properties, such as thermal conductivity, in very large systems presenting complex geometries. NEMD is an important tool for tailoring new materials, according to their thermal properties. As a conclusive remark, we should also mention that by definition NEMD can also be used for other transport properties such as viscosity and diffusion.

References

1. X. L. Feng. *Microscale Therm. Eng.* 7, 153 (2003).
2. I. Ponomareva, D. Srivastava, and M. Menon. *Nano Lett.* 7, 1155 (2007).
3. P. K. Schelling, S. R. Phillpot, and P. Keblinski. *Phys. Rev.* B 65, 144306 (2002).
4. N. Yang, G. Zhang, and B. Li. *Nano Lett.* 8, 276 (2002).

5. Y. Chen, D. Li, J. R. Lukes, Z. Ni, and M. Chen. *Phys. Rev. B* 72, 174302 (2005).
6. K. Termentzidis, P. Chantrenne, and P. Keblinski. *Phys. Rev. B* 79, 214307 (2009).
7. R. J. Stevens, L. V. Zhigilei, and P. M. Norris. *Int. J. Heat Mass Transfer* 50, 3977 (2007).
8. X. Zhang, H. Xie, M. Hu, H. Bao, S. Yue, G. Qin, and G. Su. *Phys. Rev. B* 89, 054310 (2014).
9. Q. X. Pei, Z. D. Sha, and Y. W. Zhang. *Carbon*, 49, 4752 (2011).
10. Z. Guo, D. Zhang, and X. G. Gong. *Appl. Phys. Lett.* 95, 163103 (2009)
11. D. M. Heyes. *J. Chem. Soc. Farad. Trans.* 2, 80, 1363 (1984).
12. R. Vogelsang, C. Hoheisel, and G. Ciccotti. *J. Chem. Phys.* 86, 6371 (1987).
13. D. Bedrov, G. D. Smith, and T. D. Sewell. *Chem. Phys. Lett.* 324, 64 (2000).
14. F. Müller-Plathe. *J. Chem. Phys.* 106, 6082 (1997).
15. S. K. Chien and Y. T. Yang. *Appl. Phys. Lett.* 98, 033107 (2011).
16. J. E. Turney, A. J. H. McGaughey, and C. H. Amon. *Phys. Rev. B* 79, 224305 (2009).
17. C. Oligschleger and J. C. Sch On. *Phys. Rev. B* 59, 8 (1998).
18. R. L. C. Vink, G. T. Barkema, W. F. Van der Weg, and N. Mousseau. *J. Non-Crystalline Solids*, 282, 248 (2001).
19. A. France-Lanord, E. Blandre, T. Albaret, S. Merabia, D. Lacroix, and K. Termentzidis. *J. Phys.: Condensed Matter* 26, 055011 (2014).
20. A. France-Lanord, S. Merabia, T. Albaret, D. Lacroix, and K. Termentzidis. *J. Phys.: Condensed Matter* 26, 355801 (2014).
21. J. Moon, M. Cho, and M. Zhou. *J. Appl. Phys.* 117, 214307 (2015).
22. M. Hu and D. Poulikakos. *Nano Lett.* 12, 5487 (2012).
23. P. J. Newby, B. Canut, J. M. Bluet, S. Gomès, M. Isaiev, R. Burbelo, and V. Lysenko. *J. Appl. Phys.* 114, 014903 (2013).
24. E. Blandre, L. Chaput, S. Merabia, D. Lacroix, and K. Termentzidis. *Phys. Rev. B* 91, 115404 (2015).
25. Y. He and G. Galli. *Phys. Rev. Lett.* 108, 215901 (2012).

Chapter 8

Approach-to-Equilibrium Molecular Dynamics

Evelyne Lampin, Pier Luca Palla, Hayat Zaoui, and Fabrizio Cleri
IEMN, UMR CNRS 8520 and University of Lille, CS 60069, 59652 Villeneuve d'Ascq Cedex, France
evelyne.lampin@univ-lille1.fr

8.1 Introduction

Besides steady-state methods presented in Chapters 8 and 9, molecular dynamics simulations can also be used to simulate temperature transients, by means of approach-to-equilibrium molecular dynamics (AEMD) calculations. The interest for such methods is related to the short MD simulation time required to study a transient, compared to a steady state. Temperature transients can be used to calculate thermal boundary resistances under the lumped capacitance approximation, valid when a high resistance separates two good thermal conductors (Shenogin, 2004). The approach-to-equilibrium of an instantaneously heated portion of a supercell has also been used to obtain the thermal conductivity of a system of molecules (Hulse, 2005). In the present work, we present a

Nanostructured Semiconductors: Amorphization and Thermal Properties
Edited by Konstantinos Termentzidis

ISBN 978-981-4745-64-2 (Hardcover), 978-1-315-36445-2 (eBook)
www.panstanford.com

variation of these transient methods where two separate blocks are equilibrated at different temperatures before the approach-to-equilibrium is monitored to extract a characteristic decay time (Section 8.2). In parallel, the heat equation is solved analytically under the same conditions of periodicity and initial temperature profile (Section 8.3). It is shown that the temperature profile obtained by MD has the spatial and temporal form of the Fourier series solution of the heat equation. The same relation between the transient time and the thermal conductivity is therefore used to calculate the thermal conductivity of the atomic system. An application to the case of a crystal is shortly recalled in Section 8.4. The approach is also applied to an amorphous material, amorphous silica, and the differences of time scale and length dependences are discussed (Section 8.5).

8.2 Approach-to-Equilibrium Molecular Dynamics

To obtain the thermal conductivity of a bulk material at the nanoscale, we first build a supercell where atoms are set at their equilibrium positions, and use periodic boundary conditions to artificially suppress the effect of surfaces. The length of the supercell along z, L_Z, is chosen larger than the sides of the supercell in order to induce a one-dimensional heat current (Fig. 8.1). This heat current is created subsequently to a preliminary step, in which a temperature difference is established between two parts of the supercell. A "Hot" block is heated at a temperature T_1, and the other "Cold" block at a temperature $T_2 < T_1$. For this purpose, the supercell is successively heated at T_1 and T_2 during separate NVT simulations of 10 ps. The positions and velocities of two halves are then cut and pasted to

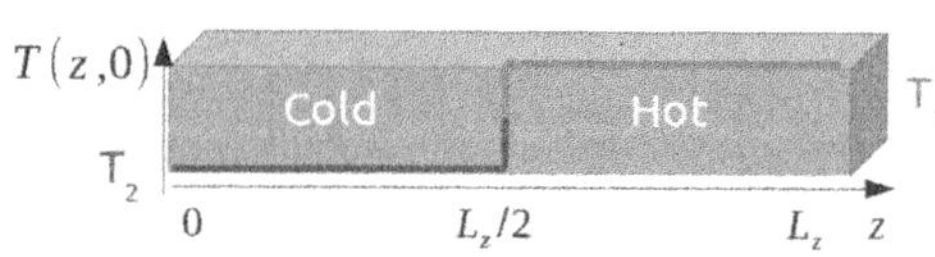

Figure 8.1 Elongated supercell.

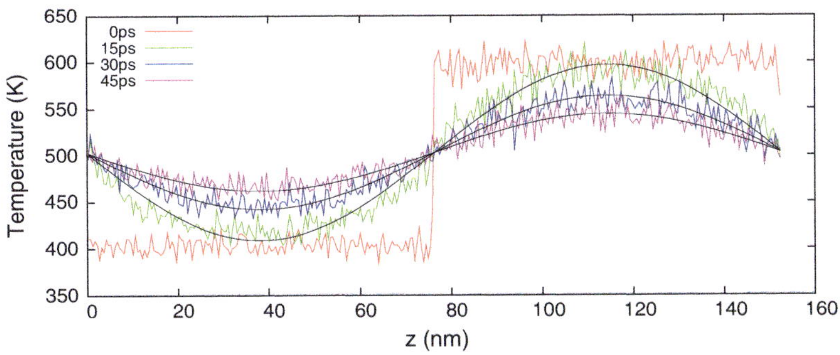

Figure 8.2 Temperature profiles during the approach-to-equilibrium: 0 ps (red), 15 ps (green), 30 ps (blue) and 45 ps (magenta). The black lines are sinusoidal fits of the form $f(z) \propto \sin(2\pi z/L_z)$. The supercell is 152 nm long and the square cross section is 4.3×4.3 nm^2.

obtain a 1D step like temperature profile as shown in red in Fig. 8.2 for crystalline silicon (c-Si). The temperatures of the two blocks are set to 400 and 600 K, respectively. A target temperature of 500 K is chosen close to the Debye temperature of silicon since quantum effects are not included in classical molecular dynamics. The initial temperature difference is taken large enough (200 K) to increase the precision of the decay time determination.

After this first step, the flat equilibrium temperature profile is approached during a NVE simulation. During the approach-to equilibrium, the step-like profile becomes sinusoidal and its amplitude decreases with time (Fig. 8.2). The temperature difference between the two initial blocks at T_1 and T_2 is monitored on the fly (Fig. 8.3). It results to be an exponential, with a characteristic decay time τ:

$$\Delta T = T_1 - T_2 = \Delta T_0 \exp\left(-\frac{t}{\tau}\right) \tag{8.1}$$

8.3 Heat Equation

We consider an homogeneous and isotropic solid of number density ρ (m^{-3}) and heat capacity C_V (J.K^{-1}). The first law of thermodynamics states that the infinitesimal change of enthalpy dH in the constant volume V of mass $M = \iiint_V \rho dv$ delimited by the

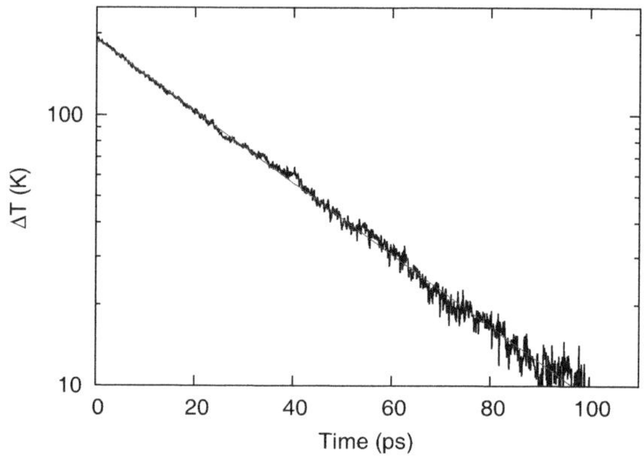

Figure 8.3 Temperature difference versus time (black curve) and exponential fit with $\tau = 32$ ps (red curve) monitored during the approach-to-equilibrium presented in Fig. 8.2.

surface S between t and $t + dt$ is given by the infinitesimal amount of heat δQ supplied to the volume element:

$$dH = \delta Q \tag{8.2}$$

The change in enthalpy is equal to

$$dH = H(t + dt) - H(t) = \frac{\partial H}{\partial t} dt. \tag{8.3}$$

Introducing C_V and ρ to obtain H in the volume V, we obtain

$$dH = \frac{\partial}{\partial t}\left(\iiint_V \rho C_V T dv\right) dt = \left(\iiint_V \rho C_V \frac{\partial T}{\partial t} dv\right) dt \tag{8.4}$$

In the absence of internal heat sources, the incoming heat δQ is given by the heat flux across S:

$$\delta Q = \left(-\iint_S \vec{j}.d\vec{S}\right) dt = \left(-\iiint_V \mathrm{div}(\vec{j}) dv\right) dt \tag{8.5}$$

by using the divergence theorem. The Fourier law states a relation between the heat flux and the temperature gradient with a proportionality coefficient to be identified with the thermal conductivity κ ($\mathrm{W.K^{-1}.m^{-1}}$) of the material:

$$\vec{j} = -\kappa.\overrightarrow{\mathrm{grad}} T \tag{8.6}$$

The equality between the received heat and enthalpy variation (Eq. 8.2), combined with the Fourier law, therefore results in the heat equation:

$$\rho C_V \frac{\partial T}{\partial t} = \kappa \Delta T \tag{8.7}$$

The equation becomes in 1D

$$\frac{\partial T(t, z)}{\partial t} = D \frac{\partial^2 T(t, z)}{\partial z^2} \tag{8.8}$$

with the thermal diffusion coefficient $D = \kappa/(C_V \rho)$. The equation can be solved analytically by separating spatial and temporal dependencies:

$$T(t, z) = \Theta(t).\zeta(z) \tag{8.9}$$

We substitute $T(t, z)$ into Eq. (8.8) and obtain

$$\frac{1}{D}\frac{\Theta'(t)}{\Theta(t)} = \frac{\zeta''(z)}{\zeta(z)} \tag{8.10}$$

Since the right hand side depends only on t and the left hand side depends only on z, both sides are equal to a constant value λ and we have to solve

$$\Theta'(t) = \lambda D \Theta(t) \tag{8.11}$$

and

$$\zeta''(z) = \lambda \zeta(z). \tag{8.12}$$

The solution of the first equation is

$$\Theta(T) = \Theta_0 \exp(\lambda D t) \tag{8.13}$$

It can be shown that a non-zero solution for the second equation only exists if $\lambda < 0$, i.e.,

$$\zeta(z) = A\cos(\alpha z) + B\sin(\alpha z) \tag{8.14}$$

with $\lambda = -\alpha^2$. Since we use periodic boundary conditions, the temperature and its derivative, i.e., the flux, are equal in $z = 0$ and $z = L_Z$.

$$\begin{aligned} \zeta(0) &= \zeta(L_Z) \\ \zeta'(0) &= \zeta'(L_Z) \end{aligned} \tag{8.15}$$

Substituting Eq. (8.14) into Eq. (8.15) results in the systems of two equations:

$$A[1-\cos(\alpha L_Z)] - B\sin(\alpha L_Z) = 0 \qquad (8.16)$$
$$A\sin(\alpha L_Z) + B[1-\cos(\alpha L_Z)] = 0$$

This imposes that the determinant $(1-\cos(\alpha L_Z))^2 + \sin^2(\alpha L_Z) = 0$, i.e., $1-\cos(\alpha L_Z) = 0$. The periodicity therefore results in a series of discrete values of α:

$$\alpha = \frac{2\pi}{L_z}n, \qquad n = 0, \pm 1, \pm 2, \ldots . \qquad (8.17)$$

and consequently of λ:

$$\lambda = -\frac{4\pi^2}{L_z^2}n^2 \qquad n = 0, 1, 2, \ldots . \qquad (8.18)$$

The total solution is therefore a Fourier series:

$$T(t,z) = A_0 + \sum_{n=1}^{\infty}\left[A_n\cos\left(\frac{2\pi}{L_z}nz\right) + B_n\sin\left(\frac{2\pi}{L_z}nz\right)\right]e^{-n^2 t/\tau} \qquad (8.19)$$

with

$$\tau = \frac{1}{4\pi^2}\frac{L_z C_V}{S\kappa}, \qquad (8.20)$$

where S is being the area of the section of the supercell. The coefficients of the Fourier series are

$$A_0 = \frac{1}{L_z}\int_0^{L_z} T(0,z)dz, \qquad (8.21)$$

$$A_n = \frac{2}{L_z}\int_0^{L_z} T(0,z)\cos\left(\frac{2\pi}{L_z}nz\right)dz \quad \text{for} \quad n > 0 \qquad (8.22)$$

and

$$B_n = \frac{2}{L_z}\int_0^{L_z} T(0,z)\sin\left(\frac{2\pi}{L_z}nz\right)dz. \qquad (8.23)$$

In the AEMD simulation, the initial temperature profile is

$$T(0,z) = T_1, \qquad L_z/2 < z < L_z \qquad (8.24)$$
$$T(0,z) = T_2, \qquad 0 < z < L_z/2$$

The integrals in the expressions of A_n, for $n > 0$ are equal to 0 for this particular initial profile, while the B_n are non-null for odd values of n only:

$$B_n = \frac{2(T_2 - T_1)}{n\pi} \tag{8.25}$$

and A_0 is equal to $(T_1 + T_2)/2$.

The analytical expression of the temperature profile corresponding to the AEMD simulation is therefore a Fourier series:

$$\begin{aligned} T(t, z) = & \frac{1}{2}(T_1 + T_2) \\ & + \sum_{m=0}^{\infty} \frac{2(T_2 - T_1)}{(2m+1)\pi} \sin\left(\frac{2\pi(2m+1)}{L_z} z\right) \\ & \times e^{-(2m+1)^2 t/\tau} \end{aligned} \tag{8.26}$$

The dominant contribution to the series for $m = 0$ is a sinusoid:

$$T(t, z) = \frac{1}{2}(T_1 + T_2) + \frac{2}{\pi}(T_2 - T_1)\sin\left(\frac{2\pi}{L_z} z\right) \times e^{-t/\tau} \tag{8.27}$$

The sine form of the temperature profile obtained during AEMD (Fig. 8.2) is therefore in agreement with the solution of the heat equation. The equivalent of the temperature difference monitored during AEMD can also be obtained from the solution of the heat equation:

$$\begin{aligned} \Delta T(t) &= \frac{2}{L_z}\int_{L_z/2}^{L_z} T(t, z)dz - \frac{2}{L_z}\int_{0}^{L_z/2} T(t, z)dz \\ &= \sum_{m=0}^{\infty} \frac{8(T_2 - T_1)}{(2m+1)^2\pi^2} e^{-(2m+1)^2 t/\tau} \end{aligned} \tag{8.28}$$

The dominant contribution for $m = 0$ is an exponential decay:

$$\Delta T(t) = \frac{8}{\pi^2}(T_2 - T_1)e^{-t/\tau} \tag{8.29}$$

The exponential decay of the temperature difference between the "Hot" and "Cold" block observed during the simulations of approach-to-equilibrium (Fig. 8.3) is therefore also obtained by solving the heat equation.

8.4 From Decay Transient to Thermal Conductivity

The transient regime obtained by AEMD has the same temporal and spatial form as the solution of the heat equation. The coefficient in the heat equation, i.e., the thermal conductivity, is typically obtained from experiments on bulk materials. Here we use molecular dynamics simulations to obtain the value of the thermal conductivity accounting for nanoscale effects in a high-purity model material that cannot be achieved experimentally. For this purpose, we fit the decay time τ of the temperature difference in AEMD and use the same relation between τ and the thermal conductivity κ, obtained from the heat equation:

$$\kappa = \frac{L_Z C_V}{4\pi^2 S}\frac{1}{\tau} \tag{8.30}$$

It will be noticed that this expression involves a non-trivial dependence on the system length L_z, in both C_V and τ. The values of the thermal conductivity of c-Si obtained with this method are plotted versus the length L_Z of the supercell in Fig. 8.4. Although simulations were performed up to very long supercells ($L_Z = 1.2\ \mu\text{m}$, $S = 76\ \text{nm}^2$, i.e., 4.5 millions of atoms), the plateau is not yet reached. The MD values can then be extrapolated to infinite length

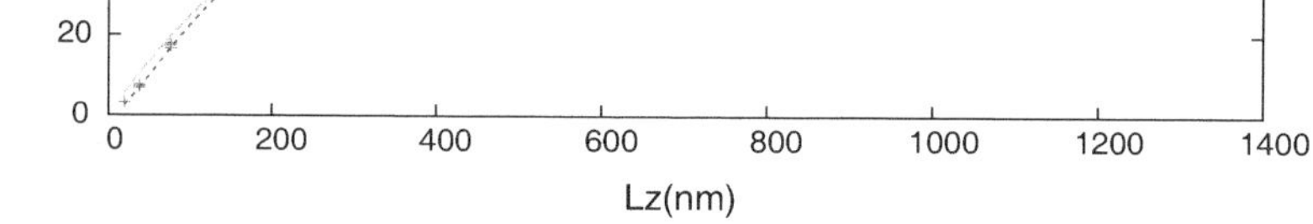

Figure 8.4 Thermal conductivity of c-Si versus the supercell length (red points). The lines are extrapolations of the first (green) and second order (blue).

according to a Taylor expansion:

$$\frac{1}{\kappa} = \frac{1}{\kappa_\infty}\left(1 + \frac{\Lambda_1}{L_Z} + \frac{1}{2}\left(\frac{\Lambda_2}{L_Z}\right)^2\right) \qquad (8.31)$$

The first-order term (green curve in Fig. 8.4) does not fit perfectly the thermal conductivities on the whole range of lengths. The thermal conductivities are too high at both short ($L_Z < 200$ nm) and long sizes ($L_Z = 1200$ nm). The agreement is better when we include the second-order term as shown in the blue curve. The thermal conductivity extrapolated to a system of infinite length κ_∞ is equal to 122 $\pm$ 2 $W.K^{-1}.m^{-1}$, in agreement with values obtained in the literature using the same interatomic potential (Abs da Cruz, 2011). The value overestimates the experimental thermal conductivity of silicon at 500 K, equal to 80 $W.K^{-1}.m^{-1}$ (Glassbrenner, 1964). This may be attributed to the purity of the crystal used in the model and to an imperfect description of the anharmonicity in the interatomic potential.

8.5 Thermal Conductivity of Crystalline and Amorphous Silica

The thermal conductivity of an amorphous material can be obtained using the same method. We present the results obtained on amorphous silica and a comparison with a crystalline phase of silica, α-quartz. The two systems were prepared for different applications, and the interatomic potential is not the same for the amorphous and crystal phase. Nevertheless, we will see that already interesting conclusions can be drawn from the comparison of the thermal conductivities.

The amorphous silica (a-SiO_2) supercell was obtained in a previous work, devoted to the study of the interface between c-Si and a-SiO_2 (Lampin, 2012). The choice of the interatomic potential was a major issue in that case, because an interatomic potential was required that could describe both structures, amorphous and crystal, and materials, Si and SiO_2, and mixed bonds at the interface. We have chosen to use an unique formulation, the Tersoff interatomic potential, using a parametrization of O–O and Si–O

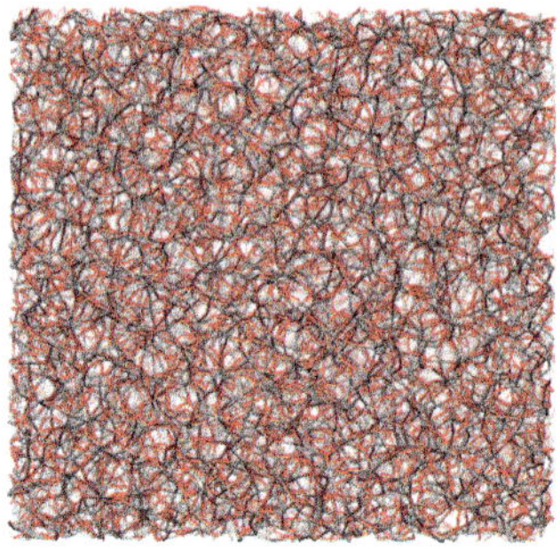

Figure 8.5 Amorphous silica.

interactions by Munetoh et al. (Munetoh, 2007). The amorphous supercell was obtained by a slow quench from melted silica. The section of the supercell is 4.4×4.4 nm^2 (Fig. 8.5). This is slightly compressed to match the c-Si cross section in the c-Si/a-SiO_2/c-Si multilayer. α-Quartz, on the other hand, was generated independently in order to test the AEMD method on a range of good to poor thermal conductors (Lampin, 2013). The interatomic potential by van Beest et al. (van Beest, 1990) was used in this case. The structure is trigonal (Fig. 8.6) and the section of the supercell is set to 4.5×7.7 nm^2.

A temperature transient is created and monitored to extract a characteristic decay time in both systems. The initial temperature

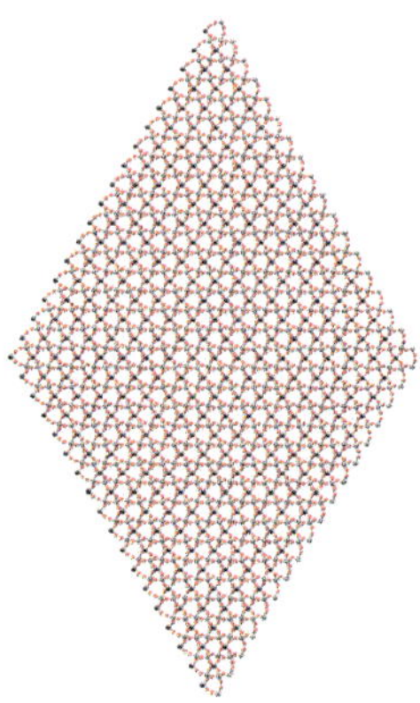

Figure 8.6 α-quartz.

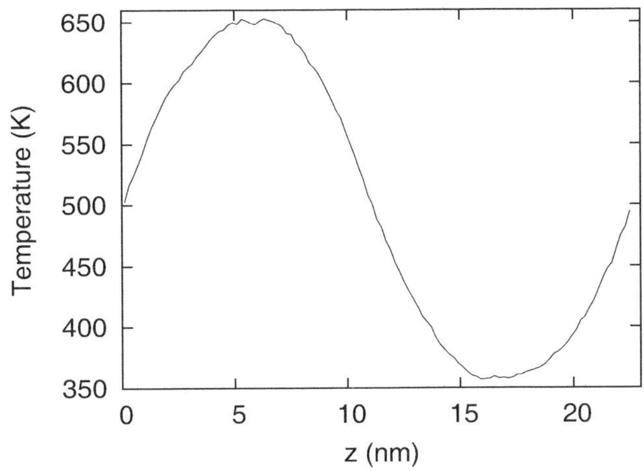

Figure 8.7 Temperature profile in amorphous silica. Thermostat at 600 K on the left half, and 400 K on the right half.

profile, however, was not be obtained from a simple cut and paste of the successive profiles at T_1 and T_2 as in the case of crystalline Si. This procedure is not adapted to the case of an amorphous, where atoms can be found very close to the boundary between the hot and cold blocks. Instead, we used two Nosé-Hoover thermostats (Nosé, 1984; Hoover, 1985) at $T_1 = 600$ K and $T_2 = 400$ K. The equilibrium temperature profile after 100 ps is shown in Fig. 8.7. It is composed of two parabolic curves that join at 500 K. The parabolic profile is the solution of the heat equation at steady state, when a constant internal heat source is added in Eq. 8.5. We have checked on c-Si that the step-like and parabolic profiles give the same decay time during the successive approach-to-equilibrium. Figure 8.8 shows the decay of the temperature difference between the "Hot" and "Cold" blocks of the supercell. The decay is exponential, and longer for a given size compared to c-Si, due to the much smaller thermal conductivity of a-SiO_2. One could think that the study of transient becomes less interesting in this case. On the other hand, the thermal conductivities obtained using Eq. (8.30) and plotted in Fig. 8.9 show a plateau already for L_z greater than 10 nm. The computational time therefore remains short. The value obtained in the plateau, ≈ 1.1 W.K^{-1}.m^{-1}, is in agreement with experiments (Cahill, 1987).

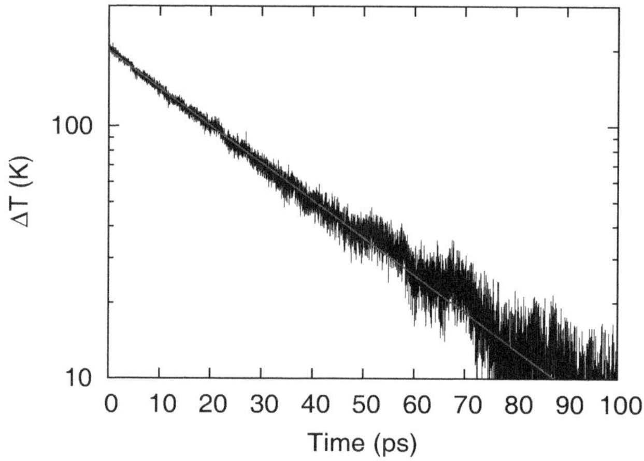

Figure 8.8 Temperature difference versus time in a-SiO_2 (black) and exponential fit with $\tau = 29$ ps (red). $L_z \approx 20$ nm.

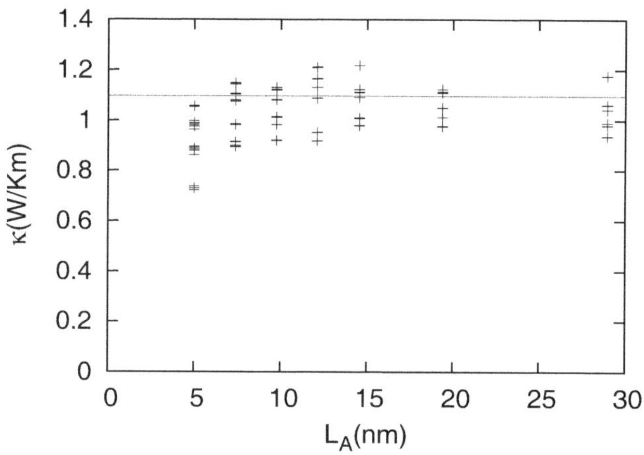

Figure 8.9 Thermal conductivity of amorphous silica (points) versus length and plateau at $\kappa_\infty = 1.10 \pm 0.01$ W.K^{-1}.m^{-1} (red line).

The difference in length dependence of the thermal conductivities of c-Si and a-SiO_2 is expected from the comparison of the respective thermal conductivities, using the kinetic theory of heat transfer by phonons: $\kappa = \frac{1}{3} C_V c M$, where c is the mean sound velocity and M an average mean free path. Although this is a simple

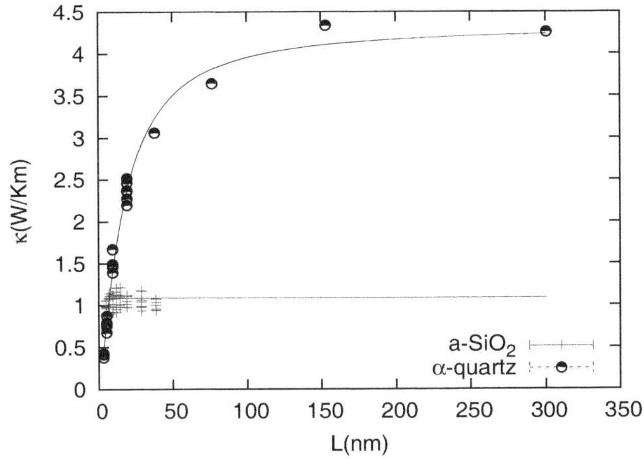

Figure 8.10 Thermal conductivity of a-SiO_2 (red crosses) and α-quartz (black dots) versus length. The bulk thermal conductivity of α-quartz is $\kappa_\infty = 4.32 \pm 0.07$ W.K^{-1}.m^{-1}.

expression, and we know that phonon mean free paths can take values ranging over many decades in some materials, it can help understand why the length dependency is strongly reduced in a-SiO_2. From Fig. 8.9, we can conclude that even for 10 nm-long supercells, all the mean free paths are taken into account in the periodic atomic structure, and the thermal conductivity has attained its bulk value.

Figure 8.10 compares the thermal conductivity of a-SiO_2 and α-quartz. In the crystal, the thermal conductivity exhibits a stronger length dependence. However, this dependence remains much less pronounced than in c-Si, and the curve saturates at $\kappa_\infty = 4.32 \pm 0.07$ W.K^{-1}.m^{-1}. As expected, the thermal conductivity is reduced in the amorphous, and the ratio between the thermal conductivities of the crystal versus amorphous is found to be around 4 in our simulations.

8.6 Conclusions

We presented in this chapter a novel computational approach, to obtain the thermal conductivity of real materials from molecular dynamics simulations of temperature transients. The approach-to-equilibrium is monitored, and a characteristic decay time is obtained. The temperature profile has a spatial and temporal evolution that closely follows the solution of the heat equation, as can be shown if the latter is written and solved under the same conditions of the molecular dynamics simulation, namely: 1D heat current, periodic boundary conditions and step-like initial temperature profile. Therefore we use the same relationship between the thermal conductivity and the temperature decay time.

This method has been applied to crystals, and a length dependence of the thermal conductivity has been evidenced, more pronounced in good conductors such as c-Si, compared to relatively poor conductors such as α-quartz. The computer simulation of temperature transients is much faster compared to the simulation of a steady-state regime, and large supercells containing up to 4.5 millions of atoms can be easily studied with this approach. The same approach can also be applied for amorphous materials, such as a-SiO_2. In this case, the decay time is typically longer for a given supercell length, but in the quasi absence of length dependence, this does not prevent to extract the conductivity of amorphous materials.

Acknowledgments

The authors acknowledge the support from French National Agency project NOODLES, equipex EXCELSIOR and the HPC resources of CINES and IDRIS.

References

Shenogin, S., Xue, L., Ozisik, R., Keblinski, P. and Cahill, D. G. (2004). Role of thermal boundary resistance on the heat flow in carbon-nanotube composites, *J. Appl. Phys.*, Vol. 95, pp. 8136–8144.

Hulse, R. J., Rowley, R. L. and Wilding, W. V. (2005). Transient nonequilibrium molecular dynamics simulations of thermal conductivity: 1. Simple fluids, *Int. J. Thermophys.*, Vol. 26, pp. 1–12.

Lampin, E., Palla P. L., Francioso, P.-A. and Cleri, F. (2013). Thermal conductivity from approach-to-equilibrium molecular dynamics, *J. Appl. Phys.*, Vol. 114, pp. 033525-1–5.

Tersoff, J. (1988). Empirical interatomic potential for silicon with improved elastic properties, *Phys. Rev. B* Vol. 38, pp. 9902–9905.

Abs da Cruz, C., Termentzidis, K., Chantrenne, P. and Kleber, X. (2011). Molecular dynamics simulations for the prediction of thermal conductivity of bulk silicon and silicon nanowires: Influence of interatomic potentials and boundary conditions, *J. Appl. Phys.*, Vol. 110, pp. 034309-1–8.

Glassbrenner, G. C. and Slack, G. A. (1964). Thermal conductivity of silicon and germanium from 3°K to the melting point, *Phys. Rev.*, Vol. 134, pp. A1058-A1069.

Lampin, E., Nguyen, Q.-H., Francioso, P.-A. and Cleri, F. (2012). Thermal boundary resistance at silicon-silica interfaces by molecular dynamics simulations, *Appl. Phys. Lett.*, Vol. 100, pp. 131906-1–4.

Munetoh, S., Motooka, T., Moriguchi, K. and Shintani, A. (2007). Force fields for silicas and aluminophosphates based on ab initio calculations, *Phys. Rev. Lett.* Vol. 64, pp. 1955–1958.

van Beest, B. W. H., Kramer, G. J., van Santen, R. A. (1990). Interatomic potential for Si-O using Tersoff parametrization, *Comput. Mater. Sci.* Vol. 39, pp. 334–339.

Nosé, S. (1984). A molecular dynamics method for simulations in the canobical ensemble, *Mol. Phys.* Vol. 52, pp. 255.

Hoover, W. G., (1985). Canonical dynamics: Equilibrium phase-space distributions, *Phys. Rev. B* Vol. 31, p. 1695.

Cahill, D. G. and Pohl, R. O. (1987). Thermal conductivity of amorphous solids above the plateau, *Phys. Rev. B* Vol. 35, pp. 4067–4073.

Chapter 9

Atomistic Simulations of Vibrational Modes of Crystalline Nanoinclusions in an Amorphous Matrix

Tanguy Damart, Yaroslav M. Beltukov, Amani Tlili, and Anne Tanguy

LAMCOS Laboratory UMR 5259, INSA-Lyon, CNRS, University of Lyon 1, Bâtiment Jean d'Alembert, 18-20 rue des Sciences F69621, Villeurbanne, France

tanguy.anne1@gmail.com

In this chapter, we present numerical methods that can be used to track the vibrational properties of atomistic systems. Among them, we detail the calculation and analysis of normal modes, also called resonant modes for the nanoparticles and the calculation of the vibrational density of states and the dynamical structure factors. These calculations are applied to the vibrational dynamics of isolated nanoparticles, of amorphous and crystalline matrix, and of nanoparticles embedded in a bulk matrix. Additionally, we compare the results previously obtained on the global system with those given by in situ propagation of a quasi-monochromatic wave packet. This comparison allows to discuss the proper definition of the acoustic mean free path in a composite system in presence of

Nanostructured Semiconductors: Amorphization and Thermal Properties
Edited by Konstantinos Termentzidis

ISBN 978-981-4745-64-2 (Hardcover), 978-1-315-36445-2 (eBook)
www.panstanford.com

disorder, as well as the effect of the coupling between the matrix and the embedded nanoparticles.

9.1 Introduction

Vibrational response of materials has important applications not only in terms of acoustic damping and acoustic resonance [DeSilva (2007)] but also in terms of thermal capacity [Kittel (2005)], thermal conductivity [Zeller (1971)] and structural analysis [Poilblanc (2006)]. It is thus important to understand what controls the vibrational response at different frequencies, and what can affect the acoustic propagation at small scales when an external signal is sent to the analyzed material.

We will briefly summarize here different techniques that can be used to infer numerically the vibrational properties of materials: calculation of normal modes and eigenfrequencies, calculation of the vibrational density of states (VDOS), calculation of the dynamical structure factor (DSF) for transverse as well as longitudinal waves, and propagation of a wave packet. The last two methods allow to get different estimations of the lifetime of phonons in the system, and of energy diffusivity. We will review the approximations made and the different data that these techniques may provide. We will then apply these techniques to the example of nanoinclusions embedded in a matrix.

The example of nanoinclusions in a matrix raises different interesting questions [Crut (2015); Saviot (2010)]. The presence of nanoinclusions with different crystallographic structures, compared with the matrix, may indeed induce an additional scattering of phonons. What happens when phonons are already strongly scattered before arriving on the inclusion? What is the role played by the interface? How does the presence of the inclusion affect the average mean free path and thermal conductivity of the composite system?

Numerically, the systems can be generated in more or less realistic way. In general, numerical simulations can give at most a qualitative description of the system and of the mechanisms studied. This is why it is important not only to focus on the numerical results but also to combine the atomistic simulation with a critical analysis allowing to build an intuition for a more general understanding of the elementary processes at play.

In this chapter, we will focus more specifically on the case of crystalline inclusions in an amorphous matrix. We will first review briefly different numerical techniques that can be used to infer the vibrational properties, and compare with experimental measurements. We will then focus first on the vibrations of isolated, crystalline and non-crystalline, clusters (Section 9.3). We will then focus on the role of disorder in the vibrational response of amorphous bulk samples (Section 9.4). In the last part, we will merge inclusions and matrix together, and discuss the role played by inclusions in the normal modes, as well as on the propagation of phonons. We will end with a general conclusion.

9.2 Atomistic Simulations of Vibrations

9.2.1 General Approximations in Molecular Dynamics Simulations

The goal of classical molecular dynamics is to solve Newton's equation of motion with an atomistic resolution. The equations of motion can be written on each atom i

$$m_i \frac{\partial^2 u_\alpha}{\partial t^2}(\mathbf{r}_i) - \eta_i \frac{\partial u_\alpha}{\partial t}(\mathbf{r}_i) = -\frac{\partial E_{total}}{\partial r_{i\alpha}} + f_\alpha(\mathbf{r}_i) \tag{9.1}$$

$$\approx -\sum_j M_{i\alpha,j\beta}.u_\beta(\mathbf{r}_j) + f_\alpha(\mathbf{r}_i), \tag{9.2}$$

where $\mathbf{u}_i$ is the Lagrangian displacement of particle i located at $\mathbf{r}_i$ and E_{total} is the potential energy describing interatomic interactions and $\mathbf{f}$ are external forces applied on each individual particle. Equation (9.2) is the harmonic approximation of the equation of motion where forces are linearized, with

$$M_{i\alpha,j\beta} = \frac{\partial^2 E_{total}}{\partial r_{i\alpha}\partial r_{j\beta}}$$

describing the harmonic interactions between atoms. The viscous damping in the left hand side of Eq. (9.1) accounts for dissipative processes, like thermalization. Depending on the coupling to the thermostat, for example, an additional equation can be written on η_i [Allen (1991); Frenkel (2002)]. It is clear from Eqs. (9.1) and (9.2) that three major approximations are made when dealing with molecular dynamics simulations.

First, considering the large number of equations ($3 \times N$ for N atoms) the system size is limited. One million particles with an interparticle distance of the order of 1 Å corresponds to a box size of 10 nm side only. This small size is not limiting when looking at amorphous samples where characteristic lengthscales are in general below 5 nm [Tanguy (2015a)], but it can be insufficient when long range stresses are involved, as for crack propagation or dislocations in crystals. Moreover, this strong limitation gives an upper limit to the size and the distance between the inclusions used in composite samples. Using periodic boundary conditions, for example, [Frenkel (2002)], the system size fixes the distance between the inclusions that are positioned in a regular array due to the repetition of the primitive cell.

Second, the interatomic interactions are described using an empirical description of the energy E_{total}. The empirical interactions are often parametrized in order to minimize the discrepancy with quantum calculations for different parameters like structural parameters [Stillinger (1985)] or vibration frequencies of well-defined cluster modes together with optimization of elastic constants [van Beest (1990); Carre (2007)]. Note, however, that for silica, for example, there exist at least five different empirical descriptions of the Si–Si, Si–O, and O–O interactions [To (2007); Tangney (2002)], but none is able at the moment to reproduce the density of vibrational states on the complete frequency range, as measured experimentally [Buchenau (2007); Price (1987)]. The major problem when using an empirical parametrization of the interactions is the definition of the time scale. The time scale $[t_s]$ is given by

$$[t_s] = \sqrt{\frac{[m].[a]^2}{[\Delta E]}}, \tag{9.3}$$

where $[m]$ is the unit for masses, $[a]$ is the length-scale and $[\Delta E]$ is the typical energy barrier of individual particles. $t_s = 1$ corresponds to the inertial time needed for one particle to reach the interparticle energy barrier. The energy barrier is fixed by the empirical potential. However, it is also fixed by the positions of all the other particles, the global energy barriers height being far smaller than the energy barrier for interparticle interactions. A reasonable choice for an empirical potential can be motivated by the ability to give realistic structures within a small amount of computational time. A time scale $[t_s] = 10^{-13}\ s = 0.1$ ps, often related to interatomic interactions from Eq. (9.3) allows to explore a temporal window of only 1 ns within 10^6 steps using a time step $\delta t = 0.01 \times [t_s] = 1$ fs. Starting the simulation in the liquid state at 1000 K, a quench up to 300 K in less than 10^6 steps thus requires an effective quenching rate larger than 7×10^{11}°K.s^{-1}, far larger than any experimental quenching rate. However, such a quenching rate gives rise to structures that compare quite well with the experiments [van Beest (1990)], and it is not clear whether a slower quenching rate would enhance the precision [Vollmayr (1996)]. The frequency range that can be covered with dynamical simulations in such an amount of time is limited by $1/\delta t \approx 10^{15}$ s^{-1} for the upper value, and $1/\Delta t \approx 10^{9}$ s^{-1} for the lower part. These results are indeed strongly dependent from the definition of the unit time and from the choice of the empirical potential.

Third, the dissipative processes appearing in the equation of motion, for example, through a viscous damping term can have different physical origins, such as phononic excitation [Sevincli (2007)] or electron transfer [Rittmeyer (2015)]. Usually, a spatially homogeneous damping is used for simplicity [Lemaitre (2006)]. Complex electron/phonon/photon processes can, however, take place [Sun (2011)], that could contribue to inhomogeneous damping as a function of local strain [Niehaus (2005)]. Quasi-static athermal simulations are a well-defined limit obtained by looking at the succession of nearest equilibrium positions, with the help of energy minimization algorithms [Rodney (2011)]. In this limit, the dissipation time τ_{diss} is supposed to be smaller than the typical time between successive strain increments, that is

$$\dot{\gamma} << \delta\gamma/\tau_{diss},$$

where $\delta\gamma$ is the elementary strain increment and $\dot{\gamma}$ the strain rate. In the same time, spontaneous structural relaxations are frozen due to the successive fast quenches (energy minimizations), and temperature cannot play a role. The elementary strain step $\delta\gamma$ must be chosen sufficiently small to ensure that the system spends some time in its initial basin when starting from any equilibrium configuration [Leonforte (2005)]. In this limit, apparent diffusion of the particles is only due to geometrical effects and mechanical instabilities [Tsamados (2010); Tanguy (2006)]. A realistic description of dissipative processes is probably very important to describe anharmonic processes [Vacher (2005)]. In this chapter we will focus only on the harmonic contribution to vibrations and acoustic propagation.

9.2.2 Resonant Modes

The resonant modes or eigenmodes of the system are obtained by solving Eq. (9.2) without the external, nor the damping force. Defining the elements of the dynamical matrix as

$$D_{i\alpha,j\beta} = M_{i\alpha,j\beta}/\sqrt{m_i.m_j}$$

the eigenmodes of the system can be obtained by the exact diagonalization of $\mathbf{D}$. Indeed, with $u_\alpha(\mathbf{r}_i, t) = v_{n,i\alpha}.\exp(i\omega_n t)$ (ω_n is an eigenfrequency), Eq. (9.2) can be rewritten

$$\omega_n^2\sqrt{m_i}v_{n,i\alpha} = \sum_{j,\beta} D_{i\alpha,j\beta}\sqrt{m_j}v_{n,j\beta}$$

and ω_n^2 are the eigenvalues of $\mathbf{D}$, while the eigenmodes $\mathbf{v}_n$ are obtained simply from the eigenvectors. Different numerical methods can be used to diagonalize the dynamical matrix. One recent method uses the MKL library based on the FEAST method for the eigenvalues [Polizzi (2009)], and using sparse solvers for the eigenvectors [Schenk (2000)]. The main difficulty in getting the eigenmodes and eigenfrequencies is the high numerical resolution needed in evaluating the dynamical matrix elements. Indeed, due to translational invariance, this matrix is poorly conditioned. Another method often used to characterize the vibrational properties of a solid material is the determination of the dynamical structure factor.

9.2.3 Dynamical Structure Factor

The dynamical structure factor (DSF) is a dynamical generalization of the static structure factor. The static structure factor can be measured from neutron, or X-ray, scattering experiments. It is defined as

$$S(\mathbf{q}) = \left\langle \left| \sum_{i=1}^{N} \exp\left(i\mathbf{q}.\mathbf{r}_i\right) \right|^2 \right\rangle = \left\langle \sum_{i=1}^{N} \sum_{j=1}^{N} \exp\left(i\mathbf{q}.\left(\mathbf{r}_i - \mathbf{r}_j\right)\right) \right\rangle$$
$$= \left\langle \rho_{\mathbf{q}}.\rho_{-\mathbf{q}} \right\rangle .$$

The DSF for longitudinal waves is defined as

$$S_L(\mathbf{q}, \omega) = \int \exp(-i\omega t) \left\langle \left(\rho(\mathbf{q}, t) - \rho_{\mathbf{q}}\right) . \left(\rho(-\mathbf{q}, t) - \rho_{-\mathbf{q}}\right) \right\rangle dt$$

with

$$\rho(\mathbf{q}, t) = \int \rho(\mathbf{r}, t) \exp(i\mathbf{q}.\mathbf{r}) d\mathbf{r},$$

where $\rho(\mathbf{r}, t)$ is the coarse-grained mass density at the position $\mathbf{r}$. It can be rewritten as a function of the equilibrium position $\mathbf{r}_i^{eq}$ and of the Eulerian displacement $\mathbf{u}_i(t)$ of the atoms i as

$$\rho(\mathbf{q}, t) = \sum_{i=1}^{N} \exp\left(i\mathbf{q}.\left(\mathbf{r}_i^{eq} + \mathbf{u}_i(t)\right)\right) \approx \sum_{i=1}^{N} \exp\left(i\mathbf{q}.\mathbf{r}_i^{eq}\right) . \left(1 + i\mathbf{q}.\mathbf{u}_i(t)\right)$$
$$\approx \rho_{\mathbf{q}} + i\mathbf{q}. \sum_i \mathbf{u}_i(t) \exp\left(i\mathbf{q}.\mathbf{r}_i^{eq}\right) . \tag{9.4}$$

Thus

$$S_L(\mathbf{q}, \omega) = \int \exp(-i\omega t) \sum_{ij} (\mathbf{q}.\mathbf{u}_i)(\mathbf{q}.\mathbf{u}_j) \exp\left(i\mathbf{q}.\left(\mathbf{r}_i^{eq} - \mathbf{r}_j^{eq}\right)\right) dt$$
$$\approx \frac{k_B T}{m\omega^2} q^2 \sum_{n=1}^{3N} \left| \sum_i \frac{\mathbf{q}}{q} . \mathbf{v}_n \exp(i\mathbf{q}.\mathbf{r}_i^{eq}) \right|^2 \delta(\omega - \omega_n) \tag{9.5}$$

after decomposition on the $3N$ eigenmodes $\mathbf{v}_n$, and assuming the classical equipartition of energy. For transverse waves, the DSF can be written analogously:

$$S_T(\mathbf{q}, \omega) = \frac{k_B T}{m\omega^2} q^2 \sum_{n=1}^{3N} \left| \sum_i \frac{\mathbf{q}}{q} \times \mathbf{v}_n \exp(i\mathbf{q}.\mathbf{r}_i^{eq}) \right|^2 \delta(\omega - \omega_n)$$

The DSF can be computed either from a dynamical run at constant energy and recording the instantaneous positions $\mathbf{u}_i(t)$ of each atom i or from the eigenmodes $\mathbf{v}_n$ of the initial equilibrium configuration in the harmonic approximation. The DSF is a description of the vibrations in terms of phonons with a well-defined wave vector $\mathbf{q}$. Its maxima define the dispersion relation $\omega_n(\mathbf{q})$, and different estimations for its width can be used to infer an approximate mean free path, or inverse lifetime of the phonons. Indeed, in amorphous materials, the eigenmodes are not phonons, because they are not characterized by a well-defined wave vector but have a more complex shape that will be discussed later in different cases [Buchenau (2014)].

9.2.4 Wave Packet Propagation

Another quantity that can be computed from molecular dynamics simulations is the mean free path, which is the attenuation length of a wave packet inside the sample [Schelling (2002)]. Indeed, in amorphous samples, phonons are not eigenmodes. The coupling between phonons is measured from the damping of wave packets. Wave packets are defined as localized excitations due to an imposed displacement of the form

$$u_\alpha^{WP}(\mathbf{r}_i, t) = A\sin(\omega t + \phi_{i,\alpha})\exp\left(-\frac{x_i^2}{2\Delta x^2} - \frac{t^2}{2\Delta t^2}\right)$$

applied to one samples side, perpendicular to the x-direction. Δx and Δt are the spatial and temporal extents of the excitation. The excitation is transmitted to the other side by solving Newton's equations of motion, without any additional external damping force. From the propagation of a wave packet, it is simple to identify a *mean free path*, and the *diffusivity*. The mean free path is obtained from the attenuation of the maximum of the wave packets amplitude [Damart (2015); Akkermans (2004)], and the diffusivity is obtained from the spatial spreading of the wave packet during its motion [Beltukov (2013, 2015)]. The spreading of the wave packet is due to its scattering along the sample, because it is not an eigenmode.

We will now discuss first what the eigenmodes of isolated (crystalline and non-crystalline) inclusions are. Then we will separately study the vibrational properties of amorphous bulk materials. Finally, we will show numerical results concerning the vibrational properties of amorphous matrices with inclusions inside.

9.3 Eigenmodes of Isolated Inclusions

9.3.1 Continum Mechanics

The continuum theory of elasticity for the vibrations of isotropic solid materials gives the well-known Navier's equation:

$$\rho\frac{\partial^2 \mathbf{u}}{\partial t^2}(\mathbf{r},t) = (\lambda+2\mu)\nabla\nabla.\mathbf{u} - \mu\nabla\times\nabla\times\mathbf{u}, \tag{9.6}$$

where λ and μ are the Lamé coefficients, and ρ is the mass density. The resolution of the Navier's equation for spheroidal objects [Eringen (1975)] involves the decomposition:

$$\mathbf{u} = \nabla\Phi + \nabla\times(\Psi\mathbf{r}) + \nabla\times\nabla\times(\chi\mathbf{r})$$

with Φ, Ψ, and χ satisfying the waves equation $\ddot{f} = v^2\nabla^2 f$, where v is the longitudinal waves velocity $v_L = \sqrt{(\lambda+2\mu)/\rho}$ for Φ, and it is the transverse waves velocity $v_T = \sqrt{\mu/\rho}$ for Ψ and χ. In spherical coordinates (r, θ, ϕ) with the condition of free stresses at the surface of the sphere with radius R, the displacements can be regrouped into *torsional modes* and *spheroidal modes*. The *torsional modes* are given by $\Psi = 0$. They have the form [Eringen (1975)]

$$\begin{aligned} u_r &= 0 \\ u_\theta &= B_{nm}\frac{m}{r\sin\theta}\sqrt{\frac{\pi}{2qr}}J_{n+1/2}(qr)P_n^m(\cos\theta)f(m\phi)\exp(-i\omega t) \\ u_\phi &= B_{nm}\frac{1}{r}\sqrt{\frac{\pi}{2qr}}J_{n+1/2}(qr)\frac{d}{d\theta}P_n^m(\cos\theta)g(m\phi)\exp(-i\omega t), \end{aligned} \tag{9.7}$$

with m integer, $J_{n+1/2}$ the Bessel function, P_n^m the associated Legendre function, $f(m\phi) = \{\sin m\phi, -\cos m\phi\}$ and $g(m\phi) = \{\cos m\phi, \sin m\phi\}$, and $q = \omega/v_T$. The eigenfrequency ω is obtained as root of the compatibility equation:

$$(n-1)J_{n+1/2}(\omega R/v_T) - \omega R/v_T J_{n+3/2}(\omega R/v_T) = 0.$$

The *spheroidal modes* do not preserve the volume of the sphere. They have a more complex expression of the form

$$\begin{aligned}
u_r &= A_{nm} U_n(r, v_L, v_T, \omega) P_n^m(\cos\theta) h(m\phi) \exp(-i\omega t) \\
u_\theta &= A_{nm} V_n(r, v_L, v_T, \omega) \frac{d}{d\theta} P_n^m(\cos\theta) g(m\phi) \exp(-i\omega t) \\
u_\phi &= -m A_{nm} V_n(r, v_L, v_T, \omega) \frac{P_n^m(\cos\theta)}{\sin\theta} f(m\phi) \exp(-i\omega t), \qquad (9.8)
\end{aligned}$$

where $U_n(r, v_L, v_T, \omega)$ and $V_n(r, v_L, v_T, \omega)$ are functions of the distance r to the center, as well as of the frequency and sound velocities [Eringen (1975)], $h(m\phi) = \{\sin m\phi, \cos m\phi\}$, and the eigenfrequency ω is obtained as root of the compatibility equations:

$$J_{n+1/2}(\omega R/v_T) = 0 \qquad (9.9)$$

or

$$\begin{aligned}
&\frac{2v_T}{v_L}\left[\frac{v_T}{\omega R} + \frac{(n-1)(n+2)v_T^2}{\omega^2 R^2}\right. \\
&\times \left.\left(\frac{J_{n+3/2}(\omega R/v_T)}{J_{n+1/2}(\omega R/v_T)} - \frac{(n+1)v_T}{\omega R}\right)\right] J_{n+3/2}(\omega R/v_L) \\
&+ \left[-\frac{1}{2} + \frac{(n-1)(2n+1)v_T^2}{\omega^2 R^2} + \frac{v_T}{\omega R}\left(1 - \frac{2n(n-1)(n+2)v_T^2}{\omega^2 R^2}\right)\right. \\
&\times \left.\frac{J_{n+3/2}(\omega R/v_T)}{J_{n+1/2}(\omega R/v_T)}\right] J_{n+1/2}(\omega R/v_L) = 0
\end{aligned}$$

Examples of torsional and of spheroidal modes are shown in Fig. 9.1 together with the stiffness dependence of the eigenfrequencies. Depending on the stiffness of the inclusion, it will thus be possible to tune its resonant frequencies.

9.3.2 Amorphous Inclusions

The numerical resolution can be done on atomistic systems by the exact diagonalization of the dynamical matrix, as discussed

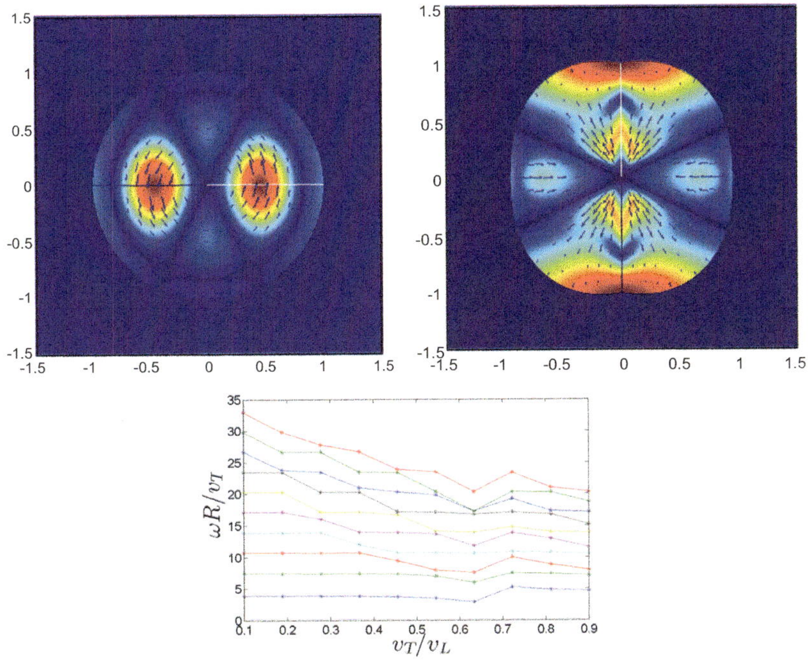

Figure 9.1 (Top) Vizualization of a torsional mode ($n = 3$, $m = 2$, second frequency) in the $\phi = 0$-plane for $v_T = 0.5v_L$. Colors indicate the amplitude of displacement. Arrows indicate the direction of displacement in the plane. The spatial map is done in the Eulerian coordinates. (Middle) Idem for a spheroidal mode ($n = 3$, $m = 2$, second frequency). (Bottom) Ten first eigenfrequencies of the spheroidal mode (with $n = 1$) as a function of the stiffness ratio v_T/v_L. The frequencies of the torsional modes are independent of v_L.

before [Wittmer (2002)]. An inclusion made with N atoms has $3N$ eigenmodes that can be described by the displacement vectors supported by each atom. For sufficiently large wavelength, the resulting displacements and the related eigenfrequencies are very close to those obtained by the continuum calculation [Wittmer (2002)], when the nanoparticle is amorphous. The minimum wavelength for the convergence to the continuum mechanics result is of the order of 30 interatomic distances. Below that wavelength, degeneracies of the modes are split due to local anisotropy due to disorder [Tanguy (2002)].

9.3.3 Crystalline Inclusions

In case of crystalline inclusions, elastic constants have the symmetry of the crystalline lattice. Equation (9.6) must be rewritten as follows:

$$\rho \frac{\partial^2 \mathbf{u}}{\partial t^2} = \nabla . \left(\mathbf{C}(x, y, z) : (\nabla \mathbf{u} + \nabla \mathbf{u}^T) \right),$$

with $\mathbf{C}(x, y, z)$ the elastic stiffness tensor with 21 independent components in the general case. In case of isotropic medium the number of independent components of $\mathbf{C}$ is only two, but it is three in case of a cubic symmetry, yielding to a refinement of the nanoparticle's vibrations as compared to the isotropic case. The crystalline case was extensively studied by Saviot et al. [Saviot (2010); Portales (2008); Saviot (1996); Hu (2004)]. It shows a splitting of the spheroidal modes, and a shift of the eigenfrequencies as a function of the degree of anisotropy inside the elastic stiffness tensor. However, the non-degenerated radial breathing modes of the spheres are only weakly altered with less than 2% modification in the mode frequency [Crut (2015)]. This is a consequence of directional averaging of the elastic properties in symmetric oscillations.

Different experiments can be used to probe the vibrational frequencies of nanoparticles embedded in a matrix, such as low-frequency resonant Raman spectroscopy [Saviot (2010)], or time-resolved optical pump-probe experiments [Crut (2015)]. Note that the vibrations of the inclusions depend on the coupling to the environment, and especially on the acoustic mismatch, or relative impedance of the inclusion to the medium, as detailed in [Crut (2015)].

We will now study the vibrational properties of the amorphous matrix embedding the nanoinclusions.

9.4 Vibrational Response of the Amorphous Matrix

The vibrational response of the amorphous matrix is studied with the different methods explained in Section 9.2. We will summarize the main results obtained in amorphous bulk materials.

Illustrations are taken from simulations of amorphous silicon-like samples [Damart (2015); Fusco (2010)]. More details can be found in ref. [Beltukov (2015)].

9.4.1 Boson Peak

In amorphous materials, the density of vibrational states is excessive in comparison with the density of vibrational states obtained from simple acoustic propagation. This excess in the low-frequency range of the spectrum is called the *Boson peak* and was widely studied in the last years [Wyart (2005); Marruzzo (2013); Baldi (2011); Gurevich (2003); Chumakov (2014)]. It is responsible for the large heat capacity of glasses and anomalous temperature dependence, in comparison with crystals [Zeller (1971)]. The vibrational density of states (VDOS) as a function of ω reads

$$g_{DOS}(\omega) = \frac{1}{3N}\sum_{i=1}^{3N}\delta(\omega - \omega_i) \tag{9.10}$$

From a numerical point of view, the VDOS can be obtained either from the exact diagonalization of the dynamical matrix, either from dynamical estimations of the density of states without exact calculation of the eigenfrequencies and eigenmodes. Recent algorithm such as the FEAST algorithms [Polizzi (2009)] can be used to get the full set of eigenfrequencies together with the eigenvectors of the dynamical matrix $\mathbf{D}$. The VDOS is then computed by the relative number of eigenmodes whose frequency belongs to $\omega \pm \Delta\omega$. For $N > 10^4$, however, this direct method requires too much time as well as random access memory. Indirect methods can be used such as the kernel polynomial method (KPM), and the velocity auto-correlation method.

The KPM method is detailed in [Beltukov (2015)]. Its principle is to give an estimation of Eq. (9.10) by replacing the δ functions by a series of second kind Chebyshev polynomials.

$$\delta(\epsilon - \epsilon_j) = \frac{2}{\pi}\sqrt{1-\epsilon^2}\sum_{k=0}^{\infty} U_k(\epsilon)U_k(\epsilon_j)$$

The VDOS can then be expressed [Beltukov (2015)] as

$$g_{DOS}(\omega) = \frac{8\omega}{\pi\omega_{max}^2}\sum_{k=0}^{\infty}\mu_k \sin\left(2(k+1)\arcsin(\omega/\omega_{max})\right),$$

where μ_k is the k-th Chebyshev moment

$$\mu_k = \frac{1}{N}\sum_{i=1}^{N} U_k(\epsilon_i) \approx \overline{\langle u_0|U_k(\mathbf{I} - 2\mathbf{D}/\omega_{max}^2)|u_0\rangle}$$
$$= \overline{\langle u_0|u_k\rangle} \quad (9.11)$$

Here μ_k is approximated by averaging over several realizations of a Gaussian random vector $|u_0\rangle$ with unit norm. The exact diagonalization is then replaced by a simple recursive summation. This method can be compared to the velocity auto-correlation function, to which it resembles.

The velocity auto-correlation function can be used to get the VDOS as well, from the numerical resolution of the equations of motion at constant energy [Dove (2005)]. Indeed, the velocity $\dot{\mathbf{u}}(i, t)$ is given as

$$\dot{\mathbf{u}}(i, t) = \frac{-i}{\sqrt{Nm_i}}\sum_n \omega_n A_n \mathbf{v}_n \exp(i\omega_n t).$$

Its correlation function reads

$$\sum_i m_i \langle|\dot{\mathbf{u}}(i, t).\dot{\mathbf{u}}(i, 0)|\rangle = \frac{1}{N}\sum_n \omega_n^2 |An|^2 \exp(i\omega_n t)$$
$$\approx \frac{k_B T}{N}\sum_n \exp(i\omega_n t)$$
$$= \frac{k_B T}{N}\int d\omega 3N g_{DOS}(\omega)\exp(i\omega_n t) \quad (9.12)$$

after using classical equipartition of energy. The VDOS is thus given by the inverse Fourier transform of the velocity auto-correlation function computed at low, but finite, temperature in the classical harmonic approximation. The VDOS of both amorphous and crystalline samples is shown in Fig. 9.2a,b. The well-defined peaks of the crystalline sample are broadened in the amorphous one. The corresponding Boson peak is shown in Fig. 9.2c. The physical origin of the Boson peak is still discussed [Beltukov (2013); Baldi (2011); Chumakov (2014)]. We have shown that its intensity is increased for more flexible bounds [Beltukov (2015)] and that it was related to transverse heterogeneous motion of atoms [Beltukov (2013, 2015)]. We will now present more information on the specificities of eigenmodes in amorphous samples.

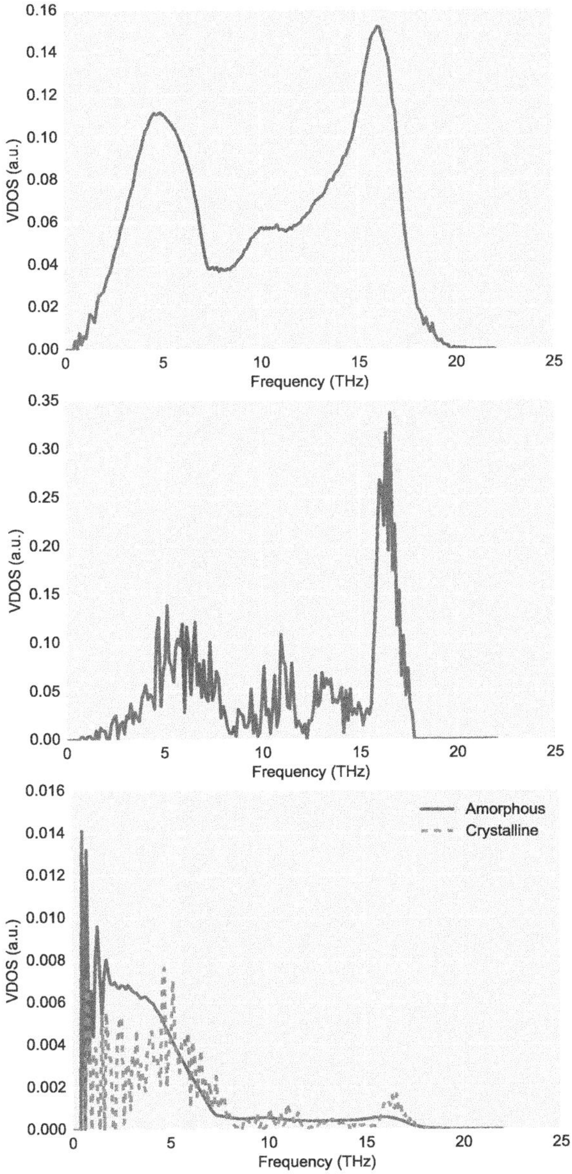

Figure 9.2 (a) VDOS of amorphous a-Si sample computed with SW potential [Stillinger (1985)] and exact diagonalization of the matrix (b) VDOS of CFC crystal of Si-like sample described with SW potential (c) VDOS divided by the Debye density of states showing the Boson peak for amorphous Si-like samples, compared to CFC Si-like samples.

9.4.2 Extended and Localized Eigenmodes

The eigenmodes are obtained from the exact diagonalization of the dynamical matrix. Contrary to plane waves that are extended modes with a well-defined wave vector $\mathbf{q}$, the extent of eigenmodes is in general limited. It can be characterized by the participation ratios of the eigenmodes. The generalized participation ratio of mode n is defined for different values of k as

$$P_k(\omega_n) = \frac{1}{N} \frac{\left(\sum_i \mathbf{v}_{n,i}^2\right)^k}{\sum_i \mathbf{v}_{n,i}^{2k}} = \frac{1}{N \sum_i \mathbf{v}_{n,i}^{2k}}.$$

It is the ratio between the norm of $\mathbf{v}_n$ and its k-th moment. For $k = 2$, P_2 quantifies the relative number of atoms moving with the same amplitude. For example, if only one atom is moving, then $P_2 = 1/N$, while $P_2 = 1$ for translational motion. Varying k allows to look for possible multifractal behavior of the mode $\mathbf{v}_n$. For that, it is necessary to introduce a scale dependence in P_k. This can be done by replacing the displacements $\mathbf{v}_{n,i}$, supported by each atom i, by their coarse-grained description [Goldenberg (2007)] at scale w:

$$\mathbf{v}_{CG,n}(\mathbf{r}, w) = \sum_i \frac{\mathbf{v}_{n,i}}{\left(2\pi w^2\right)^{3/2}} \exp\left(-(\mathbf{r} - \mathbf{r}_i)^2/2w^2\right)$$

The participation ratio with $k = 2$ is shown in Fig. 9.3. In the left-hand side of the Boson peak, there is an accumulation of eigenmodes with very low participation ratios, also called *soft modes*. These modes are precursors of plastic instabilities. At the center of the Boson peak, the participation ratio reaches a local maximum as a function of the frequency $P_2 \approx 0.5$ that correspond to the value obtained from Gaussian random noise with two degrees of freedom [Beltukov (2013)]. It decays to a secondary local minimum $P_2 \approx 1/3$, when only longitudinal waves are still present (only one degree of freedom); then it raises again and falls definitely down at the transition to optic-like modes. At this transition, modes have a multifractal shape characterized by

$$P_k(\omega_n, w) \propto w^{\tau_n(k)}$$

corresponding to Anderson's localization. These modes are also sometimes called *fractons* [Aharony (1987); Allen (1999); Faez (2009)].

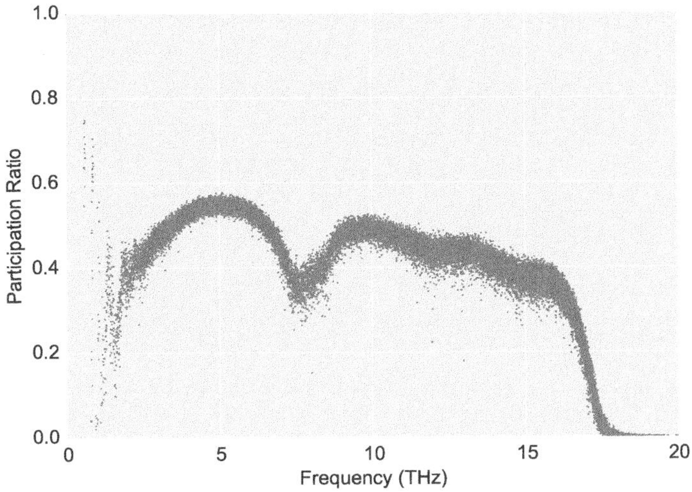

Figure 9.3 Participation ratio with $k = 2$ for the amorphous Si-like bulk sample, as a function of ω_n.

This analysis of the eigenmodes can be completed by an analysis of the spatial correlation functions

$$C_n(\mathbf{r}, w) = \frac{\langle \mathbf{v}_{CG,n}(\mathbf{r}+\mathbf{r}', w).\mathbf{v}_{CG,n}(\mathbf{r}', w)\rangle_{\mathbf{r}'}}{\langle \mathbf{v}_{CG,n}(\mathbf{r}', w).\mathbf{v}_{CG,n}(\mathbf{r}', w)\rangle_{\mathbf{r}'}}.$$

For plane waves, $C_n(r, \omega) \propto \sin(q_n r)/(q_n r)$ with a first minimum located at $r^* \approx 0.449/q_n$. The correlation function of the *soft modes* is very similar to that of plane waves. The presence of the soft spots thus not affects the long-range correlation function. However, the right-hand side of the Boson peak corresponds to a sudden change in the correlation function and to departure from plane waves (Fig. 9.4). The departure from plane waves can be studied in more details by looking now at the propagation of a wave packet.

9.4.3 Mean Free Paths and Diffusivity

The propagation of a wave packet simulates the propagation of phonons [Cohen (1977)]. It allows a direct measurement of the mean free path $L_{p,\,dir}(\omega)$, and of the diffusivity $D_{dir}(\omega)$ as detailed in Section 9.2.4. We have done the measurement of the position and of the amplitude of the maximum in the kinetic energy due to

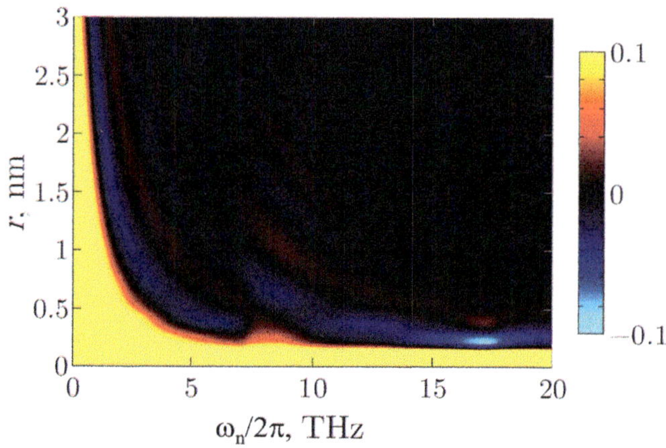

Figure 9.4 Spatial correlation function of the eigenmodes in the amorphous Si-like bulk sample, as a function of ω_n.

the wave packet, at different times. It is drawn as a function of its position in Fig. 9.5. It shows a transition between an exponential attenuation at low frequencies ω, to a power-law attenuation at higher frequencies. The exponential attenuation is characteristic of weak-scattering, while the power-law behavior is characteristic of strong-scattering [Akkermans (2004)].

The mean free path is also often obtained from the fit of the dynamical structure factor at a given frequency ω. The most often used fit is the so-called DHO fit inspired from a generalized Langevin equation written on the density correlators as done, for example, in mode coupling theory [Ruocco (2000)]. It is written as

$$S_{DHO}(q,\omega) = \frac{A}{\left(\omega^2 - \omega(q)^2\right)^2 + \omega^2\Gamma^2},$$

where $\omega(q)$ is the dispersion relation and Γ is the *inverse lifetime*. The mean free path is deduced from S_{DHO}, when it is possible, as

$$\text{Ł}_{p,DHO} = v/\Gamma,$$

where v is the group velocity of a given type of waves (L or T) at a given wave vector q. The two estimations $L_{p,\,dir}$ and $L_{p,DHO}$ of the *mean free path* have been measured in our amorphous matrix. They are shown in Fig. 9.6. As can be seen in this figure, the DHO fit is possible only at sufficiently low frequency. At higher frequencies,

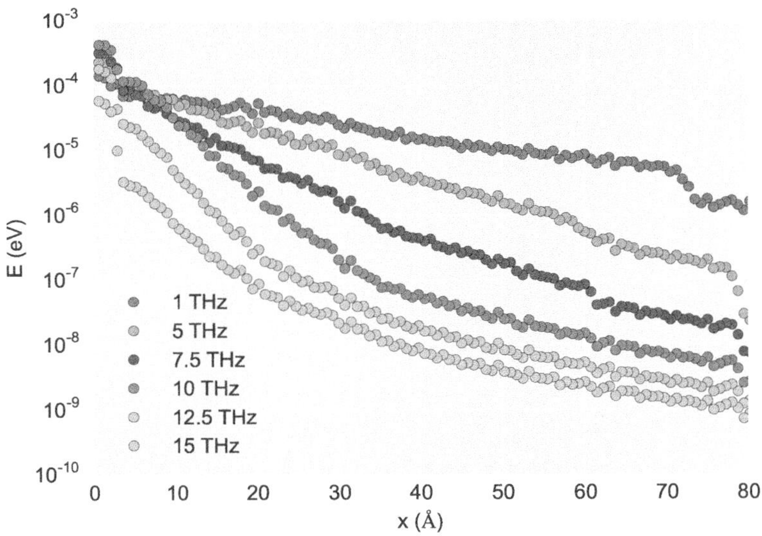

Figure 9.5 Position and amplitude of the maximum in the envelope of the kinetic energy, obtained for different times after the excitation by a wave packet with different frequencies $\omega = \{1$ THz, 5 THz, 7.5 THz, 10 THz, 12.5 THz and 15 THz$\}$ in a Si-like amorphous sample.

the sound velocity v is ill-defined and L_p is sometimes related to the diffusivity D as $L_p l = \sqrt{D/\Gamma}$ as done, for example, in [Larkin (2014)]. Quite remarkably, the two estimations $L_{p,\,dir}$ and $L_{p,\,DHO}$ differ significantly, $Ll_{p,\,dir}$ being one order of magnitude smaller than $L_{p,\,DHO}$. The DHO fit thus overestimates strongly the attenuation length of acoustic vibrations.

In the two cases, the mean free path decreases with the frequency. The Ioffe–Regel transition [Ioffe (1960)] is defined as $L_p \approx 2\pi/q$. In a-Si, this limit is reached first for transverse waves at $2\pi/q \approx 9$ Å, and then for longitudinal waves at a smaller wavelength [Beltukov (2015)]. For SiO_2, the Ioffe–Regel criterion is reached for wavelengths larger than 100 nm, that is the largest simulation box size used in actual MD simulations [Larkin (2014)]. The origin of the mean free path is due to the fact that the wave packet is not an eigenmode of the sample. Its motion involves several eigenmodes with different dynamics and its energy is not preserved because it is spread over the different eigenmodes.

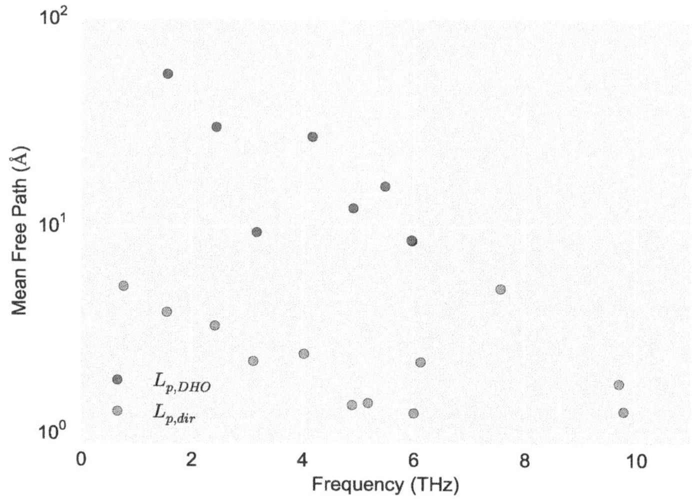

Figure 9.6 Mean free paths $L_{p,\,dir}(\omega)$ and $L_{p,\,DHO}(\omega)$ obtained from the propagation of a wave packet, and from the DHO fit of longitudinal waves, respectively.

9.5 Vibrations of the Amorphous Matrix with the Inclusions

We will now consider a 3D amorphous sample of size $L = 60\,\text{Å}$ containing spherical inclusions with radius R between 10 to 30 Å. In order to mimic waves propagation across successive inclusions, a sample with five inclusions, aligned along one direction, was used to study the propagation of a wave packet. In this part, we will discuss the vibrational changes due to the presence of crystalline inclusion(s).

9.5.1 Eigenmodes of the Matrix with the Inclusions

The eigenmodes of the amorphous matrix with the inclusions are shown in Fig. 9.7. In some of them, the vibrations of the matrix and of the inclusions are decoupled (mode 3, for example), but in many others there is no apparent separation between the inclusion and the matrix. In any cases, the vibrations of the inclusions are very different from the vibrations of isolated inclusions. The participation

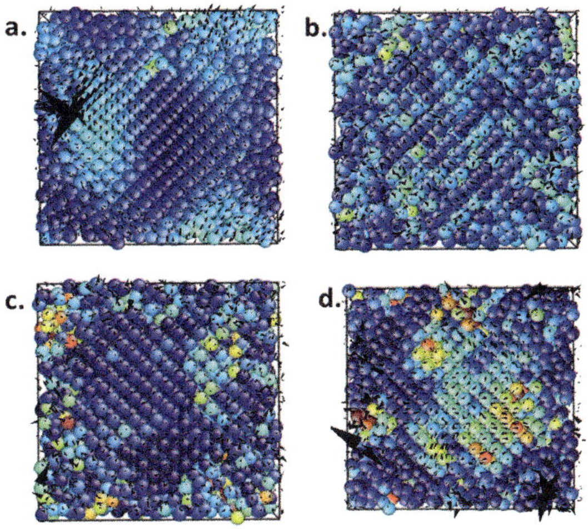

Figure 9.7 Snapshots of different eigenmodes of a system made of an amorphous matrix with a 3 nm large crystalline inclusion in Si-like sample. The colors depend on the amplitude of the displacement supported by each atom. Arrows show in-plane displacements, in a central layer with width 4 Å.

ratio P_2 of the modes is compared in Fig. 9.8 to the participation ratio of the modes of the amorphous matrix alone (Fig. 9.3). The two are very close. The main differences are located at $\omega \approx 12$ THz and $\omega \approx 15$ THz, were the participation ratio of the system with one inclusion is smaller than those of the amorphous matrix alone. In these ranges, the inclusions are mute: the vibrations are located in the matrix only. These frequency ranges correspond to gaps in the density of vibrational states of the crystal: transition from longitudinal acoustic (LA) modes to longitudinal optics (LO) modes for the later, and from LO modes to transverse optic (TO) modes of the crystal for the last one [Persans (1989)].

9.5.2 Vibrational Density of States (VDOS)

The difference between the vibrations of the composite sample with inclusions and the vibrations of the amorphous matrix is also shown in the vibrational density of states. The two VDOS are compared with that of a pure crystalline sample in Fig. 9.9. As already pointed out

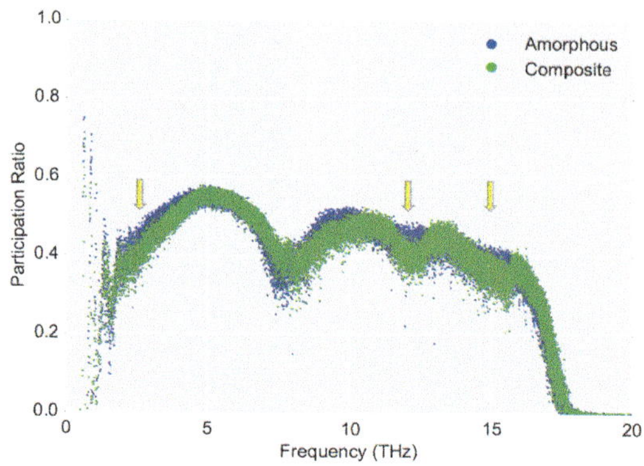

Figure 9.8 Participation ratio P_2 of the eigenmodes for the composite system, compared to the participation ratio of the eigenmodes in the bulk amorphous Si-like sample. Arrows highlight the differences between the two.

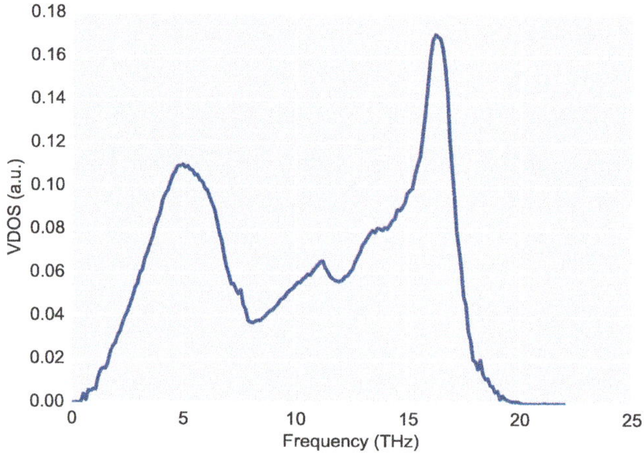

Figure 9.9 VDOS of Si-like composite samples with size $L = 50$ Å and $2R = 30$ Å.

in the study of the eigenmodes, the VDOS falls down around $\omega \approx 12$ THz and $\omega \approx 15$ THz as compared to the amorphous one. The partial density of states (computed by summing only the displacements of atoms belonging to the inclusion) confirms that the inclusions are mute in these frequency ranges. The resonant modes do not involve the inclusion and are confined in the matrix at these frequencies.

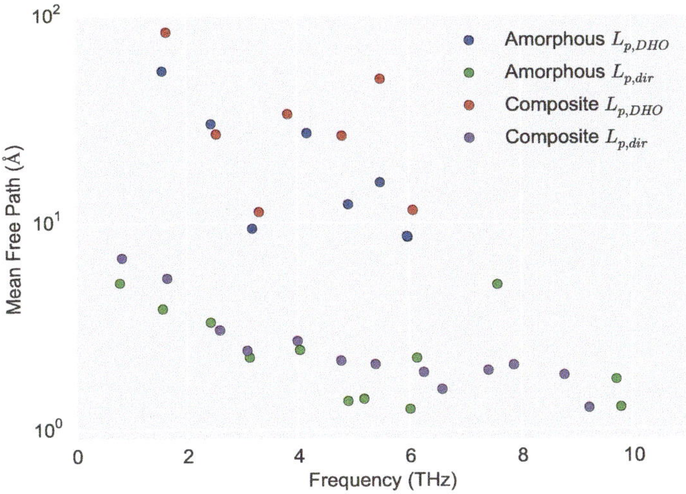

Figure 9.10 Mean free paths $L_{p,\,dir}(\omega)$ and $L_{p,\,DHO}(\omega)$ of Si-like composite samples, compared to the bulk amorphous sample. The DHO fit is given from longitudinal waves here.

The low-frequency range ($\omega < 5$ THz) is dominated as well by the motion of the amorphous matrix. After divided by the Debye density of states, the VDOS highlights a Boson peak similar to that of the amorphous matrix alone.

9.5.3 Dynamical Structure Factor

Experimentally, the measurement of the vibrational properties is given by the measurement of the dynamical structure factor through inelastic dynamic scattering methods. As can be seen in Fig. 9.10, the DHO fits of the dynamical structure factor, however, do not show any significant difference between the pure amorphous samples, and the amorphous samples with the inclusions.

9.5.4 Heterogeneous Propagation in the Presence of Nanocrystalline Inclusions

The propagation of a wave packet inside the sample is again more informative than the fit of the dynamical structure factor

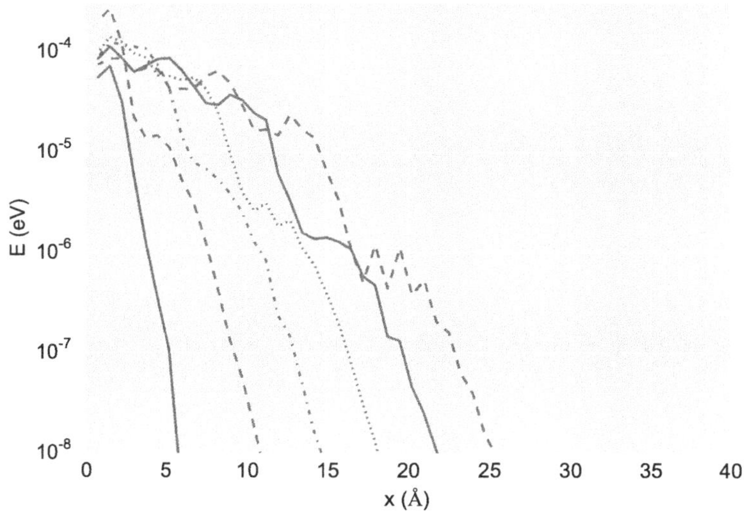

Figure 9.11 Kinetic energy envelope for a wave packet with $\omega = 5$ THz, at different times ($n \times 800$ fs), in the Si-like composite sample with inclusions radius $R = 30$ Å. The single maximum splits into two maxima at the interface with the inclusion.

alone. When the wave packet enters the crystalline inclusion, the envelope of the kinetic energy splits into two maxima (Fig. 9.11). The secondary maximum (lower kinetic energies) accelerates, while the highest maximum (larger energies) is slowed down at the interface. This low-pass filter behavior of the inclusions does not affect too much, however, the average mean free path obtained by averaging the evolution of the two maxima as a function of space.

Such heterogeneity in the spatial extent of a wave packet cannot be caught by the usual mean free path measurement but still plays an important role in thermal transport and local heating.

9.6 Conclusion

In this chapter, we have detailed different numerical methods that can be used to get information on the vibrational properties of amorphous and composite amorphous/crystals samples. We focused on molecular dynamics simulations and related tools. After discussing the main assumptions of such simulations at the atomic scale, we have compared the information obtained from the study of eigenmodes (diagonalization of the dynamical matrix), dynamical structure factor, and propagation of wave packets on silicon-like samples.

It was shown that the vibrational density of states of the amorphous matrix has a large excess in the low-frequency range as compared to the continuous Debye model, also referred to as *Boson peak* and visible in the systems with and without the inclusions. At higher frequencies, the presence of crystalline inclusions diminishes the vibrational density of states at the frequencies where the crystal is mute. In general, however, it is difficult to separate the vibrations of the inclusions from the vibrations of the amorphous matrix, unless the stiffness of the two is very different. Finally, the comparison between DHO fits of the dynamical structure factor and the propagation of wave packets shows that the mean free path is overestimated in the first case, and that the role of inclusions is to split the propagation of the wave packet into a slower propagation of high kinetic energy components that are repealed at the inclusion/matrix interface, and an enhanced propagation of low energy modes into the inclusions.

All the methods detailed in this chapter can be used in many other systems. Their combined use allows to get precise information on the propagation of vibrations, and thus on the thermal properties of the samples.

References

de Silva, C. W., ed. (2007). *Vibration, Damping, Control, Design*, CRC Press, Boca Raton.

Kittel, C. (2005). *Introduction to Solid State Physics*, John Wiley and sons.

Zeller, R. C., and Pohl, R. O. (1971). Thermal conductivity and specific heat of non-crystalline solids. *Physical Review B*, Vol. 4, p. 2029.

Poilblanc, R., and Crasnier, F. (2006). *Spectroscopies infrarouge et raman*, EDP Sciences, Les Ulis.

Crut, A. (2015). Acoustic vibrations of metal nano-objects: time-domain investigations. *Physics Reports*, Vol. 549, No. 15, pp. 1–49.

Saviot, L., Mermet, A., and Duval, E. (2010). Acoustic vibrations in nanoparticles, in *Nanoparticles and Quantum Dots: Handbook of Nanophysics*, Sattler, K. D., ed., CRC Press, Boca Raton.

Allen, M. P., and Tildesley, D. J. (1991). *Computer Simulation of Liquids*, Clarendon Press, Oxford.

Frenkel, D., and Smit, B. (2002). *Understanding Molecular Simulations*, Academic Press, New-York.

Tanguy, A. (2015). Vibration modes and characteristic length scales in amorphous materials. *JOM*, Vol. 67, No. 8, pp 1832–1839.

Stillinger, F. G., and Weber, T. A. (1985). Computer simulation of local order in condensed phases of silicon. *Physical Review B*, Vol. 31, No. 8, pp. 5262–5271.

van Beest, B. W. H., Kramer, G. J., van Santen, R. A. (1990). Force Fields for Silicas and Aluminophosphates Based on *Ab initio* calculations. *Physical Review Letters*, Vol. 64, No. 16, pp. 1955–1958.

Carré, A., Berthier, L., Horbach, J., Ispas, S., and Kob, W. (2007). Amorphous silica modeled with truncated and screende Coulomb interactions: a molecular dynamics simulation study. *The Journal of Chemical Physics*, Vol. 127, pp. 114512-1,114512-9.

To, T. T. *Etude par dynamique moléculaire de spectres vibrationnels de verres de silice* (2007). PhD Thesis, université des sciences et Technologies de Lille- Lille 1.

Tangney, P., and Scandolo, S. (2002). An *A initio* parametrized interatomic force field for silica. *Journal of Chemical Physics*, Vol. 117, No. 19, pp. 8898–8904.

Buchenau, U., Wischnewski, A., Ohl, M., and Fabiani, E. (2007). Neutron scattering evidence on the nature of the Boson peak. *Journal of Physics: Condensed Matter*, Vol. 19, No. 20, p. 205106.

Price, D., and Carpenter, J. (1987). Scattering function of viterous silica. *Journal of Non-Crystalline Solids*, Vol. 92, No. 1, pp. 153–174.

Vollmayr, K., Kob, W., and Binder, K. (1996). Cooling-rate effects in amorphous silica: a computer-simulation study. *Physical Review B*, Vol. 54, p. 15808.

Sevincli, H., Mukhopadhyay, S., Senger, R. T., and Ciraci, S. (2007). Dynamics of phononic dissipation at the atomic scale: dependence on internal degrees of freedom. *Physical Review B*, Vol. 76, pp. 205430-1, 205430-10.

Rittmeyer, S. P., Meyer, J., Inaki Juaristi, J., and Reuter, K. (2015). Electronic friction-based vibrational lifetimes of molecular adsorbates: beyond the independent-atom approximation. *Physical Review Letters*, Vol. 115, p. 046102.

Lemaitre, A., and Maloney, C. (2006). Sum rules for the quasi-static and visco-elastic response of disordered solids at zero temperature. *Journal of Statistical Physics*, Vol. 123, No. 2, pp. 415–453.

Niehaus, T. A., Heringer, D., Torralva, D., and Frauenheim, Th. (2005). Importance of electronic self-consistency in the TDDFT based treatment of nonadiabatic molecular dynamics. *European Physical Journal D*, Vol. 35, pp. 467–477.

Rodney, D. Tanguy, A., and Vandembroucq, D. (2011). Modeling the mechanics of amorphous solids at different length and time scales. *Modelling and Simulation in Materials Science and Engineering* Vol. 19, p. 083001-1,083001-34.

Sun, B.-R., Zhan, Z.-J., Liang, B., Zhang, R.-J., Wang, W.-K. (2011). Light emission and dynamic failure mechanism of hypervelociy impact on Zr-Ti-Ni-Cu-Be bulk metallic glass. *Chinese Physics Letters*, Vol. 28, No. 9, pp. 096102-1,096102-4.

Leonforte, F., Tanguy, A. Boissiere, R., Wittmer, J. R., and Barrat, J.-L. (2005). Continuum limit of amorphous elastic bodies (III): three-dimensional systems. *Physical Review B* Vol. 72, pp 224206-1, 224206-11.

Tsamados, M. (2010). Plasticity and dynamical heterogeneity in driven glassy materials. *European Physical Journal E*, Vol. 32, No. 2, pp. 165–181.

Tanguy, A., Leonforte, F., and Barrat, J.-L. (2006). Plastic response of a 2D Lennard-Jones amorphous solid: detailed analysis of the local rearrangements at very slow strain rate. *European Physical Journal E*, Vol. 20, pp. 355–364.

Buchenau, U. (2014). Evaluation of x-ray Brillouin scattering data. *Physical Review E*, Vol. 90, p. 062319.

Schelling, P. K., Philipot, S. R., and Keblinski, P. (2002). Phonon wave-packet dynamics at semiconductor interfaces by molecular-dynamics simulation. *Applied Physics Letters*, Vol. 80, No. 14, pp. 2484–2486.

Vacher, R., Courtens, E., and Foret, M. (2005). Anharmonic versus relaxational sound damping in glasses. II. Vitreous silica. *Physical Review B*, Vol. 72, pp. 214205-1,214205-11.

Polizzi, E. (2009). Density-Matrix-Based algorithms for solving eigenvalue problems. *Physical Review B*, Vol. 79, p. 115112.

Schenk, O., Gartner, K., and Fichtner, W. (2000). Efficient sparse LU factorization with left-right looking strategy on shared memory multiprocessors. *BIT*, Vol. 40, No. 1, pp. 158–176.

Damart, T., Giordano, V., and Tanguy, A. (2015). Nano-crystalline inclusions as a low-pass filter for thermal transport in amorphous silicon. *Physical Review B*, Vol. 92, p. 094201.

Beltukov, Y., Kozub, V. I., and Parshin, D. A. (2013). Ioffe-Regel criterion and diffusion of vibrations in random lattices. *Physical Review E*, Vol. 87, pp. 134203-1,134203-20.

Beltukov, Y., Fusco, C., Parshin, D. A., and Tanguy, A. (2015). Boson peak and Ioffe-Regel criterion in amorphous silicon-like materials: the effect of bond directionality. Condmat: arXiv:1510.01761

Eringen, A. C., and Suhubi, E. S. (1975). *Elastodynamics Vol. I & II*, Academic Press, New-York.

Wittmer, J. P., Tanguy, A., Barrat, J.-L., and Lewis, L. (2002). Vibrations of amorphous nanometric structures: when does continuum theory apply? *Europhysics Letters*, Vol. 57, pp. 423–429.

Tanguy, A., Wittmer, J. P., Leonforte, F., and Barrat, J.-L. (2002). Continuum limit of amorphous elastic bodies: a finite-size study of low-frequency harmonic vibrations. *Physical Review B*, Vol. 66, pp. 174205-1, 174205-17.

Fusco, C., Albaret, T., and Tanguy, A. (2010). Role of local order in the small-scale plasticity of model amorphous materials. *Physical Review E*, Vol. 82, pp. 066116-1,066116-11.

Portales, H., Goubet, N., Saviot, L., Adichtchev, S., Murray, D. B., Mermet, A., Duval, E., and Pileni, M.-P. (2008). Probing atomic ordering and multiple twinning in metal nanocrystals through their vibrations. *Proceedings of the National Academy of Sciences of the United States of America*, Vol. 105, pp. 14784–14789.

Ramirez, F., Heyliger, P. R., Rappe, A. K., and Leisure, R. G. (2008). Vibrational modes of free nanoparticles: from atomic to continuum scales. *Journal of the Acoustical Society of America*, Vol. 123, p. 709.

Saviot, L., Champagnon, B., Duval, E., Kudriavtsev, I. A., and Ekimov, A. I. (1996). Size dependence of acoustic and optical vibrational modes of CdSe nanocrystals in glasses. *Journal of Non-Crystalline Solids*, Vol. 197, p. 238.

Hu, M., Hillyard, P., Hartland, G. V., Kosel, T., Perez-Juste, J., and Mulvaney, P. (2004). Determination of the elastic constants of gold nanorods produced by seed mediated growth. *Nano Letters* Vol. 4, pp. 2493–2497.

Wyart, M., Nagel, S. R., and Witten, T. A., (2005). *Europhysics Letters* Vol. 72, p. 486 (2005).

Marruzzo, A., Schirmacher, W., Fratalocchi, A., and Ruocco, G. (2013). *Scientific Reports*, Vol. 3, p. 1407.

Baldi, G., Giordano, V. M., and Monaco, G. (2011). *Physical Review B*, Vol. 83, p. 174203.

Gurevich, V. I., Parshin, D. A., and Schober, H. R. (2003). *Physical Review B*, Vol. 67, p. 094203.

Chumakov, A. I., Monaco, G., Fontana, A., Bosak, A., Hermann, R. P., Bessas, D., Wehinger, B., Crichton, W. A., Krisch, M., Rüffer, R., Baldi, G., Carini Jr., G., Carini, G., D'Angelo, G., Gilioli, E., Tripodo, G., Zanatta, M., Winkler, B., Milman, V., Refson, K., Dove, M. T., Dubrovinskaia, N., Dubrovinsky, L., Keding, R., and Yue, Y. Z. (2014). Role of Disorder in the Thermodynamics and Atomic Dynamics of Glasses *Physical Review Letters*, Vol. 112, p. 025502.

Dove, M. T. (2005). *Introduction to Lattice Dynamics*, Cambridge university press, Cambridge.

Goldenberg, C., Tanguy, A., and Barrat, J.-L. (2007). Particle displacements in the deformation of amorphous materials: local fluctuations vs. the nonaffine field. *EuroPhysics Letters*, Vol. 80, pp. 16003-1,16003-6.

Aharony, A., Alexander, S., Entin-Wohlman, O., and Orbach, R. (1987). Scattering of fractons, the Ioffe-Regel criterion, and the (4/3) conjecture, *Physical Review Letters*, Vol. 58, p. 132.

Allen, P. B., Feldman, J. L., Fabian, J., and Wooten, F. (1999). Diffusons, locons and propagons: character of atomic vibrations in amorphous Si. *Philosophical Magazine B*, Vol. 79, No. 11/12, pp. 1715–1731.

Faez, S., Strybulevych, A., Page, J. H., Lagendijk, A., and van Tiggelen, B. A. (2009). Observation of Multifractality in Anderson Localization of Ultrasound. *Physical Review Letters*, Vol. 103, p. 155703.

Cohen-Tannoudji, C., Diu, B., and Laloe, F. (1977). *Quantum Mechanics*, Hermann, Paris.

Akkermans, E., and Montambaux, G. (2004). *Physique mesoscopique des electrons et des photons*, CNRS editions, EDP sciences, Paris.

Ruocco, G., Sette, F., Di Leonardo, R., Monaco, G., Sampoli, M., Scopigno, T., and Viliani, G. (2000). Relaxation processes in harmonic glasses. *Physical Review Letters* Vol. 84, No. 25, p. 5788.

Larkin, J. M., and McGaughey, A. J. H. (2014). Thermal conductivity accumulation in amorphous silica and amorphous silicon. *Physical Review B*, Vol. 89, p. 144303-1,144303-12.

Ioffe, A. F., and Regel, A. R. (1960). Non-crystalline, amorphous and liquid electronic semiconductors. *Progress in Semiconductors*, Vol. 4, pp. 239–291.

Persans, P. D. (1989). Raman Scattering as probe of a structure in amorphous multilayers, in *Advances in Disordered Semi-Conductors, Vol. 1, Amorphous silicon and related materials*, Fritzsche, H., ed., World Scientific, Singapore.

Chapter 10

Structural Properties and Defects of III-Nitride Semiconductors at the Nanoscale

J. Kioseoglou
Department of Physics, Aristotle University of Thessaloniki, GR-54124 Thessaloniki, Greece
sifisl@auth.gr

During the past two decades, group III-nitride heterostructures have attracted considerable interest due to their potential applicability in high efficiency optoelectronic devices. Extensive research in the area has led to a variety of optoelectronic devices, such as high-power and high-temperature transistors, optical data storage, and biochemical sensors. However, it is inevitable that their full potential is still under intense investigation. There are a variety of specific fields that should be investigated and controlled such as the impact of the defects on the optoelectronic and thermal properties the surface reconstructions and the morphological properties of the nanostructures. In this chapter, we present in detail and finally review five selected topics which span multiple bulk and nanoscale structural properties of III-nitrides [1–3]. Initially we focus on a variety of extended defects in III-nitrides. In the second part, we investigate the structural properties of non-polar interfaces as

---

Nanostructured Semiconductors: Amorphization and Thermal Properties
Edited by Konstantinos Termentzidis

ISBN 978-981-4745-64-2 (Hardcover), 978-1-315-36445-2 (eBook)
www.panstanford.com

well as on the diffusion mechanisms of the most favourable point defect (vacancy) along polar interfaces. Afterwards the analysis of surface reconstructions is presented in detail focusing on a specific semipolar surface. Finally we analyse the shape of the GaN nanowires by the means of interatomic potential simulations.

10.1 Introduction

The family of nitrides semiconductors (AlN, GaN and InN) is one of the most motivating classes of optoelectronic materials. All three binaries and their alloys exhibit some unique properties among semiconductors, such as strong interatomic bonds, high thermal conductivity. Their energy gaps cover a wide spectral range from the red to deep ultraviolet. They are also used for the fabrication of devices for high power in high-temperature operation. However, these compounds do not exist in natural crystalline form and they have to be grown, in most cases, by heteroepitaxy. Consequently, various crystallographic defects are introduced in their structures, and they alter significantly the performance of devices.

One of the most interesting areas where there is a lack of extensive research till now is the influence of the structural characteristics and defects on the thermal properties of the III-N compounds. In particular for GaN, there are several experimental [4, 5] as well as theoretical studies [6, 7] showing that the thermal conductivity ranges between 160 and 250 W/mK at room temperature (300 K). Recently, it has been found that even the finite size of the GaN layers reduces the thermal conductivity lower than the bulk value [8]. However, several studies have been focused on the influence of the extended defects on the thermal conductivity of the III-N compound semiconductors [9, 10]. Recently, extensive research has been done on nanostructures [11] and on the most dominant defects in III-N films, the threading dislocations [12, 13].

In general, concerning the theoretical studies, two methods are used for the investigation of the thermal conductivity be the use of molecular dynamics [14, 15]. The most common and accurate method in order to determine the thermal conductivity is the equilibrium molecular dynamics (EMD) method. The main idea of an EMD simulation is that the decay of the heat flux fluctuations is proportional to the thermal conductivity via the Green–Kubo formula [16]. The principle relies on the fact that the regression of the thermal fluctuations of the thermal flux follows macroscopic laws. Consequently, the time decay of the flux fluctuations is proportional to thermal conductivity. This relationship is interpreted by the Green–Kubo formulae [17], which state that the time integral of the heat flux autocorrelation function is proportional to the thermal conductivity tensor.

The non-equilibrium molecular dynamics (NEMD) method is similar to the hot-plate experimental set-up and proposed by Kotake and Wakuri [18] and has been applied successfully for nanostructured systems as superlattices [19] or NWs [12, 20]. In this method a temperature difference is imposed on the two edges of the NW. Once the steady state is attained, the temperature profile and the heat power exchange between the hot and cold thermostats are analysed. These two quantities are used in the determination of the thermal conductivity employing Fourier's law. Fixed boundary conditions are used in all directions, while along the NW, two additional sets of fixed atoms after the two thermostats are used to assure the stability of the NW. All the calculations are performed for finite temperatures T_{NEMD}, with the two thermostats at about 10% higher and lower temperatures with respect to T_{NEMD}.

The main advantage of the EMD method is that just one simulation is needed in order to compute the full conductivity tensor, which is very useful in simulations of non-uniform supercells of atoms that exhibit large thermal anisotropies. Moreover, the EMD method does not depend on finite size effects in contrast to NEMD, owing to the fact that in EMD the phonons are not strongly scattered by the boundaries of the supercell that is used in the simulations.

Equilibrium structural parameters can be calculated by the use of DFT methods. However, atomistic simulations methods have still a significant impact on studding extended defect structures in these systems. Many cases of structures and energies of planar and line defects and strain fields associated with nanostructure growth processes have been described recently by the use of empirical potential calculations [21–26]. It is mentioned [27] that while the majority of empirical potentials fall into either the Stillinger–

Weber [28] (SW) or Tersoff [29, 30] approaches, the Tersoff form is fundamentally different from the SW one in that the strength of individual bonds is affected by the presence of surrounding atoms and it is much more realistic for highly distorted configurations. However, many studies are carried out by the use of SW and Tersoff potentials for bulk properties [31, 32] and defected configurations such as planar and point defects [25, 33, 34] dislocated structures [23, 24, 35, 36], surface and interface structures [37–39], and the liquid and amorphous states [40, 41].

The defect structure of these compounds may contain different interatomic interactions. Among them metallic interactions are also present and the interatomic potential which can be used should include such interactions. An empirical bond-order interatomic potential, based on Tersoff potential, is implemented for wurtzite structure III (Al, Ga, In)–N compound semiconductors. The parameterization, made for GaN, of Nord et al. [33] is used and the study for AlN and InN is based on a previous approach for zing blende structure [42]. The potential is tuned to reproduce accurately the equilibrium lattice constant a, the axial ratio c/a, the internal parameter u, the binding energy, the formation enthalpy and the elastic constants. An environment dependent approach is developed for AlN and InN having as variables the range of the In–In and Al–Al interaction distances and compared with the standard one [1].

Using molecular dynamic calculations the interatomic potential is applied on a series of extended defects: 16 unrelaxed crystallographic planar defects corresponding to Austerman [43], Holt [44] and IDB* [45] type of $\{10\bar{1}0\}$ and $\{1\bar{2}10\}$ inversion domain boundaries (IDBs) and on two principal models Amelinckx [46] and Drum [47], of $\{1\bar{2}10\}$ prismatic translation domain boundaries (TDBs).

At the second part the structural properties of non-polar AlN/GaN interfaces taking into account initially elastic and afterwards plastic relaxation is investigated. Moreover, the diffusion mechanisms of the most favourable point defect (metal vacancy) along polar interfaces are discussed. In order to achieve a deeper understanding of surface phenomena, calculations on the atomic and electronic structure of $(10\bar{1}3)$ and $(10\bar{1}\bar{3})$ surfaces are discussed.

Finally the shape of the GaN nanowires is analysed by the means of interatomic potential simulations.

10.2 Extended Defects

10.2.1 Planar Defects

In order to obtain the relaxed structures of the observed planar defects in these materials, molecular dynamic calculations of 16 unrelaxed crystallographic models based on Austerman-, Holt- and IDB*-type $\{10\bar{1}0\}$ IDBs and on two principal Amelinckx and Drum prismatic $\{1\bar{2}10\}$ TDB models are implemented. The defects are constructed in supercells with periodic boundary conditions and are relaxed from a temperature of 600 K slowly to 0 K by the use of the quenched molecular dynamic method [48]. At least 1000 atoms are considered for the construction of the supercell since in molecular dynamic calculations strong internal relaxations may lead to finite-size effects [33] in smaller cells. In the planar defect models where "wrong" bonds are present (i.e. III–III and N–N bonds), the d-spacing of the boundary planes is also considered. Starting from the ideal lattice value, the spacing of the boundary planes is increased by steps of 0.1 Å, in order to obtain accurately the optimum conditions [49]. In each step a full relaxation procedure is implemented and finally the minimum energy position is achieved. Although the III-species environment approach contains complicated bonding conditions, forces can still be evaluated with computational speed comparable to the standard interaction approach.

10.2.1.1 The $\{10\bar{1}0\}$ planar defects

The Austerman IDB model is formed in a wurtzite bicrystal when one hcp sublattice is kept at a fixed orientation whereas the second switches to a crystallographically equivalent site taking into account an inversion operation. It is assumed that in the original Austerman model the nitrogen atoms of the two domains are in the closest positions along the boundary plane, while in the Austerman **b** model the closest sites belong to the III-type atoms. The Holt IDB is obtained through an exchange of the III (Al, Ga, In) and nitrogen atomic positions and the relative translation **p** of the two domains is 3/8[0001]. The IDB* model is related to the Holt model by a $c/2$ translation, and it is characterized by the absence of "wrong" bonds (**p** = -1/8[0001]). The IDB plane can reside on two locations for each IDB model, i.e. either cutting single bonds, or cutting double bonds; the former is designated as type 1 plane and the latter as type 2 plane, as illustrated in Fig. 10.1.

In Table 10.1, the energies of the $\{10\bar{1}0\}$ inversion domain boundaries in III-N calculated by the interatomic potential [1] are presented.

The interatomic potential calculations reveal that the energetically favourable $\{10\bar{1}0\}$ planar defects are those not including III–III "wrong" bonds, the IDB* being the most energetically favourable model. The Austerman model is found to require only 0.01 eV/Å^2 more energy than the IDB* type and consequently it is a strong candidate among IDB atomic configurations. In all cases the type 2 models are found to require much more energy than the corresponding type 1. Particularly the IDB* 2 and Austerman 2 models are found to comprise double energy than the corresponding type 1 models. The Austerman **b** models in all cases relax to the corresponding IDB* models, due to the fact that they initially include III–III-type atomic interactions.

Regarding the two approaches used, for AlN and InN the "wrong" bonded boundaries are estimated to be more energetically unfavourable using the standard approach than the III-species environment one. This finding justifies the significance of the III-species environment approach. Regarding the structures using the III-species environment approach, the equilibrium III–III bonds for the relaxed atomic configurations of the Holt-type IDBs (3.69 Å for In–In and 3.39 Å for Al–Al for the type 1 and 3.54 Å for In–In and 3.24 Å for Al–Al for the type 2) are lengthened with respect to the corresponding bulk metallic bonds. Contrary, by the standard approach, smaller values very close to their bulk metallic bond lengths are obtained (3.19 Å for In–In and 2.8 Å for Al–Al). For the case of GaN, the Ga–Ga bonds are lengthened by about 2.72 Å with respect to the values exhibited in bulk α-Ga (2.5 Å). The above results show that the III-species environment approach describes

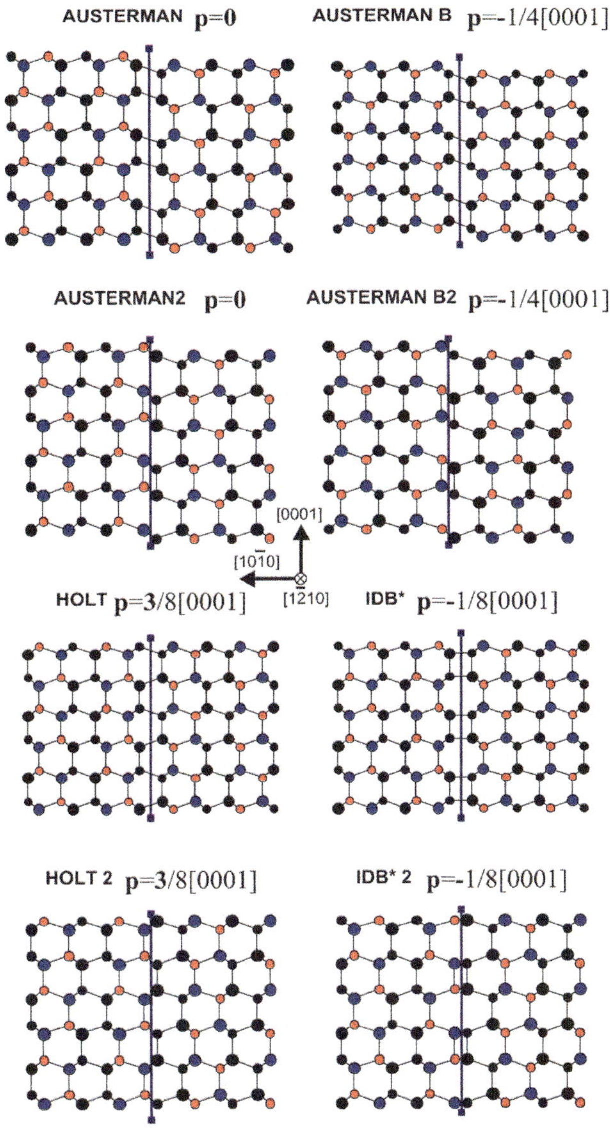

Figure 10.1 Atomic models of the $\{10\bar{1}0\}$ inversion domain boundaries in III-N, projected along $<1\bar{2}10>$. The position of the boundary plane is indicated by the solid line, large and small circles denote III and N atoms, respectively, light-shaded circles are at 0 level, and dark circles are at level $a/2$ along the projection direction.

Table 10.1 Energies (eV/Å2) of $\{10\bar{1}0\}$ inversion domain boundaries in III-N calculated by the interatomic potential [1]

Energy eV/Å^2		Inversion Domain Boundaries $\{10\bar{1}0\}$							
		Holt	**IDB***	**Holt 2**	**IDB* 2**	**Austerman**	**Austerman b**	**Austerman 2**	**Austerman b2**
GaN	Standard	0.21	0.06	0.37	0.13	0.07	IDB*	0.13	IDB* 2
AlN	III-sp. env.	0.22	0.05	0.34	0.15	0.06	IDB*	0.13	IDB* 2
InN	III-sp. env.	0.20	0.03	0.32	0.07	0.04	IDB*	0.08	IDB* 2

the "wrong" bonded $\{10\bar{1}0\}$ planar defects more realistically in AlN and InN. By the use of the Tersoff potential the same defect energy hierarchy with DFT calculations [1] is achieved for $\{10\bar{1}0\}$ planar defects in GaN.

10.2.1.2 The $\{1\bar{2}10\}$ planar defects

The Amelinckx TDB model is formed by a $1/6<20\bar{2}3>$ translation while the Drum model is formed by a $1/2$ $<10\bar{1}1>$ one, both comprising a $1/2<0001>$ component. The Amelinckx model has a basal component equal to $1/3<10\bar{1}0>$ while the basal component of the Drum model is equal to $1/2<10\bar{1}0>$. A $1/6<10\bar{1}0>$ shift converts the Amelinckx to the Drum model. Two more models are studied; both of them exhibit a total translation vector $1/2<10\bar{1}0>$ which is derived by subtracting a $1/2<0001>$ component from the translation vector of the Drum model. The first one, namely the Drum–Holt model, contains "wrong" bonds, while the second, namely the DRUM* model, comprises inversion character (Fig. 10.2).

The Austerman ($\mathbf{p} = 0$), Austerman b ($\mathbf{p} = -1/4[0001]$), IDB* ($\mathbf{p} = -1/8[0001]$) and Holt ($\mathbf{p} = 3/8[0001]$) models of $\{1\bar{2}10\}$ IDBs are illustrated in Fig. 10.3.

In Table 10.2, the energies of the $\{1\bar{2}10\}$ translation and inversion domain boundaries in III-N calculated by the interatomic potential [1] are presented.

The Drum model is found to be the most energetically favourable $\{1\bar{2}10\}$ translation domain boundary in all cases, whilst the second position in the hierarchy belongs to the Drum*. Regarding the $\{1\bar{2}10\}$ IDBs, the IDB* type is the most energetically favourable while the Austerman **b** model requires more energy than the IDB* in all cases. It is noted that the energy difference between the Drum and Drum* models, and between the IDB* and Austerman **b** models is about 0.01–0.02 eV/Å^2, very close to the minimum error of the interatomic potential calculations (0.01 eV/Å^2).

The Holt-type TDBs and IDBs are found to be the most unfavourable defects. The Amelinckx and Austerman models in all cases are relaxed to the Drum and IDB* models, respectively.

By the use of the III-species environment approach for AlN and InN, the Holt IDB is found more favourable than the Drum–Holt

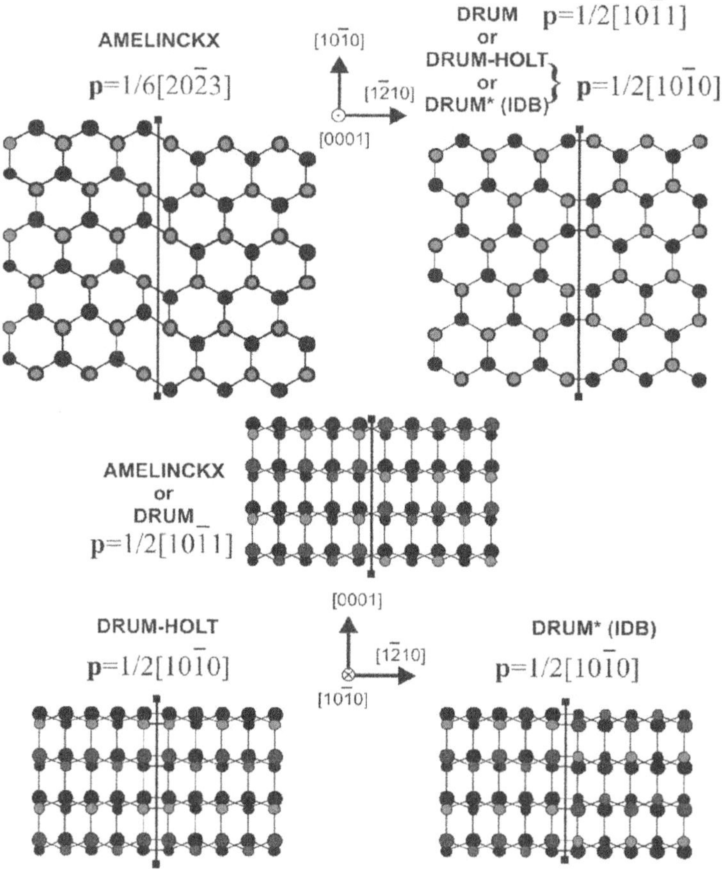

Figure 10.2 Atomic models of the $\{1\bar{2}10\}$ translation domain boundaries in III-N, projected along <0001> and <10$\bar{1}$0>. The boundary plane is indicated by the solid line, large and small circles denote III and N atoms, respectively, and shading of atoms denotes different levels along the projection direction.

TDB, whereas these are energetically equivalent when following the standard approach. GaN follows the same trend as AlN and InN on comparing the Holt IDB with Drum–Holt TDB using the III-species environment approach. This is due to the fact that the III-species environment approach represents more accurately the metal-metal interactions for (Al, In)-N. It is again found that the

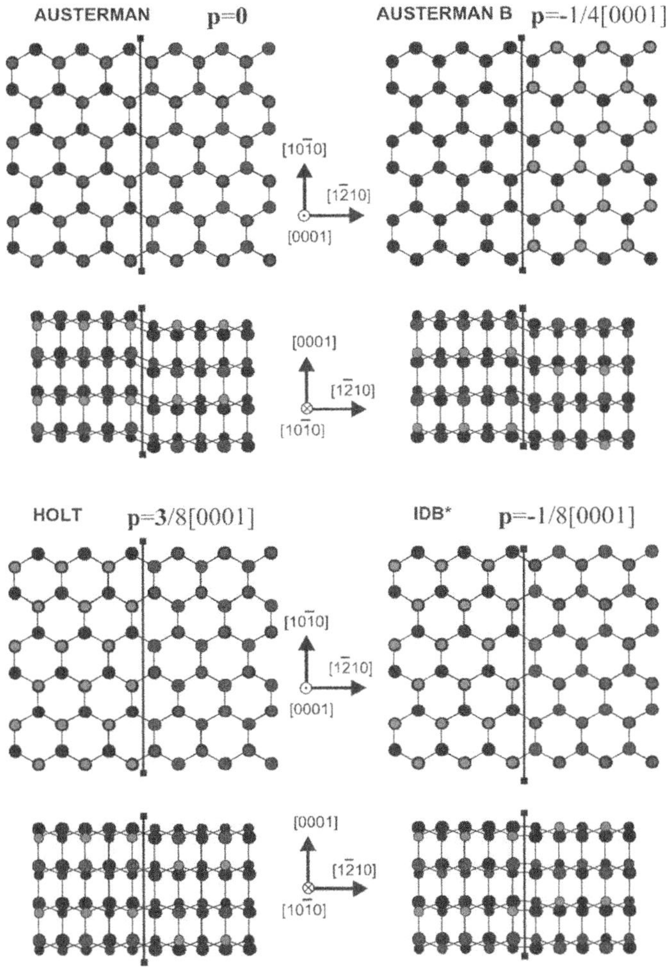

Figure 10.3 Atomic models of the $\{1\bar{2}10\}$ inversion domain boundaries in III-N, projected along $<0001>$ and $<10\bar{1}0>$. Atoms and symbols are as given in Fig. 10.2.

standard approach overestimates the "wrong" bond energies more than the III-species environment approach.

Regarding the structure in the III-species environment approach, the equilibrium III–III bonds in the relaxed structures of the $\{1\bar{2}10\}$ planar defects (3.69 Å for In–In and 3.28 Å for Al–Al) are lengthened.

Table 10.2 Energies (eV/Å^2) of $\{1\bar{2}10\}$ translation and inversion domain boundaries in III-N calculated by the interatomic potential [1]

Energy eV/Å^2		Translation Domain Boundaries Inversion Domain Boundaries $\{1\bar{2}10\}$							
		Amelinckx	**Drum**	**Drum–Holt**	**Drum***	**Austerman**	**Austerman b**	**IDB***	**Holt**
GaN	Standard	Drum	0.18	0.27	0.19	IDB*	0.08	0.07	0.23
AlN	III-sp. env.	Drum	0.12	0.28	0.16	IDB*	0.08	0.06	0.25
InN	III-sp. env.	Drum	0.07	0.20	0.09	IDB*	0.04	0.03	0.14

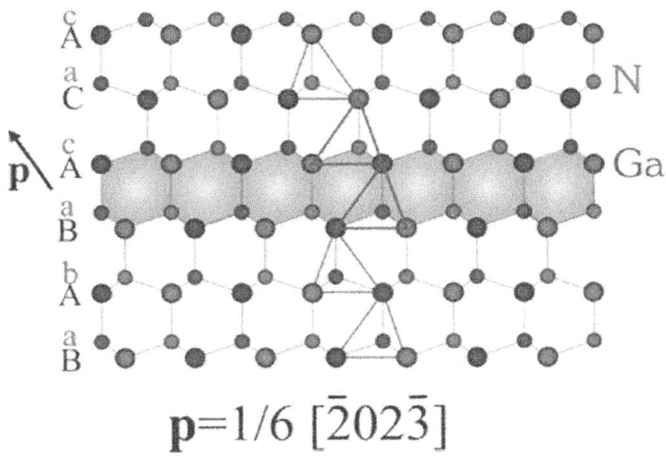

Figure 10.4 Atomistic model of the I_1 SFs in wurtzite GaN projected along the [$\bar{1}2\bar{1}0$] direction. The translation vector **p** and the stacking sequence along the SF are denoted.

In the standard approach, the bond lengths in the compounds are found to be very close to their metallic bond lengths in the bulk (3.19 Å for In–In and 2.8 Å for Al–Al).

In the present study the boundary plane spacing is also investigated. In GaN, the Ga–Ga bond along the boundary of the Drum–Holt defect was found to be 2.85 Å. This is in agreement with the DFT results in GaN (2.8 Å) [49].

The {$1\bar{2}10$} IDB* type in GaN is the most energetically favourable {$1\bar{2}10$} planar defect, in agreement with Ref. [50]. In addition, the energetic defect hierarchy is in agreement with DFT results reported for some of these defects in GaN. For the {$10\bar{1}0$} or the {$1\bar{2}10$} planar defects in AlN and InN there are no DFT results known or reported yet.

10.2.2 Stacking Faults and Point Defects

The defects in crystals are typically classified by dimension. Point defects are those that occur only around a single lattice point and are not extended in space such as vacancies, interstitials and antisite defects. Extended defects include linear defects such as dislocations

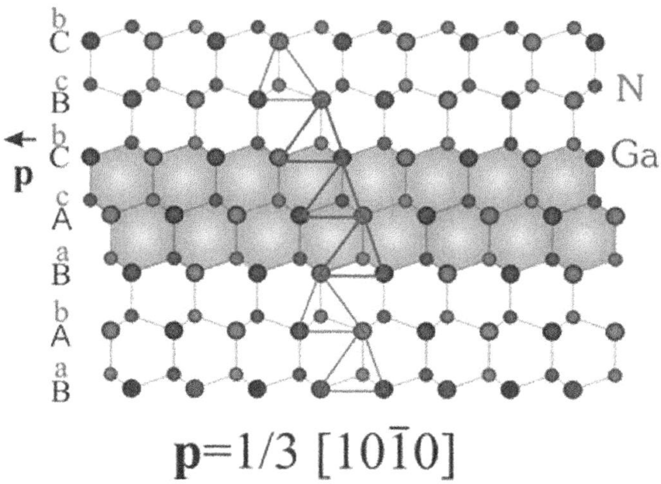

Figure 10.5 Atomistic model of the I_2 SFs in wurtzite GaN projected along the [$\bar{1}2\bar{1}0$] direction. The translation vector **p** and the stacking sequence along the SF are denoted.

and planar defects such as grain boundaries, interfaces and stacking faults. The three dimension defects include precipitates, holes and surface hillocks and pits.

Stacking faults are common defects in III-nitrides and a high density of stacking faults is usually observed in III-nitride epitaxial layers near the interface with the substrate. There are three different types of stacking faults I_1, I_2, E which contain one, two or three cubic bilayers, respectively, as they are presented in Figs. 10.4, 10.5 and 10.6. The energy of SFs in wurtzite III-nitrides is found to be almost proportional to the number of cubic bilayers in their structural units.

Among the stacking faults, I_1 has the lowest energy but its formation requires the removal of a double layer followed by slip of $\frac{1}{3}$ <1$\bar{1}$00>. As a result, it cannot be created due to a strain relaxation but must be growth related. This also applies to the E-type SF that can be created by adding an extra plane. The I_2 SF has the second lowest energy and unlike the I_1- and E-type SFs it can be created by a simple shear and therefore it can be caused by strain relaxation in the growing layer.

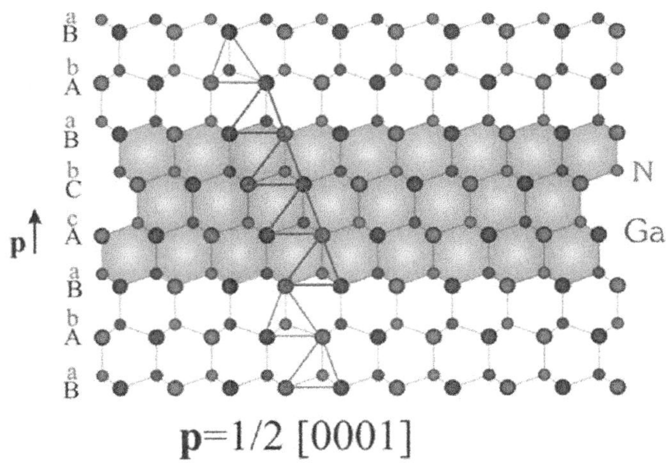

Figure 10.6 Atomistic model of the E SFs in wurtzite GaN projected along the $[\bar{1}2\bar{1}0]$ direction. The translation vector **p** as well as the stacking sequence along the SF are denoted.

10.3 Study of Nonpolar $(1\bar{1}00)$ AlN/GaN Interfaces Using ab initio and Empirical Potential Calculations

In group-III nitride semiconductor technology, nonpolar growth directions are promising for increasing the output efficiency of optoelectronic devices. Heterostructures grown on nonpolar planes like $\{11\bar{2}0\}$ (*a*-plane) or $\{1\bar{1}00\}$ (*m*-plane) are not affected by the presence of the piezoelectric fields and exhibit flat-band conditions. However, the role of the interfacial structures as source of line and extended defects must be considered, since such defects are detrimental for the radiative efficiency and device lifetimes. Moreover, defect formation mechanisms at nonpolar and semipolar interfaces, are different to those established for *c*-plane interfaces and therefore it is significant to address such issue.

Possible structural configurations of the nonpolar $(1\bar{1}00)$ AlN/GaN interfacial region were investigated by both interatomic potential and density functional theory (DFT) calculations. Firstly, pseudomorphically grown heterostructures were assumed. Atomic

Figure 10.7 Relaxed GaN/AlN non-polar interface projected along the [$\bar{1}2\bar{1}0$] direction having a planar SF along the interface.

scale simulations and band structure calculations were performed, either imposing elastic relaxation on both the epilayer and the substrate or only on the epilayer on a rigid substrate. Concerning the plastic strain relaxation two principal mechanisms were compared, i.e. the formation of a planar defect along the heteroepitaxial interface or the introduction of planar defects emanating vertically from the interface towards the GaN and AlN crystals. Such defects are intrinsic or extrinsic stacking faults.

It is concluded that the existence of stacking mismatch at the heteroepitaxial interface is energetically favoured and consequently it should be present at least at the first stages of growth. However, the sp^2 hybridization of the atoms along the stacking mismatch boundary is not a stable configuration for the wurtzite structure, thus, the introduction of I_1-type stacking faults is not avoided for the *m*-plane heteroepitaxial growth. In the latter case the relaxed structural models were analysed and the energetic results indicated the favourable cases.

Initially the existence a planar defect at the heteroepitaxial interface which is bounded by two dislocations with Burger vectors $b = \frac{1}{2}$ <0001> and $b = -\frac{1}{2}$ <0001> is considered as it is presented in Fig. 10.7. The calculations were based on the empirical interatomic potential method and were performed by using LAMMPS code. This method offered to examine large interfacial areas with thousands of atoms, thus taking into account the structural mismatch resulting to the coexistence of "good fit" and "bad fit" regions. The Tersoff potential, which is known to reproduce the lattice and elastic parameters of nitride semiconductors with great accuracy, was used.

Another interesting case is the existence of a stacking mismatch boundary (SMB)-type SF at the interface. The atomistic model of the SMB is presented in Fig. 10.8.

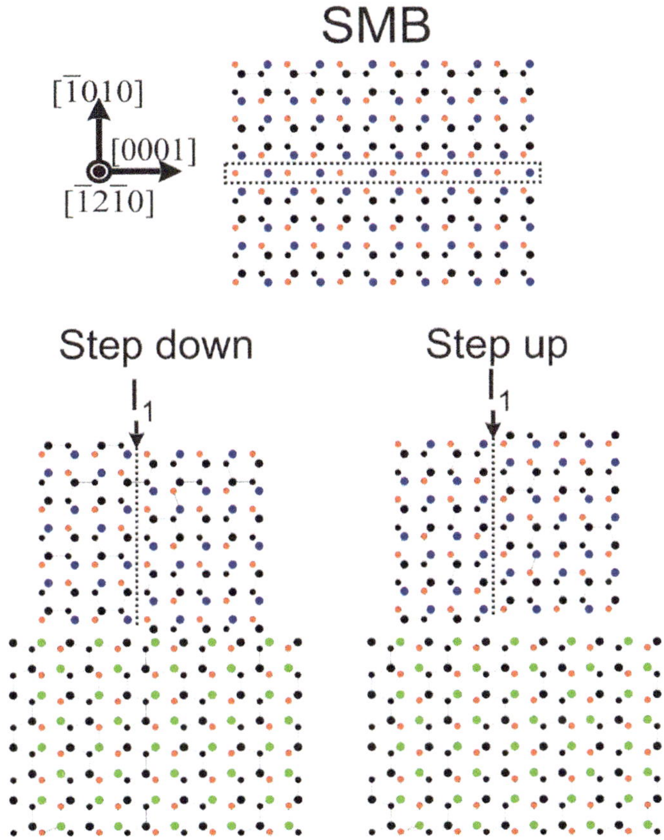

Figure 10.8 The atomistic model of the SMB as well as the two cases of the I_1 SF on the non-polar GaN/AlN interface. Two cases are considered in order to accommodate the step along the $< 10\bar{1}0 >$ at the interfacial structure: the "step down" where the $\mathbf{b}_{e_x}$ component is taken towards the interface and the "step up" case where the $\mathbf{b}_{e_x}$ component is considered to the opposite direction.

Afterwards, the existence of I_1-type SF, perpendicular to the plane of the interface is considered.

The I_1 intrinsic SF corresponds to one violation of the stacking rule and can be formed by the removal of a plane followed by shearing by $\frac{1}{3}$ $<10\bar{1}0>$.

In order to describe the I_1 SF configuration with bounding partial dislocations, we start from the perfect crystal. Upon removal or addition of a basal disc of atoms (vacancy or interstitial disc) and assuming that only displacements normal to the basal plane are associated with the precipitation, the SF is bounded by a Frank loop with Burgers vectors 1/2[0001]. A loop of Shockley partial dislocation is then nucleated in the faulted region leading to the formation of I_1 SF.

Two types of admissible partial dislocations may bound the I_1 SF [24]. An edge dislocation with Burgers vector equal to $\mathbf{b}_e = 1/6 <20\bar{2}3>$ and a mixed-type dislocation with Burgers vector $\mathbf{b}_m = 1/6 <2\bar{2}03>$. In both cases the dislocation line is taken to be $[1\bar{2}10]$. The edge dislocation can be decomposed into two normal edge components $\mathbf{b}_{e_x} = 1/3 <10\bar{1}0>$ and $\mathbf{b}_{e_y} = 1/2 <0001>$ while the mixed dislocation can be decomposed into a screw $\mathbf{b}_s = 1/6 <1\bar{2}10>$ and two normal edge components $\mathbf{b}_{e_x} = 1/6 <10\bar{1}0>$ and $\mathbf{b}_{e_y} = 1/2 <0001>$. Consequently, in the present study two cases are considered in order to accommodate the step along the $<10\bar{1}0>$ at the interfacial structure: the "step down" where the $\mathbf{b}_{e_x}$ component is taken towards the interface and the "step up" case where the $\mathbf{b}_{e_x}$ component is considered to the opposite direction as they are given in Fig. 10.8. On the other hand, the extrinsic stacking faults E are bounded by partial dislocations with Burgers vector $\mathbf{b}_{e_y} = 1/2 <0001>$.

In Fig. 10.9 the relaxed atomistic model of the non-polar GaN/AlN interface having a SMB along the interface is presented.

In Fig. 10.10 the relaxed GaN/AlN non-polar interface projected along the $[\bar{1}2\bar{1}0]$ direction having two I_1 SFs in GaN perpendicular to the interfaces is given.

In Fig. 10.11 the relaxed GaN/AlN non-polar interface having two I_1 SFs in AlN perpendicular to the interfaces is presented.

The plastically relaxed wurtzite $(1\bar{1}00)$ AlN/GaN interfaces with E-type stacking faults is also investigated.

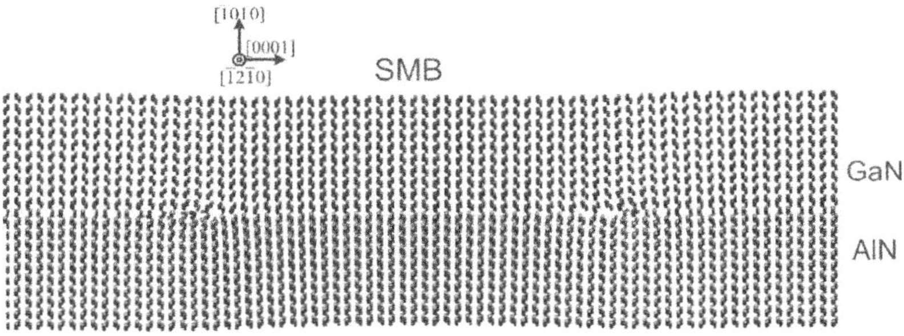

Figure 10.9 Relaxed GaN/AlN non-polar interface projected along the $[\bar{1}2\bar{1}0]$ direction having SMB along the interface.

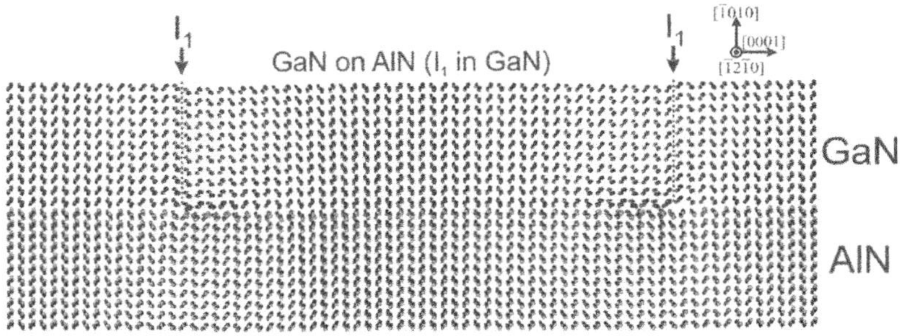

Figure 10.10 Relaxed GaN/AlN non-polar interface projected along the $[\bar{1}2\bar{1}0]$ direction having two I_1 SFs in GaN perpendicular to the interfaces.

Two different cases were examined. In the first case, AlN has grown on GaN and E-type stacking faults are introduced in AlN perpendicular to the plane of the interface and above of the extra planes of GaN. In the second case GaN has grown on AlN and E-type stacking faults are introduced in AlN. Both step up and step down configurations are examined.

The models consist of 16 monolayers of AlN and 16 monolayers of GaN along $[10\bar{1}0]$. The dimensions for GaN were $36 \cdot c_{\text{GaN}}$ along [0001] and ... along $[\bar{1}2\bar{1}0]$ while the dimensions for AlN were $37 \cdot c_{\text{AlN}}$ along [0001] and ... along $[\bar{1}2\bar{1}0]$ and the supercells consist of about 185,000 atoms totally. The relaxed models of the two aforementioned cases are presented in Figs. 10.12 and 10.13.

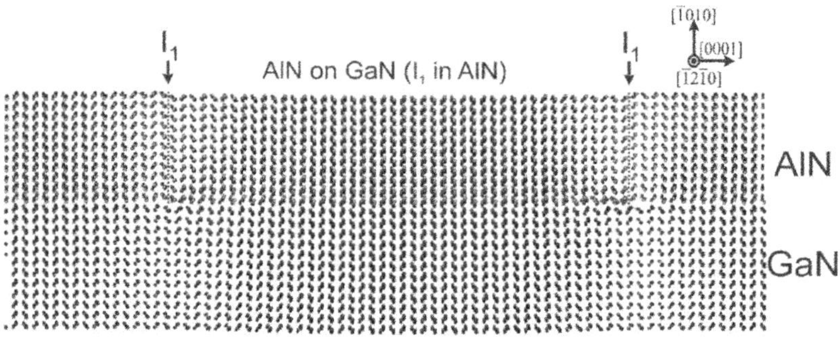

Figure 10.11 Relaxed GaN/AlN non-polar interface projected along the $[\bar{1}2\bar{1}0]$ direction having two I_1 SFs in AlN perpendicular to the interfaces.

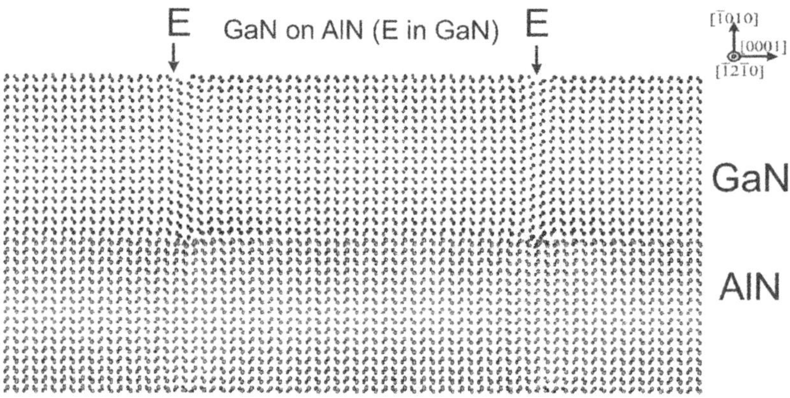

Figure 10.12 Relaxed GaN/AlN non-polar interface projected along the $[\bar{1}2\bar{1}0]$ direction having two E SFs in GaN perpendicular to the interfaces.

The interfacial energies of the relaxed models are calculated. They also compared with the interfacial energies of models with planar SF, SMB and I_1 SFs. and the results are presented in Table 10.3.

We conclude that the existence of E-type SF is the energetically favoured and consequently it should be present at least at the first stages of growth. However, since it requires almost three times the energy of the I_1-type SF, it will not be a stable configuration and a transition from the E-type to the I_1-type must be expected during

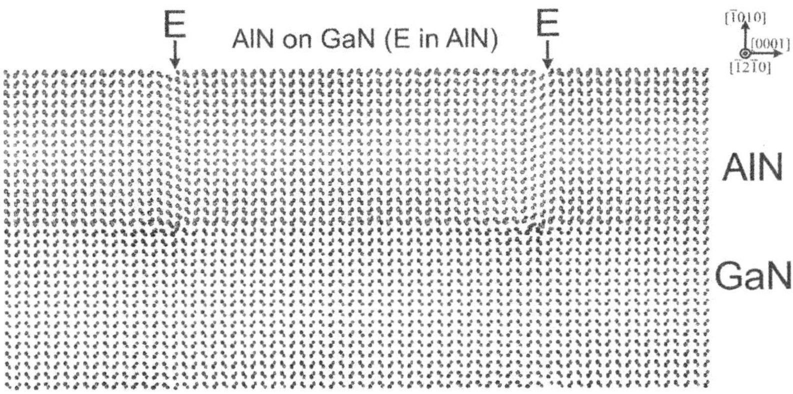

Figure 10.13 Relaxed GaN/AlN non-polar interface projected along the $[\bar{1}2\bar{1}0]$ direction having two E SFs in AlN perpendicular to the interfaces.

Table 10.3 Interfacial Energies calculated by molecular dynamic calculations

Type of SF	Interfacial energy (eV/ A^2)
E in GaN (GaN on AlN)	0.05
E in AlN (AlN on GaN)	0.07
SMB	0.09
I_1 in GaN (GaN on AlN)	0.10
I_1 in AlN (AlN on GaN)	0.17
Planar SF	0.18

growth. Also the "Step down" configuration is found energetically favourable with respect to the "Step up".

The formation of a SMB at the heteroepitaxial interface is also favoured. However, the sp^2 hybridization of the atoms along the interfacial area of SMB is not a stable configuration for the wurtzite structure. As a result in order to maintain the ground state interfacial structure, the formation of I_1-type SFs will be induced. Finally it is found that the introduction of a planar SF is the most energetically unfavourable case.

10.4 Diffusion

Diffusion is the transport of matter from one point to another by atomic motion. By this process inhomogeneous materials can convert to homogeneous and in order to take place the diffusion process the temperature should be high enough to overcome energy barriers to atomic motion. It occurs in all states of aggregation and it is relatively fast in gases, slow in liquids and very slow in solids. Generally the factors that influence diffusion are the temperature, the mechanism of diffusion, the diffusing and host species and the microstructure.

There are two different types of diffusion. In the "impurity diffusion" which occurs in response to a concentration gradient, atoms tend to migrate from regions of large concentration to regions with low concentration of impurities. In the self-diffusion, which occurs in one component material, all atoms that exchange positions are of the same type.

There are four different mechanisms of diffusion of atoms in a crystalline lattice. "Vacancy-assisted diffusion" occurs when a substitutional atom exchanges lattice positions with a vacancy. "Interstitial diffusion" occurs when an interstitial atom jumps to another interstitial position. "Interstitialcy diffusion" occurs when an interstitial atom pushes one of its nearest neighbours on a normal lattice site into another interstitial position and itself occupies the lattice site of the displaced atom. And finally the "ring diffusion".

10.4.1 Diffusion of Metal Vacancies through AlN/GaN Interfaces

The diffusion of metal vacancies in wurtzite (0001) AlN/GaN interfaces is investigated. The calculations were based on the density functional theory (DFT) within the local density approximation (LDA) and were performed by using VIENNA ab initio simulation package (VASP) [51, 52]. The Projector augmented-wave potentials were adopted for gallium, aluminium and nitrogen atoms and the gallium 3d electrons were included in the valence. The energy cut-off was set to 460 eV and the Brillouin zone was sampled by a $2\times2\times1$ Gamma centred k-point grid.

Three different cases were examined [2]. In the first case the a lattice parameter of GaN is adjusted to conform to that of AlN in order to simulate the pseudomorphic growth of GaN layer strained on AlN substrate. In the second case, the a lattice parameter of AlN is adjusted to conform to that of GaN leading to an AlN layer strained on GaN substrate and in the third case both AlN and GaN layers are allowed to freely relax. The third case represents the AlN/GaN multilayer where both the AlN and GaN layers are strained. For the third case where both AlN and GaN layers are strained, the energy variation of our optimized structure versus the strain induced along the basal plane is investigated. The optimized strained a' lattice constant in order to construct the final, stress free, supercell is calculated. In order to validate the optimized a' lattice constant the elasticity theory is considered. Following the biaxial formulation in wurtzite structure, the strain components along the basal plane (e_{xx}) and along the polar direction (e_{zz}) are defined by the expressions:

$$e_{xx} = \frac{a' - a}{a}, \quad e_{zz} = \frac{c' - c}{c}$$

while the relationship between them is

$$e_{zz} = -2\frac{C_{13}}{C_{33}}e_{xx}, \tag{10.1}$$

where a and c are the unstrained bulk lattice constants with a' and c' the strained values and C_{13} and C_{33} are elastic constants of either GaN or AlN.

As shown in Fig. 10.14, each supercell, which comprises 320 atoms, contains two different interfaces. It should be noticed that due to the fact that the <0001> direction of the wurtzite structure is a polar axis, the AlN/GaN supercells exhibit two distinct interfaces: the 3Ga–N–Al interface, where each N interface atom is bonded to one Al interface atom and three Ga interface atoms, and the Ga–N–3Al interface, where each N interface atom is bonded to three Al interface atoms and one Ga interface atom.

What is interesting, is that the 3Ga–N–Al interface repels the vacancies in contrast to the Ga–N–3Al interface which attracts them. Also it is known that gallium and aluminium vacancies can occur in

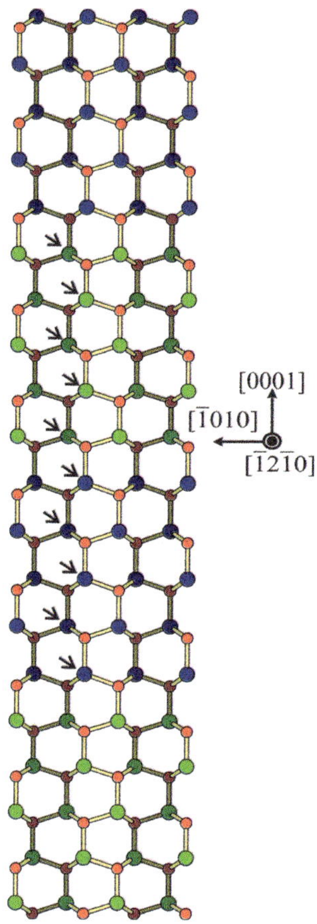

Figure 10.14 The relaxed atomistic model of the 3Ga–N–Al interface where each N interfacial atom is bonded to one Al and three Ga interface atoms and the Ga–N–3Al interface in which each N atom is bonded to three Al and one Ga interface atoms. The 10 positions of the vacancies that are investigated are also denoted.

0, –1, –2 and –3 charge states, but they have the lowest energies for the bulk AlN or GaN in the –3 charge state. Moreover, it is concluded that when the charge of the vacancies is increased then the concentration of the vacancies is increased on the Ga–N–3Al interface while it is decreased on the 3Ga–N–Al interface [2].

Therefore, we focused on triply negatively charged metal vacancies near the Ga–N–3Al interface.

Computing the formation energy of defects requires examining the modification of the chemical potentials of species upon introducing a vacancy in the computational box. In pure bulk III-N compounds, these satisfy the following relation [53]:

$$\mu^s_{\mathrm{IIIN}} = \mu_{\mathrm{III}} + \mu_{\mathrm{N}}, \tag{10.2}$$

where μ_{III} and μ_{N} represent the chemical potentials of metallic and N atoms, respectively, whereas the heat of formation of the III-N compound is given by

$$\Delta H^f_{\mathrm{IIIN}} = \mu^s_{\mathrm{IIIN}} - \mu^s_{\mathrm{III}} - \mu^g_{\mathrm{N_2}} \tag{10.3}$$

with, μ^s_{IIIN}, the chemical potential of the bulk wurtzite III-N and, μ^s_{III} and $\mu^g_{\mathrm{N_2}}$, the chemical potentials of the condensed (III metal) and of the gaseous (N2) phases, respectively. These expressions can conveniently be rewritten as

$$\begin{aligned} \mu_{\mathrm{III}} &= \mu^s_{\mathrm{III}} + (1-\lambda)\Delta H^f_{\mathrm{IIIN}} \\ \mu_{\mathrm{N}} &= \mu^g_{\mathrm{N_2}} + \lambda \Delta H^f_{\mathrm{IIIN}}, \end{aligned} \tag{10.4}$$

the parameter λ $(0 \le \lambda \le 1)$ defining N-rich $(\lambda = 0)$ or III-rich $(\lambda = 1)$ conditions. The chemical potential difference between species is then given by

$$\begin{aligned} \Delta\mu &= (\mu_{\mathrm{III}} - \mu^s_{\mathrm{III}}) - (\mu_{\mathrm{N}} - \mu^g_{\mathrm{N_2}}) \\ &= (\mu_{\mathrm{III}} - \mu_{\mathrm{N}}) - (\mu^s_{\mathrm{III}} - \mu^g_{\mathrm{N_2}}) = (1-2\lambda)\Delta H^f_{\mathrm{IIIN}} \end{aligned} \tag{10.5}$$

and should satisfy the obvious condition

$$\begin{matrix} -\Delta H^f_{\mathrm{IIIN}} & \le \Delta\mu \le & \Delta H^f_{\mathrm{IIIN}} \\ (\lambda = 1) & & (\lambda = 0)\;. \\ \text{III-rich} & & \text{N-rich} \end{matrix} \tag{10.6}$$

The formation energy of an electrically neutral defect at $T = 0$ K is then given by [54]

$$\begin{aligned} \Omega_{\mathrm{D}}(\mu_{\mathrm{III}}, \mu_{\mathrm{N}}) &= E_{\mathrm{D}} - \frac{1}{2}(n_{\mathrm{III}} + n_{\mathrm{N}})\mu^s_{\mathrm{IIIN}} \\ &\quad -\frac{1}{2}(n_{\mathrm{III}} - n_{\mathrm{N}})(\mu^s_{\mathrm{III}} - \mu^g_{\mathrm{N_2}}) - \frac{1}{2}(n_{\mathrm{III}} - n_{\mathrm{N}})\Delta\mu, \end{aligned} \tag{10.7}$$

where n_{III} and n_{N} represent the numbers of III-type and N atoms, respectively. By combining the relations above, one easily obtains

$$\Omega_{\mathrm{D}}(\mu_{\mathrm{III}}, \mu_{\mathrm{N}}) = E_{\mathrm{D}} - E^{\mathrm{total}}_{\mathrm{ref}} - n'_{\mathrm{III}}\mu^s_{\mathrm{IIIN}} - (n'_{\mathrm{N}} - n'_{\mathrm{III}})\mu_{\mathrm{N}}, \tag{10.8}$$

where $E_{\text{ref}}^{\text{total}}$ is the total energy of the reference structure, and n'_{III}, n'_{N} are the differences between the numbers of III and N atoms in the defected supercells and in the reference structures, respectively. Conversely, the formation energy of a defect in a specific charge state, Q, is given by

$$\Omega_{\text{D}}(\mu_{\text{III}}, \mu_{\text{N}}) = E_{\text{D}} - E_{\text{ref}}^{\text{total}} - n'_{\text{III}}\mu_{\text{IIIN}}^{s} - (n'_{\text{N}} - n'_{\text{III}})\mu_{\text{N}} + Q(E_{\text{D}}^{\text{Fermi}} + E_{\text{ref}}^{\text{VBM}}), \tag{10.9}$$

where $E_{\text{D}}^{\text{Fermi}}$ is the Fermi level energy of the defective system with respect to the valence band maximum (VBM), $E_{\text{ref}}^{\text{VBM}}$, in the reference system, shifted by the difference between average electrostatic potentials in the defected and the reference systems.

It is found that when the average a lattice constant of AlN and GaN was used we have the lowest formation energy of the aluminium and gallium vacancy for both N-rich and III-rich conditions. Also it is found that the formation energies are changed considerably depending on growth condition (N-rich or III-rich). More specifically, under N-rich conditions the formation energies are smaller for all the cases examined.

It is found that when we have aluminium or gallium vacancy the lowest strain corresponds to the AlN/GaN interface where the AlN is strained on the a GaN bulk lattice constant.

The structural analysis has been also performed for neutral metal vacancies near the Ga–N–3Al interface.

As before it is found that when we have aluminium or gallium vacancy the lowest strain corresponds to the AlN/GaN interface where the AlN is strained on the a GaN bulk lattice constant.

In Fig. 10.15 the neighbouring N atoms to vacancies bond strain for the neutral and the -3 charge state is presented with respect to bulk bond length. An outward relaxation of the N atoms surrounding the vacancy is observed and this behaviour is more pronounced for the Al vacancies. Moreover, it is also noticeable on the atomistic models that when we impose a smaller a lattice constant, i.e. for the case of both AlN and GaN strained and for the GaN layer strained on AlN, the strain of the N atoms surrounding the vacancy is increased. The lowest strain corresponds to the AlN strained case on bulk GaN. It is established that the strain is increased with the charge state of the vacancy.

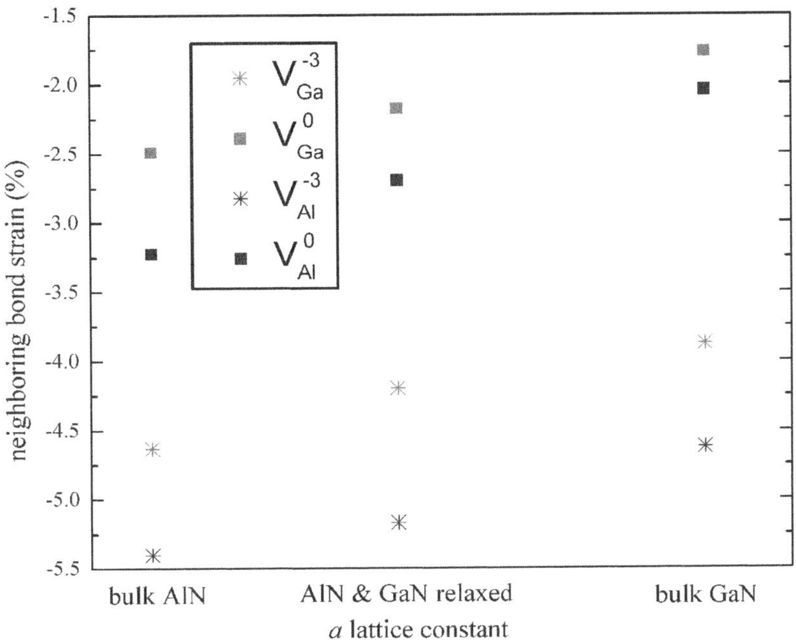

Figure 10.15 Bond strain of neighbouring to vacancies N atoms, with respect to bulk bond length versus the imposed *a* lattice constant of the three aforementioned cases.

In order to investigate the diffusion of a metal vacancy across the strained AlN/GaN interface a well-established method [55–57] is implemented. Starting from the relaxed atomistic configuration of the energetically favourable positions of the metal vacancies at the AlN/GaN interface, the migration path of a metallic atom through the interface towards the vacancy position is investigated. In Fig. 10.16a the schematic illustration of the migration path of a metallic atom (Al or Ga) across the AlN/GaN interface between the energetically favourable positions for Al and Ga vacancies is presented. Consequently, the final positions of the migration paths are an Al atom in the position of the Ga vacancy and an Al vacancy in the AlN and vice versa. Hence, the diffusion of a Ga vacancy across the interface is simulated by the diffusion of an Al atom towards the Ga vacancy and vice versa. The charge state of the vacancy in all the cases is considered $Q = -3$. In Fig. 10.16b the total energy

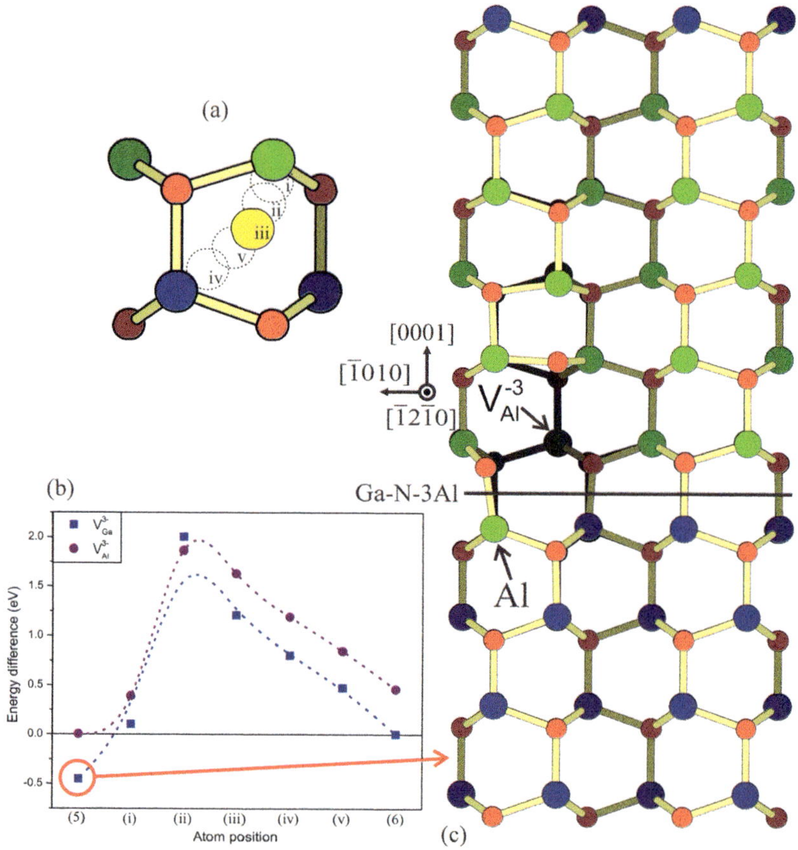

Figure 10.16 (a) Illustration of the migration path of a metallic atom (Al or Ga), across the GaN/AlN interface through their energetically favourable positions. The various atomic positions describing the diffusion process of a metal vacancy are denoted. (b) Total energy differences, referenced to the initial configurations (Al vacancy at the fifth position and Ga vacancy at the sixth position) during vacancy diffusion process. The final positions are an Al atom in the position of the Ga vacancy and vice versa. The charge state of the vacancy in all the cases is $Q = -3$. (c) The final configuration for Ga vacancy diffusion process is an Al atom in the position of the Ga vacancy and an Al vacancy in the AlN. This specific structural configuration is found energetically favourable with respect to the initial configuration by about 0.45 eV.

differences, referenced to the initial configurations of metallic atoms during the diffusion process is presented. It is concluded that the diffusion barrier for the Ga and Al vacancies across the interface is about 2.0 and 1.8 eV, respectively. Taking into account that the diffusion barrier of Ga vacancy in bulk GaN is calculated about 1.9 eV [55], it is concluded that the interfusion of the vacancies across the AlN/GaN interface does not require more energy. However, since the diffusion barrier is close to 2 eV, the metal vacancies will be immobile bellow about 800 K.

An interesting outcome has emerged concerning the final positions of the diffusion process. The energy difference of the final position of the diffusion process for the Al vacancy, with respect to the initial configuration is about 0.46 eV. On the contrary, the final position of the Ga vacancy diffusion is an Al atom in the position of the Ga vacancy and an Al vacancy in the AlN. This specific structural configuration is presented in Fig. 10.16c, and has been found energetically favourable with respect to the initial configuration by about 0.45 eV. This means that an Al atom in GaN close to the AlN/GaN interface in combination with an Al vacancy is considerable energetically favourable with respect to the most energetically favourable configuration of a Ga vacancy in GaN.

10.5 Surface Reconstructions: The Semipolar $\{10\bar{1}3\}$ GaN Surfaces

III-Nitrides are well established components of the opto- and micro- electronics industry, being implemented in light-emitting diodes (LEDs), laser diodes (LDs) as well as high electron mobility transistors for chemical and biological sensors [58]. Nevertheless, their common growth along the polar, *c*-axis of the wurtzite structure results in polarization induced electric fields, which hinder the realization of their full potential. To that end growth along semipolar orientations, i.e. inclined relative to the *c*-axis by angles $\theta \neq 90^\circ$, is acknowledged as a highly prominent approach. Recently proof has been provided that heterostructures' growth on $\{10\bar{1}1\}$

or $\{11\bar{2}2\}$ substrates reduces the quantum confined stark effect, while on $\{10\bar{1}3\}$ exact cancelation could be achieved [59]. Semipolar growth is also considered a key approach in resolving the "green gap" issue; fabrication of green LEDs (e.g. [60]) and LDs [61] has been up to date achieved through exploitation of semipolar InGaN/GaN quantum wells whereas in the particular case of $\{10\bar{1}3\}$ GaN interest is further exemplified since it comprises the primary facet orientation bounding polar quantum dots [62].

However, control over the growth and properties of semipolar material has proven to be an elaborate issue: Considering $\{10\bar{1}3\}$ GaN, consecutive growth of $(11\bar{2}2)$ parasitic crystallites and extensive twinning have been reported under both metal-organic vapour phase epitaxy (MOVPE) [63] and hydride vapour phase epitaxy (HVPE) [64, 65]. In the case of plasmaassisted molecular beam epitaxy (PAMBE), spontaneous formation of $(10\bar{1}3)$ crystallites was evidenced during $(11\bar{2}2)$ GaN growth [66].

This study is presented in detail in Ref. [3], in which a comprehensive analysis is performed on the atomic geometry, surface morphology and electronic structure of both polarity senses' $\{10\bar{1}3\}$ surfaces. DFT calculations are accomplished in two different cell periodicities, namely the 1×1 and 2×1. Surfaces were modelled under the slab geometry with each slab containing nine GaN double layers and fractionally charged pseudohydrogens to saturate dangling bonds on the bottom slab surfaces. Both ideal (cleaved) surfaces are stoichiometric, i.e. involving equal numbers of Ga and N atoms. Each Ga atom on $(10\bar{1}3)$ has one dangling bond (db) whereas N atoms have alternately one or two dbs. The situation is reversed on $(10\bar{1}\bar{3})$.

Total energy calculations were performed under the Generalized Gradient Approximation (GGA) using Projector Augmented Wave (PAW) pseudopotentials [67] and the VASP code [68]. The cut off energy was 500 eV and the Monkhorst-Pack mesh was $4\times4\times1$ for the 1×1 surface unit cells. A vacuum area of 12 Å and Periodic Boundary Conditions (PBC) were used normal to the surface and on the two perpendicular directions respectively. In order to explore the electronic structure, exchange and correlation were treated using the local density approximation under modified, norm conserving Troullier–Martins pseudopotentials (MPP-LDA), which result in a band gap value of 3.40 eV for bulk GaN thus overcoming the band gap underestimation problem [69].

Surface reconstructions were considered where Ga replace N atoms and vice versa. These structures hereafter referred to as Ga and N ML could be described as Ga–Ga and N–N dimers respectively, on top of the cleaved surface. Metal bilayer (BL) structures were also investigated, where N atoms in the topmost and first subsurface layers are replaced by Ga atoms. Various Ga and N adatom/vacancy reconstructions were examined. Finally, specific geometries of these surfaces comprise equal numbers of cation and anion derived dbs and satisfy the electron counting rule (ECR) [70].

It is revealed that a Ga adatom may be stabilized on the $(10\bar{1}3)$ GaN surface if bonded to a two-fold coordinated N atom (Ga adatom-D) while bonding to a triply coordinated N (Ga adatom-T) is less favourable by $\sim$0.026 eV/Å^2. A single N adatom bound to a surface Ga could be stabilized on the surface under extreme N-rich conditions. Nevertheless, a N adatom close to a surface N results in strong N–N bonds formation and desorption from the surface in the form of nitrogen molecules; the same effect was recorded for higher N content morphologies.

The two considered ECR morphologies are energetically degenerate and more stable than the cleaved surface (Fig. 10.17a) independent of growth stoichiometry. However, they are not the primary reconstructions favoured on $(10\bar{1}3)$. For extreme N-rich conditions, they are degenerate with the Ga adlayer structure (Fig. 10.17b), which remains favourable for the entire range of the Ga chemical potential. It is only at the Ga-extreme that formation of the Ga BL (Fig. 10.17c) could be observed.

In contrast to the remarkable stability of the Ga adlayer on $(10\bar{1}3)$ GaN, various transitions are predicted to take place on the opposite polarity surface depending on growth stoichiometry. The cleaved surface (Fig. 10.18a) is stabilized on $(10\bar{1}\bar{3})$ GaN for growth under extreme N-rich conditions and up to $\mu_{\text{Ga}} - \mu^s_{\text{Ga}} \leq -0.68$ eV. The Ga adatom reconstruction is favoured in the $-0.68 \leq \mu_{\text{Ga}} - \mu^s_{\text{Ga}} \leq -0.60$ eV range and formation of a Ga adlayer (Fig. 10.18b) is predicted for growth between $-0.60 \leq \mu_{\text{Ga}} - \mu^s_{\text{Ga}} \leq -0.23$ eV. From that point and up to the Ga-extreme, the Ga BL (Fig. 10.18c)

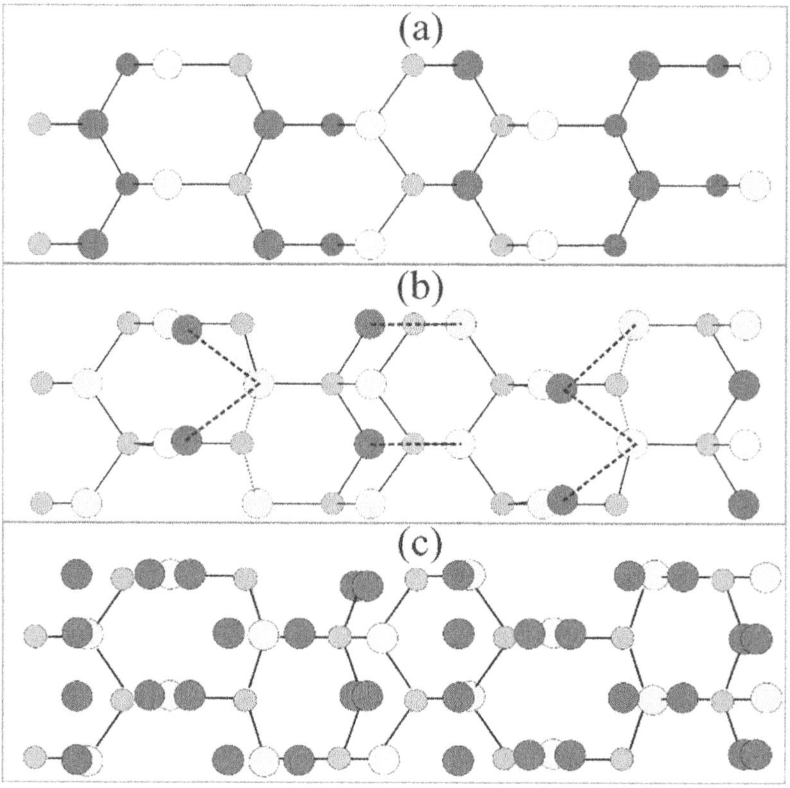

Figure 10.17 (a) Cleaved, relaxed $(10\bar{1}3)$ GaN surface. In (b) the Ga adlayer and in (c) the Ga BL are illustrated. Reconstructions are presented in top view, normal to $(10\bar{1}3)$. In all cases, the 2×1 periodicity is given for higher clarity. Large/ small spheres correspond to Ga/N atoms. Darker spheres designate topmost surface atoms.

is promoted. In that case as well, N-rich morphologies were found to be highly unstable inducing strong N–N bonds and desorption from the surface in the form of nitrogen molecules. The considered $(10\bar{1}\bar{3})$ ECR morphologies, i.e. single Ga vacancy or N adatom in 2×1 periodicity, are not favoured even under extreme N-rich conditions.

At this point it should be noted that although 1×1 and 2×1 periodicities of the $(10\bar{1}\bar{3})$ reconstructions were energetically degenerate, an exception was evidenced for the cleaved surface, where the 2×1 reconstruction is favoured. This feature is attributed

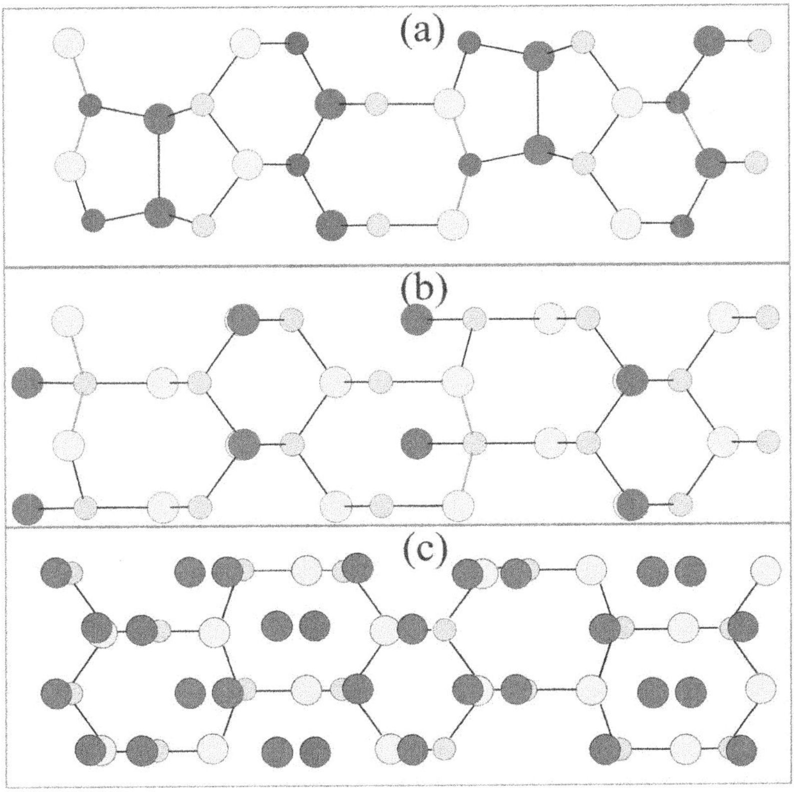

Figure 10.18 (a) Cleaved, relaxed (10$\bar{1}\bar{3}$) GaN surface. In (b), (c) the Ga adlayer and Ga BL structures, respectively, are presented. Illustrations are in accordance with Fig. 10.17.

to the doubly coordinated Ga surface atoms, which tend to minimize their free charge by forming Ga–Ga bonds as evident in Fig. 10.18a. For the 1×1 cell, these atoms form two bonds with surface N, with average bond length equal to 1.88 Å and their separation distance to neighbouring Ga atoms is 3.22 Å. In the 2× cell, the corresponding Ga–N bond length is 1.90 Å while the Ga–Ga distance is 2.49 Å, i.e. equal to the dimer-like covalent bond and ~9% contracted relative to the metallic bonds in a-Ga [71] and references therein]. These low coordinated Ga atoms relax towards sp^2 hybridization, exhibiting average bond angles equal to 116.3°. Triply coordinated Ga atoms are unaffected, residing at 1.90 Å from neighbouring N atoms.

The stabilization of metal-rich reconstructions on both polarity surfaces for a wide range of the Ga chemical potential, is a common trend of semipolar orientations as attested by recent studies [72–75] and appoints them a character closer to polar material ([76] and references therein). It is only under extreme N-rich conditions and up to $\mu_{\mathrm{Ga}} - \mu_{\mathrm{Ga}}^{s} \leq -0.68$ eV that the cleaved N-polarity surface is promoted, resembling *m*-plane GaN [77]. The similar metallic morphology of $\{10\bar{1}3\}$ and $\{11\bar{2}2\}$ orientations shall hence but result in competitive growth rates and pronounced interplay, as experimentally attested. However, an efficient scheme emerges for growth of $\{10\bar{1}3\}$ material isolated from $\{11\bar{2}2\}$ parasitic orientations, i.e. growth of $(10\bar{1}\bar{3})$ under moderate or extreme N-rich conditions. Taking into account previous results on $(11\bar{2}2)$ GaN [72, 73], the stabilization of Ga–N dimers shall result in enhanced incorporation probability of III- and N-species on $(10\bar{1}\bar{3})$ relative to $(11\bar{2}\bar{2})$ thus promoting the former growth front.

Furthermore metallic species on the $\{10\bar{1}3\}$ surfaces will play a key role in the enhancement of adatom mobility and the accomplishment of two-dimensional (2D) layer formation with smooth surface morphology. This behaviour could be augmented in extreme III-rich growth of $(10\bar{1}3)$ GaN and certainly in a wider stoichiometry range for N-polarity material, since a more efficient diffusion channel of metallic species is expected when metallic bilayers are formed.

10.6 Analysis on the Shape of the GaN Nanowires

It is well known that there are several types of nanostructures that may have several applications in sensors, optoelectronics and biomedical sciences.

The nanostructures can by classified in several categories. Taking into account that the growth of bulk materials is characterized as 3D growth mode and the layered structures as two-dimensional (2D) growth, the nanostructures can be considered in two categories:

(i) the one-dimensional (1D) nanostructures, i.e. mainly the nanowires (NWs) and their variations nanobelts, nanosprings and nanospirals
(ii) the well-known zero-dimensional (0D) nanostructures, the nanodots (NDs) or quantum dots (QDs)

Quantum-sized structures such as optimized NWs are considered to be highly attractive in optoelectronic as well as sensing applications, provided they can be electrically activated.

The LAMMPS molecular dynamics simulator [78] with a bond-order many body Tersoff potential [29, 30] is used for the molecular dynamics (MD) simulations. For Ga–N, Ga–Ga and N–N interactions the parameters by Nord et al. [33] are used.

In order to investigate the effect of the shape on the energetics of the GaN NWs several models are relaxed by MD. Specifically, NWs grown along the *c*-axis of trigonal, hexagonal and dodecagonal shapes with facets along the $\{10\text{–}10\}$ (*m*-plane) or $\{11\text{–}20\}$ (*a*-plane) planes (both in the case of the dodecagonal NW) are constructed (Fig. 10.19).

MD models consist of several thousands of atoms and have diameters from 0.2 to 80 nm. In all the cases, the NWs are surrounded by a vacuum, sufficiently big to ensure that each NW is far away from its nearest image. All the models are relaxed and the total energies are calculated. The total energies are divided by the number of pairs and the energy per Ga–N is extracted (Fig. 10.20).

The hexagonal NW with facets along the *m*-plane and the dodecagonal NW present the lowest energies by the use of MD calculations. This result is in agreement with the calculation results in Table 10.4 that show that the energy of the *m*-plane surface of GaN presents a lower energy than the *a*-plane surface. It is worth noting that the two shapes are almost energetically degenerate, especially in the small-diameter area, which corresponds to the initial stages of growth.

During this phase the bulk portion of the NW is small compared to the surface area. The bulk can increase with the reduction of the surface area, thus the NW prefers a more circular shape (dodecagonal and hexagonal). As the nanowire gets bigger, the

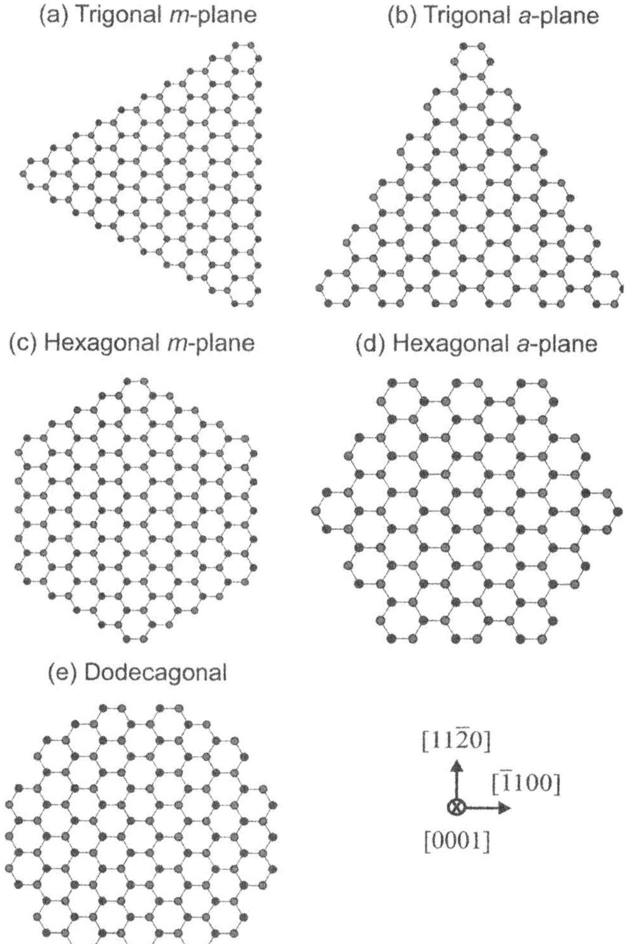

Figure 10.19 Atomistic models constructed: (a) trigonal NW with facets along the m-plane, (b) trigonal NW with facets along the a-plane, (c) hexagonal NW with facets along the m-plane, (d) hexagonal NW with facets along the a-plane and (e) dodecagonal NW with facets along both the m- and a-planes.

Table 10.4 Energy of GaN {10–10} (*m*-plane) and {11–20} (*a*-plane) surfaces in eV/Å^2

Surface	Northrup et al. [62]	Do et al. [63]	Kioseoglou et al. [64]
m-plane	0.117	0.133	0.123
a-plane	0.122	0.139	0.143

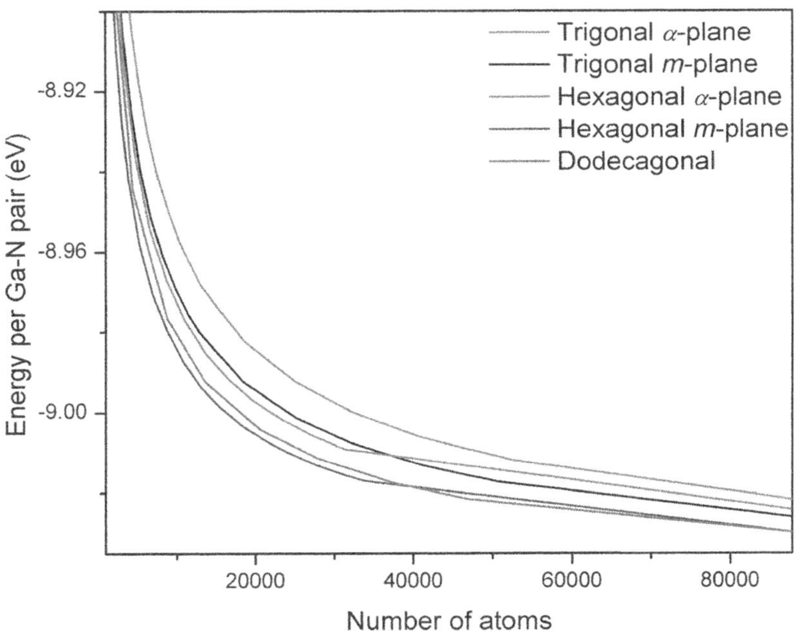

Figure 10.20 The energy per Ga–N pair versus the number of atoms for all the examined NW shapes, by the use of molecular dynamics calculations.

bulk portion increases and the facets switch to more energetically preferable ones (i.e. *m*-plane facets and hexagonal shape).

Acknowledgements

Part of this research has been co-financed by the European Union (European Social Fund—ESF) and Greek national funds through the Operational Program "Education and Lifelong Learning" of the National Strategic Reference Framework (NSRF)—Research Funding Program: ARCHIMEDES III Investing in knowledge society through the European Social Fund. Part of the computations in this work were supported by the LinkSCEEM-2 project, funded by the European Commission under the 7th Framework Program through Capacities Research Infrastructure, INFRA-2010-1.2.3 Virtua Research Communities, Combination of Collaborative Project and Coordination and Support Actions (CP-CSA) under grant agreement no RI-261600. The author is grateful to Professors Th. Karakostas, Ph. Komninou, Th. Kehagias, G. P. Dimitrakopulos and Dr. E. Kalesaki for valuable discussions, and Mr. M. Zaxaropoulos, Mrs. K. Iordanidou and Mr. Th. Pavloudis for their help.

References

1. J. Kioseoglou, Ph. Komninou and Th. Karakostas, *Phys. Stat. Sol. (b)*, 245(6), 1118–1124 (2008).
2. J. Kioseoglou, V. Pontikis, Ph. Komninou, Th. Pavloudis, J. Chen, and Th. Karakostas, *J. Phys.: Condens. Matter*, 27, 125006 (2015).
3. J. Kioseoglou, E. Kalesaki, L. Lymperakis, Th. Karakostas, Ph. Komninou, *J. Phys.: Condens. Matter*, 25(4), 045008 (2013).
4. A. Jezowski, B. A. Danilchenko, M. Bockowski, I. Grzegory, S. Krukowski, T. Suski, and T. Paszkiewicz, Solid State Commun. 128, 69 (2003).
5. A. Jezowski, O. Churiukova, J. Mucha, T. Suski, I. A. Obukhov, and B. A. Danilchenko, Mater. Res. Express 2, 085902 (2015).
6. X. W. Zhou, S. Aubry, R. E. Jones, A. Greenstein, and P. K. Schelling Phys. Rev. B 79, 115201 (2009).
7. L. Lindsay, D. A. Broido, and T. L. Reinecke, Phys. Rev. Lett. 109, 095901 (2012).
8. T. E. Beechem, A. E. McDonald, E. J. Fuller, A. A. Talin, C. M. Rost, J.-P. Maria, J. T. Gaskins, P. E. Hopkins and A. A. Allerman, *J. Appl. Phys.*, 120, 095104 (2016).
9. D. Florescu, V. Asnin, F. H. Pollak, R. Molnar and C. Wood, *J. Appl. Phys.*, 88, 3295 (2000).
10. R. B. Simon, J. Anaya and M. Kuball, *Appl. Phys. Lett.*, 105, 202105 (2014).
11. T. Pavloudis, K. Termentzidis, P. Komninou, C. D. Latham, P. R., Briddon, J., Kioseoglou, *J. Appl. Phys.*, 119 (7), art. no. 074304 (2016).

12. K. Termentzidis, M. Isaiev, A. Salnikova, I. Belabbas, D. Lacroix, J. Kioseoglou, Impact of screw and edge dislocations on the thermal conductivity of GaN nanowires.
13. Z. Su, L. Huang, F. Liu, J. P. Freedman, L. M. Porter, R. F. Davis and J. A. Malen, *Appl. Phys. Lett.*, 100, 201106 (2012).
14. K. Termentzidis and S. Merabia, Molecular dynamics simulations and thermal transport at the nano-scale, in *Molecular Dynamics: Theoretical Developments and Applications in Nanotechnology and Energy* (ed. L. Wang), DOI: 10.5772/2383 (2002).
15. S. Merabia and K. Termentzidis, *Phys. Rev. B*, 86, 094303 (2012).
16. D. J. Evans, *Phys. Rev. A*, 34(2), 1449–1453 (1986).
17. D. Frenkel and B. Smit, *Understanding Molecular Simulation*, Academic Press, San Diego, pp. 75, 90 (2002).
18. S. Kotake and S. Wakuri, *JSME Int. J., Ser. B*, 37, 103 (1994).
19. K. Termentzidis, P. Chantrenne and P. Keblinski, *Phys. Rev. B*, 79, 214307 (2009).
20. K. Termentzidis, T. Barreteau, Y. Ni, S. Merabia, X. Zianni, Y. Chalopin, P. Chantrenne and S. Volz, *Phys. Rev. B*, 87, 125410 (2013).
21. I. Belabbas, P. Ruterana, J. Chen, and G. Nouet, *Philos. Mag.*, 86(15), 2241 (2006).
22. J. Chen, P. Ruterana, and G. Nouet, *Mater. Sci. Eng. B*, 82, 117 (2001).
23. A. Bere and A. Serra, *Phys. Rev. B*, 65, 205323 (2002).
24. J. Kioseoglou, G. P. Dimitrakopulos, Ph. Komninou, and Th. Karakostas, *Phys. Rev. B*, 70, 035309 (2004).
25. J. Kioseoglou, G. P. Dimitrakopulos, Ph. Komninou, H. M. Polatoglou, A. Serra, A. Béré, G. Nouet, and Th. Karakostas, *Phys. Rev. B*, 70, 115331 (2004).
26. M. Kohyama, *Modell. Simul. Mater. Sci. Eng.*, 10, R31 (2002).
27. M. Z. Bazant, E. Kaxiras, and J. F. Justo, *Phys. Rev. B*, 56, 8542 (1997).
28. F. Stillinger, T. A. Weber, *Phys. Rev. B*, 31, 5262 (1985).
29. J. Tersoff, *Phys. Rev. B*, 39, 5566 (1989).
30. J. Tersoff, *Phys. Rev. B*, 41, 3248 (1990).
31. K. Albe, K. Nordlund, J. Nord, and A. Kuronen, *Phys. Rev. B*, 66, 035205 (2002).
32. W. H. Moon, H. J. Hwang, *Phys. Lett. A*, 315(3–4), 319 (2003).
33. J. Nord, K Albe, P. Erhart, and K. Nordlund, *J. Phys.: Condens. Matter*, 15, 5649 (2003).

34. J. Nord, K. Nordlund, J. Keinonen, and K. Albe, *Nucl. Instr. Meth. Phys. Res. B*, 202, 93 (2003).
35. A. S. Nandedkar and J. Narayan, *Philos. Mag. A*, 56, 625 (1987).
36. A. S. Nandedkar and J. Narayan, *Philos. Mag. A*, 61, 873 (1990).
37. K. Nordlund, J. Nord, J. Frantz, and J. Keinonen, *Comp. Mater. Sci.*, 18, 283 (2000).
38. L. Nurminen, F. Tavazza, D. P. Landau, A. Kuronen, and K. Kaski, *Phys. Rev. B*, 67, 035405 (2003).
39. A. Bere and A. Serra, *Philos. Mag.*, 86, 2159 (2006).
40. J. Tersoff, *Phys. Rev. B*, 49, 16349 (1994).
41. M. Ishimaru, K. Yoshida, and T. Motooka, *Phys. Rev. B*, 53, 7176 (1996).
42. F. Benkabou, M. Certier, and H. Aourag, *Mol. Simulation*, 29(3), 201 (2003).
43. S. B. Austerman and W. G. Gehman, *J. Matter. Sci.*, 1, 249(1966).
44. D. B. Holt, *J. Phys. Chem. Solids.*, 30, 1297 (1969).
45. J. E Northrup, J. Neugebauer, and L. T. Romano, *Phys. Rev. Lett.*, 77, 103 (1996).
46. H. Blank, P. Delavignette, R. Gevers, and S. Amelinckx, *Phys. Stat. Sol.*, 7, 747 (1964).
47. C. M. Drum, *Philos. Mag. A*, 11, 313 (1965).
48. L. Verlet, *Phys. Rev.*, 159, 98 (1967).
49. J. Elsner, M. Kaukonen, M. I. Heggie, M. Haugk, Th. Frauenheim and R. Jones, *Phys. Rev. B*, 58, 15347–15323 (1998).
50. J. E. Northrup, *Phys. B*, 273, 130 (1999).
51. G. Kresse and J. Furthmüller, *Comput. Mater. Sci.*, 6, 15–50 (1996).
52. G. Kresse and J. Furthmüller, *J. Phys. Rev. B*, 54, 11169–11186 (1996).
53. S. Limpijumnong and C. G. Van de Walle, *Phys. Rev. B*, 69, 035207 (2004).
54. G. X. Qian, R. Martin, and D. Chadi, *Phys. Rev. B*, 38, 7649 (1998).
55. S. Limpijumnong and C. G. Van de Walle, *Phys. Rev. B*, 69, 035207 (2004).
56. C. G. Van de Walle and J. Neugebauer, *J. Appl. Phys.*, 95, 3851 (2004).
57. T. Trevethan, C. D. Latham, M. I. Heggie, M. J. Rayson, and P. R. Briddon, *Phys. Rev. B*, 90, 174108 (2014).
58. R. F. Service, *Science*, 327, 1598 (2010).
59. P. Bigenwald, B. Gil, F. Benharrats, K. Zitouni, and A. Kadri, *Semicond. Sci. Technol.*, 27, 024009 (2012).
60. M. Funato, M. Ueda, Y. Kawakami, Y. Narukawa, T. Kosugi, M. Takahashi, and T. Mukai, *Jpn. J. Appl. Phys.*, 45, L659 (2006).

61. Y. Enya, Y. Yoshizumi, T. Kyono, K. Akita, M. Ueno, M. Adachi, T. Sumitomo, S. Tokuyama, T. Ikegami, K. Katayama, and T. Nakamura, *Appl. Phys. Exp.*, 2, 082101 (2009).
62. B. Daudin, *J. Phys. Condens. Matter*, 20, 473201 (2008).
63. S. Ploch, M. Frentrup, T. Wernicke, M. Pristovsek, M. Weyers, and M. Kneissl, *J. Cryst. Gr.*, 312, 2171 (2010).
64. T. J. Baker, B. A. Haskell, F. Wu, J. S. Speck, and S. Nakamura, *Jpn. J. Appl. Phys.*, 45, L154 (2006).
65. T. J. Baker, B. A. Haskell, F. Wu, P. T. Fini, J. S. Speck, and S. Nakamura, *Jpn. J. Appl. Phys.*, 44, L920 (2005).
66. L. Lahourcade, J. Renard, B. Gayral, E. Monroy, M. P. Chauvat, and P. Ruterana, *J. Appl. Phys.*, 103, 093514 (2008).
67. G. Kresse and D. Joubert, *Phys. Rev. B*, 59, 1758 (1999).
68. G. Kresse and J. Furthmüller, *Phys. Rev. B*, 54, 11169 (1996).
69. D. Segev, A. Janotti, and C. G. Van de Walle, *Phys. Rev. B*, 75, 035201 (2007).
70. M. D. Pashley, *Phys. Rev. B*, 40(10) 481 (1989).
71. K. Albe, K. Nordlund, J. Nord, and A. Kuronen, *Phys. Rev. B*, 66, 035205 (2002).
72. T. Yamashita, T. Akiyama, K. Nakamura, and T. Ito, *Jpn. J. Appl. Phys.*, 48, 120201 (2009).
73. T. Akiyama, D. Ammi, K. Nakamura, and T. Ito, *Jpn. J. Appl. Phys.*, 48, 100201 (2009).
74. J. E. Northrup, *Appl. Phys. Lett.*, 95, 133107 (2009).
75. T. Yamashita, T. Akiyama, K. Nakamura, and T. Ito, *Jpn. J. Appl. Phys.*, 49, 018001 (2010).
76. J. Neugebauer, *Phys. Stat. Sol. (b)*, 227, 93 (2001).
77. J. E. Northrup and J. Neugebauer, *Phys. Rev. B*, 53, R10477 (1996).
78. S. Plimpton, *J. Comput. Phys.*, 117(1), 1–19 (1995).
79. d J. Neugebauer, *Phys. Rev. B*, 53, R10477 (1996).
80. E. C. Do, Y.-H. Shin, and B.-J. Lee, *J. Phys.: Condens. Matter*, 21, 325801 (2009).
81. Pavloudis, Th. Kehagias, Ph. Komninou, Th. Karakostas, C. D. Latham, M. J. Rayson, P. R. Briddon, and M. Eickhoff, *J. Appl. Phys.*, 118, 034301 (2015).

Part II

FABRICATION AND MEASUREMENT TECHNIQUES

Chapter 11

Amorphization of Porous Silicon Nanostructures by Heavy Ion Irradiation

B. Canut,[a] P. Newby,[a,b] M. Massoud,[a] V. Lysenko,[a] P. O. Chapuis,[c] and J. M. Bluet[a]

[a] *Université de Lyon, Institut des Nanotechnologies de Lyon (INL), CNRS, INSA de Lyon, F-69621 Villeurbanne, France*

[b] *C2MI 45 Boulevard de l'Aéroport, Bromont, QC J2L 1S8, Canada*

[c] *Université de Lyon, Centre d'Energétique et de Thermique de Lyon (CETHIL), CNRS, INSA de Lyon, F-69621 Villeurbanne, France*

bruno.canut@insa-lyon.fr, jean-marie.bluet@insa-lyon.fr

11.1 Introduction

The control of silicon thermal conductivity (TC) from the high bulk value of 156 $W.m^{-1}.K^{-1}$ down to the very low values of insulating materials, like glass (1–2 $W.m^{-1}.K^{-1}$) became a crucial issue for the micro-electronics industry. Indeed, new technologies development, compatible with CMOS process, in order to control thermal flux in highly integrated systems (3D integration) are highly desired by the industry. Not only heat dissipation issue has to be addressed but also heat conversion for the fast growing market of wireless application. This thematic is defined as a major one to be studied

Nanostructured Semiconductors: Amorphization and Thermal Properties
Edited by Konstantinos Termentzidis

ISBN 978-981-4745-64-2 (Hardcover), 978-1-315-36445-2 (eBook)
www.panstanford.com

in the forthcoming years by the Nanoelectronics Initiative Advisory Council (ENIAC) [1].

One of the most efficient approaches to reduce the TC of Si is to fabricate porous silicon (PSi) by electrochemical etching of single crystal silicon wafers (c-Si). By this way, TC values 2–3 orders of magnitude lower than that of bulk c-Si have been demonstrated in the past [2, 3]. However, reducing TC by increasing porosity, is limited by the detrimental effect of porosification on the mechanical properties of PSi [4]. A trade-off must therefore be found between TC and mechanical performance. One approach, explored in the past, is to slightly oxidize PSi. By this way, a reduction of TC by a factor of two has been achieved [5, 6]. Here again this process has a limitation due to swelling and stress in the PSi layer which are detrimental for further processing [7, 8]. Furthermore, a too important oxidation implies oxygen incorporation which reduces porosity. This makes TC to increase again beyond a given limit [5, 6]. Consequently, alternative techniques for reducing the TC of PSi without changing its porosity or damaging it are desirable. The chosen way is to amorphize Si as disorder is known to reduce thermal transport in crystalline solids [9]. In the case of Si this gives a two orders of magnitude reduction with values ranging between 1 and 5 $W.m^{-1}.K^{-1}$ depending on the fabrication process [10].

In this study, we evaluate ion beam treatment onto PSi wafers in view of an additional decrease of their TC due to partial or total amorphization of the material. For this purpose, we have irradiated PSi targets of different porosities (41%, 56%, and 75%) with projectiles accelerated in the electronic regime (energy per mass unit above 0.5 MeV/u). Up to now, the radiation-induced effects in porous materials have been mainly studied in porous silicon by using a large variety of incident ions (from H to Au) and energies (10 keV–100 MeV). As an example, the results obtained for columnar and spongy type porous Si irradiated with 4 MeV He^+ ions evidence a densification which occurs preferentially in a narrow depth region around the mean projected range of the ions. This effect was ascribed to the displacement cascades resulting from collisional interaction processes [11]. To our knowledge, the influence of the inelastic processes (electronic slowing-down) on possible defect creation or structural changes in PSi was not yet investigated. In most insulating materials irradiated at sufficient electronic slowing-down, defects are created in the wake of each projectile (the so-called "latent track") [12]. Besides, it is well established that the usual single crystalline semiconductors (Si, Ge, SiC) irradiated with monoatomic heavy ions accelerated in the GeV range cannot be damaged via electronic processes [13]. Many experiments, conducted on high energy accelerators at the highest electronic stopping powers (S_e) available (i.e., $\approx$25 keV.nm^{-1} with ^{238}U ions of 1 GeV) evidenced that the radiation-induced disorder in bulk Si is only due to nuclear collisions and thus occurs at the end of range of the projectiles [14, 15]. Amorphous latent tracks in Si single crystals have been only registered when using C_{60} clusters as projectiles [16–18]. The insensitivity of bulk Si to irradiation in the electronic regime can be mainly ascribed to its rather high thermal diffusivity (D = 0.9 cm^2.s^{-1}). According to the thermal spike model [19–21], the prompt energy received by the target electrons after the passage of the projectile (10^{-17} s per unit cell) gives rise to a local thermalization of the electron gas within a period of $\approx 10^{-15}$ s. This step is followed by an electron–phonon coupling, causing a local temperature increase (thermal spike). For a few 10^{-12} s, a molten phase develops around the ion path and produces the so-called track after ultrafast quenching. Such a scenario does not occur in materials having a high thermal conductivity, like metals and semiconductors, as the lattice temperature cannot reach the melting temperature. The situation should be completely different in PSi samples whose TC is by two orders of magnitude lower than that of bulk Si. One can thus expect a local temperature rise up to the melting point around the projectile path, inducing the amorphization of the crystalline skeleton and consequently a significant lowering of the thermal conductivity. The aim of this study is then to investigate the "response" of PSi targets to ion irradiation dominated by electronic slowing-down processes.

11.2 Experiments

11.2.1 Porous Silicon Formation

Porous silicon samples were formed by electrochemical etching of p+ monocrystalline (100) Si wafers (resistivity 0.01–0.02 Ωcm). This was carried out in a Teflon cell (Fig. 11.1), using a (1:1)

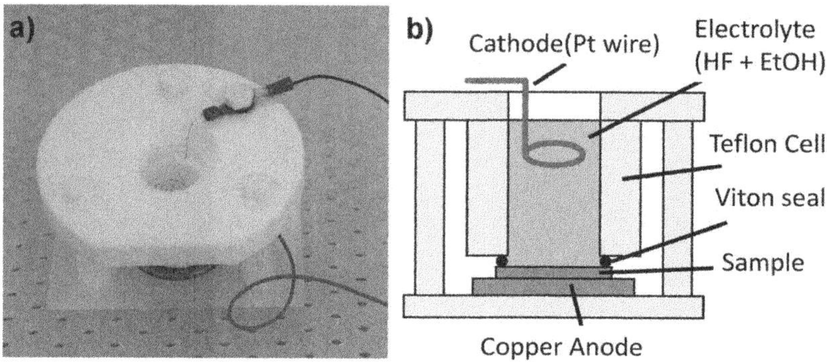

Figure 11.1 (a) Photography of a Teflon used for electrochemical etching of Si. (b) Transverse schematic drawing of the cell.

48% hydrofluoric acid and anhydrous ethanol mixture. A pulsed current was used, with zero-current etch-stop periods to improve uniformity of the PSi layers [22]. Their thickness was 10 μm, and three different porosities were used: 41%, 56%, and 75%. This parameter was controlled via the anodization current density and the effective porosity was measured by infrared reflectivity using an effective medium model for the refractive index.

A scanning electron microscopy (SEM) image of PSi fabricated in this way is presented in Fig. 11.2. The PSi formed is of dendritic silicon quasi-columns, with a diameter on the order of 10 nm, perpendicular to the wafer surface, and separated by pores. These quasi-columns remain monocrystalline and retain the crystalline orientation of the substrate. This morphology is typical of the so-called mesoporous structure of PSi.

11.2.2 Swift Heavy Ion Irradiation Experiments

The PS targets were irradiated at the SME and IRRSUD beamlines of the GANIL accelerator [23], using ^{130}Xe ions (91 MeV and 29 MeV energies), ^{208}Pb ions (545 MeV energy) or ^{238}U ions (110 MeV and 850 MeV energies). The fluences ranged from 10^{12} to 7×10^{13} cm^{-2}. All the irradiations were performed at room temperature with a flux limited to 10^9 cm^{-2}.s^{-1} in order to avoid any overheating of the targets. The main irradiation parameters, calculated in bulk

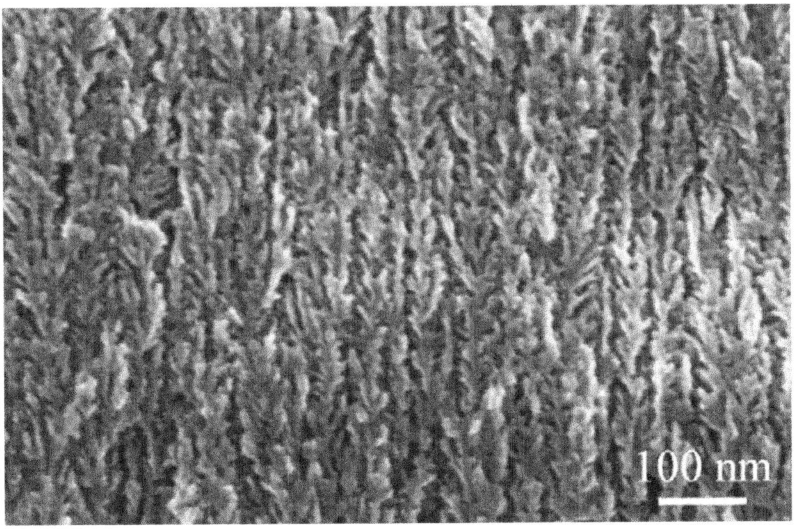

Figure 11.2 SEM image of mesoporous Si with a porosity of 56%.

Si from the SRIM2013 code [24], are listed in Table 11.1. The nuclear damage cross sections A_n indicated in the last column of Table 11.1 correspond to elastic displacements induced per one incident projectile at the target surface. They were calculated assuming a displacement energy of 21 eV [25] for silicon atoms.

As illustrated in Fig. 11.3 (irradiation of Si with ^{238}U ions), the electronic slowing-down exceeds the nuclear one by two orders of magnitude.

Table 11.1 Main features of the ion irradiations in Si: incident energy (E), projected range (R_p), electronic (S_e) and nuclear (S_n) stopping powers

Projectile	E (MeV)	R_p (μm)	S_e (keV.nm^{-1})	S_n (keV.nm^{-1})	A_n (nm^2)
^{130}Xe	29	6.8	6.4	0.24	0.032
^{130}Xe	91	13.7	10.9	0.10	0.012
^{208}Pb	545	37.2	22.2	0.07	0.009
^{238}U	110	14.2	12.3	0.35	0.040
^{238}U	850	47.6	27.0	0.07	0.008

Note: A_n is the cross section for disordering the target surface via elastic processes (see text).

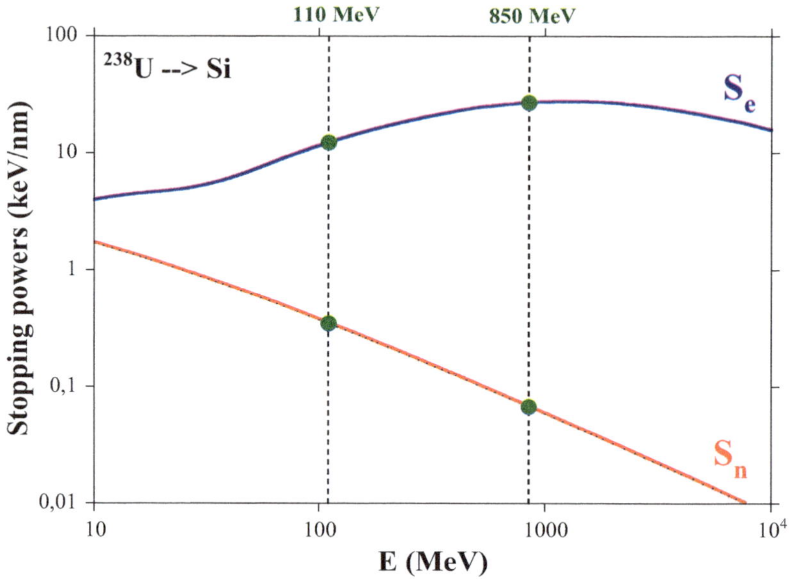

Figure 11.3 Energy dependence of the electronic (S_e) and nuclear (S_n) stopping powers in the case of silicon irradiated with ^{238}U ions.

11.2.3 Characterization Experiments

The samples composition and surface chemistry were studied by infrared spectroscopy (Bruker Vertex 80 spectrometer) in attenuated total reflectance (ATR) mode using a Ge prism (Specac). Their morphology was evaluated using SEM in cross section on cleaved samples and Raman spectroscopy (Renishaw RM 1000 spectrometer) in micro Raman mode (objective magnification 50) with a laser excitation at 532 nm. During the Raman experiment, the laser power was reduced in order to prevent sample heating by the laser beam. Indeed, as the 56% PSi TC is about 20 times lower than the one of bulk material, usual laser powers in the 10 mW range cause heating and consequently spectral shifting to lower energies and broadening.

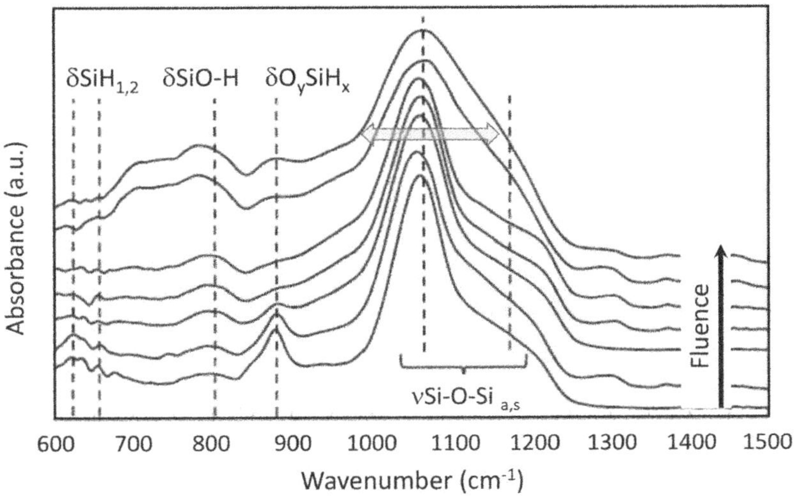

Figure 11.4 Infrared absorbance spectra obtained in ATR mode for 56% porosity Si irradiated with ^{238}U at 110 MeV. For easier observation, the spectra have been shifted vertically starting from the non-irradiated sample up to the sample irradiated with the highest fluence (10^{12} cm^{-2}; 3×10^{12} cm^{-2}; 7×10^{12} cm^{-2}; 10^{13} cm^{-2}; 3×10^{13} cm^{-2} and 7×10^{13} cm^{-2}). The horizontal arrow shows the peaks broadening occurring for the highest fluences.

11.3 Results

11.3.1 Surface Chemistry

Considering the huge surface to volume ratio of mesoporous Si, the surface chemistry may play a key role for different physical properties like electrical and thermal conduction. This surface chemistry may also be modified by ion irradiation. The infrared absorbance in the middle infrared range has been measured for the different samples. As the same feature was observed for the different ions and energy, only the case of ^{238}U irradiation is displayed in Fig. 11.4 as an example.

With increasing ion fluence, the principal observations are the following:

(i) The SiH and SiH_2 bending modes at 625 cm^{-1} and 660 cm^{-1} respectively vanish for fluencies higher than 10^{12} cm^{-2}. The same is observed for the SiH, SiH_2 stretching modes at 2082 cm^{-1} and 2112 cm^{-1} (not presented in Fig. 11.4).

(ii) A pronounced peak at 880 cm^{-1} attributed to O_ySiH_x bending mode in PSi [26] also vanishes quickly. For the highest fluence, a broad peak with small low energy shift, which may be due to Si–O–Si bending, seems to reappear.

(iii) The broad peak at 800 cm^{-1} attributed to SiO–H bending slightly increases with irradiation fluence. The same is observed for the broad stretching mode band of SiO–H between 3100 cm^{-1} and 3600 cm^{-1} (not presented in Fig. 11.4).

(iv) The symmetric and asymmetric Si–O–Si stretching modes at 1070 cm^{-1} and 1180 cm^{-1} respectively are broadening especially for the highest fluence (3×10^{13} cm^{-2} and 7×10^{13} cm^{-2}).

These observed features indicate a global reduction but not total disappearance of the surface H bonding accompanied by a slight oxidation of the sample for all fluences higher than 10^{12} cm^{-2}.

The Si–O–Si stretching bands are not growing in intensity as it would be expected for a volume oxidation of the crystallites but are broadening due to the formation of Si–O surface modes and also due to partial amorphization of the sample. This amorphization of the sample is clearly evidenced by the broadening of all the peaks in the spectrum (horizontal arrow in Fig. 11.4 for the Si-O-Si peak) for the 3×10^{13} cm^{-2} and 7×10^{13} cm^{-2} fluences.

11.3.2 Morphology and Amorphous Fraction

The evolution of PSi morphology with increasing ion fluence is observed by SEM. In Fig. 11.5, an example of the typical evolution is given for ^{238}U ion irradiation in a PSi of 75% porosity. For the low irradiation fluence (10^{12} cm^{-2}) as for the non-irradiated sample (Fig. 11.2), the ramification of the dendritic structure ends with small spike structures with a width of few nanometers. As the fluence is increased up to 7×10^{13} cm^{-2} the PSi skeleton remains unchanged and the sample porosity does not change. This

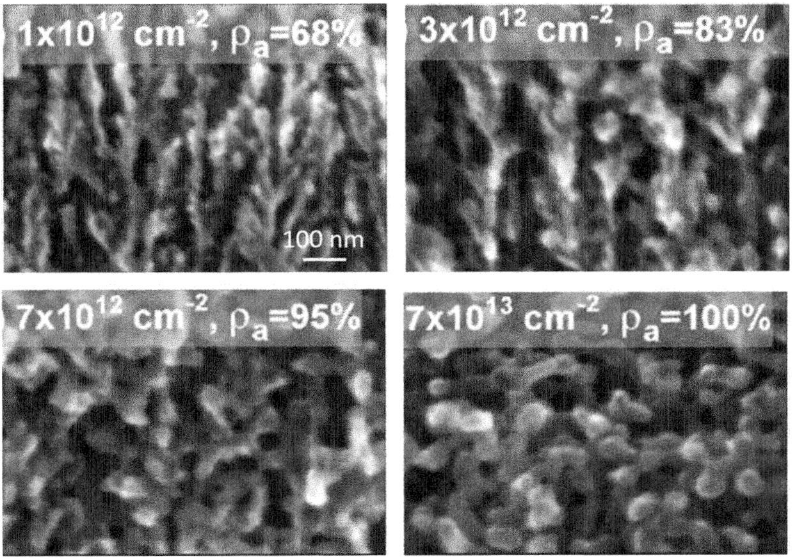

Figure 11.5 SEM images of PSi with 75% porosity with increasing ^{238}U at 110 MeV irradiation fluence. ρ_a in the images legend is the amorphous fraction determined by Raman Spectroscopy (see text).

confirms that the interaction regime is mainly electronic. Indeed, in the case of nuclear interaction with heavy ions like ^{238}U or ^{130}Xe, the cascade of collisional damage is expected to crush down the porous structure as explained in the introduction and for light $^4He^+$ ion irradiation in the MeV range [11]. Nevertheless, for 3×10^{13} cm^{-2} and 7×10^{13} cm^{-2} fluences, the small ramifications ending the dendrite are turning into beads. This morphology at the ramification ending calls to mind solidification of molten Si droplet, which may have been formed during irradiation. Indeed, according to the thermal spike model [19] discussed in the following, the small PSi terminations can melt locally, if the deposited energy density is high enough. If PSi crystallites are not totally melted, liquid phase epitaxial regrowth occurs during cooling down and the crystalline nature is kept. In the opposite case, for higher fluences, if the wall crystallite is molten, amorphous Si will form as it appears in infrared measurement. In order to obtain a more accurate evaluation of the

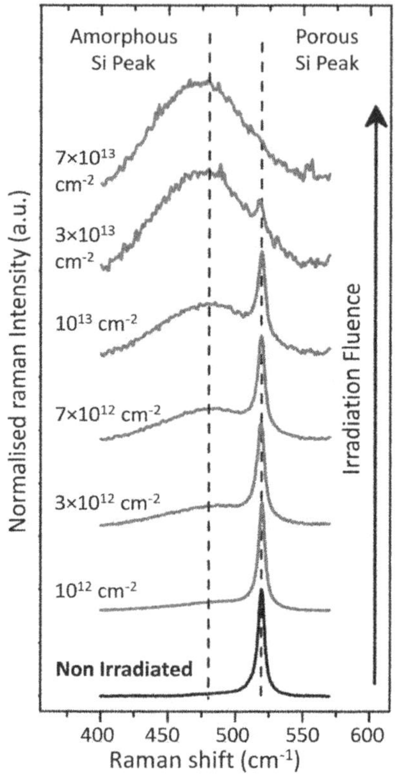

Figure 11.6 Raman spectra for Psi with 56% porosity irradiated with ^{238}U at 110 MeV. The vertical dashed lines at 480 cm^{-1} and 520 cm^{-1} are guide for the eyes to follow the intensity evolution of amorphous Si Raman peak and PSi Raman peak respectively.

amorphous fraction (ρ_a in the following) present in the sample, Raman spectroscopy is used.

Toward this end, Raman spectra were collected for the different ions and different fluences. The spectra obtained with increasing the ion fluence are given in Fig. 11.6 in the case of ^{238}U irradiation at 110 MeV. As can be observed, a broad band centered at 480 cm^{-1} increases in intensity as the fluence is enhanced while the PSi sharp peak vanishes and finally totally disappears for the highest fluence (7×10^{13} cm^{-2}). This qualitative observation demonstrates the appearance of an amorphous phase in the PSi samples as the fluence is increased. Indeed, Si amorphization results in selection rules breakdown, which makes all phonon modes observable in Raman spectroscopy and not only the optical mode in Brillouin zone center ($k = 0$). Consequently a complex spectrum in which the intensity is mainly proportional to phonon density of states, as all phonons can be seen, is extending from 0 to 521 cm^{-1}. In order to extract more quantitative information, an empirical way to determine the amorphous fraction from this spectrum is to compare the PSi crystalline Raman peak area (I_c) to the area of the TO branch peak appearing at 480 cm^{-1} (I_a) in amorphous material. The amorphous fraction is calculated using the following expression [27]:

$$\rho_a = \frac{yI_a}{(I_c + yI_a)} \quad \text{with } y = \frac{\Sigma_c}{\Sigma_a},$$

where y is the ratio of the crystalline and amorphous Raman cross section (Σ_c and Σ_a, respectively). More details on I_a and I_c extraction can be found in ref. [28].

The correct value for y is still debated in the literature, with values anywhere between 0.1 and 1.7 having been published [27, 29–33]. The simple value $y = 1$ was chosen (i.e., the Raman cross sections of amorphous and crystalline (or porous) Si are considered to be the same) in this work in order to compare all the different samples.

Figure 11.7 displays, for the different studied porosities, the fluence evolution of the amorphous fraction ρ_a in PS targets irradiated with ^{238}U ions of 110 MeV energy. In the three cases, the damage kinetics were fitted using Poisson's law:

$$\rho_a = 1 - \exp(-A_d.\Phi),$$

where Φ is the fluence and A_d is the amorphization cross section. Although not entirely satisfactory to account for the experimental data at all fluences, these fittings allowed to extract the following values for A_d: 14 nm^2, 26 nm^2, and 95 nm^2 for 41%, 56%, and 75% porosities, respectively. Such damage cross sections exceed by three orders of magnitude the cross sections A_n for displacement events via nuclear processes reported in Table 11.1. This confirms that the disorder creation in PS targets is necessarily due to the electronic slowing-down of the projectiles. The increase of A_d with the porosity

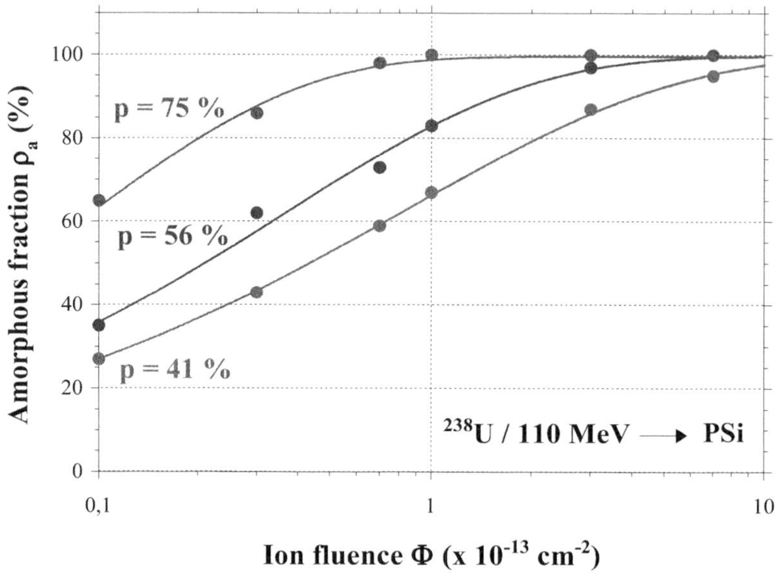

Figure 11.7 Amorphous fraction (ρ_a) versus the fluence of PS targets irradiated with ^{238}U ions of 110 MeV energy. Three different porosities were used: 41% (green), 56% (blue), and 75% (red). The continuous curves are the best fits of the data using Poisson's law (see text).

P can be ascribed to the thermal conductivity K_{PS} of the mesoporous target which, in the case of a columnar morphology, decreases with P according to the following relation [6]:

$$K_{PS} = \frac{K_{Si}}{1 + \frac{4\Lambda_{Si}}{3d_{cr}}}(1 - P)^3,$$

where K_{Si} is the thermal conductivity of bulk Si ($K_{Si} = 150$ W.m^{-1}.K^{-1}), Λ_{Si} is the phonon mean free path in single crystalline Si ($\Lambda_{Si} = 43$ nm at room temperature) and d_{cr} is the mean size of the nanocrystallites. The obtained values ($K_{PS} = 7.7$ W.m^{-1}.K^{-1}, 2.8 W.m^{-1}.K^{-1} and 0.4 W.m^{-1}.K^{-1} for $P = 41\%$, 56% and 75%, respectively) are by two orders of magnitude lower than K_{Si}. In these conditions and according to the thermal spike model, the target temperature after electron–phonon coupling should reach the melting temperature thus inducing amorphous track formation along the projectile path. It is worthwhile to notice that in the case of the highest porosity (75%), the damage cross section corresponds to

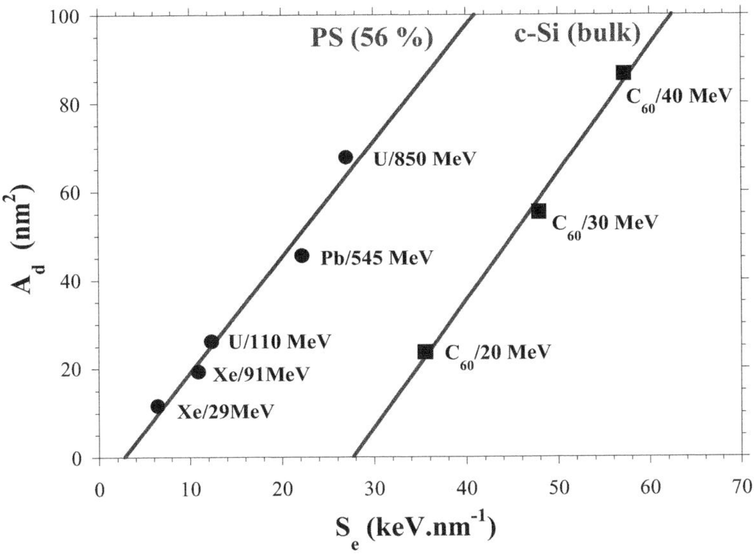

Figure 11.8 Damage cross sections (A_d) versus the incident electronic stopping power (S_e) for 56% porosity PS targets (circles). Previous results obtained in bulk Si irradiated with C_{60} clusters (squares) are also given for comparison. In both cases, the continuous straight lines are the best fits to the data.

a track diameter $d_{tr} = 11$ nm ($A_d = \pi d_{tr}^2/4$) which is very close to the crystallite size ($d_{cr} = 12$ nm for $P = 75\%$). This indicates that only one projectile impact onto the crystallite is sufficient to amorphize it entirely in the first micrometers below the surface.

From Raman spectroscopy and its quantitative exploitation described above, the damage cross sections were measured for the different target porosities and irradiation conditions [34, 35]. The evolution of A_d versus the electronic stopping power S_e for a target of medium porosity (56%) is plotted in Fig. 11.8. A linear fitting of the data intercepts the horizontal axis at $S_{eth} \approx 3$ keV.nm^{-1}. Within the experimental errors and by neglecting the possible influence of the different projectile velocities, this value may be regarded as the threshold for damage creation in PS of 56% porosity irradiated in the electronic regime. As a comparison, from C_{60} bombardment in the 10 MeV range, a much higher threshold $S_{eth} \approx 30$ keV.nm^{-1} was estimated in bulk single crystalline Si [16]. The low threshold evidenced in the studied targets, which is comparable to that obtained in some insulating materials like SiO_2 [36] and $LiNbO_3$ [37], confirms the increased sensitivity of columnar porous silicon submitted to irradiation in the electronic stopping power regime.

11.4 Conclusions

We have shown that swift heavy ion irradiation with 110 MeV and 850 MeV ^{238}U ions, 91 MeV and 29 MeV ^{129}Xe ions, and 545 MeV ^{208}Pb ions can create an amorphous phase in mesoporous silicon, while no effect is observed in bulk material. Combined with a partial surface oxidation observed in FTIR, this amorphization results in a considerable reduction of the PSi TC (see following chapters). This amorphization is assumed to result from the following mechanism: the PSi small ramification melt due to electronic interaction, as can be explained by the thermal spike model, and lose their crystallinity during re-solidification. Nevertheless, the PSi keeps its global structure with no change in porosity. As a consequence, it is possible to adjust the amorphous volume fraction and therefore the TC by varying the ion fluence.

References

1. http://www.eniac.eu/web/downloads/SRA2007.pdf.
2. G. Gesele, J. Linsmeier, V. Drach, J. Fricke, R. Arens-Fischer, *J. Phys. D* 30, 2911 (1997).
3. V. Lysenko, S. Perichon, B. Remaki, D. Barbier, *Sens. Actuators A* 99, 13 (2002).
4. C. Populaire, B. Remaki, V. Lysenko, D. Barbier, H. Artmann, T. Pannek, *Appl. Phys. Lett.* 83, 1370 (2003).
5. S. Périchon, V. Lysenko, P. Roussel, B. Remaki, B. Champagnon, D. Barbier, P. Pinard, *Sens. Actuators A* 85, 335 (2000).
6. V. Lysenko, S. Perichon, B. Remaki, D. Barbier, B. Champagnon, *J. Appl. Phys.* 86, 6841 (1999).
7. K. Imai, H. Unno, *IEEE Trans. Electron Devices* 31, 297 (1984).
8. K. Barla, R. Herino, G. Bomchil, *J. Appl. Phys.* 59, 439 (1986).

9. D. G. Cahill, S. K. Watson, R. O. Pohl, *Phys. Rev. B* 46, 6131 (1992).
10. Y. He, D. Donadio, G. Galli, *Appl. Phys. Lett.* 98, 144101 (2011).
11. A. Simon, F. Paszti, A. Manuaba, A. Z. Kiss, *Nucl. Instrum. Phys. Res., Section B* 158, 658 (1999).
12. M. Toulemonde, S. Bouffard, F. Studer, *Nucl. Instrum. Phys. Res., Section B* 91, 108 (1994).
13. W. Wesh, A. Kamarou, E. Wendler, *Nucl. Instrum. Phys. Res., Section B* 225, 111 (2004).
14. M. Levalois, P. Bogdanski, M. Toulemonde, *Nucl. Instrum. Phys. Res., Section B* 63, 14 (1992).
15. P. Mary, P. Bogdanski, M. Toulemonde, R. Spohr, J. Vetter, *Nucl. Instrum. Phys. Res., Section B* 62, 391 (1992).
16. B. Canut, N. Bonardi, S. M. M. Ramos, S. Della-Negra, *Nucl. Instrum. Phys. Res., Section B* 146, 296 (1998).
17. A. Dunlop, G. Jaskierowicz, S. Della-Negra, *Nucl. Instrum. Phys. Res., Section B* 146, 302 (1998).
18. A. Colder, "Contribution à la compréhension des effets d'irradiation dans les semi-conducteurs germanium, silicium et arséniure de gallium," PhD thesis, Caen University (2001).
19. M. Toulemonde, C. Dufour, E. Paumier, *Phys. Rev. B* 46, 14362 (1992).
20. C. Dufour, A. Audouard, F. Beuneu, J. Dural, J. P. Girard, A. Hairie, M. Levalois, E. Paumier, M. Toulemonde, *J. Phys.: Condens. Matter* 5, 4573 (1993).
21. C. Dufour, F. Beuneu, E. Paumier, M. Toulemonde, *Europhys. Lett.* 45, 585 (1999).
22. S. Billat, M. Thönissen, R. Arens-Fischer, M. G. Berger, M. Krüger, H. Lüth, *Thin Solid Films* 297, 22 (1997).
23. Grand Accélérateur National d'Ions Lourds, Bd Henri Becquerel, BP 55027-14076 CAEN Cedex, France.
24. J. F. Ziegler, J. P. Biersack, U. Littmark, *Stopping Power and Ranges of Ions in Matter*, vol. 1 (Pergamon, New York, 1985).
25. J. Bourgoin, M. Lannoo, *Point Defects in Semiconductors*, vol. 1 (Springer, Berlin, 1983).
26. Douglas B. Mawhinney, John A. Glass Jr., John T. Yates Jr., *J. Phys. Chem. B* 101 (7), 1202–1206 (1997).
27. E. Bustarret, M. A. Hachicha, and M. Brunel, *Appl. Phys. Lett.* 52, 1675 (1988).

28. P. Newby, B. Canut, S. Gomes, K. Termentzidis, P. Chantrenne, V. Aimez, L. Fréchette, J. M. Bluet, V. Lysenko, *J. Appl. Phys.* 114 (1), 014903 (2013).
29. R. Tsu, J. Gonzalez-Hernandez, S. S. Chao, S. C. Lee, K. Tanaka, *Appl. Phys. Lett.* 40, 534 (1982).
30. E. Vallat-Sauvain, C. Droz, F. Meillaud, J. Bailat, A. Shah, and C. Ballif, *J. Non-Cryst. Solids* 352, 1200 (2006).
31. C. Smit, R. A. C. M. M. van Swaaij, H. Donker, A. M. H. N. Petit, W. M. M. Kessels, M. C. M. van de Sanden, *J. Appl. Phys.* 94, 3582 (2003).
32. D. Han, J. D. Lorentzen, J. Weinberg-Wolf, L. E. McNeil, and Q. Wang, *J. Appl. Phys.* 94, 2930 (2003).
33. M. Ledinsky, A. Vetushka, J. Stuchlık, T. Mates, A. Fejfar, J. Kocka, J. Stepanek, *J. Non-Cryst. Solids* 354, 2253 (2008).
34. B. Canut, M. Massoud, P. Newby, V. Lysenko, L. Fréchette, J. M. Bluet, I. Monnet, *Nucl. Instrum. Methods Phys. Res., Sect. B* 327, 99–102 (2014).
35. M. Massoud, B. Canut, P. Newby, L. Frechette, P. O. Chapuis and J. M. Bluet, *Nucl. Instrum. Phys. Res., Section B* 341, 27–31 (2014).
36. A. Meftah, F. Brisard, J. M. Costantini, E. Dooryhee, M. Hage-Ali, M. Hervieu, J. P. Stoquert, F. Studer and M. Toulemonde, *Phys. Rev. B* 49, 12457 (1994).
37. B. Canut, S. Ramos, N. Bonardi, J. Chaumont, H. Bernas, E. Cottereau, *Nucl. Instrum. Phys. Res., Section B* 122, 335 (1997).

Chapter 12

Crystalline/Amorphous Porous Ge Nanostructures by Electrochemical Etching

Sergii Tutashkonko[a] **and Tetyana Nychyporuk**[b]

[a]*Graduate School of Engineering, Nagoya University, 464-8603, Furo-cho, Chikusa-ku, Nagoya-shi, Aichi-Ken, Japan*

[b]*INL, Site INSA, Bât. Blaise Pascal, 7 Avenue Jean Capelle, 69621 Villeurbanne, France*

s.tutashkonko@gmail.com

12.1 Introduction

Electrochemical porosification is a frequently used method to realize semiconducting nanostructures. The method consists of the electrochemical etching of a semiconductor material, such as Si, Ge, SiC, and GaAs, resulting in the formation of the cavities (pores) in bulk and the crystalline skeleton in-between the pores. The advantages of this method are simplicity, compatibility with the CMOS process, ability to control the morphology of the porous layers, and preservation of the long-range crystalline order in the resulting nanostructures.

Nanostructured Semiconductors: Amorphization and Thermal Properties
Edited by Konstantinos Termentzidis

ISBN 978-981-4745-64-2 (Hardcover), 978-1-315-36445-2 (eBook)
www.panstanford.com

The porous Silicon (PGe) was discovered in 1956 by Arthur Uhlir Jr. and Ingeborg Uhlir from Bell Labs while developing the process to polish and to texture the surface of Si and Ge. Following the discovery in 1990 of the intense room-temperature photoluminescence emission from nanostructured porous Si (Canham, 1990; Cullis and Canham, 1991), porous semiconductors have attracted considerable interest owing to their potential application in nanoelectronics, nanophotonics, photocatalysis, energy conversion, and biological sensing (Föll et al., 2002). Besides Si, significant attention has been also paid to Ge. Indeed, Ge, like Si, is a semiconductor with a broad range of properties that could push the breadth of applications to new levels. It has one of the highest dielectric constants (ε(Ge) $\approx$ 16.2) among the conventional semiconductors, a very high refractive index (n(Ge) $\approx$ 4.14 in the near infrared), and thus is of a noticeable interest for photonic crystal applications (El Kurdi et al., 2008). The lack of a stable oxide makes the functionalization of its surface quite easy (Loscutoff and Bent, 2006). Dimensional reduction of Ge from the bulk to the nanoscale size induces the appearance of its new complementary functionalities such as quantum confinement effects (Nataraj et al., 2010), highly developed internal surface area and size selective absorption (Gaponenko, 2005). Porous layers with crystallographic pores (for ex. with the columnar morphology) are of particular interest for their birefringence properties (Golovan' et al., 2007).

The porous materials are characterized by multiple quantitative parameters (Rouquerol et al., 1994). The **porosity** P is a ratio of the total volume of the pores V_p to the total volume of the porous layer V. In case of a powder of porous particles, V is the apparent volume of the particles excluding the inter-particle voids.

$$P = \frac{V_p}{V} \tag{12.1}$$

The **size of the pores** L_p is the distance between two opposite pore walls. The value of L_p has a meaning when the geometric shape of pores is known and well defined (for example cylindrical, spherical, slit-like, etc.) Usually, L_p refers to the smallest measure of a pore (the width of a slit-like pore, the diameter of a cylindrical pore, the diameter of a spherical pore). By analogy, L_{cr} is the **size of a crystallite** in-between the pores. Finally, the pore packing density is given by the parameter L_{ip}—**the interpore distance**, defined as a distance between the centers of two adjacent pores. According to the International Union of Pure and Applied Chemistry (IUPAC) classification, there are three types of pores (Sing et al., 1985):

- micropores ($L_p < 2$ nm)
- mesopores ($L_p = 2$–50 nm)
- macropores ($L_p > 50$ nm)

A homogenous material, which contains most of the pores of the same class, is called a *micro/nanoporous, mesoporous* or *macroporous* material. The **mean values** of L_p, L_{cr}, and L_{ip} in a uniform porous layer are represented with the parameters d_p, d_{cr}, and d_{ip}, respectively. The crystallites separated by pores are interconnected by crystalline necks of mean size d_c. This latter parameter is particularly important, since it determines the capacity of the porous material to transfer electric current or heat.

The shape of individual pores and their arrangement determine the morphology of the porous material. The pores can be classified with respect to their shape (Fig. 12.1). They can be spherical (a), cylindrical (c,f), bottle-shaped (b), funnel-shaped (d) or slit-shaped. A rough surface (g) cannot be considered as porous unless the depth or the pores is more that their width. The pores can also be classified

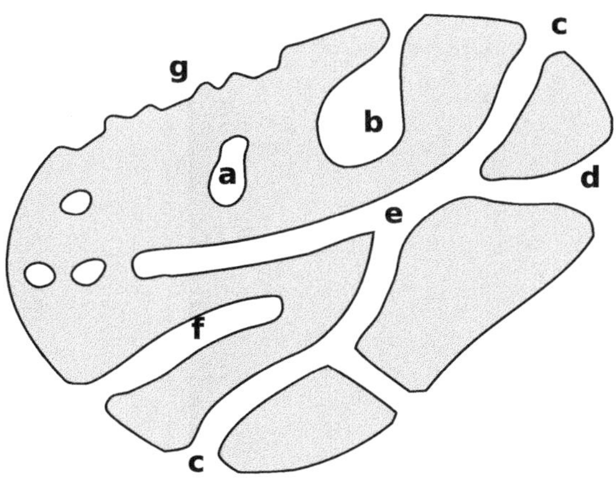

Figure 12.1 Schematic cross-sectional view of a porous solid (Rouquerol et al., 1994).

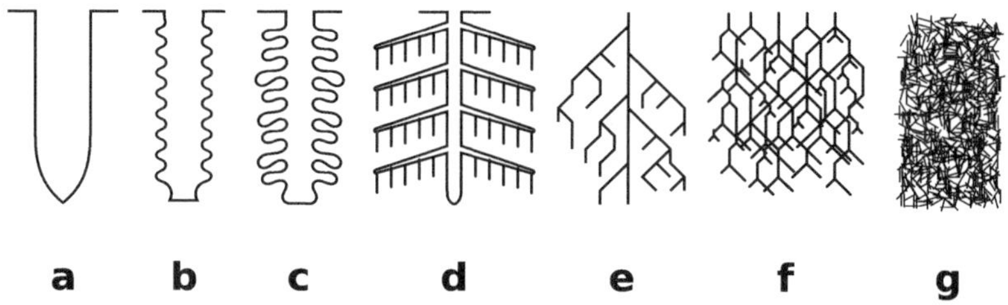

Figure 12.2 Classification of pores with respect to their branching, inspired by the reference (Zhang, 2006): (a) Smooth pore walls. (b) Branches are shorter than the diameter of a main pore. (c) Second level branches only. (d) Main pores with the branches of the second and of the third level. (e) Dendritic branches: recursive branching of the lateral branches. (f) Dendritic branches without a clearly defined central trunk. (f) Short dense random branches.

depending on their accessibility to the electrolyte (Rouquerol et al., 1994): the closed (a) and the open pores (b,c,d,e,f). The closed pores strongly affect the macroscopic properties of a porous material, such as the bulk density, mechanical resistivity, and thermal conductivity, but they are inactive for interaction with fluids or gas absorption.

In case of the pores formed by electrochemical etching, two major classes of pores can be distinguished: crystallographic pores (those which propagate along the crystallographic planes of the substrate independently on the substrate orientation) and current-line pores (following the direction of the electric current flow).

Finally, the pores can be classified with respect to branching of individual pores and interconnection of adjacent pores (Fig. 12.2).

Hereafter, the naming of the pores will refer to their visual appearance ("columns," "dendrites," "sponge," "trees," "fish-bones," etc.) along with an explanation in terms of characteristics presented above.

12.2 Experimental Setup

Porosification is performed in an electrochemical cell schematically represented in Fig. 12.3. The housing of the cell is made of Teflon^{TM}, resistant to the action of acids. The wafer of Si or Ge is made in direct

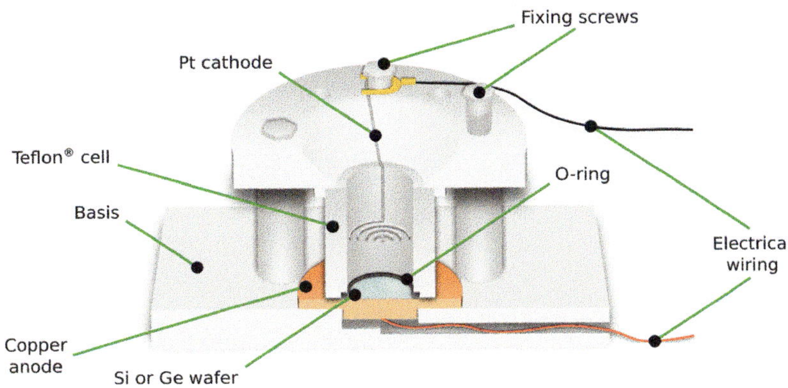

Figure 12.3 Schematic representation of a cross-sectional view of the electrochemical cell used for porosification (Tutashkonko, 2013).

contact with the copper electrode by its rear face. The surface area to porosify is delimited by a viton O-ring. A cylindrical cell containing the electrolyte is located above. The counter-electrode made of Pt or a Pt/Rh alloy is located at mid-height of the cell. The shape of the counter-electrode must ensure the lateral uniformity of the electric field in-between two electrodes At the same time, as the electrolyte is stirred in the upper part of the cell, the counter-electrode should not affect the flow of electrolyte in the cell. If a large wafer is used, the cell can be modified to make a wet contact with the rear face of the wafer to avoid the potential drop across the wafer and so to ensure the lateral uniformity of the potential distribution.

In case of Si, the etching is performed by applying DC current of a density J within the range of 0.1–100 mA/cm^2. The current can also be applied in form of a square wave function with the high level J and the low level of “0” to ensure evacuation of the molecular hydrogen generated in-depth the pores during the etching. Applying even more complicated current profiles, like a saw wave function, enables modulation of the shape of pores (Wang et al., 2014). In case of Ge, the range of the current density is much more narrow (0.1–10 mA/cm^2). Moreover, the current is applied in form of a square wave with the length of positive t_+ and negative t_- pulses of 0.1–3 s.

Finally, the electrolyte is typically composed of hydrofluoric acid (HF) diluted with ethanol to decrease the surface tension. In case of

Si, the HF content in the electrolyte can be varied in a wide range. However, to realize mesoporous Ge layers, the HF concentration should be no less than 40%.

Once the etching is finished, the cell is drained, the sample is rinsed in ethanol and dried in air. Thick porous layers can be rinsed in hexane or dried using the supercritical drying process to ensure its mechanical stability and avoid cracks.

12.3 Mesoporous Ge

In general, all structural and morphological properties of porous semiconductors, such as the thickness, the porosity, the diameter of pores, the size of crystallites are controlled by the following anodization process parameters:

- anodic current density (J_+)
- anodization time (t_{etch})
- composition of the electrolyte ($HF_{\%}$)
- substrate type and doping level
- irradiation intensity

PSi is a well-studied material. The influence of the above-mentioned parameters on the morphology and the structure of the PSe layers can be found elsewhere (Föll et al., 2002; Lehmann, 2002). In the simplest case of a DC etching of Si, its morphology is mainly determined by the substrate doping, whereas the size of the pores (and so do the porosity) depend on the electrolyte composition and the applied current density.

While PGe can only be formed at a very restricted range of experimental conditions, the use of bipolar electrochemical etching adds more degrees of freedom. Three more parameters should be considered:

- duration of anodic and cathodic pulses (t_+, t_- respectively)
- cathodic current density (J_-)

The unique feature of PGe is that its morphology depends on the applied current density and the passivation degree, allowing to change the morphology on the fly and to engineer complex multilayer nanostructures. Here we present examples of different morphologies of mesoporous Ge. Some of these morphologies (like "fish-bones") are unique to PGe and have not been seen in other porous semiconductors.

12.3.1 Morphology Tuning: Anodic Current Density

Figure 12.4a–f presents SEM images of the interface between bulk Ge and the porous region for different current densities and two etching times: (i) 3 h and (ii) >6 h. The pulse duration t_+/t_- was fixed to 1 s/2 s. For "higher" current densities (>1.8 mA/cm^2) the pores show a clear orientation and separation (Fig. 12.4a,b). When the total etching time is "short" enough (≈3 h), the individual pores propagate straight in the preferential direction ⟨100⟩ (Fig. 12.5) with very small levels of branching (Fig. 12.4a), thus forming a columnar morphology with short second level branches only. The growth of numerous lateral and branched pores was observed for longer etching times (>6 h) (Fig. 12.4b). In this case, the lateral pores even had second, third, or further levels of branches. Contrary to the main pores, the branches do not appear to be oriented toward clear crystal directions. Independently of the etching time, the application of the "low" current densities (<1.8 mA/cm^2) produced a highly interconnected pore network without a clearly defined growth direction, as illustrated in Fig. 12.4c–f. The current densities below 1.6 mA/cm^2 are characterized by a dendritic morphology (Fig. 12.4e,f) while the current density range of 1.6–1.8 mA/cm^2 results in a sponge-like morphology (Fig. 12.4c,d).

Table 12.1 lists the values of the average pore diameter of PGe layers presented in Fig. 12.4. Some general trends can be observed. First, the average pore diameter of PGe layers as obtained from SEM images shows a monotonic slightly increases with current density. Second, all the PGe layers formed at "low" current densities (<1.8 mA/cm^2) (Fig. 12.4c–f) are characterized by small pores with a narrow size distribution (4–5 nm). In the case of "higher" current densities (>1.8 mA/cm^2) the pore diameter strongly depends on the etching time. For the total etching time ≈3 h, the diameter of the

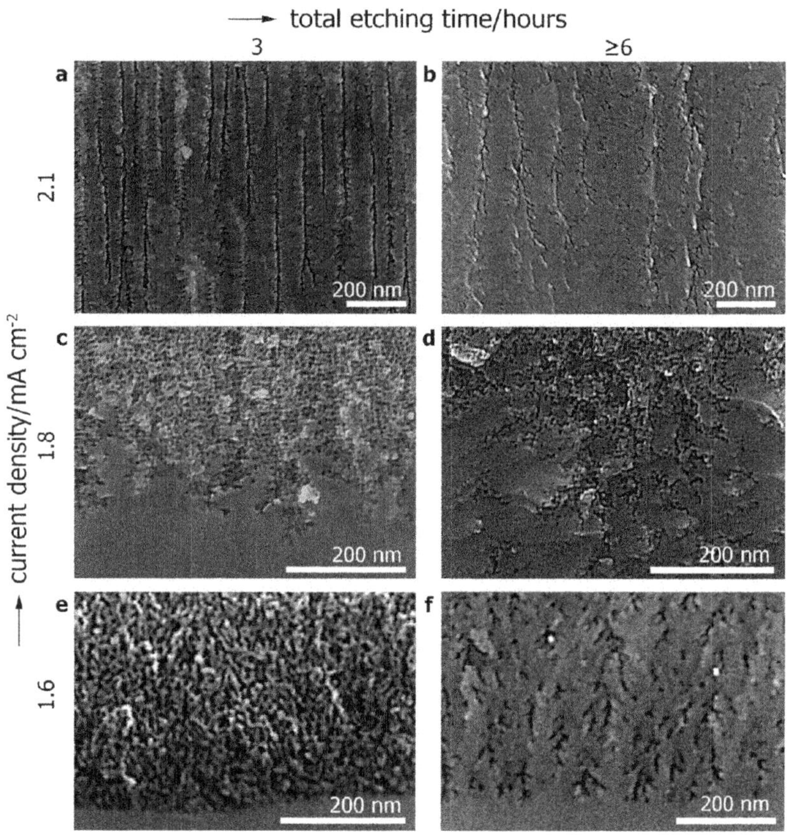

Figure 12.4 SEM images of the interface between bulk Ge and the porous region for different current densities and two etching times: (i) 3 h and (ii) >6 h. Six morphologies could be identified: (a) "Columnar" morphology with short second level branches only, (b) "columnar" morphology with second and third level branches, (c) high-porosity "sponge-like" morphology with close, random, and short branches, (d) low-porosity "sponge," (e) high-porosity "dendritic," and (f) low-porosity "dendritic" morphologies (Tutashkonko et al., 2013).

branched pores (4–6 nm) is about two times smaller than that of the main pores (8–15 nm) (Fig. 12.4a). With depth the diameter of the main pores decreases and becomes comparable to that of one of the branches (Fig. 12.4b).

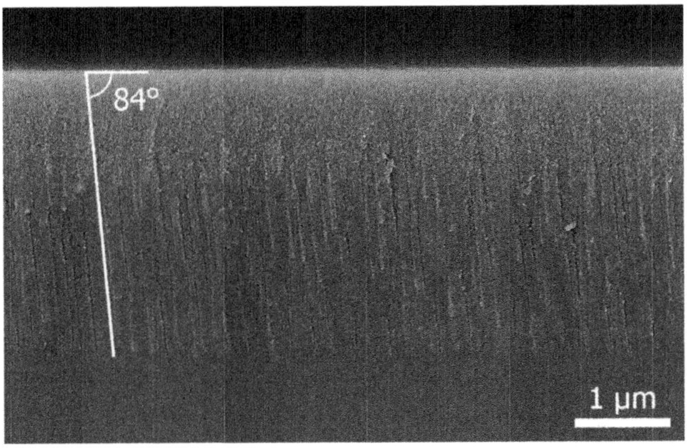

Figure 12.5 Cross-sectional SEM image of the PGe layer with columnar pores formed on a Ge wafer with (1 0 0) orientation and 6° disorientation toward the nearest (1 1 1) direction. The pore propagation direction forms an 84° angle with the surface, which corresponds to the (1 0 0) orientation.

Table 12.1 The etching parameters and the nanopore diameter of the samples presented in Fig. 12.4

Current density (mA/cm^2)	Total etching time (h)	Nanopore diameter (nm)
1.6 mA/cm^2	3	5–8
1.6 mA/cm^2	8	5–9
1.8 mA/cm^2	3	4–10
1.8 mA/cm^2	6	4–10
2.1 mA/cm^2	3	8–15 (main pores); 4–6 (branches)
2.1 mA/cm^2	20	5–8

12.3.2 Morphology Tuning: Surface Passivation

The morphology of the PGe layers can be controlled by adjusting the degree of the surface passivation with hydrogen. The latter depends on the duration of the passivating cathodic pulse t_{-}. Three ranges of cathodic pulse durations will be discussed, corresponding to high (i) $t_{off} > 1$ s, intermediate (ii) $0.38 \text{ s} \le t_{off} \le 1$ s and low (iii) $t_{off} < 0.38$ s passivation degrees.

Typical SEM views at low and high magnifications of the mesoporous Ge layer prepared under experimental conditions

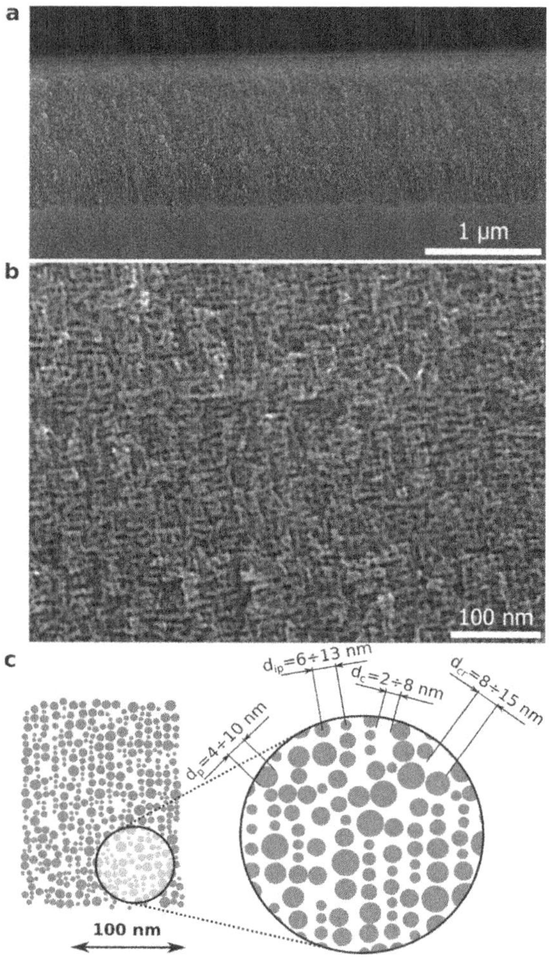

Figure 12.6 Low (a) and high (b) magnification cross-sectional SEM images of the sponge-like PGe layer ($J_+ = J_- = 2$ mA/cm^2, $t_{on}/t_{off} = 1$ s/2 s, where J_- is the applied cathodic current density); (c) its schematic presentation. The blue regions correspond to the pores (Tutashkonko et al., 2015).

corresponding to the high degree of surface passivation ($J_+ =$ 2 mA/cm^2, 1 s/2 s) are presented in Fig. 12.6a,b, respectively. As one can see, the 1.5 μm-thick sponge-like PGe layer is similar to the one presented in Fig. 12.4c. This layer is composed of highly interconnected mesopores, separated by a crystalline skeleton. The pores are disordered at short range; however, some ordering can be observed at long range (Fig. 12.1b). The schematic view of the sponge-like PGe layer is presented in Fig. 12.6c. The blue regions correspond to the pores. The mean diameters of the pores (d_p) and of the nanocrystallites (d_{cr}) are 4–10 nm and 2–12 nm, respectively. The mean interpore distance (d_{ip}) is 6–13 nm.

The same morphology is observed for all $t_{off} > 1$ s; this result is attributed to a high degree of passivation. However, at cathodization times greater than 2 s, the release of molecular hydrogen was observed, which leads to a progressive degradation of the porous layer (Tutashkonko, 2013). Indeed, the hydrogen bubbles accumulating inside the pores can create important mechanical stresses within the structure. On the other hand, they obstruct the diffusion of reactive species deep into the pores. While the PGe layers formed with $t_{off} \in [3\text{ s}, 5\text{ s}]$ are significantly amorphized, the samples prepared with $t_{off} > 5$ s were completely destroyed because of these two phenomena.

Another extreme case—the samples fabricated with cathodic pulse shorter than 0.38 s. Under these experimental conditions corresponding to the low degree of the surface passivation, the formation of a novel morphology, named "a pine-tree" was observed (Fig. 12.7). As one can see, a 3 μm thick porous layer (Fig. 12.7a) consists of the vertical pores that follow the [100] direction and the lateral branches propagating at $[1 \pm 1 \pm 1]$ directions (Fig. 12.7b). The diameter of the main pores varies from 4 to 11 nm, whereas the thickness of the lateral pores is in the range from 4 to 7 nm (Fig. 12.7c). It should be underlined that the structure is well periodic. The vertical spacing between the lateral pores is between 20–25 nm and the distance between the principal pores is in the range from 80 to 110 nm. This last distance was not found to depend on the etching current density, which lets us suppose that the nucleation of the porous layer occurs on the periodical defects of the surface of the crystal. In addition, the branches of two adjacent trees do not intersect and the tips of the pores are located on the $\{111\}$-family planes.

All the samples fabricated under experimental conditions corresponding to the low degree of the surface passivation were laterally and vertically homogeneous. However, it was not possible to create

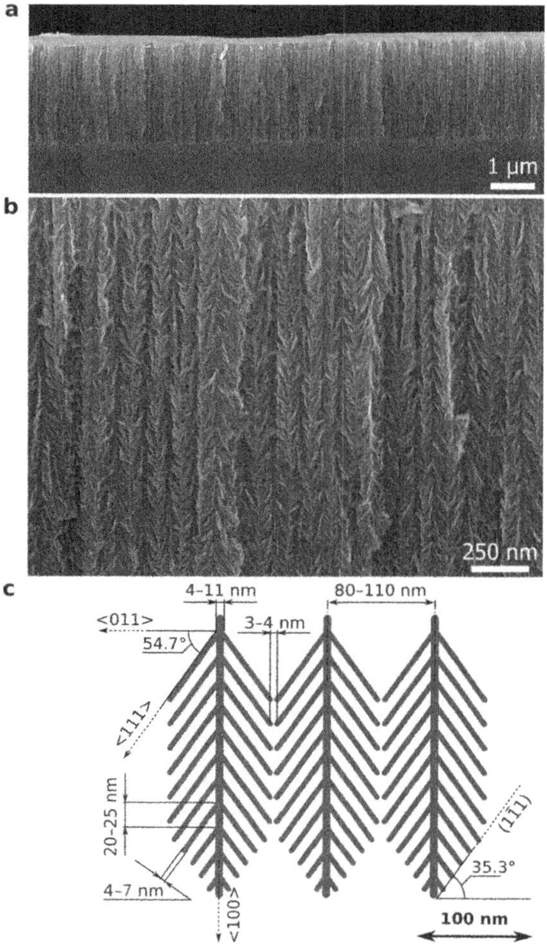

Figure 12.7 Low (a) and high (b) magnification cross-sectional SEM images of the "pine-tree"-like PGe layer ($J_+ = J_- = 2$ mA/cm^2, $t_{on}/t_{off} = 1$ s/0.38 s); (c) its schematic presentation. The pore network is presented in green (Tutashkonko et al., 2015).

the porous layers with cathodization times less than 0.25 s. These layers crumbled during the etching, leaving a textured surface on the sample.

The cross-sectional SEM view of the PGe layer prepared under experimental conditions corresponding to the intermediate degree

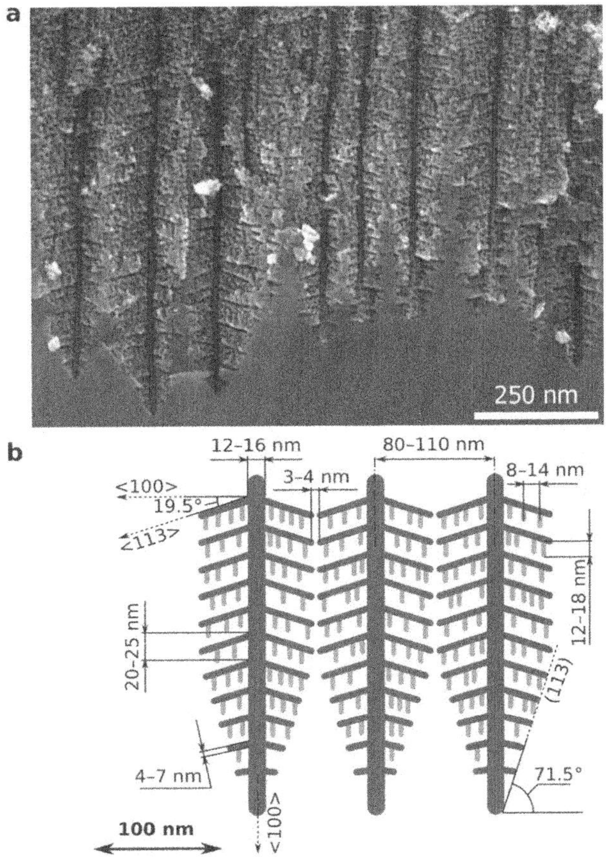

Figure 12.8 Cross-sectional SEM image (a) and the structural view (b) of the "fish-bone"-like PGe monolayer ($J_+ = J_- = 2$ mA/cm^2, $t_{on}/t_{off} = 1$ s/0.43 s). The pore network is presented in red (Tutashkonko et al., 2015).

of the surface passivation ($J_+ = 2$ mA/cm^2, $t_{on}/t_{off} = 1$ s/0.43 s) is presented in Fig. 12.8a. A novel morphology, named the "fish-bone," is observed. Similarly to the "pine-tree" morphology, the "fish-bone" morphology is well periodic. It consists of the main vertical trunks, which follow the [100] direction, and the lateral branches propagating in the [$1 \pm 1 \pm 3$] directions, thus creating an obtuse angle of $70.5° \pm 0.5°$ with the trunks (Fig. 12.8b). The diameters of the main trunks are in the range of 12–16 nm, almost two times larger than the diameter of the main pores of the "pine-tree" morphology. On the other hand, the thickness of the lateral pores as well as the period of the main trunks and the spacing between the lateral pores are the same as in the "pine-tree" morphology and constitute 4–7 nm, 80–110 nm, and 20–25 nm, respectively. The branches of two adjacent "fish-bones" do not intersect and are separated by the tiny isthmuses of 3–4 nm. Finally, the second-order pores of 12–18 nm in length are observed beneath of the lateral branches.

12.4 Structural Analysis of Mesoporous Ge

A typical cross-sectional transmission electron microscope (TEM) view of the sponge-like meso-PGe layer is shown in Fig. 12.9a. Homogeneous and highly interconnected sponge-like porous network can be observed. The average pore diameter estimated from the SEM image is found to be in the range between 3 and 8 nm. Interconnected Ge nanocrystals constituting the porous layer are shown on the high-resolution TEM picture in Fig. 12.9b. Log-normal-like size distribution of the nanocrystals centered at 4.9 nm with standard deviation of 0.3 nm is presented in Fig. 12.9c.

In order to get further insight into the structural properties of the fabricated meso-PGe layers, micro-Raman scattering spectroscopy measurements were carried out. In our study, Raman spectra were recorded at room temperature on (001)-oriented cross-sectional plane of the meso-PGe layers with the use of an excitation wavelength of 514 nm and an excitation power of 0.5 mW/cm^2 applied through an objective $\times 100$. The obtained characteristic Raman spectrum (opened circles) of the meso-PGe layer is shown in Fig. 12.9d. As one can see, the width, shape, and spectral position of the Raman peak strongly differ from well-known quite narrow symmetric and centered at 300 cm^{-1} peak of original monocrystalline Ge substrate (Roodenko et al., 2010). Indeed, the former is wider, red shifted, and asymmetrically broadened towards the lower frequency because of partial selection rule breakdown for the backscattered Stokes Raman signal recorded in these conditions (Fauchet et al., 1991; Kartopu et al., 2008; Ren and Cheng, 2002;

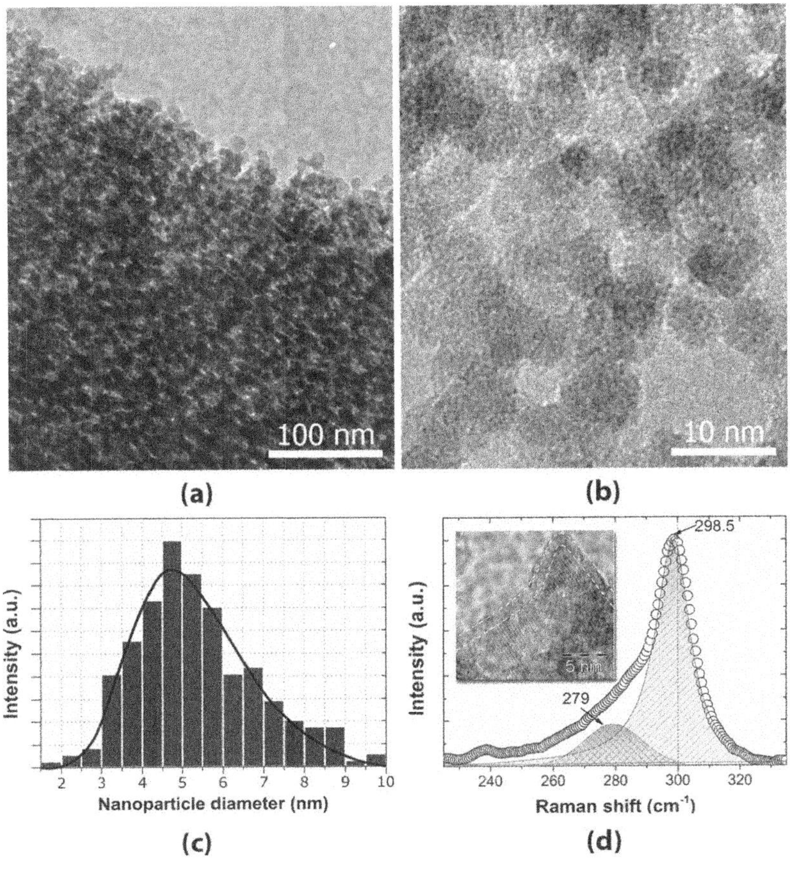

Figure 12.9 (a) Cross-sectional SEM view of the meso-PGe layer, (b) TEM picture of the sponge-like porous network, (c) size distribution of the nanocrystals, (d) characteristic Raman spectrum of the meso-PGe layer (Isaiev et al., 2014; Tutashkonko, 2013).

Roodenko et al., 2010), A phenomenological phonon-confinement model initially developed for nano-Si (Tutashkonko et al., 2013) allows estimation of a mean diameter of the Ge nanocrystallites from the spectral position and the asymmetric shape of the Raman spectrum. The Raman profile calculated from the phonon-confinement model (Tutashkonko, 2013) is presented as a hatched blue area in Fig. 12.9d. The mean crystallite diameter of 3.1 nm with standard deviation of 0.68 nm was found from the model to ensure the best fitting of the high energy tail and position of the peak maximum. However, in order to perfectly fit the whole spectrum, an intense band corresponding to amorphous Ge (a-Ge) which is centered near 279 cm^{-1} (hatched red area) has to be taken into account. This amorphous phase can correspond to the disordered $\approx$1 nm-thick shell covering the Ge nanocrystallites, which is quite well visible in the high-resolution TEM picture shown as insert in Fig. 12.9d. Taking in to account thickness of the amorphous shell and diameter of the crystalline core, a good agreement with the size distribution shown in Figure 12.9c can be stated.

12.5 Thermal Conductivity of Mesoporous Germanium

The thermal transport simulation on the crystalline core/amorphous shell germanium nanoparticles shown in Fig. 12.10 by non-equilibrium molecular dynamic was performed. For the experimental evaluation of the meso-PGe thermal conductivity, the PA gas-microphone technique already described by M. Isaiev et al. (Isaiev et al., 2014) was applied. Two light-emitting diodes (LEDs) with maximum spectral density wavelengths at $\lambda_1 = 470$ nm and $\lambda_2 = 670$ nm were chosen as sources for excitation of PA signals. Different absorption coefficients of the meso-PGe layer at these

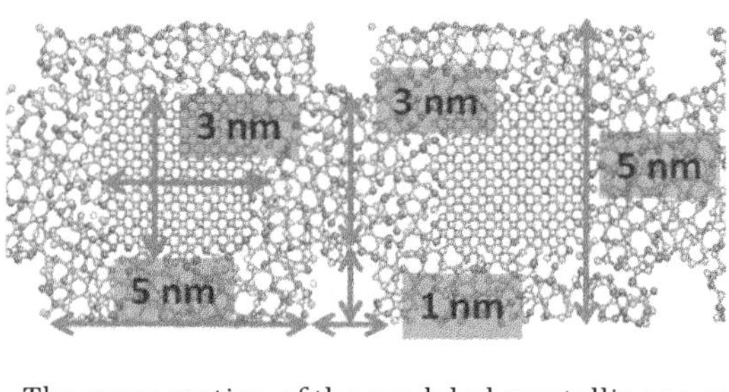

Figure 12.10 The cross section of the modeled crystalline core/amorphous shell germanium nanoparticles with molecular dynamics is depicted. The characteristic lengths of the geometry are given. Gray atoms indicate four coordinated atoms, blue with one, yellow with two and green with three (Isaiev et al., 2014).

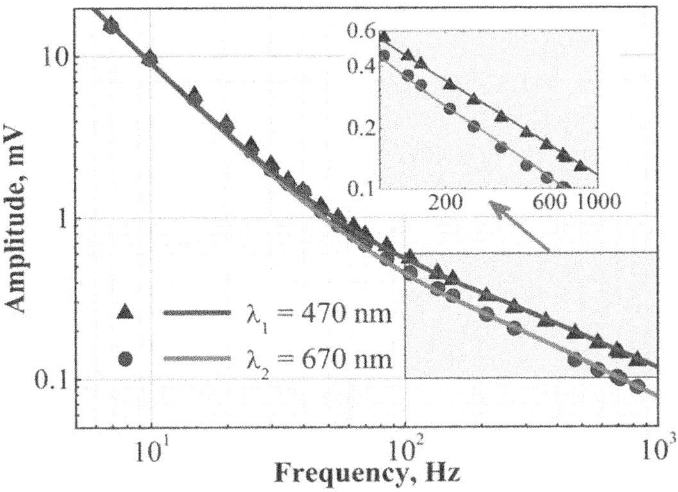

Figure 12.11 Experimental amplitude frequency dependences (symbols) for the meso-PGe layers at different excitation wavelengths (Isaiev et al., 2014).

wavelengths (2.7 μm^{-1} and 0.48 μm^{-1}, respectively) were obtained from ellipsometry measurements and was taken into account in order to accurately set spatial distribution of the absorbed light intensities for the both wavelengths. The obtained experimental amplitude-frequency characteristics (see Fig. 12.11) were fitted with development model for the both excitation wavelengths with thermal conductivity of the meso-PGe layers as a fitting parameter. The best fitting (solid lines) was obtained for $K \approx 0.6$ W/(mK) (Isaiev et al., 2014), which is in excellent agreement with the values estimated from solving the Boltzmann transport equation with the use a Monte Carlo algorithm (Jean et al., 2014).

12.6 Conclusions

Bipolar electrochemical etching, which consists of periodical switching from cathodic to anodic bias during the etching process, is a powerful tool to form mesoporous Ge layers with controlled morphology. By precisely controlling the surface passivation degree and the etching duration, it is possible to engineer porous layers of different morphologies. The structural analysis at nanoscale level revealed that each Ge nanocrystallite, constituting mesoporous Ge layer, is covered by disordered 1 nm amorphous shell. On the basis of these observations, Ge mesoporous thermal conductivity was simulated and measured by the photoacoustic technique. It was found that at room temperature the thermal conductivity of mesoporous Ge is of two orders of magnitudes lower (0.6 W/(m K)) than that of bulk Ge substrates (58 W/(m K)). The authors believe that the obtained crystalline/amorphous porous Ge structures, besides their visual beauty, being perfectly periodical and self-organized, will find a large spectrum of applications.

References

Canham, L. T. (1990). Silicon quantum wire array fabrication by electro-chemical and chemical dissolution of wafers, *Applied Physics Letters* **57**, 10, pp. 1046–1048, doi:10.1063/1.103561, URL http://link.aip.org/link/?APL/57/1046/1.

Cullis, A. G. and Canham, L. T. (1991). Visible light emission due to quantum size effects in highly porous crystalline silicon, *Nature* **353**, 6342, pp. 335–338, URL http://dx.doi.org/10.1038/353335a0.

El Kurdi, M., David, S., Checoury, X., Fishman, G., Boucaud, P., Kermarrec, O., Bensahel, D. and Ghyselen, B. (2008). Two-dimensional photonic crystals with pure germanium-on-insulator, *Optics Communications* **281**, 4, pp. 846–850, URL http://dx.doi.org/10.1016/j.optcom.2007.10.008.

Fauchet, P., Lockwood, D. and Young, J. (1991). Light Scattering in Semiconductor Structures and Superlattices, *by DJ Lockwood and JF Young (Plenum, NY)*, pp. 229–245.

Föll, H., Christophersen, M., Carstensen, J. and Hasse, G. (2002). Formation and application of porous silicon, *Materials Science and Engineering: R: Reports* **39**, 4, pp. 93–141, doi:10.1016/S0927-796X(02)00090-6, URLhttp://www.sciencedirect.com/science/article/pii/S0927796X0 20%00906.

Gaponenko, S. (2005). *Optical Properties of Semiconductor Nanocrystals*, Cambridge Studies in Modern Optics (Cambridge University Press), ISBN 9780521019231.

Golovan', L., Kashkarov, P. and Timoshenko, V. (2007). Form birefringence in porous semiconductors and dielectrics: A review, *Crystallography Reports* **52**, 4, pp. 672–685, doi:10.1134/S1063774507040153, URL http://dx.doi.org/10.1134/S1063774507040153.

Isaiev, M., Tutashkonko, S., Jean, V., Termentzidis, K., Nychyporuk, T., Andrusenko, D., Marty, O., Burbelo, R., Lacroix, D. and Lysenko, V. (2014). Thermal conductivity of meso-porous germanium, *Applied Physics Letters* **105**, 3, p. 031912.

Jean, V., Fumeron, S., Termentzidis, K., Tutashkonko, S. and Lacroix, D. (2014). Monte carlo simulations of phonon transport in nanoporous silicon and germanium, *Journal of Applied Physics* **115**, 2, p. 024304.

Kartopu, G., Sapelkin, A. V., Karavanskii, V. A., Serincan, U. and Turan, R. (2008). Structural and optical properties of porous nanocrystalline Ge, *Journal of Applied Physics* **103**, 11, pp. 113518–113524, URL http://dx.doi.org/10.1063/1.2924417.

Lehmann, V. (2002) (Wiley-VCH Verlag GmbH), ISBN 9783527600274, pp. 51–75, doi:10.1002/3527600272.ch4, URL http://dx.doi.org/10.1002/3527600272.ch4.

Loscutoff, P. W. and Bent, S. F. (2006). Reactivity of the germanium surface: chemical passivation and functionalization, *Annu. Rev. Phys. Chem.* **57**, pp. 467–495.

Nataraj, L., Xu, F. and Cloutier, S. G. (2010). Direct-bandgap luminescence at room-temperature from highly-strained Germanium nanocrystals, *Opt. Express* **18**, 7, pp. 7085–7091, doi:10.1364/OE.18.007085, URL http://www.opticsexpress.org/abstract.cfm?URI=oe-18-7-7085.

Ren, S.-F. and Cheng, W. (2002). Calculations of surface effects on phonon modes and Raman intensities of Ge quantum dots, *Phys. Rev. B* **66**, pp. 205328–205333, doi:10.1103/PhysRevB.66.205328, URL http://link.aps.org/doi/10.1103/PhysRevB.66.205328.

Roodenko, K., Goldthorpe, I., McIntyre, P. and Chabal, Y. (2010). Modified phonon confinement model for Raman spectroscopy of nanostructured materials, *Physical Review B* **82**, 11, pp. 115210–115220.

Rouquerol, J., Avnir, D., Fairbridge, C., Everett, D., Haynes, J., Pernicone, N., Ramsay, J., Sing, K. and Unger, K. (1994). Recommendations for the characterization of porous solids (Technical Report), *Pure and Applied Chemistry* **66**, 8, pp. 1739–1758.

Sing, K., Everett, D., Haul, R., Moscou, L., Pierotti, R., Rouquerol, J. et al. (1985). International Union of Pure and Applied Chemistry: reporting physisorption data for gas/solid systems with special reference to the determination of surface area and porosity, *Pure Appl Chem* **57**, pp. 603–619.

Tutashkonko, S. (2013). *Elaboration du Ge mesoporeux et etude de ses proprietes physico-chimiques en vue d'applications photovoltaiques*, Ph.D. thesis, Université de Sherbrooke / INSA de Lyon.

Tutashkonko, S., Alekseev, S. and Nychyporuk, T. (2015). Nanoscale morphology tuning of mesoporous ge: electrochemical mechanisms, *Electrochimica Acta* **180**, pp. 545 – 554, doi:http://dx.doi.org/10.1016/j.electacta.2015.08.112, URL http://www.sciencedirect.com/science/article/pii/S0013468615303625.

Tutashkonko, S., Boucherif, A., Nychyporuk, T., Kaminski-Cachopo, A., Arès, R., Lemiti, M. and Aimez, V. (2013). Mesoporous Germanium formed by bipolar electrochemical etching, *Electrochimica Acta* **88**, pp. 256–262, URL http://dx.doi.org/10.1016/j.electacta.2012.10.031.

Wang, K., Liu, G., Hoivik, N., Johannessen, E. and Jakobsen, H. (2014). Electrochemical engineering of hollow nanoarchitectures: pulse/step anodization (si, al, ti) and their applications, *Chem. Soc. Rev.* **43**, pp. 1476–1500, doi:10.1039/C3CS60150A, URL http://dx.doi.org/10.1039/C3CS60150A.

Zhang, G. (2006). Porous silicon: Morphology and formation mechanisms, in C. G. Vayenas, R. White and M. Gamboa-Adelco (eds.), *Modern Aspects of Electrochemistry*, *Modern Aspects of Electrochemistry*, Vol. 39 (Springer US), ISBN 978-0-387-23371-0, pp. 65–133, doi:10.1007/978-0-387-31701-4_2, URL http://dx.doi.org/10.1007/978-0-387-31701-4_2.

Chapter 13

Amorphous/Nanocrystalline Composites: Recrystallization

Valentina M. Giordano and Ameni Tlili
ILM Institute of Light and Matter, UMR5306 Université Lyon 1-CNRS, 69622 Villeurbanne cedex, France
valentina.giordano@univ-lyon1.fr

13.1 Introduction

In this chapter, we present an experimental protocol for obtaining amorphous/crystalline composites by means of controlled annealings. More specifically, here we are interested in composites in which nanometric crystals are embedded in an amorphous matrix. Such materials indeed are of relevant technological interest because of the change in the physical properties due to the presence of ordered/disordered interfaces at the nanoscale (Hebert, 2011). We will focus on amorphous/crystalline metallic composites, relatively novel materials that can be produced quite easily by controlled thermal protocols starting from metallic glasses (MG).

Nanostructured Semiconductors: Amorphization and Thermal Properties
Edited by Konstantinos Termentzidis

ISBN 978-981-4745-64-2 (Hardcover), 978-1-315-36445-2 (eBook)
www.panstanford.com

13.1.1 Thermal Transport in Amorphous/Nanocrystalline Composites

The effects of nanocrystalline inclusions on the thermal properties of a metallic glass start from their effect on the electric conductivity. Indeed, even if poorer metals with respect to their crystalline counterpart, MG remain good electric conductors, which means that the most of heat transport is assured by electrons.

Generally speaking, heat in a crystalline system is carried by electrons and lattice vibrations, described as (pseudo)particles, the phonons. In insulators and semi-conductors these latter dominate the thermal transport. Conversely, in a metal, electrons carry most of the heat. The repartition of the heat transport can be written as

$$K_T = K_L + K_e, \tag{13.1}$$

where K_T is the total thermal conductivity, K_L and K_e the lattice and electron contributions, respectively. Intuitively it is clear that the latter will be proportional somehow to the electrons ability to move, i.e., the electric conductivity σ. This proportionality is described by the Wiedemann–Franz law, which has been developed for a simple metal with only one conduction band, but is largely used in all kinds of systems:

$$K_e = \sigma LT, \tag{13.2}$$

where T is the temperature in K and L the Lorentz number,

$$L = 2.45 \times 10^{-8} W\Omega/K^2.$$

The same separation can be adopted in a disordered system, where the lattice thermal contribution turns out to be very weak and weakly temperature dependent. As an example, the thermal conductivity of the prototype silica glass, SiO_2 ranges between 1.3 and 1.5 W/mK (Baldi et al., 2008; Regner et al., 2013). In a non-metallic system like SiO_2, heat can be transported only by atomic vibrations. Indeed, despite the lack of periodicity, we can still identify collective excitations, reminiscent of lattice vibrations, phonons. Like crystals, the low-energy phonons are plane-waves and propagate, thus carrying heat. However, they are characterized by a very short lifetime, compared to crystalline systems, due to scattering from disorder, and which decreases very rapidly with the phonons energy. As a consequence, the distance traveled by the phonon during the space of its lifetime, the mean free path, is very short, turning out to be of the order of the inter-particle distance, and the heat transport efficiency very poor.

It is, however, important to underline that propagating modes contribute only very little to the heat transport in glasses. Indeed three kinds of atomic vibrations can be generally identified in glasses: low-energy propagating plane-wave like modes, called *propagons*, high-energy localized modes, called *locons*, and finally delocalized non-propagating modes in an intermediate energy range, called *diffusons*. These latter are found to dominate the vibrational spectrum and to be the predominant heat carriers via a diffusive transport mechanism at high temperature (Beltukov et al., 2013).

While the phononic contribution ranges between 0.1 and 2 W/mK, in metallic glasses the actual thermal conductivity can be much higher, due to the electronic contribution. Despite the disordered structure, the Wiedemann–Franz law has been found to hold also in metallic glasses (Mizoguchi et al., 1980) with a value for the Lorentz number very close to the theoretical one and thus can be used for estimating the relative importance of the two contributions. With typical electric resistivity in the 60–250 μ Ohm cm (MIZUTANI, 2001), metallic glasses have an electronic thermal conductivity which is in the 3–13 W/mK range, thus covering almost the whole thermal conductivity measured value.

The presence of nanometric inclusions in a bulk amorphous material can affect both phonon and electron transport.

Phonon transport at an amorphous/crystalline macroscopic interface has been addressed in quite recent studies (Lampin et al., 2012). The effects of an interface at the nanoscale is, however, largely unknown, and of great interest. Intuitively, dramatic phonon scattering can be expected when the inclusion size is comparable to the Boson Peak wavelength (usually in the nm range) (Beltukov et al., 2013; Tanguy et al., 2010). More interestingly, a very recent molecular dynamics investigation has reported unexpected filtering effects on high-energy vibrational modes (see Chapter 4) (Damart et al., 2015). More details have already been given in Chapters 4 and 5.

On the other hand, effects of nanometric amorphous/crystalline interfaces on the electron transport are still completely unknown and not investigated yet. The nanoparticles can create potential variations at the nanoscale due to the hetero-interfaces, altering the electrons transport in the matrix. This topic has been largely investigated for crystalline nanoinclusions in a crystalline bulk material.

Electronic transport is generally described by the Boltzmann transport theory, which calculates the non-equilibrium electron distribution function, due to perturbations introduced by collisions of the electrons with external elements, such as impurities, defects, grain boundaries, acoustic and optical phonons. In the relaxation time approximation, the electric conductivity can be written as

$$\sigma = e^2 \int \frac{d^3k}{4\pi^3} \tau(k) v^2(k) \left(\frac{-\delta f_0}{dE} \right), \tag{13.3}$$

where the integral runs over all the electrons states defined by the electron momentum k, v is the electrons velocity and τ the electrons relaxation or scattering time. τ is a function of k, and thus of the energy of the electron, and can be calculated starting from the scattering times of the different collision sources τ_i as $\frac{1}{\tau} = \sum \frac{1}{\tau_i}$ (Mathiessen's rule).

The theoretical treatment proposed in (Zebarjadi et al., 2009) considers the nanoparticle as an additional scattering source and treats it in the effective medium theory. When the nanoparticles concentration is high enough, the electron wave does not see individual scatterers, but an effective medium made of the host matrix and the nanoparticles. It has been shown that the nanoparticle scattering can dramatically affect electron transport properties, with a scattering time strongly decreasing with the nanoparticle size.

In the following we will focus on the experimental preparation of an amorphous/nanocrystalline composite, for investigating the transport properties through order/disorder interfaces at the nanoscale.

The structure of the chapter is the following: In Section 13.2 we will rapidly remind how a glass is produced from the liquid and the meaning of the glass transition temperature, we will then introduce metallic glasses as a peculiar family of glasses, particularly suited for the production of amorphous/crystalline composites. Section 13.3 gives a rapid summary of the relaxation mechanisms giving raise to aging and finally to crystallization in glasses. Section 13.4 focuses on the crystallization of metallic glasses as obtained by controlled thermal protocols.

13.2 From a Liquid to a Glass

Glasses are known and used since many centuries. However, despite that, this class of materials continues to fascinate researchers due to the continuous discoveries of new properties and applications. In general, a glass can be viewed as a liquid in which a huge slowing down of the diffusive motion of the particles has destroyed its ability to flow on experimental time scales, thus remaining frozen in a metastable state.

The most common way to produce a glass consists on cooling a liquid fast enough to avoid crystallization. Other ways may also be used, such as

(i) low-temperature gas condensation,
(ii) application of hydrostatic pressure,
(iii) chemical reactions, such as polymerization or condensation of chemically reacted vapors,
(iv) high-energy milling, etc.

It is clear that the possibility to obtain a glass is not strictly related to the material itself, but it rather depends on the available time for the system to crystallize. If time is short enough, the atoms will not manage to arrange in an ordered structure.

In order to get a picture of the glass transition phenomenon, let's follow the rapid cooling of a liquid below its melting temperature T_m: two distinct processes can take place, either crystallization or vitrification. The occurrence of these transitions can be observed by following the temperature dependence of an extensive thermodynamic variable such as volume (V), entropy (S) or enthalpy (H) at constant pressure, as sketched in Fig. 13.1.

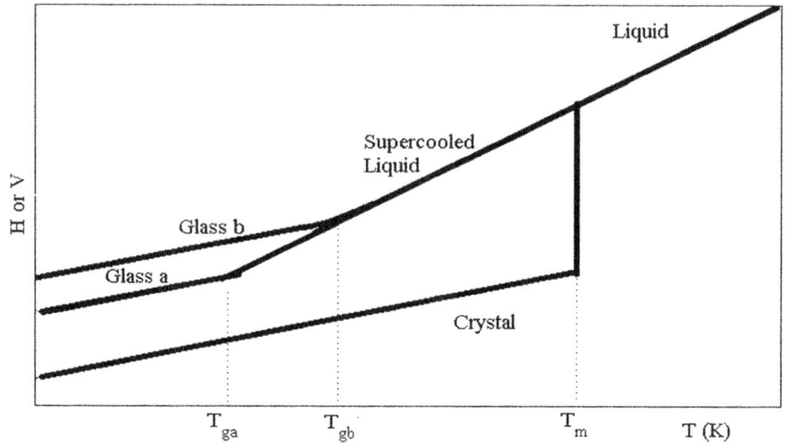

Figure 13.1 Volume or enthalpy temperature dependence while cooling at constant pressure. The glassy state is characterized by a gradual change in slope in the region around T_g, while crystallization is accompanied by an abrupt decrease at T_m. T_{ga} and T_{gb} represent the effect on the glass transition of a slower or faster cooling rate, respectively.

The crystallization process corresponds to an abrupt jump in volume or enthalpy at the melting temperature T_m. If, however, the liquid is cooled below T_m rapidly enough so that detectable nucleation and crystal growth cannot occur, it falls out of equilibrium and becomes supercooled. Upon further cooling the viscosity of the liquid increases until the system becomes solid-like around T_g. The glass transition temperature corresponds then to a smooth change in the $V(T)$ $(H(T))$ slope.

This implies a discontinuity in the corresponding derivative: the thermal expansion coefficient α_T for $V(T)$ or heat capacity C_p for $H(T)$, which abruptly decreases in a rather narrow temperature range.

At first sight, the vitrification would seem similar to a second-order phase transition, in the sense of the Ehrenfest criterion (discontinuity in the second order derivative of an extensive thermodynamic variable). However, this criterion is not adequate for glasses, as the glass transition temperature is found to depend on the thermal history and the cooling rate. Indeed, whether or not a given liquid crystallizes during cooling is strictly a kinetic problem involving the rate of nucleation and growth of the crystal on one hand, and the rate at which the thermal energy can be extracted from the cooling liquid on the other (J. Wong and Angell, 1976).

Cooling down the supercooled liquid, it becomes more and more dense and viscous and eventually it fails to reach the equilibrium structure for a given temperature. The slowing down of the diffusive motion in the liquid is expressed through the relaxation time τ_α, that represents, generally speaking, the characteristic time for the system to equilibrate its microscopic structure. This process, called structural relaxation, is related to cooperative atomic rearrangements (Jackle, 1986). At T_g τ_α, or the viscosity proportional to it, suddenly grows by several orders of magnitude in a relatively small temperature interval (Angell, 1995). It naturally follows that the material will behave as a liquid or solid depending on the time necessary to the system to relax into a new equilibrium configuration with respect to the experimental time scale, fixed by the typical time of change of an external variable, like temperature, or the observation time.

For experimental times $\tau_{\rm exp} \gg \tau_\alpha$ the material will have the time to equilibrate in a new configuration and will appear liquid-like, while for $\tau_{\rm exp} \ll \tau_\alpha$ it will appear as frozen, its change rate being too slow, and it will behave as if it were solid-like.

The slower the cooling rate of the liquid, the later the liquid will not manage to equilibrate, and thus the larger the supercooled region. The glass transition temperature will therefore be lower, as illustrated by the two lines "a" and "b" in the figure.

The glass transition can be thus classified as a kinetic transition, where everything is ruled by the comparison between an internal timescale (atomic diffusion) and an external one (the experimental observation timescale). In this way it is possible to define T_g as the temperature where $\tau_\alpha = 1000$ s, or where the viscosity reaches a value of 10^{13} poise.

The dependence of T_g on the cooling rate is rather weak (in some systems T_g changes by 3–5° when the cooling rate changes by an order of magnitude), and the transformation range is narrow, so that T_g becomes an important material characteristic.

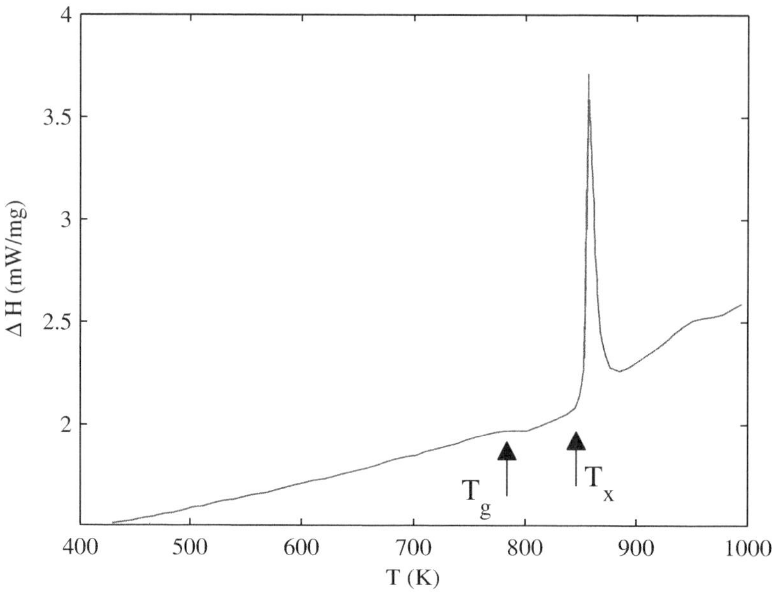

Figure 13.2 Enthalpy change measured by isochronal DSC during heating of a $Cu_{49}Hf_{42}Al_9$ metallic glass with constant heating rate of 20 K/min. The glass transition temperature $T_g = 789$ K and the onset of crystallization $T_x = 851$ K are indicated.

In general, the glass transition temperature is measured by means of differential scanning calorimetry (DSC). This technique allows performing a thermal analysis of the system by monitoring the enthalpy variations during a thermal protocol and thus to measure $C_p = (H/T)_P$. T_g is usually defined as the temperature corresponding to the change of curvature in the rise of C_p obtained by heating the system at a constant rate of 10 K/min, from the amorphous solid state. We report in Fig. 13.2 as an example the enthalpy change measured by DSC on a metallic glass $Cu_{49}Hf_{42}Al_9$, where the glass transition temperature T_g as well as the onset of crystallization T_x are indicated, although the ramp rate was in that case 20 K/min.

The mechanisms underlying the glass transition have been studied for decades and a vast variety of theories have been proposed. A general view of some of the different approaches can be found in Refs. (Adam and Gibbs., 1965; Cohen and Turnbull, 1964a,b; Reichman and Charbonneau, 2005).

13.2.1 Metallic Glasses

In the following, we will focus onto a specific family of glassy materials: the metallic glasses (MG). These represent a revolutionary new class of materials with outstanding mechanical, physical and chemical properties which place them at the cutting-edge of metal and glassy research (Wang, 2012). Their disordered structure, together with the dominant metallic atomic bonding result in a macroscopic behavior that strongly differs from that of the polycrystalline state, and which leads to unique and excellent mechanical properties like high strength, high elasticity and a very good corrosion resistance. Above their glass transition temperature, T_g, MG can be easily deformed in very precise complex shapes, making them promising candidates for many small-scale devices for advanced technical and medical equipment (Kumar et al., 2011; Zberg et al., 2009).

The first MG has been produced in the 1960s, by fast quench of the binary liquid alloy $Au_{80}Si_{20}$ (W. Klement and Duwez, 1960). The very high quenching rate required (10^6 K/s) to overcome crystallization has limited for longtime the development of these materials, until ternary alloys able to vitrify with quenching rates as low as 10^2–10^3 K/s have been discovered. Such a low cooling rate can be achieved by the simple injection of the liquid in a macroscopic water-cooled copper mold, thus allowing to produce millimeter or even centimeter-sized MG, called then bulk metallic glasses (BMG). Since then, many multi-component alloys (from 3 to more than 10 elements) have been found that can be vitrified as BMG, opening the way to new perspectives (A. Inoue and Masumoto, 1989; H. W. Kui and Turnbull, 1984; W. H. Wang, 2009).

At this day, despite the already large technological use of MG, a significant challenge stands on the way of an even more widespread use: their intrinsic out-of-equilibrium nature, which is reflected in material properties that depend on the processing history used for their fabrication, as well as on the previous thermal history, and that evolve in time. We will discuss the details in the next section.

13.3 The Metastable Nature of a Glass: Aging and Crystallization

13.3.1 Physical Aging

It is important to underline the metastable nature of the glassy state.

Indeed, the system slowly evolves in time, heading to a more equilibrium state. Such evolution is what is called physical aging of the glass and has been largely investigated through the temporal evolution of macroscopic quantities, like enthalpy and elastic moduli (Evenson et al., 2014; Hachenberg et al., 2008; Pineda et al., 2013).

The microscopic mechanisms behind aging are still not completely understood and actively investigated as this is the most important obstacle to a large spread of glasses as materials for technological applications.

In particular, many investigations have been carried on structural relaxation and aging in metallic glasses, which are among the most technologically relevant glasses at this day.

It is generally agreed upon that structural relaxation implies atomic movements heading to release internal stresses accumulated during the fast quench used to produce the glass (Evenson et al., 2014; Hachenberg et al., 2008; Pineda et al., 2013; Wang, 2012). The higher the quenching rate, the larger the amount of internal stresses present in the glass. For this reason, model systems to investigate the relationship between structural relaxation, aging, and internal stresses are metallic glasses obtained by melt spinning: a quenching technique where cooling rates as high as 10^{5-6} K/s can be reached.

As atomic mobility increases with temperature, controlled annealings can be used to accelerate the internal stress relaxation until the system is more equilibrated. A slow cooling down after the annealing will insure that the glass will not pile up new stresses. This procedure has been largely used to reduce aging in metallic glasses, which are annealed above T_g in order to release all stresses: The glass, after annealing in the supercooled liquid regime, is said to be *rejuvenated* and all its properties do not depend anymore on the thermal history previous to that annealing.

Following by x-ray diffraction the tiny structural changes during controlled annealings, two microscopic mechanisms involved in the structural relaxation and aging have been unveiled: (i) radial atomic movements causing a density increase related to the annihilation of structural defects induced by the fast quenching (*topological short-range ordering, TSRO*) and (ii) local atomic movements increasing the short-range chemical order with no density change (*short-range chemical ordering SRCO*) (Mao et al., 1995; Taub and Spaepen, 1980).

Annealings can not only boost structural relaxation but also promote crystallization: Indeed, as the system has been frozen in a disordered configuration thanks to a fast cooling down, with temperature atoms will eventually get back energy enough to rearrange in an ordered structure and the system will crystallize. How and when this will happen depends on the complex interplay of the two microscopic mechanisms cited above: On one hand, the density increase causes a reduction of the atomic diffusion (Zumkley et al., 2001), which can improve the stability of the amorphous structure against a temperature-promoted ordering, and on the other, the increased chemical short-range ordering clearly favors crystallization.

13.3.2 Crystallization

Figure 13.2 shows that continuous heating leads to the full transformation from amorphous to crystalline, which shows up as a peak in the DSC signal, whose area is proportional to the crystallization enthalpy of the material. In the case of a multi-component alloy, several crystallization peaks can be present, due to the devitrification in different crystalline phases, implying different stochiometries and compositions.

Due to the kinetic origin of all transformations in a glass, the heating ramp rate heavily affects the crystallization. As it can be seen in Fig. 13.3, increasing the ramp rate shifts both the glass transition and the crystallization peak to higher temperatures. Moreover, it can also change the crystallization mechanism, as it is the case for the widely known BMG called Vitreloy 1 ($Zr_{41}Ti_{14}Cu_{12.5}Ni_{10}Be_{22.5}$) reported in the figure, where two distinct and well-separated

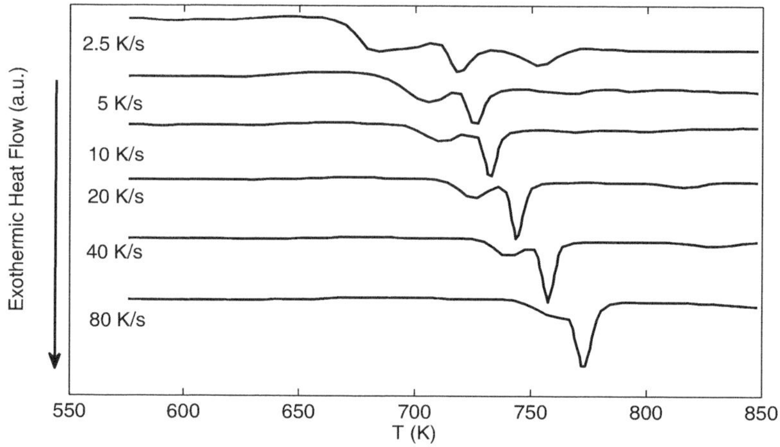

Figure 13.3 Enthalpy change measured by isochronal DSC during heating of the BMG Vitreloy 1 ($Zr_{41}Ti_{14}Cu_{12.5}Ni_{10}Be_{22.5}$) as a function of the heating rate (Adapted from (Zhuang and Wang, 2000).

crystallization peaks are visible at the lowest ramp rate, while at the highest they almost merge.

Complete crystallization can also be achieved by annealing the MG at temperatures between T_g and T_x for a time which strongly depends on temperature. This is shown for the $Mg_{83}Ni_{17}$ MG in Fig. 13.4. Here the DSC shows a single crystallization reaction (a single peak) corresponding to the crystallization of Mg_6Ni. Following the crystallization along several isotherms approaching T_x from 418 K to 438 K, it is found that increasing temperature by only 20° the crystallization peak moves down in time from about 25 to less than 5 min (Spassov et al., 2009).

13.4 Amorphous/Crystalline Composites

From the discussion in the previous section, it is clear that complete devitrification can be achieved in a MG through two time–temperature pathways: linear heating or isothermal annealing.

Irrespective of the chosen annealing strategy, if the crystallization process is abruptly interrupted by a fast quench, before

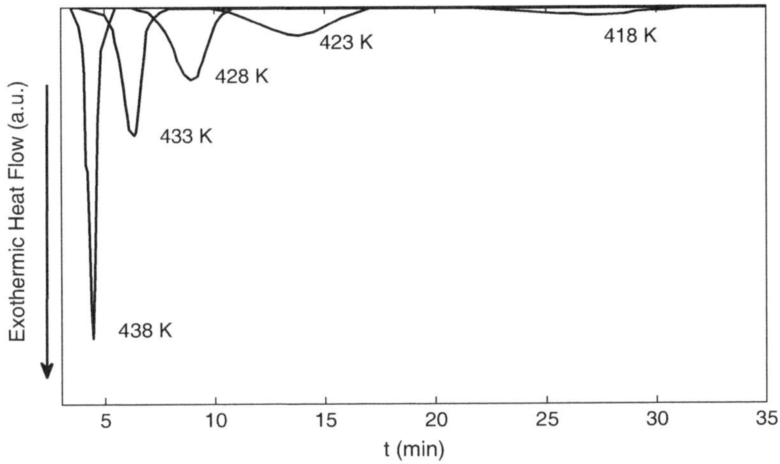

Figure 13.4 Enthalpy change measured by isothermal DSC on a $Mg_{83}Ni_{17}$ MG as a function of the annealing temperature. Adapted from (Spassov et al., 2009).

the crystallization peak is completed, an amorphous/crystalline composite will be obtained, meaning an amorphous matrix with embedded crystalline inclusions. Shape, size, and volume fraction of these latter will depend on the time–temperature path followed. Amorphous/nanocrystalline composites are of great interest for materials scientists, as their properties are usually different from both their amorphous and crystalline counterparts. They have been long investigated for the enhancement of mechanical properties, such as improved plasticity; however, much needs to be understood as for the effect of the interfaces on many properties, like thermal and electronic transport (Wang, 2007).

Classical nucleation theory describes crystallization as a two-step process: nucleation of a crystalline nucleus and then crystalline grains growth. The ability of controlling these two processes is precious for developing new technological applications, as it would allow to attain products with the required crystallized fraction and microstructure (e.g., nanocrystalline or quasicrystalline) or to avoid the material degradation at high processing or operating temperature.

Generally speaking, there exist three important crystallization types involving nucleation and growth processes, depending on the composition of a particular alloy: primary crystallization, polymorphous crystallization, and eutectic crystallization (Hsiao et al., 2002). In primary crystallization, the primary phase of the elements composing the alloy crystallizes first, with a different composition with respect to the amorphous phase. The crystallites so formed are dispersed in the amorphous matrix and may serve as nucleation sites for secondary or tertiary crystallization, taking place through the precipitation of different crystalline phases. Polymorphous crystallization is the crystallization of the amorphous phase to a crystalline one with basically the same composition: In this case at the crystal/amorphous interface there is no discontinuity in the local composition. Eutectic crystallization is simultaneous crystallization of two crystalline phases by a discontinuous reaction, which needs longer than polymorphous crystallization to proceed because a diffusive phase separation must take place in the crystallized region (Minic, 2006).

In multi-component alloys primary crystallization is generally the dominating process.

This is also the most promising for the fabrication of a nanocrystalline/amorphous composite. Indeed, in this case, the composition difference between the crystalline nucleus and the amorphous matrix inhibits the grain growth, as the nucleus is surrounded by an amorphous region depleted of its constituents and thus an important atomic diffusion will be needed for the nucleus to grow. It follows that crystallites originated by primary crystallization have usually sizes in the 5–30 nm and only temperature can promote their growth.

The small grain size and large grain densities ($\approx 10^{22}/m^3$) suggest very high nucleation rates and low growth rates, difficult to understand within the classical nucleation theory. It was first ascribed to the existence of a phase separation on a nanoscale prior to crystallization, which has been indeed observed in some BMG (Busch et al., 1995; Schneider et al., 1996). However, it has been found that not all BMG able to nanocrystallize are subject to spinodal decomposition. The nanocrystallization can indeed be understood in terms of an interplay between interfacial growth and long-range atomic diffusion in an amorphous matrix which gets depleted of the crystallized elements (Kelton et al., 2003).

As a counter-example, if polymorphous crystallization takes place, the crystallites can grow with no need of a substantial atomic diffusion. This is the case of $Mg_{60}Cu_{30}Y_{10}$, whose primary crystalline product is Mg_2Cu, very close to the amorphous matrix composition, precipitating in 100 nm sized crystallites (Wolff et al., 2004).

It is important to underline that the complex stoichiometry of multi-component MG, while being the main responsible of a nanometric crystallization, is also at the origin of the nucleation of a number of different crystalline phases, whose relative proportion and temporal sequence will depend on temperature and heating rate through the competition of their nucleation rates. For example, the crystallization of the already cited Vitreloy 1 takes place via the nucleation of several phases, the first of which seems to be a metastable quasi-crystalline phase which decomposes successively in inter-metallic crystalline compounds (Gao et al., 2003b; Wang et al., 2002).

Another final piece of useful information concerns the crystallites morphology: Wei et al. (Wei et al., 1998) have shown that devitrification from the liquid gives rise to dendritic crystallites, while crystallization from the amorphous leads to nanometric spherical nuclei.

13.4.1 The Crystalline Fraction

The crystallized volume fraction at the time t and temperature T can be expressed by the Kolmogorov–Johnson–Mehl–Avrami (KJMA) equation (Al-haj and Barry, 1997; Lu and Wang, 1991):

$$x(t) = 1 - \exp[-(k_T(t-\tau))^n], \tag{13.4}$$

where τ is an incubation time, n is the Avrami exponent and k_T contains all the kinetic information and is a function of absolute temperature:

$$k_T = k_0 \exp[-E_c(T)/RT] \tag{13.5}$$

Here $E_c(T)$ is the apparent activation energy for the crystallization process.

In a polymorphous crystallization $E_c(T)$ is constant during the whole transformation; however, this is not true anymore if there is a primary crystallization. Factors such as the amorphous matrix compositional change and the different elements diffusivity, the interplay between the interfacial attachment and growth, and the atomic diffusion need to be taken into account for a new nucleation theory more appropriate to these systems, as proposed by Kelton in his kinetic model for nucleation in partitioning systems (Kelton, 2000).

The experimental determination of the crystalline volume fraction is quite complex and usually done indirectly. The technique the most widely used to this aim is the DSC. Knowing that the area underlying the crystallization peak in a DSC scan corresponds to the crystallization enthalpy, it is assumed the proportionality between the mass crystalline fraction and the transformation enthalpy (Gravier et al., 2010; Illekovfi et al., 1995). On this basis, the crystalline fraction produced during an isothermal annealing at a time t, $x(t)$, can be calculated as follows:

$$x(t) = \frac{\int_{t_1}^{t} H(t')dt'}{\int_{t_1}^{t_2} H(t')dt'}, \tag{13.6}$$

where t_1 and t_2 mark the time at which the crystallization peak starts and ends. If a linear heating is used instead, $x(t)$ becomes

$$x(T(t)) = \frac{\int_{T_1}^{T(t)} H(t')dt'}{\int_{T_1}^{T_2} H(t')dt'}, \tag{13.7}$$

where, again, T_1 and T_2 correspond to the starting and ending temperatures of the crystallization peak, as illustrated in Fig. 13.5.

It is then possible to obtain curves of the crystalline fraction as a function of time or temperature, of the kind reported in Fig. 13.6.

Equations (13.6) and (13.7) are in fact strictly justified only for a polymorphous crystallization, as they implicitly assume that the enthalpy of crystallization is constant over the whole crystallization process, or that one kind of crystal is mostly nucleated. It also assumes that the alloy is fully crystallized at the end of the transformation. On the other hand, multi-component MG usually undergo a primary crystallization followed by secondary, etc.,

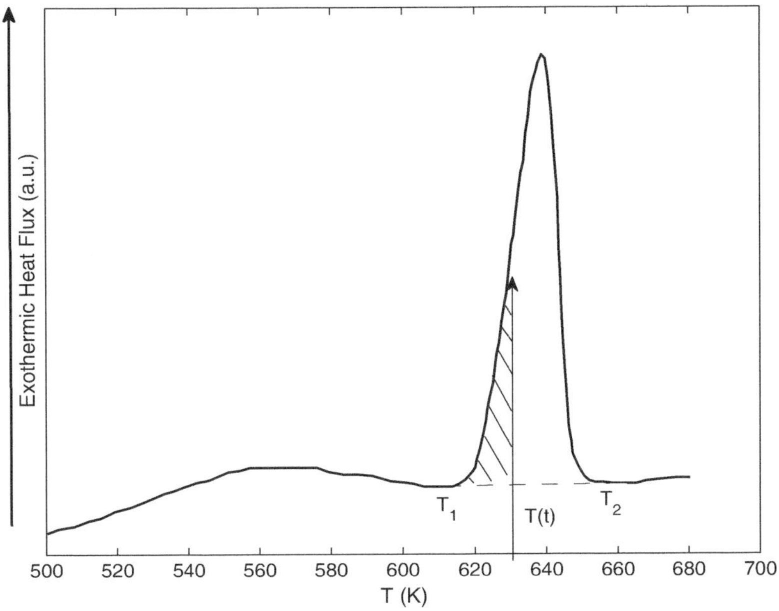

Figure 13.5 Illustration of the calculation method of the crystalline fraction starting from the crystallization peak in the heat flux measured in a DSC experiment.

with the nucleation of several, competing, phases. As said above, during this kind of crystallization the amorphous matrix changes too, as it is depleted of the crystallized elements. It follows that the free enthalpy difference between the crystalline and the amorphous phases also changes, so that the corresponding driving force for crystallization is then dependent on the fraction already transformed and the enthalpy released during the transformation cannot be proportional to the transformed fraction anymore. Moreover, the kinetic parameters, such as diffusion coefficients and other factors, are composition-dependent and will change as the crystallization proceeds.

This point has been developed by Barandiaran et al. (2003) in a study about the nanocrystallization of a Finemet-like alloy. In their model, the authors condense the different effects of the composition change of the amorphous matrix in a variable activation energy E_c appearing in the crystallization rate k_T (see Eq. 13.5), and which

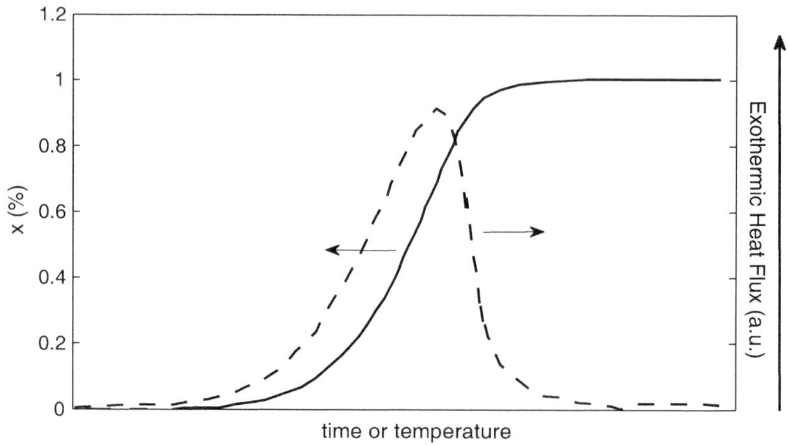

Figure 13.6 A typical crystallization fraction curve (solid line) as a function of time or temperature, obtained from the integration of the crystallization peak (dashed line).

increases all along during the transformation. This model gives a pretty good fit of the experimental crystallization peak and a better, although not exact yet, estimation of the real crystallization fraction, as deduced by Mőssbauer spectroscopy.

Other indirect methods often used in alternative to DSC are density or resistivity measurements, which, however, require the knowledge of the associated characteristics of the crystalline phases. These methods too are based on a linear approximation, thus limited in their accuracy as they do not take into account the temporal evolution of the matrix nor, in the case of resistivity, interface or percolation effects.

For this reason, direct visualization of the crystalline grains by high-resolution transmission electron microscopy (TEM) can be preferable for deducing a crystalline fraction and size distribution by image analysis techniques. Nevertheless, an uncertainty exists here too, as there are many sources of possible errors, like a too weak contrast hiding the smallest crystallites or the superposition of different crystals in the probed sample thickness (Wesseling et al., 2003).

13.4.2 The Avrami Index

The knowledge of the crystalline fraction evolution is also useful for extracting other parameters describing the crystallization kinetics, such as the Avrami index n, which is related to the dimensionality of the crystalline growth, allowing to distinguish the growth mechanism between the propagation of an interface or the atomic diffusion.

If the nucleation rate is constant, the classical Avrami index should be 4 for a spherical interfacial growth and 2.5 for a spherical diffusive growth. If the nucleation rate depends on time, then it is $n = m + 1$ where m is equal to the dimensionality of the phase transformation in case of interfacial growth or to half the dimensionality for the diffusive growth.

Classically, 4 should be the maximum value for n; it is, however, not uncommon to find Avrami indices larger than 4 in glasses, asking for more complex modeling taking into account, for example, a variable nucleation and growth rate (Kelton, 1993; Pradell et al., 1998). It is interesting to note that in their work Barandiaran et al. (2003) find that in their binary alloy simulation, the Avrami index estimation is independent of the linear or non-linear approximation for the enthalpy release.

Moreover, it is generally found that the Avrami index changes with time during crystallization in a MG, meaning that the growth mechanism changes. This can be clearly seen in Fig. 13.7, where $\ln[\ln(1 - x)^{-1}]$ is reported against $\ln(t - \tau)$ in the case of the MG $Zr_{55}Al_{10}Ni_5Cu_{30}$ (Gao et al., 2003a). From Eq. (13.4), it follows that we should obtain a straight line with a slope equal to n for a single crystallization and growth mechanism. However, it appears that n changes with time, going from 2 to 4, as can be better seen from the local Avrami index, defined as

$$n(x) = \frac{\delta \ln[-\ln(1 - x)]}{\delta \ln(t - \tau)} \tag{13.8}$$

and reported in Fig. 13.8. This means that in the case of this MG, and for the two temperatures here reported, the crystallization starts as a one-dimensional nucleation and linear growth ($n = 2$) and goes on as a three-dimensional spherical interfacial growth with constant rate ($n = 4$).

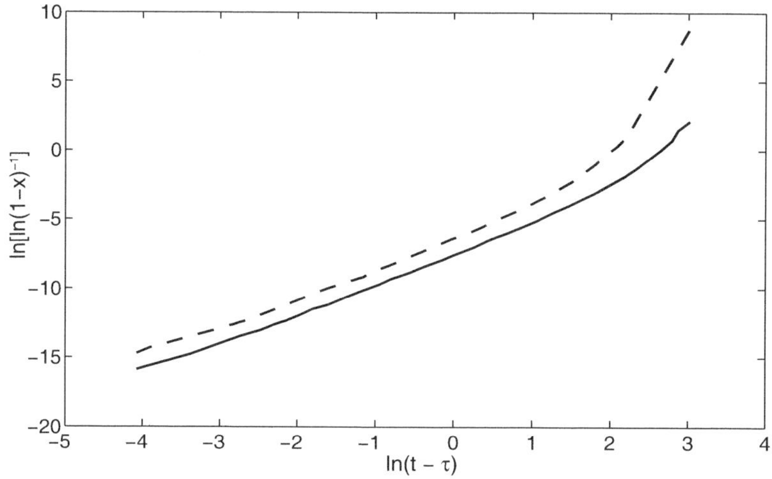

Figure 13.7 $\ln[\ln(1-x)^{-1}]$ is reported against $\ln(t-\tau)$ for the MG $Zr_{55}Al_{10}Ni_5Cu_{30}$ at $T = 713$ K (solid line) and $T = 723$ K (dashed line). The slope corresponds to the Avrami index for the transformation. Adapted from (Gao et al., 2003a).

13.4.3 The Determination of Crystalline Products

In order to model and understand the nanocrystalline/amorphous composite properties the knowledge of the involved crystalline phases is necessary. This is in fact not trivial, as many phases are often simultaneously nucleated in a multi-component MG. The most widely used technique for the phase recognition is the x-ray or electron diffraction. The former allows to follow in situ the crystallization process, during a thermal protocol. This is especially true if data are collected using a synchrotron radiation source. In this case the intense signal obtained even with 3 s acquisition time, allows to detect the appearance of crystalline phases with a very small volume fraction (less than 1%). The large q range accessible in these sources together with the high-resolution allows for a better overview of the final crystalline states, as more peaks are visible and resolved.

In Fig. 13.9 we report as an example some spectra collected in the BMG $Zr_{52.5}Cu_{27}Al_{10}Ni_8Ti_{2.5}$, also called D″, during a linear heating with a ramp rate of 5 K/s. Data were collected at the ID15

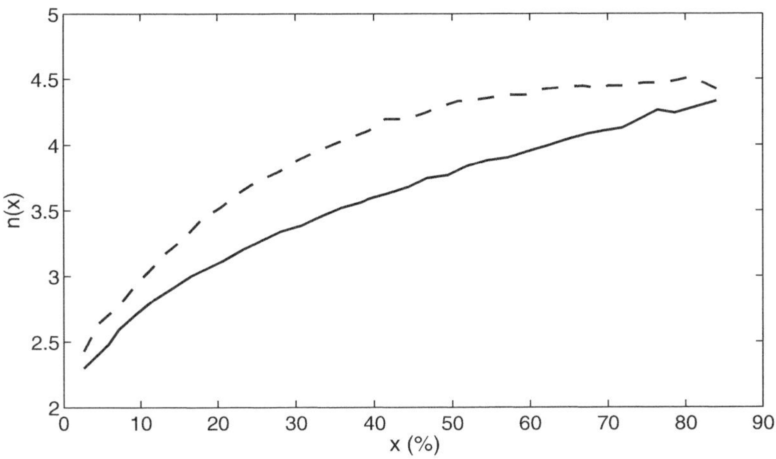

Figure 13.8 Local Avrami exponent $n(x)$ as a function of crystalline fraction x for the MG $Zr_{55}Al_{10}Ni_5Cu_{30}$ at $T = 713$ K (solid line) and $T = 723$ K (dashed line). Adapted from (Gao et al., 2003a).

beamline of the European Synchrotron Radiation Facility (ESRF), with a wavelength $\lambda = 0.14$ Å and an acquisition time of 30 s. In the figure, **a** is the amorphous sample at room temperature, **b** shows already some incipient crystalline peaks, during the heating ramp and **c** is the spectrum of the fully crystallized sample at $T = 740$ K. The theoretical spectra of the main nucleated crystalline phases, Zr_7Cu_{10} and Zr_2Cu, are also reported.

13.4.4 The Crystalline Size Distribution

The crystalline size can be deduced both from XRD data and TEM observation.

The first method is based on the Scherrer law, which relates the width Γ of a Bragg reflection (FWHM), measured in radiants, to the crystallite size d:

$$\Gamma(2\theta) = \frac{K\lambda}{d cos\theta}. \tag{13.9}$$

Here λ is the x-ray wavelength, 2θ the scattering angle and K the shape factor. The latter has a value usually close to 1 for spherical crystallites; however, it has been found to vary in the quite large

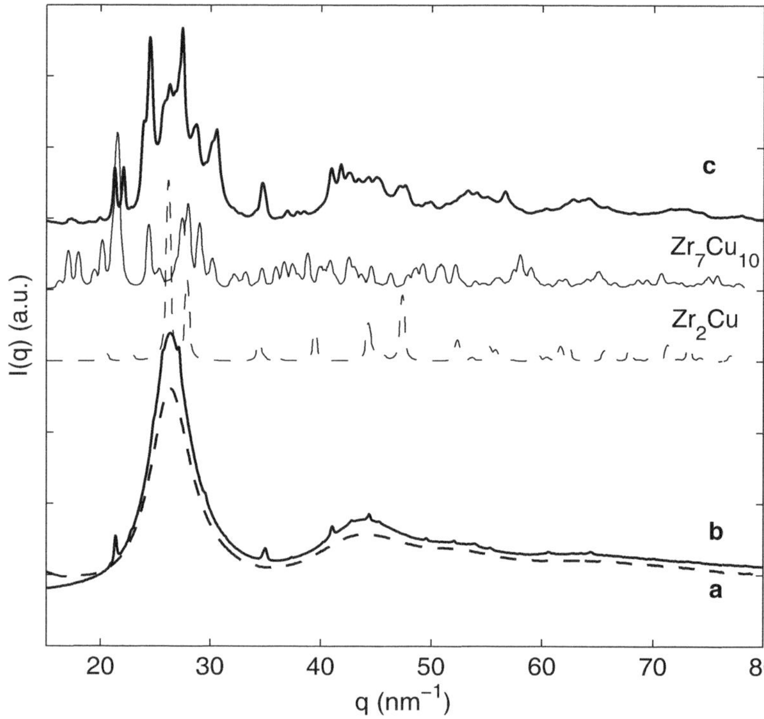

Figure 13.9 XRD spectra collected at the ID15 beamline of ESRF with incident $\lambda = 0.14$ Å on the Zr-based BMG called D″ (see text). (a) Room temperature amorphous phase; (b) beginning of crystallization during a linear heating thermal protocol (ramp rate 5 K/s); c: fully crystallized sample at 740 K. The theoretical spectra for the two main crystalline phases are also reported.

range [0.68–2], increasing thus the uncertainty on the grain size determination with this method (Langford and Wilson, 1978).

The reliability of this method remains limited also for other reasons, the main two being 1) that the quality of the peaks fit for extracting the FWHM can be poor for overlapping peaks, and 2) that crystalline size is not the only contribution to the peaks width. The presence of micro-strain gives indeed raise to a stress-induced broadening too, $\Gamma_s = 4\epsilon \tan(\theta)$ where ϵ is the microstrain. While the different angular dependence of the two contributions can help in disentangling them, it is, however, quite difficult, and the result uncertain. Some better insight can be obtained when the composition and structure of all present crystalline phases is known, as then it is possible to perform a Rietveld refinement, which takes into account also the different contributions to the broadening. It is, however, worth reminding here that the crystallites size determined by XRD analysis is in fact the size of coherently scattering crystalline domains. If grains are not single crystals, the size determined will be the size of a single crystal within a grain, and not the size of the grain itself.

The second method for extracting the particles size distribution is the image analysis from TEM measurements. High-resolution TEM indeed allows to distinguish crystalline nuclei in an amorphous matrix, as the alignment of lattice planes can be recognized (e.g. in Fig. 13.10). Image analysis allows then to obtain both the crystalline volume fraction, from the crystalline surface fraction observed in the bidimensional image, and the crystalline size distribution, built by measuring the size of all the observed grains. The limitation of this

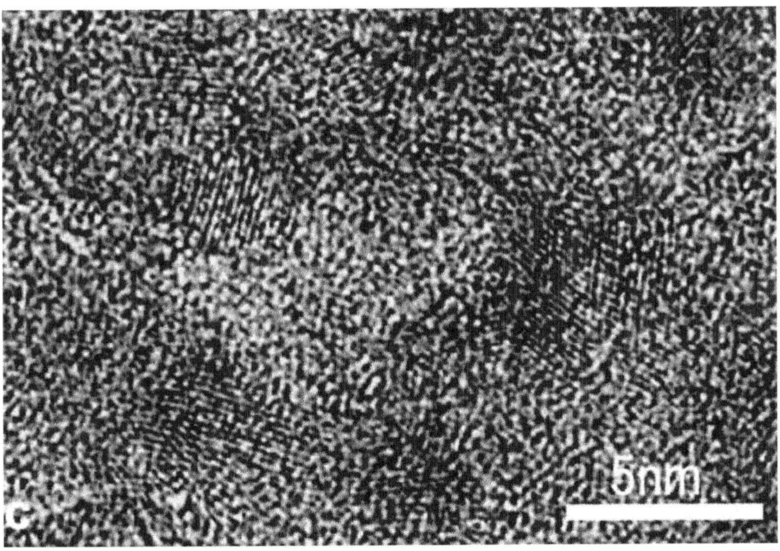

Figure 13.10 High-resolution TEM image of a $Zr_{50}Cu_{50}$ ribbon, adapted from (Inoue et al., 2005). The crystalline grains embedded in the amorphous matrix are clearly visible by the alignment of lattice planes.

method resides mainly in the local character, in the sense that only a very small portion of the sample is probed, and in the possibility that crystals are hidden by a weak contrast or overlap in the probed thickness. It is, however, still considered the best direct method for the estimation of crystalline fraction and size distribution.

13.5 A Practical Example: $Zr_{52.5}Cu_{27}Al_{10}Ni_8Ti_{2.5}$

To summarize and conclude we report here the production and characterization of a nanocrystal/amorphous composite starting from the BMG $Zr_{52.5}Cu_{27}Al_{10}Ni_8Ti_{2.5}$ (also called D″) by means of an isothermal annealing.

The sample was synthesized at the SIMaP laboratory (Grenoble, France) by arc-melting the pure elements in the right atomic proportions and suction casting the liquid alloy in a water-cooled cylindrical copper mold with an internal diameter of 3 mm. The whole procedure was done under Argon atmosphere to avoid oxidation of the elements and the alloy.

Half of the copper mold and the obtained rod are shown in Fig. 13.11. Rods of up to 8 cm length can be obtained, although the extremities are usually partially or totally crystalline because of the different quenching rate and flow during the processing. Indeed, in the figure the left extremity of the rod is clearly not shiny, indication of its polycrystalline nature.

After production, thin slices from different parts of the rod are cut to check its amorphous nature and homogeneity by x-ray diffraction: Figure 13.12 reports as an example the XRD pattern recorded at the Diffraction Center Henri Longchambon (Lyon, France) using a Bruker D8 Advance diffractometer in a Bragg-Brentano geometry and with a copper source (λ = 1.54 Å). The measured pattern is clearly non-crystalline, as no Bragg peak can be seen, at least within the quality of the signal.

The DSC analysis is then performed during a linear heating to characterize the glass transition temperature and the crystallization peak, which are found to be $T_g = 687$ K and $T_x = 755$ K (Fig. 13.13).

Once the amorphous sample has been characterized, we can proceed to prepare an amorphous/crystalline composite by means

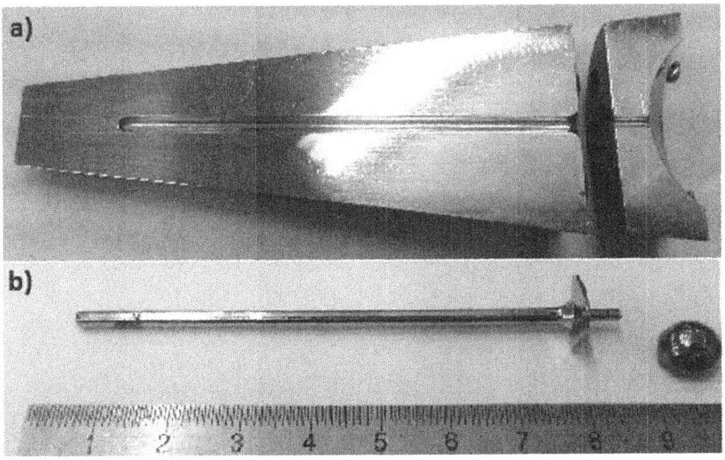

Figure 13.11 (a) Half of the copper mold on the left and crucible on the right used for casting 3 mm diameter cylinders. The hole in the crucible has a diameter of 2 mm to avoid the flow of molten alloy into the mold before suction. No direct connection exists between crucible and mold, allowing for melting at very high temperature without heating the mold too. (b) The as casted rod.

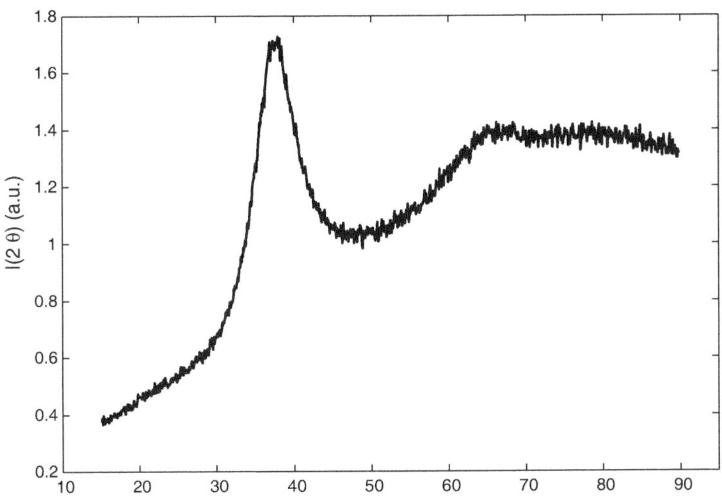

Figure 13.12 Diffraction pattern collected on a thin slice cut from the produced rod of D″, with an x-ray wavelength $\lambda = 1.54$ Å. The sample is clearly amorphous, as no crystalline peaks are visible.

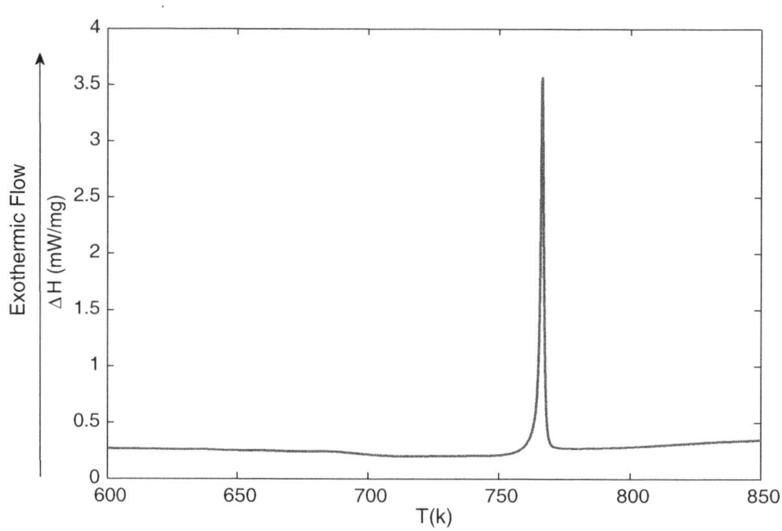

Figure 13.13 Enthalpy change measured by isochronal DSC during heating of the D″ BMG with a ramp rate of 10 K/min under argon flow. The equipment used was a Perkin Elmer Pyris Diamond DSC.

of an isothermal annealing, choosing an annealing temperature between T_g and T_X. The highest the annealing temperature, the fastest will be the crystallization process, and likely this will promote nucleation but not crystalline growth, being the available time too short for allowing an important atomic diffusion. For this reason we choose $T = 740$ K and we characterize the crystallization kinetics for an isotherm at this temperature. If the heating ramp rate is too slow, crystallization will already start during the ramp. This is indeed what can be observed in Fig. 13.9, curve **b**.

For this reason we prepare our sample in a DSC equipment, using a very small mass, in order to minimize the thermal inertia of the sample, and a ramp rate as high as 100 K/min. We do not detect any crystallization during the ramp, while we do during the isotherm. Figure 13.14 reports the crystallization peak along the isotherm reached with such a high heating rate, and the crystalline fraction as calculated using Eq. (13.6).

We can now fit the crystallization fraction evolution with Eq. (13.4) to get the Avrami index and the incubation time τ.

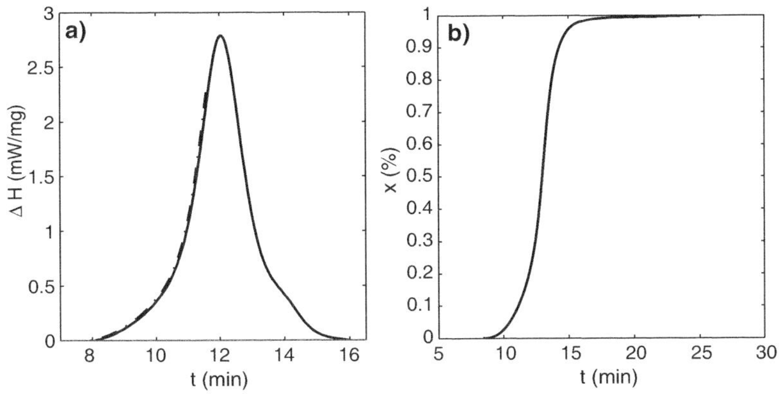

Figure 13.14 (a) Crystallization enthalpy peak measured on a D″ BMG during an isotherm at 740 K (thin solid line), and enthalpy exchange measured during a partial crystallization thermal protocol (dot-dashed thick line). (b) Crystallization fraction evolution as a function of time in a D″ BMG during an isothermal annealing at 740 K.

A closer inspection of Fig. 13.14 reveals us that there are two distinct crystallization processes, as indicated by the presence of a small shoulder on the large time-side of the main peak. This means that it will not be possible to fit the crystallization fraction with a single curve. On the other hand, the existence of two distinct crystallization processes hinders the applicability of the Barandiaran method (Barandiaran et al., 2003), as the activation energy does not smoothly change with the crystallization fraction. We therefore decide to study separately the two regimes corresponding to the two crystallization processes. The primary crystallization is mostly the only one for $t < \approx 13$ min, while longer times are dominated by the secondary crystallization, although there still is a residual primary crystallization. If we separately fit these two regimes, we get the two curves reported in Fig. 13.15, with an intermediate regime between the two. The fit gives the same incubation time for the two regimes, $\tau = 484$ s, but different Avrami indexes and kinetic parameters k: $n = 3.3$, $k = 0.0028\ \mathrm{s}^{-1}$ for small times and $n = 2.8$, $k = 0.0036\ \mathrm{s}^{-1}$ for large times.

Once the time evolution of crystalline fraction is known, we can repeat the same thermal protocol on another amorphous slice of

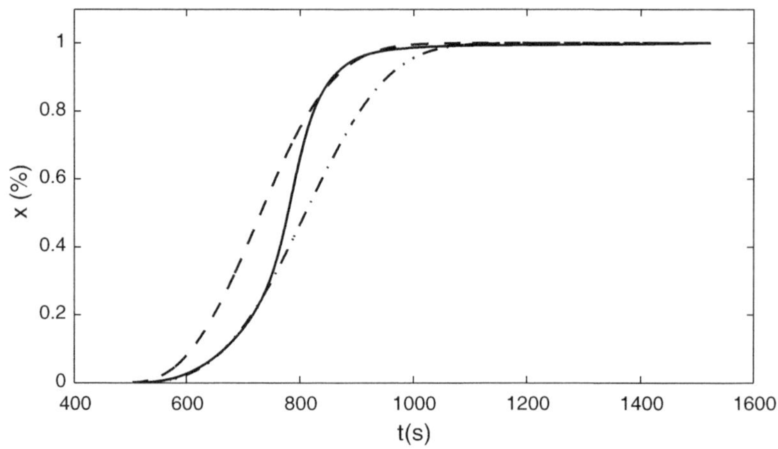

Figure 13.15 Crystallization fraction evolution of the BMG D″ during an isothermal annealing at 740 K as a function of time in seconds from Fig. 13.14 (b). The fit with Eq. (13.4) is also reported for $t < \approx 12.5$ min (dot-dashed curve) and $t > \approx 13.5$ min (dashed curve).

similar mass, interrupting it at the time t^* corresponding to the desired crystalline fraction, here 30%. The DSC signal measured during this shorter annealing is reported in Fig. 13.14a as a dot-dashed thick line. It is important to notice that the partial crystallization signal perfectly overlaps with the crystallization peak. This is a sign of the quality and homogeneity of the amorphous rod from which the two samples have been cut, and the good reproducibility of the thermal protocol. Indeed, if the rod was not homogeneous, a shift in the crystallization peak, or a change in its shape because of different crystallization processes would have been detected. The same would be true if the thermal protocol was not reproducible and the two samples were actually heated with different ramp rates.

The crystalline fraction and size distribution is finally measured by TEM, using a JEOL 2100 F microscope. We report in Fig. 13.16a a bright field image of the sample, where the crystalline nuclei are clearly detectable from the color contrast, and confirm the crystalline fraction, as estimated from DSC. The diameter distribution is found to be centered at 11 nm and with a standard deviation of 3 nm, thus confirming that in this BMG the combination

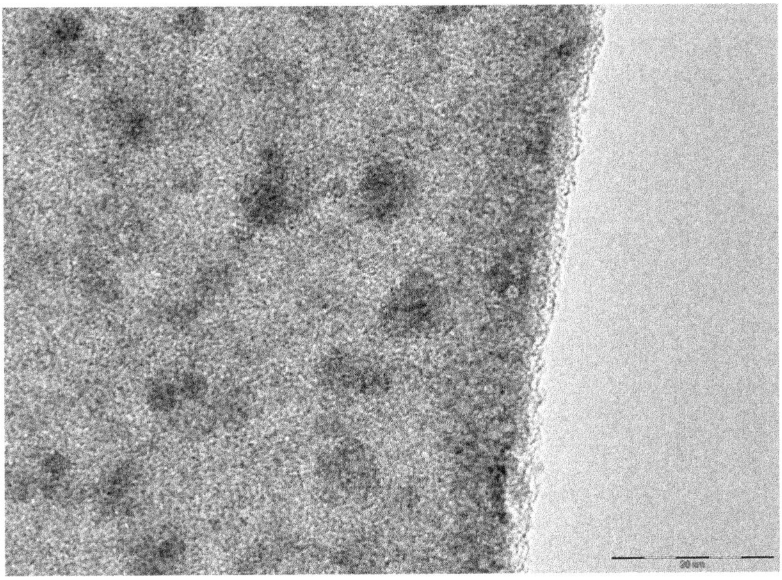

Figure 13.16 Bright field image of the amorphous/crystalline composite of the D″ BMG. The darker zones correspond to the crystalline nuclei embedded in the amorphous matrix. The crystalline fraction estimated is 30% and the average diameter is 11 nm.

of a high ramp-rate and a high-temperature isothermal annealing allow to obtain nanometric sized crystalline inclusions. The thermal protocol having been characterized and identified, samples with different crystallization fraction can be prepared for investigating their transport properties.

References

A. Inoue, T. Z. and Masumoto, T. (1989). Al–la–ni amorphous alloys with a wide supercooled liquid region. *Mater Trans JIM* **30**, p. 965–972.

Adam, G. and Gibbs., J. H. (1965). On the temperature dependence of cooperative relaxation properties in glass-forming liquids. *J. Chem. Phys.* **43**, 139.

Al-haj, M. and Barry, J. (1997). *Journal of Materials Science Letters* **16**, p. 1641.

Angell, C. A. (1995). Formation of glasses from liquids and biopolymers. *Science* **267**, p. 1924.

Baldi, G., Giordano, V. M., Monaco, G., Sette, F., Fabiani, E., Fontana, A. and Ruocco, G. (2008). Thermal conductivity and terahertz vibrational dynamics of vitreous silica, *Phys. Rev. B* **77**, p. 214309.

Barandiaran, J., Telleria, I., Garitaonandia, J. and Davies, H. (2003). Kinetic aspects of nano-crystallization in finemet-like alloys, *J. Non-Cryst. Sol.* **329**, pp. 57–62.

Beltukov, Y. M., Kozub, V. I. and Parshin, D. A. (2013). Ioffe-regel criterion and diffusion of vibrations in random lattices, *Phys. Rev. B* **87**, p. 134203.

Busch, R., Schneider, S., Peker, A. and Johnson, W. L. (1995). Decomposition and primary crystallization in undercooled zr41.2ti13.8cu12.5ni10.0be22.5 melts, *Appl. Phys. Lett.* **67**, pp. 1544–1546.

Cohen, M. H. and Turnbull, D. (1964a). Metastability of amorphous structures. *Nature* **203**, p. 964.

Cohen, M. H. and Turnbull, D. (1964b). Molecular transport in liquids and glasses. *J. Chem. Phys.* **31**, 1164.

Damart, T., Giordano, V. M. and Tanguy, A. (2015). Nano-crystalline inclusions as a low-pass fillter for thermal transport in a-si, *arXiv:1507.02112 [cond-mat.mtrl-sci].*

Evenson, Z., Naleway, S. E., Wei, S., Gross, O., Kruzic, J. J., Gallino, I., Possart, W., Stommel, M. and Busch, R. (2014). β relaxation and low-temperature aging in a au-based bulk metallic glass: From elastic properties to atomic-scale structure, *Phys. Rev. B* **89**, p. 174204.

Gao, Y. L., Shen, J., Sun, J. F., Wang, G., Xing, D. W., Xian, H. Z. and Zhou, B. D. (2003a). Crystallization behavior of zralnicu bulk metallic glass with wide supercooled liquid region. *Materials Letters* **57**, p. 1894–1898.

Gao, Y.-L., Shena, J., Suna, J.-F., Chenb, D.-M. and Wanga., G. (2003b). Nanocrystallization of zr-ti-cu-ni-be bulk metallic glass, *Materials Letters* **57**, p. 2341–2347.

Gravier, S., Donnadieu, P., Lay, S., Doisneau, B., Bley, F., Salvo, L. and Blandin, J. (2010). Evaluation of the crystal volume fraction in a partially nanocrystallized bulk metallic glass, *J. Alloys Compd.* **504S**, pp. S226–S229.

H. W. Kui, A. G. and Turnbull, D. (1984). Formation of bulk metallic glasses by fluxing, *Appl. Phys. Lett.* **45**, pp. 615–616.

Hachenberg, J., Bedorf, D., Samwer, K., Richert, R., Kahl, A., Demetriou, M. D. and Johnson, W. L. (2008). Merging of the α and β relaxations and aging via the johari–goldstein modes in rapidly quenched metallic glasses. *Appl. Phys. Lett.* **92**, p. 131911.

Hebert, R. J. (2011). *Nanocrystals in Metallic Glasses, Nanocrystal.*

Hsiao, A., McHenry, M. E., Luaghlin, D. E., Kramer, M. J., Ashe, C. and Ohkubo, T. (2002). The thermal, magnetic, and structural characterization of the crystallization kinetics of fe88zr7b4cu1, an amorphous soft magnetic ribbon, *IEEE Transactions on Magnetics* **38**, 5, pp. 3039–3044.

Illekovfi, E., Kuhnast, F.-A., Fiorani, J.-M. and Naguet, C. (1995). Peculiarities of the 'crystallization' kinetics of fe 73.5 cu 1 nb 3 si 135 b 9 ribbon to the nanostructured phase, *J. Non-Cryst. Sol.* **192 & 193**, pp. 556–560.

Inoue, A., Zhang, W., Tsurui, T., Yavari, A. R. and Greer, A. L. (2005). Unusual room-temperature compressive plasticity in nanocrystal-toughened bulk copper-zirconium glass, *Philos. Mag. Lett.* **85**, 5, pp. 221–229.

Jackle, J. (1986). Models of the glass transition. *Rep. Prog. Phys.* **49**, p. 171.

J. Wong and Angell, C. A. (1976). *Glass, Structure by Spectroscopy* (M. Dekker).

Kelton, K. (1993). Numerical model for isothermal and non-isothermal crystallization of liquids and glasses. *Journal of Non-Crystalline Solids* **163**, p. 283–296.

Kelton, K. (2000). Kinetic model for nucleation in partitioning systems, *J. Non-Cryst. Sol.* **274**, pp. 147–154.

Kelton, K., Croat, T., Gangopadhyay, A., Xing, L.-Q., Greer, A., Weyland, M., Li, X. and Rajan, K. (2003). Mechanisms for nanocrystal formation in metallic glasses, *J. Non-Cryst. Sol.* **317**, pp. 71–77.

Kelton, K. F. (2000). Time-dependent nucleation in partitioning transformations, *Acta mater.* **48**, pp. 1967–1980.

Kumar, G., Desai, A. and Schroers, J. (2011). Bulk metallic glass: The smaller the better, *Adv. Mater.* **23**, pp. 461–476.

Lampin, E., Nguyen, H.-Q., Francioso, P. and Cleri, F. (2012). *Appl. Phys. Lett.* **100**, p. 131906.

Langford, J. and Wilson, A. (1978). Scherrer after sixty years: A survey and some new results in the determination of crystallite size, *J. Appl. Cryst.* **11**, pp. 102–113.

Lu, K. and Wang, J. (1991)., *Materials Science & Engineering A* **133**, p. 501.

Mao, M., Altounian, Z. and Bruning, R. (1995). X-ray diffraction study of structural relaxation in metallic glasses, *Phys. Rev. B* **51**, p. 2798.

Minic, D. M. (2006). Synthesis, characterization and stability of amorphous alloys, *Science of Sintering* **38**, pp. 83–92.

Mizoguchi, T., Kudo, T. and Takayama., S. (1980). Lorenz number in metallic glasses. *Journal de Physique Colloques* **41 (C8)**, pp. C8-501–C8-502.

Mizutani, U. (2001). *Introduction to the Electron Theory of Metals* (Cambridge University Press).

Pineda, E., Bruna, P., Ruta, B., Gonzalez-Silveira, M. and Crespo, D. (2013). Relaxation of rapidly quenched metallic glasses: Effect of the relaxation state on the slow low temperature dynamics, *Acta Mater.* **61**, p. 3002.

Pradell, T., Crespo, D., Clavaguera, N. and M. T. Clavaguera-Mora (1998). Diffusion controlled grain growth in a primary crystallization : Avrami exponent revisited. *Journal of Physics :Condensed Matter* **10**, pp. 3833–3844.

Regner, K. T., Sellan, D. P., Su, Z., Amon, C. H., McGaughey, A. J. H. and Malen, J. A. (2013). Broadband phonon mean free path contributions to thermal conductivity measured using frequency domain thermoreflectance, *Nat. Commun.* **4**, p. 1640.

Reichman, D. R. and Charbonneau, P. (2005). Mode-coupling theory, *J. Stat. Mech.* , p. P05013.

Schneider, S., Thiyagarajan, P. and Johnson, W. L. (1996). Formation of nanocrystals based on decomposition in the amorphous s. zr41.2ti13.8cu12.5ni10be22.5 alloy, *Appl. Phys. Lett.* **68**, 4, pp. 493–495.

Spassov, T., Todorova, S. and Petkov, V. (2009). Kinetics of mg6ni nanocrystallization in amorphous mg83ni17, *J. Non-Cryst. Sol.* **355**, pp. 1–5.

Tanguy, A., Mantisi, B. and Tsamados, M. (2010). *Europhysics Letters* **90**, p. 16004.

Taub, A. and Spaepen, F. (1980). The kinetics of structural relaxation of a metallic glass, *Acta Metallurgica* **28**, pp. 1781–1788.

W. Klement, R. W. and Duwez, P. (1960). Non-crystalline structure in solidified gold-silicon alloys, *Nature* **187**, pp. 869–870.

Wang, W. H. (2007). Roles of minor additions in formation and properties of bulk metallic glasses, *Progress in Material Science* **52**, p. 540–596.

Wang, W. H. (2009). Bulk metallic glasses with functional physical properties, *Adv. Mater.* **21**, p. 4524–4544.

Wang, W. H. (2012). The elastic properties, elastic models and elastic perspectives of metallic glasses, *Prog. Mater. Sci.* **57**, pp. 487–656.

Wang, W. H., Wu, E., Wang, R. J., Kennedy, S. J. and Studer., A. J. (2002). Phase transformation in a zr41.2ti13.8cu12.5ni10be22.5 bulk amorphous alloy upon crystallization. *PRB* **66**, p. 104205.

Wei, S., Ding, B., Lei, T. and Hu, Z. (1998). Crystallization of amorphous zr65cu27al8 alloy with wide supercooled liquid region. *Materials Letters* **37**, p. 263–267.

Wesseling, P., Ko, B. C. and Lewandowski, J. J. (2003). Quantitative evaluation of a-al nanoparticles in amorphous al87ni7gd6: comparison of xrd, dsc, and tem. *Scripta Materialia* **48**, p. 1537–1541.

Wolff, U., Johnson, E. and Wert., J. A. (2004). The effect of partial crystallization on elevated temperature flow stress and room temperature hardness of a bulk amorphous mg60cu30y10 alloy. *Acta Materialia* **52**, p. 1989–1995.

Zberg, B., Uggowitzer, P. J. and Lffler, J. F. (2009). Mgznca glasses without clinically observable hydrogen evolution for biodegradable implants. *Nature Mater.* **8**, p. 887.

Zebarjadi, M., Esfarjani, K., Shakouri, A., Bian, Z., Bahk, J.-H., Zeng, G., Bowers, J., Lu, H., Zide, J. and Gossard, A. (2009). Effect of nanoparticles on electron and thermoelectric transport, *J. Electron. Mater.* **38**, p. 7.

Zhuang, Y. X. and Wang, W. H. (2000). Effects of relaxation on glass transition and crystallization of zrticunibe bulk metallic glass, *J. Appl. Phys.* **87**, 11, pp. 8209–8211.

Zumkley, T., Naundorf, V., Macht, M.-P. and Frohberg, G. (2001). Effect of reversible structural relaxation on diffusion in a zrticunibe bulk glass. *Scripta Materialia* **45**, pp. 471–477.

Chapter 14

Ball Milling–Induced Nanocrystallization and Amorphization

Eyripides Hatzikraniotis,[a] Theodora Kyratsi,[b] and Konstantinos M. Paraskevopoulos[a]

[a]*Department of Physics, Aristotle University of Thessaloniki, 54124 Thessaloniki, Greece*
[b]*Department of Mechanical and Manufacturing Engineering, University of Cyprus, 1678 Nicosia, Cyprus*
evris@physics.auth.gr

In this chapter, the use of ball milling technique for amorphization and nanocrystallization is outlined. Amorphization is the conversion of a crystalline material into an amorphous one, while nanocrystallization is the incorporation of nanometer-scale structural features in a bulk matrix. The ball milling process has two types of uses: one for *mechanical alloying* that involves the mixing of elements/compounds and chemical transformation and *mechanical milling (or grinding)* when there is no chemical transformation and only grain size reduction occurs. The case studies presented in this chapter show that amorphization may occur in both cases.

Nanostructured Semiconductors: Amorphization and Thermal Properties
Edited by Konstantinos Termentzidis

ISBN 978-981-4745-64-2 (Hardcover), 978-1-315-36445-2 (eBook)
www.panstanford.com

14.1 Introduction

In line with the book's theme, in this chapter we shall outline the use of ball milling techniques for amorphization and nanocrystallization. Nanostructuring is one of the most efficient ways to reduce thermal conductivity, without much affecting the electrical properties, and is especially applied in thermoelectric materials. We recall here that a good thermoelectric material should have high electrical conductivity, large Seebeck coefficient, and poor ability to conduct heat. More discussion concerning this kind of materials can be found elsewhere in this book [1]. Following the pioneering work of Hicks and Dresselhaus [2], who considered the role of dimensionality on the thermoelectric properties, the lower-dimensional structures can have significantly enhanced figures of merit in comparison with the usual bulk thermoelectric; in nanostructured materials, when one or more spatial dimensions are comparable to the characteristic length scale of electrons and phonons, the thermoelectric properties are greatly improved. Phenomena such as carrier confinement and energy filtering [3] may alter band structure, energy dependence of the density of states, or preferably filter out the more energetic charge carriers and thus play a positive role in enhancing the Seebeck coefficient. At the same time, high surface-to-volume ratio associated with nanostructural features and numerous interfaces that inevitably arise in structures built up of nanometer-scale structural units greatly enhance opportunities for phonon scattering with a consequent reduction in lattice thermal conductivity [4].

The term "nanostructuring" means the incorporation of nanometer-scale structural features in the otherwise bulk matrix. In general, bulk nanostructured materials can be fabricated by two principal approaches [4]:

(a) using thermodynamically driven phase segregation/rerfomulation phenomena during cooling of the melt such as spinodal decomposition [5], nucleation and growth [6], matrix encapsulations [7], and compound formation [8].
(b) relying on extrinsic processes of preparing nanosized powders by grinding, milling, or wet chemistry routes that are then compacted into ingots/pellets by pressing [9]

Amorphization is the conversion of a crystalline material into an amorphous one. The structure of an amorphous solid is usually described as possessing a crystal-like short-range molecular arrangement, but lacking a long-range order. Amorphous materials have higher Gibbs free energy than their crystalline counterparts and as a result have higher apparent solubility and faster dissolution rates. Glassy metals not only exhibit technologically interesting properties, e.g., high fracture strength, excellent soft magnetic behavior, and good corrosion resistance. They also constitute ideal materials for the studies of low-temperature transport, critical behavior, and electrical properties of disordered metals [10].

Similar to nanostructuring, two routes are used for manufacturing amorphous materials, namely *intrinsic* process by rapid solidification (rapid cooling) from molten phase, and *extrinsic* one, post processing of a crystalline solid by ball milling or other mechanical treatment. Until recently amorphous metallic alloys were almost exclusively synthesized by rapid cooling. Either the material was cooled below the crystallization temperature from the molten phase with fast cooling rates (10^6 to 10^{10} K/s) or the material was condensed from the vapor onto a cold substrate [10]. The nucleation and growth of crystalline phases could be kinetically bypassed in certain molten alloys using these high cooling rates, which yields a configuration to a frozen-in liquid, namely, a metallic glass [11]. Rapid solidification can be achieved by the spinning technique [12] to produce long and continuous ribbons and by the planar flow casting method to produce wide and thin ribbons [13].

Ball milling is a way (extrinsic) of producing amorphous alloys as well as nanostructured materials. It is a powder metallurgy processing technique involving cold welding, fracturing, and rewelding of powder particles in a ball mill and has now become an established commercial technique to produce oxide dispersion strengthened nickel- and iron-based materials. The ball milling process has two types of uses: (1) *mechanical alloying*, which involves the mixing of elements/compounds and in which chemical transformation occurs during milling and (2) *mechanical milling (or grinding)*, in which there is no chemical transformation and only grain size reduction occurs. However, amorphization can occur in both cases (sometimes referred as *mechanical disordering*). The connection between nanosizing and amorphization has been recently established as it was found that nanoparticles have more propensities for amorphization than their bulk counterparts, opening the opportunity to achieve the amorphous states in well-known poor glass-former compounds. Classical size effects are often invoked to explain such phenomenon [14].

In this chapter, the fundamentals of ball milling technique and the mechanism of amorphization by ball milling will be outlined and, then, the effect of ball milling on selected materials as case studies will be explored.

14.2 Fundamentals of Ball Milling

Typically, in a ball milling process, the starting powders are loaded in the milling container with the grinding balls and the process is continued for several hours using various conditions (type of mill, speed, duration, type and size of grinding media, ball-to-material weight ratio, process control agent, etc).

The ball milling process is used mainly for three reasons:

(a) Mechanical alloying where elements are milled together to form alloys and/or to achieve chemical transformations:

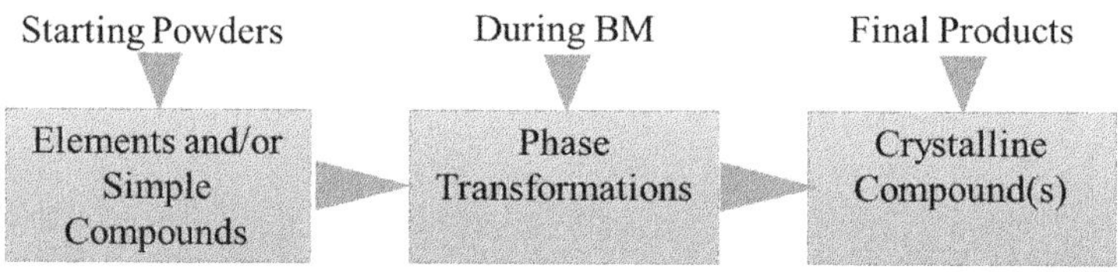

(b) Production of nanostructured materials following the steps:

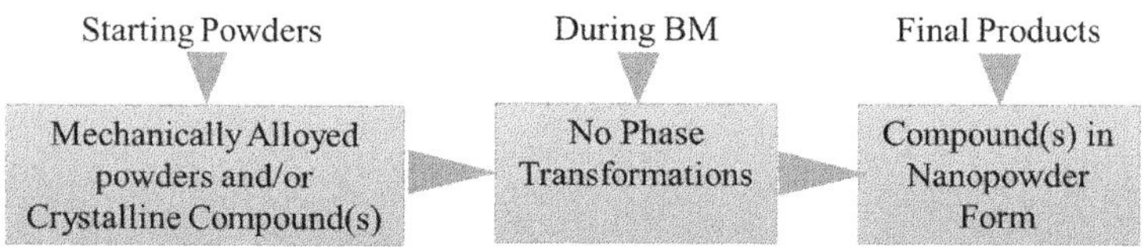

(c) Production of amorphous materials following the steps:

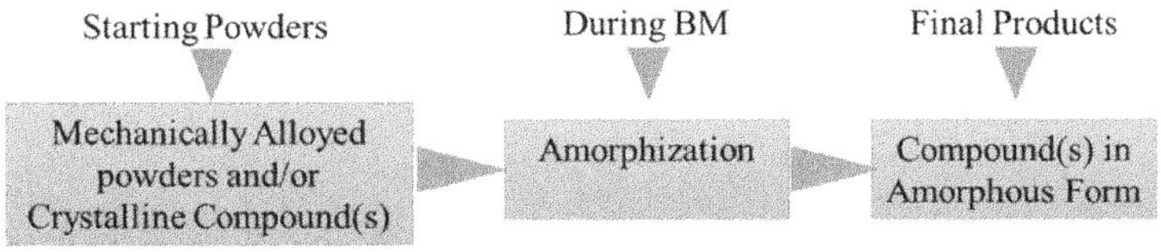

14.2.1 Mechanical Alloying

During high-energy milling, the particles of the powder are flattened, cold welded, fractured and rewelded (see the flowchart below). This sequence is repeated again and again since the powders are trapped between the balls as well as between the balls and the inner walls of the vial. The particles of the powder are plastically deformed due to the force of the impact, and this leads to work hardening and fracture. The newly created surface weld again, leading to composite-type materials that consist of a layered structure with the starting elements/compounds (mainly when ductile component participates in the process). At this stage, the cold welding predominates the other mechanisms and the size of the particles increases. While the deformation continues, the particles continue to work hardened, but now the fracture is due to fatigue failure and/or fragmentation of fragile flakes. At this stage, the fracture predominates over cold welding and the size of the particles tends to decrease. As the process continues, the structure of the particles reaches a steady-state equilibrium and their size remains the same. However, the interlayer distance keeps decreasing. Heavy deformation results in high concentration of crystal defects (dislocations, vacancies, grain boundaries, etc.) that consequently enhance the diffusion mechanism for phase transformation. More details on milling of ductile-ductile, ductile-brittle, and brittle-brittle systems are discussed in [15].

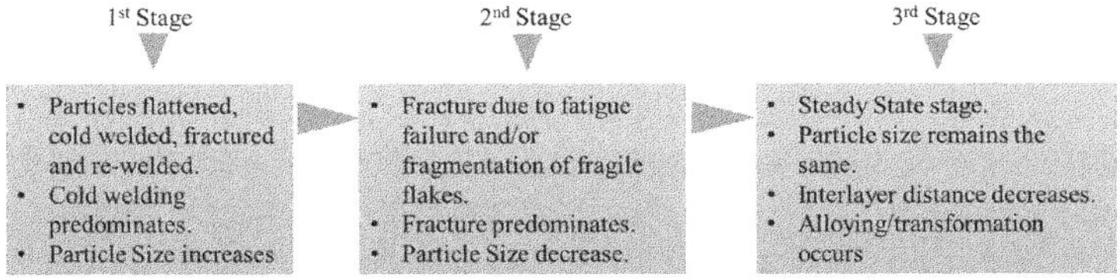

14.2.2 Production of Nanostructured Materials

The production of nanostructured materials by milling process can be described in three steps [16]:

- First, there is deformation localized in shear bands that contain high dislocations density.
- Second, the dislocations annihilate and recombine to small-angle grain boundaries separating subgrains that are already in the nanometer size (typical diameters of 20–30 run).
- Finally, the orientation of the grains becomes completely random since small-angle grain boundaries become high angle and then sliding/rotation occurs.

Almost all mechanical milled metallic materials have nanosized grains, and the advantage of this process is the ability to produce large amount of nanopowders, at room temperature with simple equipment [15].

14.2.3 Production of Amorphous Materials

The transformation of the powder from crystalline to amorphous phase during milling depends on the ability of the starting material to remain in the nanostructured state. According to experiments, two critical grain sizes have been found: d_1 is the grain size below which the surface energy becomes larger than the volume energy, while d_2 is the grain size at which the crystalline-to-amorphous transformation happens. Lattice expansion is the mechanism to compensate the energy difference below d_1, while this cannot be effective when grains are smaller than d_2. Therefore, a different mechanism, namely transformation from the crystal to the amorphous phase, exists below d_2 [17].

Amorphous materials can be prepared by the ball milling process starting from elements, alloys and/or compounds, and mixture of compounds and/or elements, all in powder form. There are two ways to make amorphous materials starting from the elements: (a) A crystalline phase is first formed and then the amorphous phase and (b) there is direct formation of the amorphous phase. The relative free energies of each case will determine whether crystalline or amorphous material will be first taken.

In addition, the milling process variables can determine the crystallinity of the product. Overall, the "soft" milling conditions such as small ball size, lower energies, and lower BMR, favor the amorphization of the powders [15], while "hard" milling conditions favor the crystallinity.

14.3 Experimental Setup

Mechanical alloying is a complex process, and finding the appropriate conditions for a number of variables is necessary to achieve the desired product. Suryanarayana [15] refers to the most important parameters as the type of mill, the milling container, the speed, the time, the type, size, and size distribution of the grinding medium, the ball-to-material weight ratio (BMR), the extent of filling the vial, the milling atmosphere, the process control agent, and the temperature of milling.

14.3.1 The Type of Mill

There are different types of mills in terms of efficiency of milling, type and speed of motion, and capacity. Due to the limited space, a brief overview of the most popular types of mills will follow, while an extended overview can be found in [18]. Figures 14.1 and 14.2 show the mills and vials of each type.

SPEX shaker mills are high-energy mills that are used at the laboratory scale. They follow a back-and-forth shaking motion combined with movements of the ends of the vial resulting in a motion that appears as number "8" [15]. They are manufactured by SPEX CertPrep Matuchen, NJ [19].

- Planetary mills (also known as pulverisette) are popular for milling experiments of a few hundred grams of powder at once. The milling vials are arranged on a rotating disk and a special mechanism rotate them at the same time around their own axis. The balls and the material act

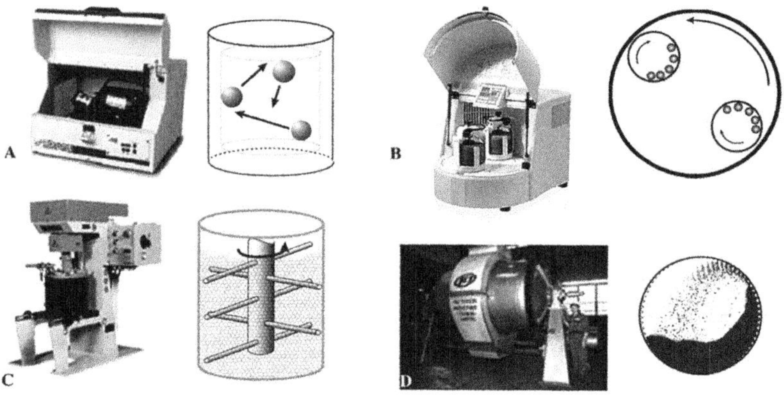

Figure 14.1 Small-scale (A) SPEX mixer/mill (65 mL), medium-scale (B) planetary mill (250 mL), (C) attritor (4 L), production-scale (D) rolling ball mill (hundreds of L). The motion of the milling media is indicated in schematics next to each mill. Reproduced from Ref. [23], with permission of the Royal Society of Chemistry.

under the centrifugal forces created by both the rotation around the vial's axis as well as the rotation on the rotating disk. This combined motion results in the grinding of the material while the grinding balls run down the inside wall of the vial (the friction effect) and then lifting off, traveling freely and colliding against the opposing inside wall (impact effect) [15]. Well-known companies that manufacture and sell planetary ball mills in Europe are Fritsch GmbH [20] and Retsch GmbH [21]. Planetary mills had been used at the lab scale, while currently industrial planetary mills are also commercially available.

- Attritor mills are high-speed mills that can be used for the lab or the production scale since 0.5–40 kg can be milled at the same time [15]. They have a stationary vessel charged with the grinding media and the material to be ground and a motor driven shaft with horizontal arms (see Fig. 14.1) is rotated at high speed causing both shearing and impact forces on the material [22].

The choice of the type of mill is usually based on the capacity (lab or industrial scale experiments) and then on the speed of operation,

Figure 14.2 From left to right, a microvial for a SPEX 5100 mixer/mill, a vial for a SPEX 8000 mixer/mill, a vial for a Fritsch Pulverisette 6 planetary mill, and a tank for a Union Process 01-HDDM attrition mill. Reproduced from Ref. [23], with permission of the Royal Society of Chemistry.

the ability to control the milling temperature, and the extent of contamination of the powders.

14.3.2 Milling Medium

The milling container/vial and the balls used for the process are chosen based on the type of the material to be ground as well as the energy that needs to be transferred to the powder [15]. Ideally, the grinding media should be made of the same material with the milled powder to avoid contamination of the product with the material of the grinding media. When this is not possible, the user can choose between the grinding media of different types of material: hardened steel, tool steel, tempered steel, stainless steel, hardened chromium steel, tungsten carbide, zirconia, sintered corundum, agate, sapphire, etc. It is crucial that both vial and balls are made of the same material; contamination of the product is always a major issue that the user should take into account.

The shape of the container has been found to be important, with flat-ended containers being more efficient than the round-ended containers. Regarding the material selection, the density of the balls is crucial since high density means high impact force on the powder. Moreover, the size of the balls varies and affects the process. Although large balls with high density transfer higher impact energy, the combination of small and large balls at the same time was found to be more efficient [15]. On the other hand, it seems that small balls promote the formation of amorphous phases.

14.3.3 Milling Speed and Time

The milling speed (or intensity) is one of the major parameters in the ball milling process that needs to be selected [15]. In general, the higher speed results in higher transferred energy. However, there are limitations on the milling speed that are related to the efficiency of the process. First, there is a maximum speed value that depends on the design of the mill. Below this maximum value and when the milling speed increases, the speed of the balls also increases. When the maximum value (critical speed) is exceeded, the balls will be pinned to the inner walls and do not fall down to exert any impact force, therefore, no milling process occurs.

Another limitation is the potential temperature increase of the vial due to the high speed. Although this could positively affect the process via diffusion enhancement achieving better homogenization and/or alloying, major drawbacks exist. The higher temperature could increase the contamination via the extensive wear of the grinding media as well as result on different types of transformations than desired.

The milling time, which is the most crucial parameter in the ball milling process, varies depending on the aim of the process (alloying, transformation, amorphization etc), type of mill, speed, temperature, BMR, etc. It is selected to achieve a steady state between fracturing and cold welding of the powder. The milling time also adheres to limitations that are basically related to the contamination; contamination increases with time, and therefore, minimum time required for the process is desired.

14.3.4 Ball-to-Material Ratio (BMR) and Extent of filling the vial

Another important parameter is the ball-to-material weight ratio. Ratios from 1:1 [24] up to 220:1 [25] have been studied in different works. For a small vial at laboratory-scale experiments, the typical BMR is 10:1. The BMR affects the total energy transferred to the powder (higher weight proportion of the balls, more collisions per unit time), therefore the ratio affects the other parameters; usually, the higher the BMR, the shorter time.

The total amount of the balls and the material affects the extent of filling of the vial, which is also an important parameter. It is required that there is enough space for the content of the vial to move around freely during milling. Typically, 50% of the vial needs to be empty.

14.3.5 Milling Atmosphere

It is common to have contamination of the powders if the milling atmosphere is not appropriate. Milling under air, for example, can create oxides and nitrides in the powder. Therefore, the milling atmosphere is crucial for the type and the purity of the products. High-purity argon is typically used as an inert gas to avoid the oxidation and contamination of the powders. Nitrogen is not that common since nitrides of various metal powders can be easily formed. Usually, the milling vials are loaded/unloaded in an argon-filled glove box, while special lids of the containers that allow filling with inert gases are also available in the market (e.g., Fig. 14.3).

14.3.6 Process Control Agents

Process control agents are commonly used during milling aiming to reduce the effect of cold working, since they adsorbed on the surface of the powder particles and lower the surface tension of the solid material. The required energy for the size reduction is given by $E = \gamma \cdot \Delta S$, where γ is the specific surface energy and ΔS is the increase of the surface area; therefore, the reduction of γ due to the existence of PCA results in shorter milling time.

Figure 14.3 Special lid that allows filling the container with inert gases [26].

Typical PCAs are stearic acid, methanol, ethanol, and hexane. Most of PCAs decompose during milling and incorporate into the powder particles. The selection of the type and the quantity of PCA depends on the nature of the powder being milled and the desired purity of the products and determine the final size and shape of the particles.

14.3.7 Milling Temperature

Since diffusion is the main mechanism participating in a milling process, the milling temperature is another important parameter. There are few reports where the temperature was intentionally changed [15]. In most cases, the milling temperature is unintentionally changed or is intentionally kept closed to room temperature (i.e., the ball milling process is automatically interrupted to avoid heating).

14.4 The Mechanism of Amorphization by Ball Milling

Amorphization is one of the most frequently reported phenomena in mechanically alloyed powder mixtures. As previously stated, amorphous phases have been obtained starting from blended elemental powders or mixtures of intermetallics (mechanical alloying), or stoichiometric compounds (pre-alloyed intermetallics or even pure elements), referred to as mechanical milling.

There is a large number of amorphous phases, which are synthesized by *mechanical alloying;* a complete list may be found in [15, 27]. In fact, it has been reported that any alloy can be made amorphous under the appropriate milling conditions. However, it should be noted that powder contamination during milling could be an important contributing factor in the formation of an amorphous phase by mechanical alloying [28].

The mechanism of amorphization by *mechanical alloying* (MA) is not clearly understood, but it is now believed that a solid-state reaction similar to that observed in thin films occurs in mechanically alloyed powders. Phenomenologically, amorphous phases can form from blended elemental powders either directly or via the formation of an intermediate phase. In some instances, it has been reported that a solid solution forms in the initial stages of mixing of blended elemental powders, which on continued milling becomes amorphous. In other instances, the sequence of phase formation is: blended elemental powder → intermediate phase → amorphous phase [28].

Amorphization by *mechanical milling* (MM), on the other hand, is believed to be due to the destabilization of the crystalline phase by the accumulation of structural defects such as vacancies, dislocations, grain boundaries, and anti-phase boundaries. The continuous decrease in grain size (and consequent increase in grain boundary area) and a lattice expansion would also contribute to the increase in free energy of the system. The complete crystal-to-amorphous conversion upon milling often requires milling times of several hours [29]. Short milling times are thus expected to induce a size reduction of both the crystallites (small single crystals) and the particles (a particle can be composed of several crystallites) without generating any noticeable amorphous content. A very important point is to determine when and how the accumulation of crystalline defects (crystal surfaces, dislocations, vacancies, etc.) upon milling gives rise to a genuine amorphous state, which is, in turn, characterized by chemical disorder. During the course of the milling process, the material cannot be described as a biphasic system made of both perfect crystalline matter and genuine amorphous matter. It appears to be constituted of a wide panel of structural states more or less disordered and ranging from the crystalline state to the amorphous state [30].

14.5 Case Studies of Ball-Milled Materials

In this part, the effect of ball milling into selected materials is explored. Since nanostructuring is one of the most effective approaches to increase the figure of merit of thermoelectric materials [31, 32] and recent works have shown that small dimensionally confined systems can exhibit a figure of merit (ZT) well in excess of unity [33, 34], the application of powder techniques is of great interest for fabrication of such nanocomposites. In this part, the materials that are discussed as case studies refer to the use of (a) *mechanical alloying* in the synthesis of nanocrystalline PbTe and (b) *mechanical milling* in the postgrowth processing of Mg_2Si and $K_2Bi_8Se_{13}$ for nanocrystallization and amorphization.

Lead telluride (PbTe) crystallizes in the rock salt (NaCl) structure and is well known as a good thermoelectric material, generally used for thermoelectric power generation in a medium temperature range between 400 and 800 K [35]. Magnesium silicide (Mg_2Si) crystallizes in the FCC CaF_2 type structure [36]. Mg_2Si has drawn much attention in the field of near-infrared optoelectronic applications [37] as a candidate thermoelectric material [38, 39] and as an alternative anode material in rechargeable lithium batteries [40, 41]. Alkali bismuth chalcogenide $K_2Bi_8Se_{13}$ has a more complex and anisotropic structure [42], consisting of 3D interconnected blocks of Bi_2Te_3-type and NaCl-type. This complex and low symmetry crystal structure with weakly bound K^+ ions in cages and structural disorder possesses low thermal conductivity, which makes it

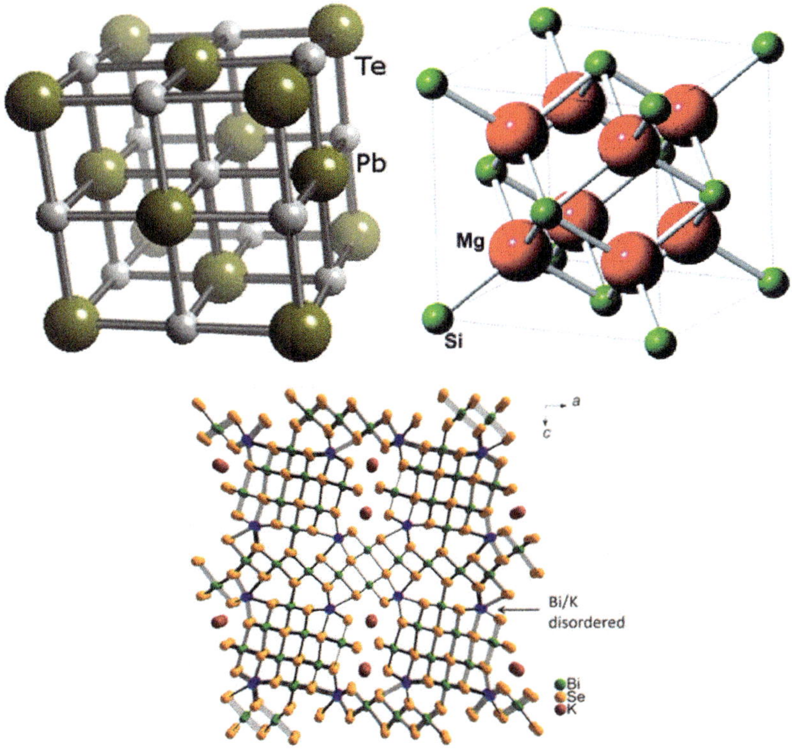

Figure 14.4 Crystal structure of PbTe, Mg_2Si and $K_2Bi_8Se_{13}$.

a candidate for thermoelectric applications. Differences in the structure (see Fig. 14.4) between materials result in differences in the ball milling features.

Ball milling was performed using a planetary mill (Fristch pulverisette P6). The vial was sealed in a glove box under an argon atmosphere to prevent the powders from undergoing oxidation during milling. Ball milling was interrupted several times and the powder was taken from the batch to be examined by X-Ray diffraction (XRD) (Shimadzu and Rigaku Miniflex), scanning electron microscopy (SEM) (TESCAN and JEOL 120CX); transmission electron microscopy (TEM) (JEOL) and high-resolution TEM (HRTEM) (JEOL). The density was evaluated either by mass/volume ratio on cold pressed pellets or by direct measurements using a Micromeritics AccuPyc 1340 gas pycnometer at room temperature.

Infrared reflectivity (IR) studies were performed on cold pressed pellets using a Bruker 113V spectrophotometer. The reflection coefficient was determined by the sample-in, sample-out method with a gold mirror as the reference. Infrared measurements were taken at near normal incidence, with a resolution of 2 cm^{-1}.

14.5.1 Nanocrystalline PbTe Synthesized by Mechanical Alloying

Granules of Pb and Te (both purity 5 N) were mechanically alloyed to give an equiatomic composition (PbTe), with speed at 400 rpm and the initial ball-to-material ratio 10:1 [43].

Figure 14.5a shows the XRD patterns for different ball milling times. As shown, the produced phase is pure PbTe, without any traces of the starting materials, at least after 6 h of ball milling, as also reported by Bouad et al. [44]. Clearly, the mechanical alloying process advances through the formation of an intermediate state: PbTe. The SEM image shown in Fig. 14.5b indicates that the typical size of the particles in ball-milled samples is found to be less than 2 μm. From XRD patterns, it is obvious that there is broadening of the peaks compared with the reference sample prepared from the melt. This is attributed to the crystalline size effect as well as the strain introduced by the ball milling process.

In order to separate the two contributions due to crystalline size and introduced strain, the 200 peaks were analyzed with the Voigt function, which is a convolution of a Gaussian and a Lorentzian component. The two components were calculated as reported by De Keijser [45]. In this method, after correction for the instrumental profile, the integral breadth of the Lorentzian component (β_L) is related to the crystalline size (t), while the Gaussian one (β_G) is attributed to the introduced misfit strain (ε):

$$t = \frac{K \cdot \lambda}{\beta_L \cos(\theta)}, \quad \varepsilon = \frac{\beta_G}{4\tan(\theta)},$$

where λ is wavelength of the Cu Kα radiation ($\lambda = 1.54184$ Å), and K is constant, typically 0.9. The left-hand equation is usually termed as "Scherrer formula" [46]. The analysis yields the crystalline about 30 nm and the strain of the order of 10^{-3}.

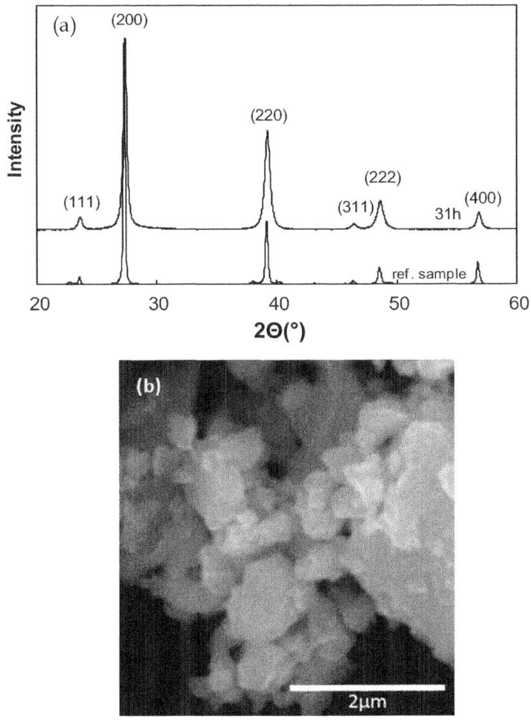

Figure 14.5 (a) XRD pattern evolution for different ball milling times. (b) SEM micrograph for material after 31 h ball milling. Th. Hasapis, Ch. Papageorgiou, E. Hatzikraniotis, Th. Kyratsi, K. M. Paraskevopoulos, IR reflectivity studies of mechanically alloyed PbTe nanocrystals, *Mater. Res. Soc. Symp. Proc.*, 1166, 65 (2009), reproduced with permission.

Figure 14.6a shows the material density during ball milling, which is smoothly decreased from 7.6 to 6.6 g/cm^3 [43]. Crystalline materials have in general higher density than their amorphous counterparts since atoms in the crystal lattice are located at a minimum possible distance from each other. An increase in lattice disorder (decreasing crystallinity) usually results in an increase in volume and therefore a decrease in density. Changes in crystallinity should therefore be accompanied by gradual, progressive changes in density [47]. Figure 14.6b shows the evolution of two phases (micro and nano) with ball milling time. The evaluation of the relative volume fractions was determined by IR measurements, using the

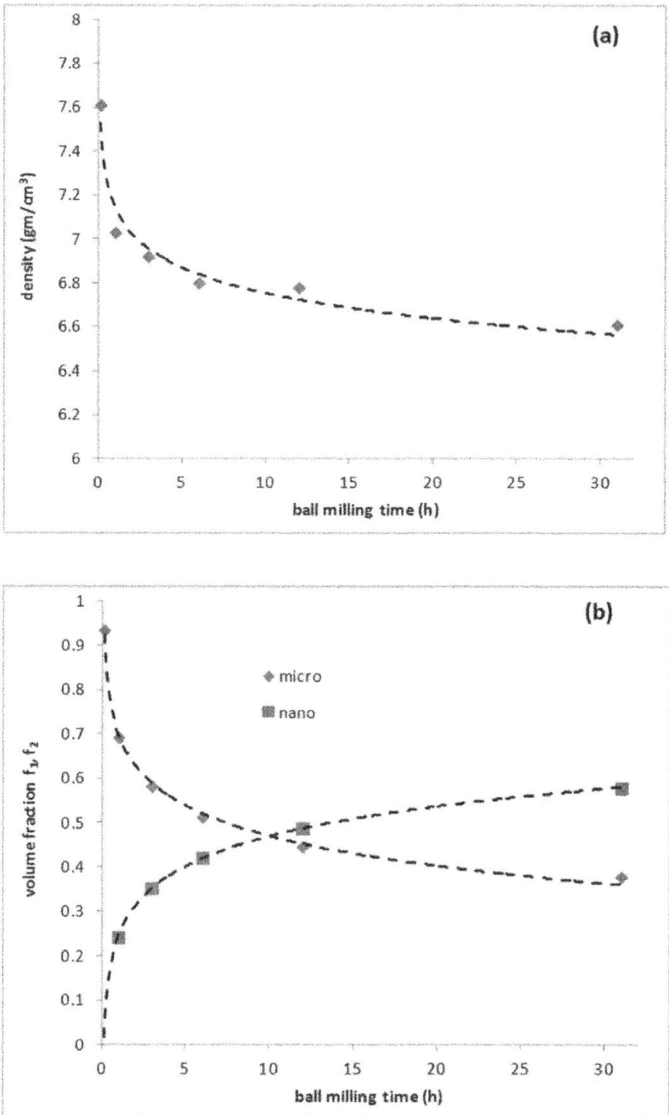

Figure 14.6 (a) Material density and (b) volume fractions for the different phases (macro and nano) during ball milling. Dashed lines serve as guide to the eye. Th. Hasapis, Ch. Papageorgiou, E. Hatzikraniotis, Th. Kyratsi, K. M. Paraskevopoulos, IR reflectivity studies of mechanically alloyed PbTe nanocrystals, *Mater. Res. Soc. Symp. Proc.*, 1166, 65 (2009), reproduced with permission.

Bruggerman approximation [48]. Infrared spectra were measured IR spectroscopy results are compared to those from the material prepared from melt; poly-crystalline material was first grounded using mortar and pestle and then cold pressed at the same pressure [43].

14.5.2 Fabrication of Mg_2 Si Nanocomposites by Mechanical Milling

Mechanical milling process was applied to commercially available powder Mg_2Si (provided by Alfa Aesar Gmb & Co, CAS#22831-39-6) using oxygen-free n-hexane as process control agent for wet milling. The speed was 300 rpm, the initial ball-to-powder mass ratio 40:1 and the fluid-to-powder mass ratio 4:1, up to 40 h [49].

In Fig. 14.7a, the powder XRD patterns of ball-milled samples (for 40 h) in comparison to the pristine material are presented. The patterns confirm the Mg_2Si as the major phase in the pristine as well as the ball-milled materials. Figure 14.7b shows the evolution of the crystalline size, with ball milling time. As can be seen, there is a saturating trend, at about 20 nm, for prolonged milling times. The arrows in Fig. 14.7a indicate a second phase: MgO. MgO peaks are present in pristine material, however at a very low concentration. The relative intensity of the MgO peak is increased in the ball-milled material indicating increase of the MgO concentration. These

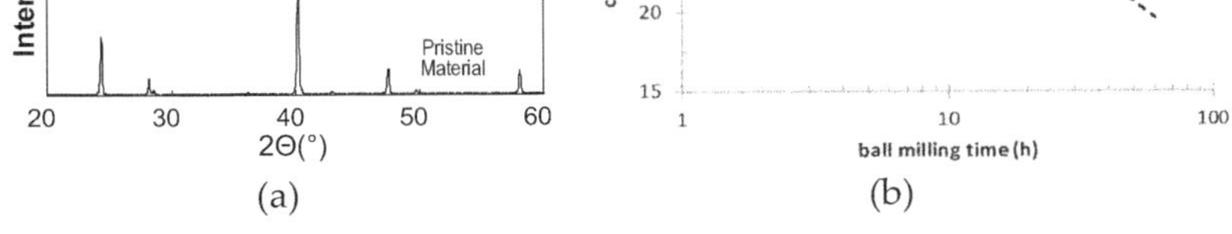

Figure 14.7 (a) Powder XRD patterns for Mg_2Si after ball milling in comparison with the pristine material. The arrows indicate the MgO peaks. (b) Crystalline size vs. ball milling time [49].

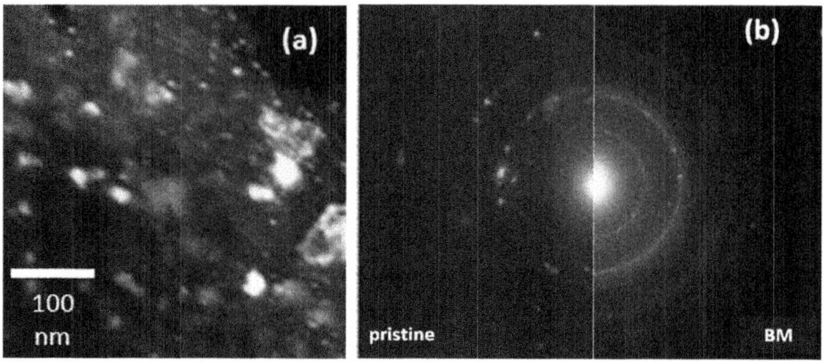

Figure 14.8 (a) Dark-field image and (b) electron diffraction for the ball-milled material (compared with the pristine one).

results suggest the in-situ formation of a two-phase nanocomposite material consisting of Mg_2Si and MgO as the minor phase.

Transmission electron micrographs and diffraction images (ED) for pristine and ball milled (for 40 h) are shown in Fig. 14.8. The ED patterns from the commercial Mg_2Si (pristine material) suggest that the material consists of two phases: one of small crystallites giving typical single crystal EDs (inset) and a polycrystalline one consisting of rings of discrete spots. Tilting experiments confirm that the main phase has FCC structure presenting characteristic extinctions of the NaCl type structure and a lattice parameter $a = 0.635$ nm in very good agreement to that of Mg_2Si (PDF#35-0773). The d-spacings deduced for the polycrystalline material indicate the appearance of cubic MgO with a lattice constant $a = 0.421$ nm (PDF#45-0946). The dark-field image from a rather small Mg_2Si crystallite makes clear that before ball milling there are crystallites of several microns at the pristine material.

The ball-milled material shows ED patterns with rings of very thin discrete spots verifying the polycrystalline character of the material. There are two sets of rings consistent with the two phases identified in the pristine material: Mg_2Si and MgO. Moreover, the spreading of the spots indicates the nanodimensions of the crystalline powder. Dark-field images clearly show nanocrystalline material. Nanocrystals with dimensions up to 90 nm were rarely found, although the mean size was estimated to be nearly 20–25 nm (receiving all crystallites) and this is in agreement with the XRD analysis.

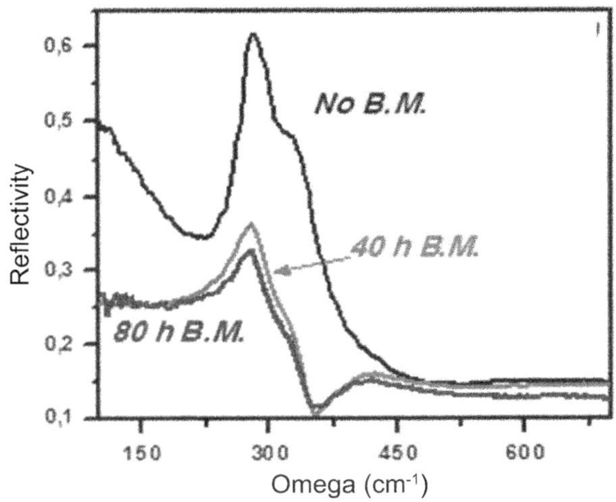

Figure 14.9 IR reflectivity spectra for pristine (0 h) and 40 h ball-milled samples.

Infrared (IR) methods can reflect significant spectral differences between crystalline and nanocrystalline (or amorphous) phases and are hence used to quantify the crystalline content, since the intensity of the vibrational bands is directly proportional to the concentration of the concerned phase. Nanocrystalline (or amorphous) forms give rise to IR spectra that differs from its crystalline counterparts. The origin of these differences relates to both the wider range of conformations, which normally leads to the presence of broader peaks relative to those found in the crystalline spectrum. The IR reflectivity spectra for pristine materials (0 h) and ball-milled material (0–40 h) are shown in Fig. 14.9. The reflection coefficient was determined using a Bruker 113V spectrophotometer on cold pressed pellets by the typical sample-in, sample-out method with a gold mirror as the reference. Spectra for both pristine and ball-milled material are mainly characterized by a multiple phonon band at 275 cm^{-1} and 321 cm^{-1}, which are attributed to Mg_2Si [50, 51]. These two bands appear to be successively lower in height and much wider in the ball-milled samples, which is indicative of the lower range order and the much smaller crystalline size. A wide band at 434 cm^{-1}, which appears in the ball-milled samples and is absent in the pristine one, may be attributed to the increased (during ball milling) concentration of nano-MgO phase [52].

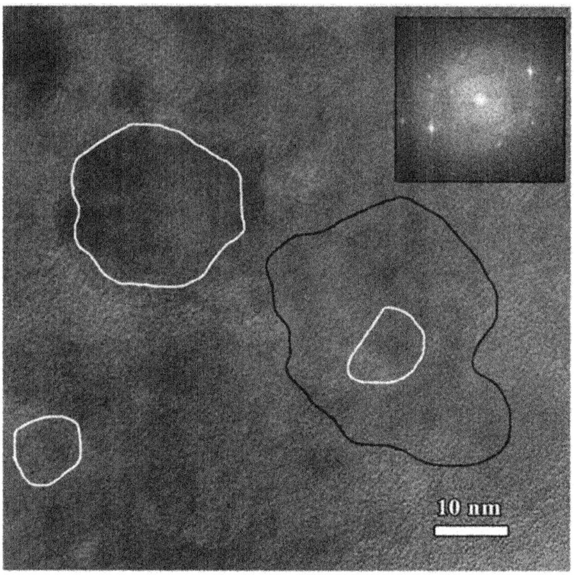

Figure 14.10 HRTEM micrograph from 40 h ball-milled Mg_2Si. The borders of the crystals denoted by white lines are deduced comparing the micrograph to the filtered image. Black line border shows a large Mg_2Si crystal and the inset presents the corresponding FFT. Reprinted from Ref. [53], Copyright (2012), with permission from Elsevier.

HRTEM observations were performed in order to distinguish the crystallites of the two phases, namely Mg_2Si and MgO. In Fig. 14.10, a typical HRTEM micrograph from the 40 h ball-milled material is presented.

The main characteristic is the presence of overlapping lattice fringes, all over the area. The inset shows the FFT of the digitized image. The distinction of the spots belonging to the two phases was obtained by comparing the calculated FFT image to the experimental ED pattern. Inverse FFT reconstruction using masks with different spot pairs reveals the nanocrystals and the orientation of the lattice planes. White lines denote the borders of the nanocrystals. Moreover, it is easy to recognize another large crystallite denoted by black line border in Fig. 14.10 and the inset display the corresponding FFT. The periodicity of the lattice fringes was found to be in very good agreement to the (200) interplanar distance of Mg_2Si. These results prove that the two phases are grown in separate nanocrystals and there is no coexistence of the two materials in the same cluster. Moreover, HRTEM images suggest the presence of an amorphous area located at the border of the particles of the material [53].

14.5.3 The Effect of Ball Size on Nanocrystallinity

In order to examine the effect of the ball-size to the crystallinity, two sets of samples have been fabricated: the *L-series*, where the material was initially ball milled with grinding balls of rather large diameter (10 mm), and the *S-series*, where the milled material (from L-series) was further milled using balls with smaller diameter (1.6 mm). Two case studies are presented; one on synthesized PbTe by mechanical alloying [54] and the other on Mg_2Si nanocomposite prepared by mechanical milling [53].

Nanocrystalline PbTe was initially milled for 31 h with 10 mm balls (L-series) and subsequently milled for additional 3 h with 1.6 mm balls (S-series). Typical XRD patterns for L-series and S-series samples are shown in Fig. 14.11a. As previously mentioned, the broadening of the peaks is evident, compared to the material grown from melt. The broadening is more pronounced in the S-series.

Figure 14.11b presents the crystalline size versus ball milling time, as estimated from the XRD data. For the L-series, the crystalline size is almost constant with ball milling time, while is greatly reduced for the S-series [54]. The crystalline size that resulted from the XRD analysis was confirmed by TEM studies performed on selected samples.

Dark-field images from L-series and S-series are shown in Fig. 14.12. The decrease of the nanocrystalline size between the two series is obvious. The nanocrystals in the powder of L-series were measured to be mainly between 11 nm and 50 nm; crystallites of 90 nm were also randomly observed. The average size was estimated to be 29 nm. In the powder from S-series, the size was clearly decreased and measured between 5 nm and 23 nm, and an average value of 12 nm was estimated. These values are in very good agreement with those that resulted from the XRD analysis.

Figure 14.13 shows the electron diffraction pattern for the series-L powder (ball milling for 31 h) and series-S powder (ball milling for additional 3 h). The ball-milled samples were identified to be the

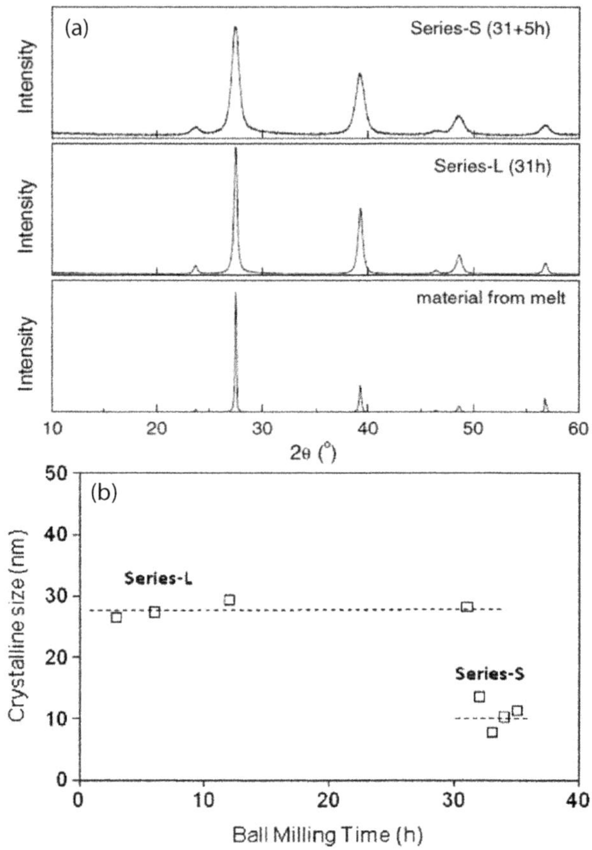

Figure 14.11 (a) Typical XRD patterns for L-series and S-series mechanically alloyed PbTe. *y*-Axis scale varies for better clarity of peak broadening. (b) Crystalline size versus ball milling time, as estimated from XRD data. Reprinted from Ref. [54], with permission of Springer.

PbTe cubic phase (PDF#38-1435) by means of electron diffraction patterns. In all cases, the patterns showed rings of discrete spots indicating the presence of nanocrystalline PbTe. HRTEM image from a sample of series S, where the lattice fringes observed in the nanocrystals allow their identification as PbTe nanocrystals. Moreover, HRTEM images permit accurate estimation of the size of small nanocrystals. The corresponding fast Fourier transform, shown as an inset, confirms that all the crystals are of the same

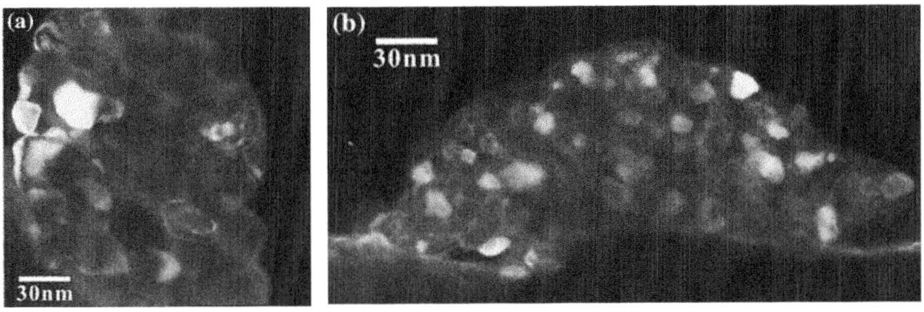

Figure 14.12 Dark-field images (a) from series L powder (ball milling for 31 h) and (b) from series S powder (ball milling for additional 3 h) of mechanically alloyed PbTe. Reprinted from Ref. [54], with permission of Springer.

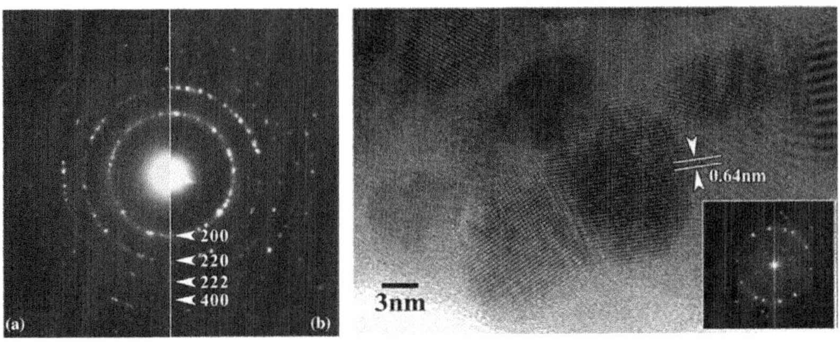

Figure 14.13 Electron diffraction patterns of mechanically alloyed PbTe from (a) series-L powder (ball milling for 31 h) and (b) series-S powder (ball milling for additional 3 h). HRTEM micrograph from series S powder of mechanically alloyed PbTe. Lattice fringes corresponding to 0.32 nm interplanar distance allow identification of the grown material. Reprinted from Ref. [54], with permission of Springer.

crystalline PbTe phase and allows, after filtering, distinction of overlapping nanocrystals and accurate estimation of the size of small nanocrystals. Traces of amorphous areas may be located at the borders of the particles of the material, in prolonged ball milling.

Nanocrystalline Mg_2Si was milled for a total of 80 h. Two types of balls were used in the milling process: (a) large balls, with 10 mm diameter and (b) small balls, with 1.6 mm diameter. Several combinations have been tried, namely, 80 h with 10 mm balls, 40 h with 10 mm balls and subsequently 40 h with 1.6 mm balls, 70 h with 10 mm balls, and additional 10 h with 1.6 mm balls, etc.

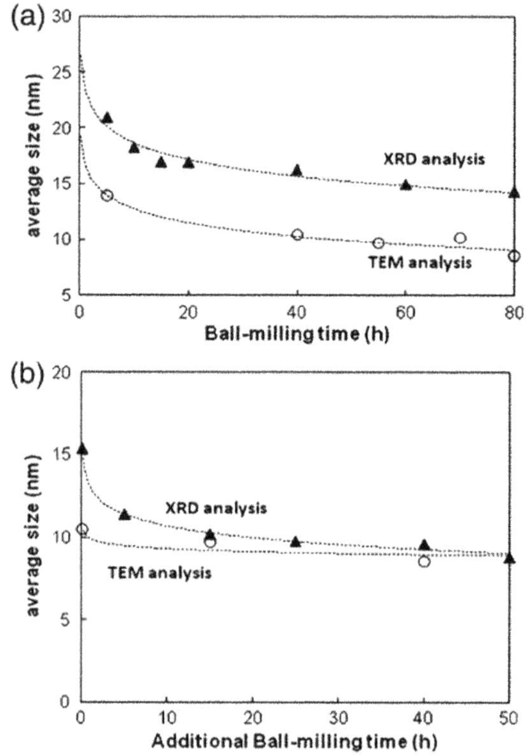

Figure 14.14 Crystalline size vs. ball milling time as calculated based on PXRD and TEM studies. Milling was carried out using (a) large balls and (b) small balls for the additional time. Reprinted from Ref. [53], Copyright (2012), with permission from Elsevier.

The evolution of the crystalline size, estimated from the Scherrer formula (for powder XRD) and from direct measurements (for TEM), is presented in Fig. 14.14. Both ways, show a decreasing trend in crystalline size, with the TEM estimation less than that of XRD. We believe that this deviation is due to the area examined; the determined size by XRD corresponds to an average on a by far much larger area, than in TEM where the edges of a single grain are examined.

Figure 14.14b shows the evolution of the crystalline size for a series of ball milling conditions of 40 h with 10 mm balls and subsequent milling with 1.6 mm balls up to additional 50 h.

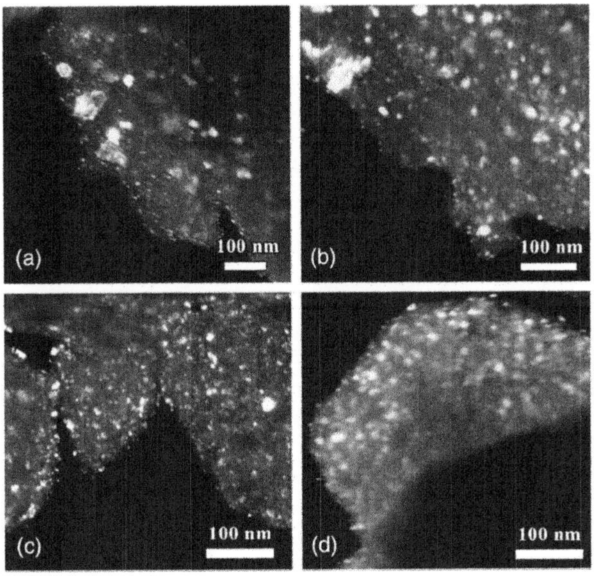

Figure 14.15 Dark-field images from 40 (a), 70 (b), 70 and 10 (c) and 40 and 40 h (d) ball-milled material. Reprinted from Ref. [53], Copyright (2012), with permission from Elsevier.

The calculations clearly show that during the ball milling process, a significant reduction of the crystalline size is being observed. According to the estimations, the crystallite size decreases down to ~15 nm for the material milled with large balls and down to ~9 nm, for the material processed with small balls. Interestingly, the values obtained by both TEM and XRD methods coincide, indicating that the nanoparticle distribution is more homogeneous in size.

Figure 14.15 shows typical dark-field TEM images from 40 h, 70 h, 70 and 10 h, and 40 and 40 h ball-milled material, using the 220 strong reflection of Mg_2Si. As in the case of mechanically alloyed PbTe, the decrease of the nanocrystalline size is evident.

14.5.4 Grain Size and Crystalline Size

SEM images were taken from several ball-milled Mg_2Si samples as shown in Fig. 14.16. The as-received pristine material had large particles with size ~0.841 mm (~20 mesh). Ball milling clearly

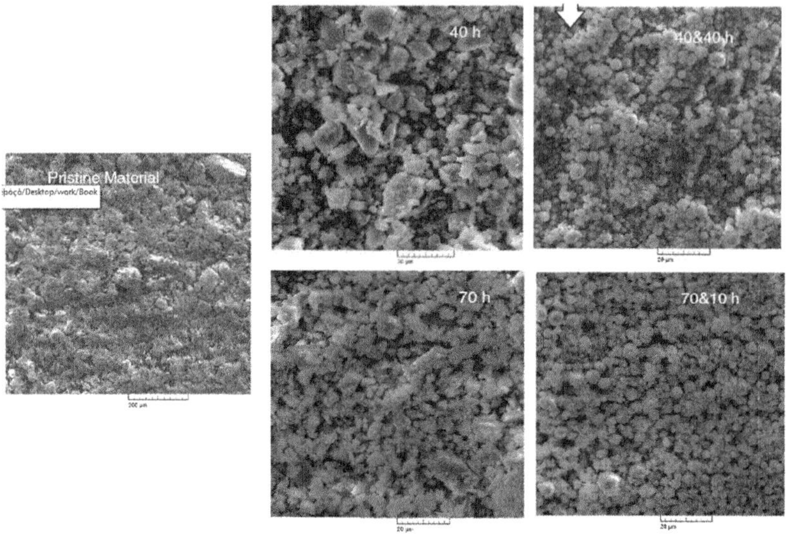

Figure 14.16 SEM micrographs for the pristine Mg_2Si and ball milled using only large balls (40 h and 70 h) and small balls (40 and 40 h and 70 and 10 h). Reprinted from Ref. [53], Copyright (2012), with permission from Elsevier.

changes the morphology of the particles, which become more irregular when milled only with large balls (i.e., 40 h and 70 h). Further milling with smaller balls makes the particles smoother, more spherical, and uniform. Image analysis in SEM pictures results in the particle (grain) size, shown in Fig. 14.17. As expected, there is a significant decrease in the mean grain size with the milling process down to ~1.6 μm. The balls' diameter affects the mean size for the same milling time; the sample 40 and 40 has lower mean size and lower standard deviation compared (1.63 ± 0.98 μm) to the 70 and 10 sample (2.04 ± 1.10 μm) though the total milling time is the same.

Here, we should distinguish between the particle size and the crystallite size. Extended ball milling, with large and small balls, shrinks down the particle size to about 1.6 μm. The crystalline size, however, is of the order of 10 nm, as results both from TEM and XRD measurements. It is the induced strain, during ball milling, which reduces the size of the crystallites inside a particle to nanoscale.

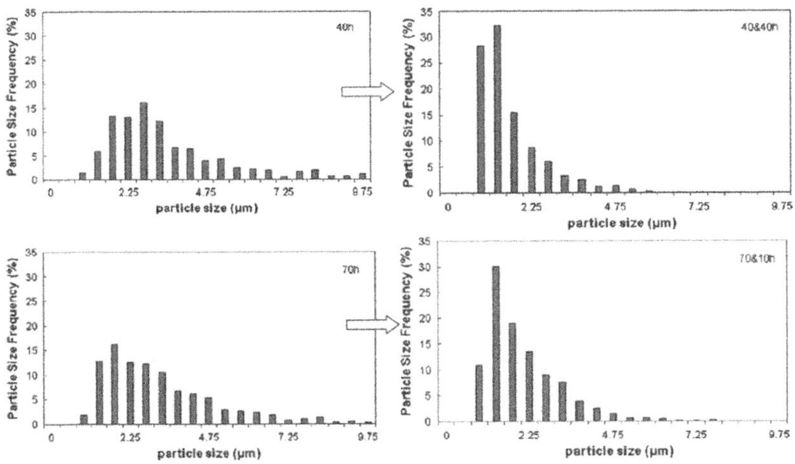

Figure 14.17 Particle size distribution as estimated based on SEM images. Reprinted from Ref. [53], Copyright (2012), with permission from Elsevier.

Figure 14.18 presents the crystalline size frequency for different ball milling times. All curves, regardless the ball milling parameters, have their maximum at about 10 nm. The distribution changes with ball milling time are the following:

- The distribution becomes sharper and higher with ball milling time.
- The maximum size of the observed crystallites is reduced from 212 nm and 92 nm for 5 and 40 h of ball milling process respectively, to 25.6 nm and 19.3 nm for 55 and 80 h.
- The frequency of crystallites which fall in the range of 5–10 nm significantly increased from 41.5%, to 50.9% to 58.6% and 78.6% for the 5, 40, 40 and 15 and 40 and 40 h, respectively.
- Two log-normal curves were needed to describe the crystallite distribution when Mg_2Si was ball milled with large balls, while, additional milling with small balls leads to a single-modal distribution.

14.5.5 Ball Milling Amorphization

Ball milling–induced amorphization was applied to β-$K_2Bi_8Se_{13}$ [55]. Ball milling was applied up to 2 h (120 min). Figure 14.19

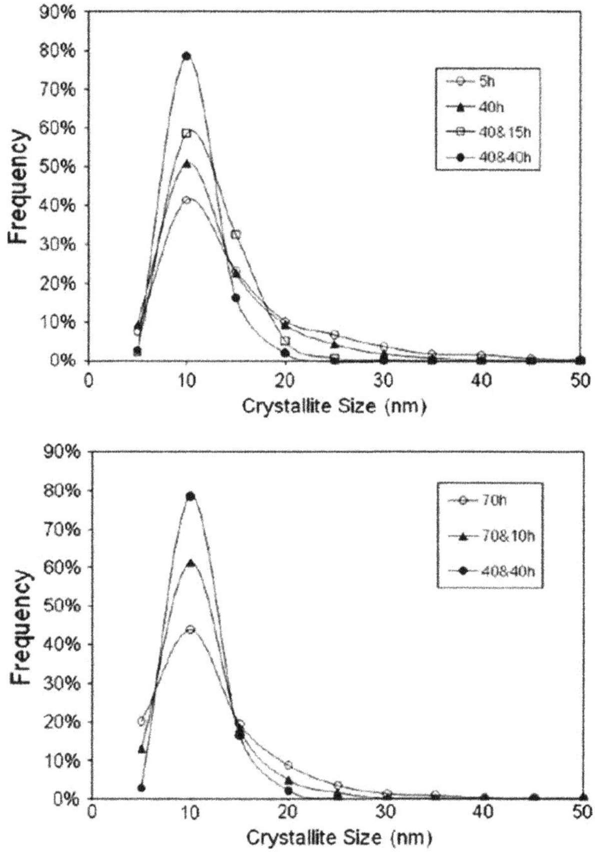

Figure 14.18 Crystallite size distribution for different ball milling times. FFT. Reprinted from Ref. [53], Copyright (2012), with permission from Elsevier.

shows the patterns of selected samples, ball milled for 10, 30, and 90 min, compared to the pristine material. As can be seen, the diffraction peaks become less intense and were significantly broadened with ball milling time, indicating that serious disorder in the crystallinity of the material is caused.

The gradual broadening can be attributed to the introduced strain and/or to the contribution of amorphous phase of the material caused by the ball milling process as well as the crystalline size effect.

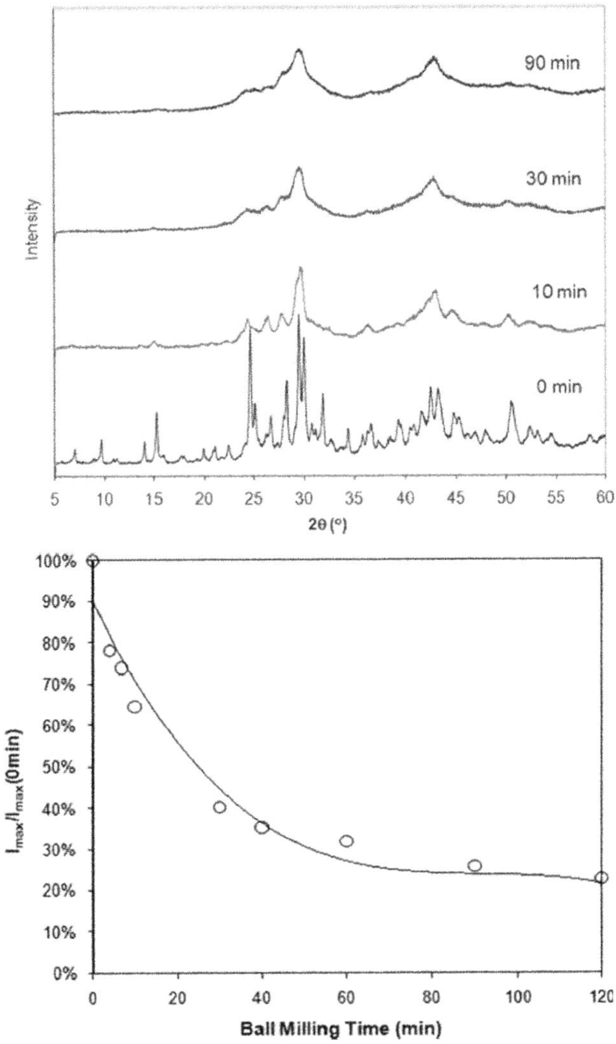

Figure 14.19 Evolution of powder XRD patterns and maximum peak intensity for β-$K_2Bi_8Se_{13}$ vs. ball milling time. The y-axis shows the relative intensity compared to the pristine material. Reprinted with permission of Ref. [55]. Copyright (2011), AIP Publishing LLC.

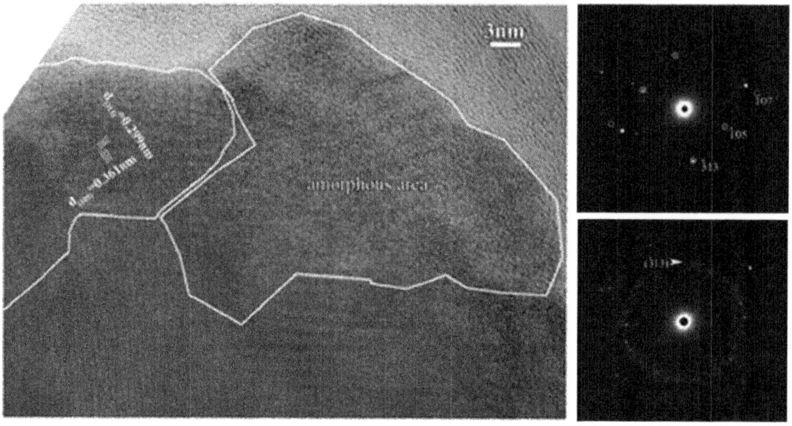

Figure 14.20 HRTEM image of $K_2Bi_8Se_{13}$ material after 30 min ball milling and electron diffraction patterns from microcrystalline area and amorphous phases. Reprinted with permission of Ref. [55]. Copyright (2011), AIP Publishing LLC.

HRTEM images (Fig. 14.20) show the existence of 15–30 nm nanocrystals in different orientations surrounded by amorphous matrix. The microcrystalline phase was identified to be β-$K_2Bi_8Se_{13}$ by indexing the corresponding diffraction pattern. The characteristic d-spacing for this system is identified by measuring distances of the reflections corresponding to (−313), (−105), and (107) planes. The diffuse rings in the diffraction pattern taken from a larger area confirm the in situ formation of an amorphous/nanocrystalline composite material during ball milling.

Figure 14.21a shows the density in two types of ball-milled $K_2Bi_8Se_{13}$: (a) in powder form and (b) pressed pellets made by the ball-milled powder. As can be seen, the measured density in powders is higher than in the pressed pellets, and this is attributed to the air inclusions existing in the pellets. The observed decrease in density, both in powder and pellets, is mainly attributed to the amorphization of the material. Longer ball milling times than 90 min did not significantly affect the density. Following the analysis presented so-far, the ball-milled $K_2Bi_8Se_{13}$ powder corresponds to a composite material that consists of a crystalline (micro- and nano-) as well as an amorphous phase. The ball milling time evolution

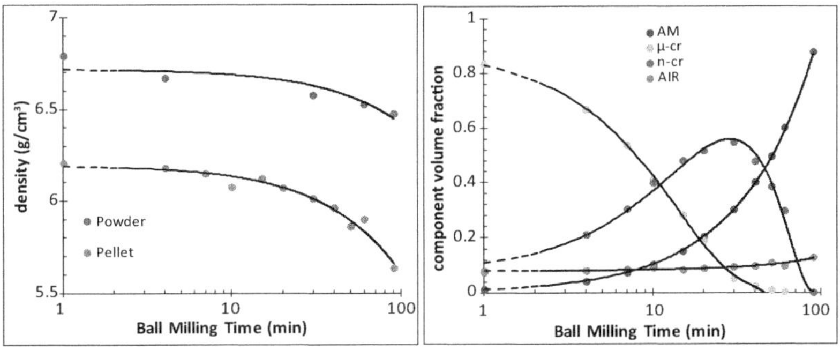

Figure 14.21 (a) Density in powder and in cold-pressed pellets and (b) volume fractions of the crystalline (micro and nano), amorphous phase and the air inclusions of the cold-pressed pellets vs. ball milling time. Reprinted with permission of Ref. [55]. Copyright (2011), AIP Publishing LLC.

of the four components (including air inclusions) is presented in Fig. 14.21b. As can be seen, $K_2Bi_8Se_{13}$ microcrystals are first converted into nanocrystals and then into amorphous material.

Amorphous $K_2Bi_8Se_{13}$ was re-crystallized by pressure-less sintering at 400°C for 48 h. SEM micrographs are shown in Fig. 14.22. The surface morphology changes from needle-like material (before ball milling) to spherical grains (after ball milling), while sintering process brings the material back to its crystalline state. Both 30 min and 60 min ball-milled samples show similar powder XRD patterns when sintered, with sharp peaks that match with that of $K_2Bi_8Se_{13}$. Relative intensities in XRD pattern of sintered material could be interpreted as due to a possible partial decomposition of $K_2Bi_8Se_{13}$ to $KBiSe_2$ during the ball milling/sintering, however decomposition is low ($<5\%$) [56]. This was confirmed by selected ED patterns; ball-milled material shows areas of discrete concentric spots, as well as the appearance of a diffuse intensity ring revealing the growth of an amorphous phase. In ED images from the sintered specimens (c and d) the rings of discrete spots show the polycrystalline character of the material. Moreover, the presence of periodic arrays of the spots, shown by small white arrows, is clear, suggesting the growth of rather large size crystals. The calculated inter-planar distances match to β-$K_2Bi_8Se_{13}$.

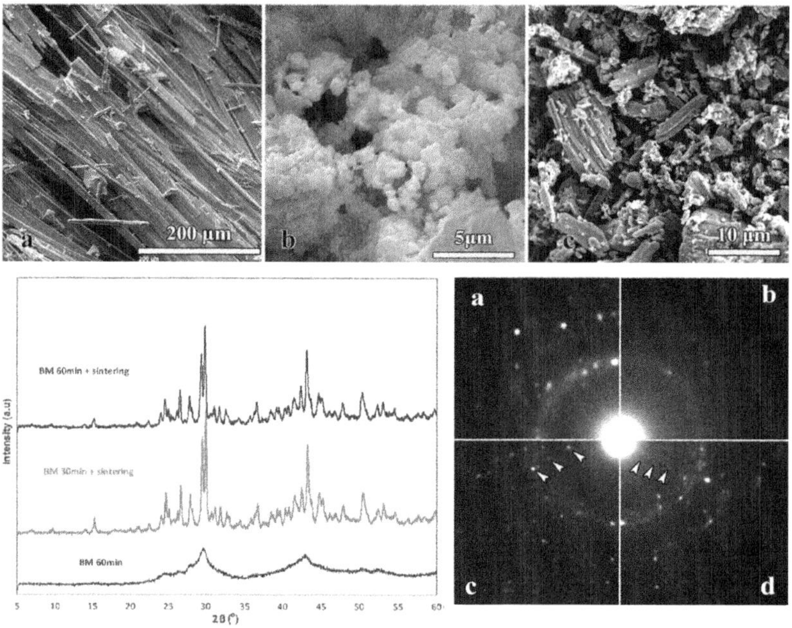

Figure 14.22 SEM micrographs for pristine (a), material milled for 30 min (b) and sintered material (c). XRD patterns for amorphized $K_2Bi_8Se_{13}$ and pressure-less sintered materials at 400°C. ED patterns from the 30 min ball-milled material before (a and b) and after (c and d) sintering. Reprinted from Ref. [56], Copyright (2011), with permission from Elsevier.

14.6 Summary

In this chapter, the use of ball milling techniques for amorphization and nanocrystallization is outlined. Amorphization is the conversion of a crystalline material into an amorphous one, while nanocrystallization is the incorporation of nanometer-scale structural features in a bulk matrix. The ball milling process has two types of uses: one for *mechanical alloying*, which involves the mixing of elements/compounds and chemical transformation occurs during milling, and *mechanical milling (or grinding)*, when there is no chemical transformation and only grain size reduction occurs. However, amorphization may occur in both cases (*mechanical disordering*). Connection between nanosizing and amorphization has been recently established as it was found that nanoparticles have more propensities for amorphization than their bulk counterparts.

Nanocrystalline PbTe was successfully synthesized by *mechanical alloying*. Granules of high-purity Pb and Te were mechanically alloyed to give an equiatomic composition for PbTe. Clearly, the mechanical alloying process advances through the formation of an intermediate PbTe nanophase. The *mechanical milling* process was applied to commercially available powder Mg_2Si, for the formation of Mg_2Si and MgO nanocomposites. In both cases, the nanocrystallinity of the materials was examined by XRD and TEM/HRTEM studies. Electron diffraction patterns show rings of very thin discrete spots verifying the polycrystalline character of the material. Dark-field images clearly show nanocrystalline material with mean size less than 30 nm for both types of materials. Reducing the milling ball size from 10 mm to 1.6 mm strongly affects the average nanocrystalline size. The average value of nanocrystals is reduced to $\sim$10 nm for both materials. At prolonged ball milling, traces of amorphous areas were located at the borders of the particles of the material.

Ball milling–induced amorphization was applied to crystalline β-$K_2Bi_8Se_{13}$. HRTEM images show the existence of 15–30 nm nanocrystals in different orientations surrounded by amorphous matrix. The diffuse rings in the diffraction pattern taken from a larger area, confirm the in situ formation of an amorphous/nanocrystalline composite material during ball milling. Analysis of the milling time evolution shows the following pattern for the amorphization process: The initial crystalline structure is partially transformed into nanocrystalline and amorphous phase, presumably at the borders of the nanocrystals. This amorphous phase then grows in volume at the expense of the nanocrystals. The amorphous $K_2Bi_8Se_{13}$ can re-crystallize by sintering.

References

1. D. Lacroix, K. Termentzidis, Applications, in *Nanostructured Semiconductors: Amorphization and Thermal Properties*, ed., Konstantinos Termentzidis, Pan Stanford Publishing Pte. Ltd. (2017).

2. L. D. Hicks, M. S. Dresselhaus, Effect of quantum well structures on the thermoelectric figure of merit, *Phys. Rev. B*, 47, 12727–12731 (1993); L. D. Hicks, M. S. Dresselhaus, Thermoelectric figure of merit of a one-dimensional conductor, *Phys. Rev. B*, 47, 16631–16634 (1993).
3. D. Vashaee, A. Shakouri, Improved thermoelectric power factor in metal-based superlattices, *Phys. Rev. Lett.*, 92, 106103-1–106103-4 (2004); T. C. Harman, P. J. Taylor, M. P. Walsh, B. E. Laforge, Quantum dot superlattice thermoelectric materials and devices, *Science*, 297, 2229–2232 (2002).
4. C. Uher, Recent advances in the development of efficient N-type skutterudites, in *Modules, Systems, and Applications in Thermoelectrics*, chapter 10, D. M. Rowe, ed., Taylor & Francis Group LLC (2012).
5. J. Androulakis, C.-H. Lin, H. Kong, C. Uher, C.-I. Wu, T. Hogan, B. A. Cook, T. Caillat, K. M. Paraskevopoulos, M. G. Kanatzidis, Spinodal decomposition and nucleation and growth as a means to bulk nanostructured thermoelectric, *J. Am. Chem. Soc.*, 129, 9780–9788 (2007).
6. J. P. Heremans, C. M. Thrush, D. T. Morelli, Thermopower enhancement in PbTe with Pb precipitates, *J. Appl. Phys.*, 98, 063703-1–063703-6 (2005).
7. J. R. Sootsman, R. J. Pcionek, H. Kong, C. Uher, M. G. Kanatzidis, Strong reduction of thermal conductivity in nanostructured PbTe prepared by matrix encapsulation, *Chem. Mater.*, 18, 4993–4995 (2006).
8. H. Li, X. F. Tang, Q. Zhang, C. Uher, High performance $In_xCe_yCo_4Sb_{12}$ thermoelectric materials with insitu forming nanostructured InSb phase, *Appl. Phys. Lett.*, 94, 102114-1–102114-3 (2009).
9. Poudel, B., Hao, Q., Ma, Y., Lan, Y., Minnich, A., Yu, C., Yan, X. et al., High thermoelectric performance of nanostructured bismuth antimony telluride bulk alloys, *Science*, 320, 634–638 (2008); W.-S. Liu, B.-P. Zhang, J.-F. Li, H.-L. Zhang, L.-D. Zhao, Enhanced thermoelectric properties in $CoSb_{3-x}Te_x$ alloys prepared by mechanical alloying and spark plasma sintering, *J. Appl. Phys.*, 102, 103717-1–103717-7 (2007).
10. H. S. Chen, Glassy metals, *Rep. Prog. Phys.*, 43, 353 (1980).
11. A. W. Weeber, H. Bakker, Amorphization by ball milling. *A Rev. Phys. B*, 153, 93–135 (1988).
12. R. W. Cahn, *Physical Metallurgy*, 3rd ed., Elsevier Science Publishers B. V. (1983); H. Liebermann, C. Graham, Production of amorphous alloy ribbons and effects of apparatus parameters on ribbon dimensions, *IEEE Trans. Magn.*, 12, 921–923 (1976).
13. P. D. Wilde, E. F. Matthys, Experimental investigation of the planar flow casting process: Development and free surface characteristics of the solidification puddle, *Mater. Sci. Eng. A*, 150, 237–247 (1992).

14. L. Piot S. Le Floch Cornier, S. Th. Daniele D. Machon, Amorphization in nanoparticles, *J. Phys. Chem. C*, 117, 11133–11140 (2013).
15. C. Suryanarayana, Mechanical alloying and milling. *Prog. Mater. Sci.*, 46, 1 (2001).
16. H.-J. Fecht, *Nano structured materials*, 6, 334 (1995).
17. P. Balaz et al., *Chem. Soc. Rev.*, 42, 7571 (2013).
18. C. Suryanarayana In: *Powder Metal Technologies and Applications. ASM Handbook, vol. 7.* Materials Park, OH: ASM International, pp. 80–90 (1998).
19. http://www.spexsampleprep.com/.
20. www.fritsch-milling.com/.
21. www.retsch.com.
22. M. Aliofkhazraei, ed., *Handbook of Mechanical Nanostructuring*, Wiley-VCH, chapter 12 (2015).
23. R. G. Blair, K. Chagoya, S. Biltek, S. Jackson, A. Sinclair, A. Taraboletti, D. T. Restrepo, *Faraday Discuss.*, 170, 223 (2014).
24. Z.-H. Chin, T.-P. Perng, *Mater. Sci. Forum*, 235–238, 121 ± 6 (1997).
25. K.-Varga, D. L. Beke. *Mater. Sci. Forum*, 225–227, 465 ± 70 (1996).
26. http://www.fritsch-milling.com/products/milling/planetary-mills/pulverisette-6-classic-line/download/.
27. C. Suryanarayana, *Bibliography on Mechanical Alloying and Milling*, Cambridge International Science Publishing, Cambridge (1995).
28. C. Suryanarayana, E. Ivanov, V. V. Boldyrev, The science and technology of mechanical alloying, *Mater. Sci. Eng. A*, 304–306, 151–158, 155 (2001).
29. I. Tsukushi, O. Yamamuro, T. Matsuo, Solid state amorphization of organic molecular crystals using a vibrating mill, *Solid State Commun.*, 94, 1013–1013 (1995).
30. T. Eifalt, O. Planinek, K. Hrovat, Methods of amorphization and investigation of the amorphous state, *Acta Pharm.*, 63, 305–334 (2013).
31. M. Dresselhaus, G. Chen, M. Y. Tang, R. Yang, H. Lee, D. Wang, Z. Ren, J. P. Fleurial, and P. Gogna, New directions for low-dimensional thermoelectric materials, *Adv. Mater.*, 19, 1043 (2007).
32. J. P. Heremans, V. Jovovic, E. S. Toberer, A. Saramat, K. Kurosaki, A. Charoenpakdee, S. Yamnaka, G. J. Snyder, Enhancement of thermoelectric efficiency in PbTe by distortion of the electronic density of states, *Science*, 321, 554 (2008).
33. T. C. Harman, P. J. Taylor, M. P. Walsh, B. E. LaForge, Quantum dot superlattice thermoelectric materials and devices, *Science*, 297, 2229 (2002).

34. K. F. Hsu, S. Loo, F. Guo, W. Chen, J. S. Dyck, C. Uher, T. Hogan, E. K. Polychroniadis, M. G. Kanatzidis, Cubic $AgPb_mSbTe_{2+m}$: Bulk thermoelectric materials with high figure of merit, *Science*, 303, 818 (2004).
35. C. Wood, Materials for thermoelectric energy conversion, *Rep. Prog. Phys.*, 51, 459 (1988).
36. D. M. Rowe (ed.), *Thermoelectrics Handbook Macro to Nano*, CRC Press (2006).
37. H. G. Lipson, A. Kahan, Infrared absorption of magnesium stannide, *Phys. Rev.*, 133, A800–A810 (1964).
38. V. K. Zailtsev, M. I. Fedorov, E. A. Gurieva, I. S. Eremin, P. P. Konstantinov, A. Y. Samunin, M. V. Vedernikov, Highly effective $Mg_2Si_{1-x}Sn_x$ thermoelectric, *Phys. Rev. B Condensed Matter*, 74, 045207 (2006).
39. J. Tani, H. Kido, Thermoelectric properties of Bi-doped Mg_2Si semiconductors, *Phys. B Condensed Matter*, 364, 218–224 (2005).
40. S.-W. Song, K. A. Striebel, X. Song, E. Cairns, Amorphous and nanocrystalline Mg_2Si thin film electrodes, *J. Power Sources*, 119–121, 110–112 (2003).
41. H. Kim, J. Choi, H. J. Sohn, T. Kang, The insertion mechanism of lithium into Mg_2Si anode material for Li-ion batteries, *J. Electrochem. Soc.*, 146, 4401–4405 (1999).
42. D. Y. Chung, K.-S. Choi, L. Iordanidis, J. L. Schindler, P. M. Brazis, C. R. Kannewurf, B. Chen, S. Hu, C. Uher, M. G. Kanatzidis, *Chem. Mater.*, 9, 3060 (1997).
43. Th. Hasapis, Ch. Papageorgiou, E. Hatzikraniotis, Th. Kyratsi, K. M. Paraskevopoulos, IR reflectivity studies of mechanically alloyed PbTe nanocrystals, *Mater. Res. Soc. Symp. Proc.*, 1166, 65 (2009).
44. N. Bouad, R. M. Marin-Ayral, G. Nabiasa, J. C. Tedenac, Phase transformation study of Pb–Te powders during mechanical alloying, *J. All. Comput.*, 353, 184–188 (2002).
45. Th. H. De Keijser, J. I. Langford, E. J. Mettemeijer, A. B. P. Vogels, Use of the Voigt function in a single-line method for the analysis of X-ray diffraction line broadening, *J. Appl. Crystallogr.*, 15, 308 (1982).
46. B. D. Cullity S. R. Stock, *Elements of X-Ray Diffraction*, 3rd ed., Prentice Hall (2001).
47. R. Suryanarayanan, Evaluation of two concepts of crystallinity using calcium gluceptate as a model compound, *Int. J. Pharm.*, 24, 1–17 (1985).

48. D. A. G. Bruggeman, Berechnung verschiedener physikalischer Konstantenvon heterogenen Substanzen, *Ann. Phys. (Leipzig)*, 24, 636–679 (1935).

49. M. Ioannou, E. Hatzikraniotis, Ch. B. Lioutas, K. M. Paraskevopoulos, Th. Kyratsi, Structural characterization of nano-crystalline Mg_2Si prepared by ball milling, *Adv. Sci. Technol.*, 74, 48–53 (2010).

50. D. McWilliams D. W. Lynch, Infrared reflectivities of magnesium silicide, Germanide, and Stannide, *Phys. Rev.*, 130, 2248 (1963).

51. W. J. Scouler, Optical properties of Mg_2Si, Mg_2Ge, and Mg_2Sn from 0.6 to 11.0 eV at 77°K, *Phys. Rev.*, 178, 1353 (1969).

52. A. M. Hofmeister, E. Keppel, A. K. Speck, Absorption and reflection infrared spectra of MgO and other diatomic compounds, *R. Astron. Soc.*, 345, 16 (2003).

53. M. Ioannou, E. Hatzikraniotis, Ch. B. Lioutas, Th. Hassapis, Th. Altantzis, K. M. Paraskevopoulos, Th. Kyratsi, Fabrication of nanocrystalline Mg_2Si via ball milling process: Structural studies, *Powder Technol.*, 217, 523–532 (2012).

54. Ch. Papageorgiou, E. Hatzikraniotis, Ch. B. Lioutas, N. Frangis, O. Valassiades, K. M. Paraskevopoulos, Th. Kyratsi, Thermoelectric properties of nanocrystalline PbTe synthesized by mechanical alloying, *J. Electronic Mater.*, DOI: 10.1007/s11664-010-1234-0 (2010).

55. Th. Kyratsi, E. Hatzikraniotis, M. Ioannou, D. Y. Chung, I. Tsiaoussis, Seebeck and thermal conductivity analysis in amorphous/crystalline β-$K_2Bi_8Se_{13}$ nanocomposite materials, *J. Appl. Phys.*, 110, 033713 (2011).

56. Ch. B. Lioutas, E. Hatzikraniotis, M. Ioannou, K. M. Paraskevopoulos, Th. Kyratsi, Structural features of ball-milled nanostructured $K_2Bi_8Se_{13}$ thermoelectric material, *Mater. Sci. Eng. B*, 176, 1568–1573 (2011).

Chapter 15

Physical Vapor Deposition for Tin-Induced and Laser Crystallization

Vlad Neimash
Institute of Physics NAS, Ukraine
neimash@gmail.com

Thin films of nanocrystalline silicon (nc-Si) are considered among the most attractive materials for the third- and next generations of solar cells and other optoelectronic devices [1]. Such properties of nc-Si, as direct bandgap (E_g) optical transitions, E_g dependence on the nanocrystals' size and accessibility of the volume processing on flexible substrates allow creating of highly efficient photovoltaic devices based on quantum dots cascade heterostructures [2, 3] and reducing processing costs due to advantages of the thin-film and roll-to-roll fabrication technologies [4, 5]. Among the problems for a wide-scale commercialization of nc-Si is high production cost and the lack of methods to efficiently control the size and concentration of Si nanocrystals during the films formation. This facilitated considerable research efforts to improve the existing and to develop advanced nc-Si technologies [6–12]. One approach is to use metal-induced crystallization (MIC) of amorphous Si (α-Si), manufactured by physical vapor deposition [13–17]. In particular, recent studies on tin (Sn)-induced crystallization of

Nanostructured Semiconductors: Amorphization and Thermal Properties
Edited by Konstantinos Termentzidis

ISBN 978-981-4745-64-2 (Hardcover), 978-1-315-36445-2 (eBook)
www.panstanford.com

α-Si demonstrated transformation of up to 90% of α-Si volume into a nanocrystalline phase with crystal sizes of 2–5 nm [18–20]. This is the range of strongest bandgap width depending on the size of silicon nanocrystals. On the other hand, physical vapor deposition method is the easiest and cheapest way to manufacture amorphous silicon.

15.1 Introduction

A controlled crystallization of amorphous silicon films from the amorphous state was studied intensively during the past decades. Recently, those studies gained an additional momentum due to advances with metal-assisted transformation of amorphous Si (α-Si) into a crystalline state [13, 14, 21, 22] Silicon doping with tin (Sn) proved to be an effective and convenient method of influencing various properties of Si. As the isovalent dopant, Sn does not create energy levels in the bandgap of Si. Therefore, doping with Sn within its solubility level does not affect electrical, optical and recombination properties of *crystalline Si*. At the same time, the presence of Sn atoms inhibits effectively the performance degradation of crystalline Si under various harmful conditions, such as heat treatment [23–25] and ionizing radiation [26–28]. In amorphous Si, doping with Sn significantly reduces the optical bandgap (E_g) [29, 30] and changes some other physical characteristics of the material. In particular, in was shown that at low concentrations of Sn, the electrical conductivity of Si:Sn alloy is characterized with a reduced activation energy of about $E_g/2$ [29, 30]. Moreover, with increasing Sn concentration in the range from 1 to 2 at.%, the hopping conductivity mechanism prevails. Light absorption in thin film alloys $Si_{1-x}Sn_x$ occurs via the direct bandgap transition mechanism [31]. However, when adding the other isovalent impurity, carbon (C) to this alloy, a reverse to indirect bandgap transitions may occur [32].

Despite a significant number of works studying Si:Sn alloys, the effects of Sn-assisted crystallization of amorphous Si were not characterized comprehensively. Tin atoms in α-Si can substitute the host atoms either in the tetragonal [33] or non-tetragonal

[32] configuration forming either a regular or non-regular grid. Under certain conditions, the precipitates of Sn atoms are formed in the amorphous matrix of Si [31, 34]. Tin is one of the eutectic forming metals that may induce a transition of amorphous Si into the crystalline state due to layer exchange effects on the metal-silicon interface [14, 17]. Such metal-induced crystallization (MIC) was observed at a relatively low-temperature (300–500°C) anneal of sandwiches of thin films of amorphous Si and metallic Sn deposited on a glass substrate [18, 19]. Raman spectroscopy results demonstrated the ability of Sn impurities to accelerate significantly crystallization of α-Si when growing thin films of $Si_{1-x}Sn_x$ by thermal evaporation of the components in vacuum [20].

The purpose of this chapter is to show the mechanism of Tin-induced crystallization of amorphous silicon manufactured by physical vapor deposition method and possibility of crystallization control during laser annealing.

15.2 Experimental Setup

Si:Sn alloy films with the thickness of 300–500 nm were prepared by thermal evaporation in vacuum of the powder mixtures of high-purity Si and Sn. Sn content in the mixtures varied from 0.1 to 30 weight %. Quartz and mono-crystalline Si wafers were used as substrates to form Si:Sn films at a temperature of 300°C and a residual pressure of 10^{-3} Pa. The deposition parameters were tightly controlled to ensure a uniform distribution of Sn in the ~100 nm-thick surface layer of the film [23]. The total thickness of the deposited Si:Sn films varied from 250 to 800 nm.

The planar α-Si/Sn/α-Si structures were prepared in three steps during a single load in a vacuum chamber at a residual pressure of 10^{-3} Pa by a sequential use of three different evaporators. First, the α-Si (99.99% electronic-grade) layer of 100 nm thickness was deposited on a borosilicate glass substrate (BSG). During the second step, a 100 nm thickness layer of metallic Sn (99.92%) was formed by deposition through the 60×60 mm^2 opening mask. The second α-Si layer of 100 nm thickness was deposited last. All depositions were made at 150°C. The central portion of the engineered samples contained three deposited layers (α-Si/Sn/α-Si), whereas the peripheral parts of the samples contained α-Si only.

The amorphous versus crystalline phase composition of the films and structures was analyzed using room temperature Raman spectroscopy (RS) activated with laser at the wavelength of 514.5 nm and 633 nm. The impurity distribution profiles across the films were characterized by Auger electron spectroscopy of reactive ion etched (RIE) films. Scanning electron microscopy (SEM) and X-ray fluorescence micro-probing (energy dispersive X-ray spectroscopy, EDX) were used for studying the surface topology and impurities distribution on the surface.

15.3 Tin-Induced Crystallization of Si in the Si–Sn Alloys

15.3.1 The Microstructure of Thin Film Si–Sn Alloys Produced by Vapor Deposition

Figure 15.1 shows typical concentration profiles in the alloy film Si:Sn (a) and amorphous silicon control sample without Sn (b) as measured by Auger electron spectroscopy of reactive ion etched films. The films were grown on crystalline silicon substrates with

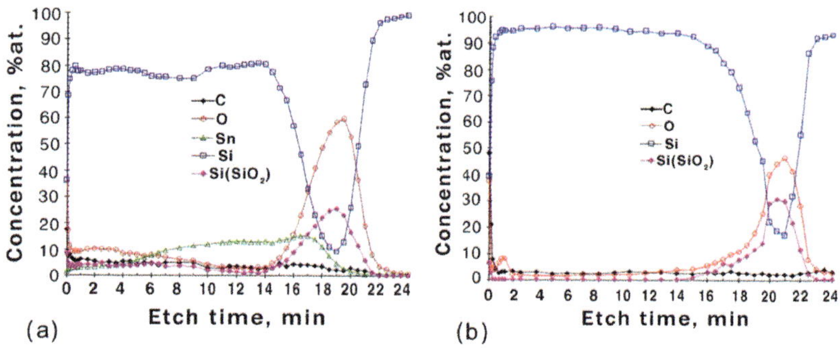

Figure 15.1 Typical concentration profiles in grown films: (a) Si:Sn film containing 2.5 at.% of Sn; (b) control film of amorphous Si grown with no Sn dopant. The etch rate was 20 nm/min.

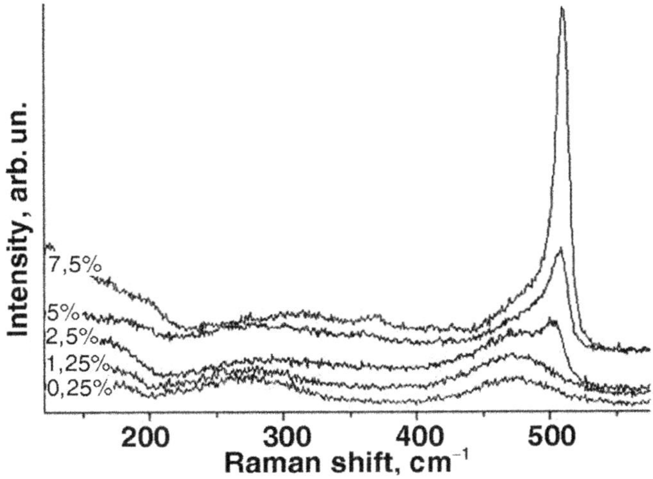

Figure 15.2 Raman spectra of Si:Sn alloys with Sn content within the range 0.25–7.5 at.%. The laser beam diameter was ~3 mm.

the thermal oxide thickness of ~100 nm. The total thickness of deposited films described in Fig. 15.1 was ~300 nm.

The phase structure of the samples with various Sn content was analyzed using Raman spectroscopy data. Figure 15.2 shows changes in Raman spectra of Si:Sn alloys with increasing Sn content. The spectra were recorded using a large diameter laser beam (3 mm) to measure the spectra integrated over a large surface. Care was taken to provide a uniform excitation across the 3 mm spot of the sample surface. A broad band at around 470 cm^{-1} corresponds to amorphous silicon. A narrow band centered around 515 cm^{-1} corresponds to Si in the crystalline state. The amplitude of this narrow band is defined by the portion of the material in the crystalline phase, whereas the peak's frequency position is defined by the dominant size of the crystallites [35–37]. The characteristic band at around 515 cm^{-1} emerged in the samples having Sn content of 2.5 at.%. With increasing Sn concentration beyond 2.5 at.%, the amplitude of this band increased significantly, but the frequency of its peak shifted only slightly to higher frequencies. The above observations indicated that a higher Sn content stimulated crystallization of amorphous Si by increasing the number of (nano)-crystals without significant changes in their dominant size.

The partial volume of the crystalline phase in the film was estimated using the expression [35]

$$X_c = (I_c/I_a)/(y + I_c/I_a), \tag{15.1}$$

where I_c and I_a are the integrated intensities of the crystalline and amorphous components in Raman spectra, respectively. The ratio y of the scattering cross sections of the crystalline and amorphous Si was determined using [36]

$$y\,(L) = 0.1 + \exp\,(-L/25)\,, \tag{15.2}$$

in which L is an average size of nanocrystals (in nanometers) that can be estimated using the expression for the scattering intensity I at frequency ν:

$$I\,(\nu) = \int \exp\left[-(q^2L^2)/(16\pi^2)\right]/\left[(\nu - \nu(q))^2 + (\Gamma_0/2)^2\right]\, d^3q, \tag{15.3}$$

where Γ_0 is the half-width of the Raman band of a crystalline Si and $\nu(q)$ the dispersion relation for the phonon vibrations in crystalline Si.

The intensity of Raman scattering originating from the α-Si phase is well described by the Gaussian distribution [24], $I\,(\nu) = B\exp\left[-(\nu - \nu_a)^2/(2\,\delta^2)\right]$, where B is a constant, $\delta = \Gamma_a/\left(2\,\sqrt{2\ln 2}\right)$ and ν_a and Γ_a are the frequency of the maximum of the α-Si scattering band and its full width at half-maximum, respectively.

Deconvolution of Raman spectra of Fig. 15.2 in the range 400–540 cm^{-1} into Gaussian and asymmetric (Eq. 15.3) components allowed estimating the partial volumes of the two phases in the thin films. We used the Levenberg–Marquardt nonlinear least square deconvolution algorithm to analyze Raman spectra of Fig. 15.2. In all cases, best fitting results were obtained using two components. Adding more asymmetric or Gaussian components caused deconvolution to converge with negligible weights of additional components, proving rather uniform distribution of nanocrystal sizes.

Figure 15.3 shows deconvolution results for Si:Sn alloys containing 2.5, 5.0 and 7.5 at.% of Sn. Table 15.1 summarizes the calculation results using Eqs. 15.1–15.3.

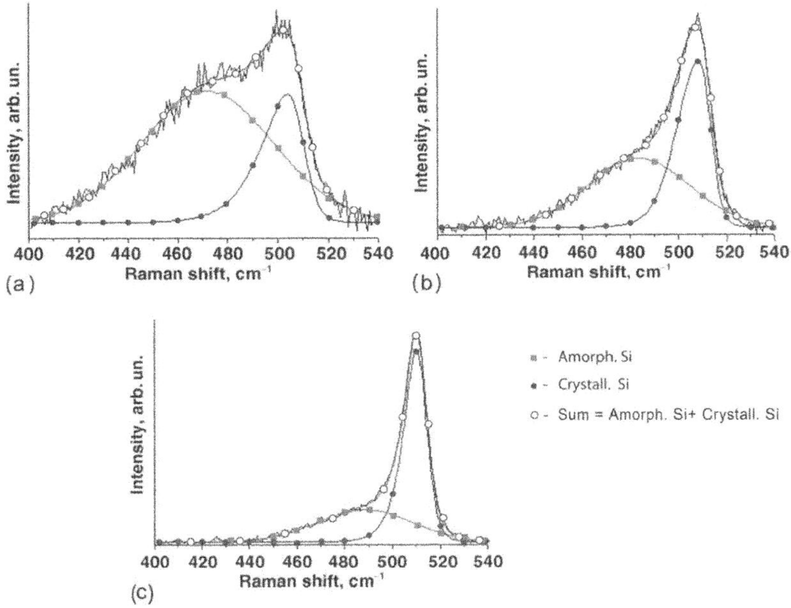

Figure 15.3 Raman spectra deconvolution into the Gaussian (amorphous, line with solid squares) and asymmetric (crystalline, line with solid circle) components for the films with Sn content (a) 2.5 at.%, (b) 5.0 at.%, (c) 7.5 at.%. Line with open circles is the sum of two components shown with solid circles and squares.

Table 15.1 The crystal average (dominant) size L and crystalline phase partial volume X_c for the samples containing 2.5, 5.0, and 7.5 at.% of Sn

Sn content	I_c	I_a	I_c/I_a	L	X_c
2.5 at.%	25.7	75.7	0.34	2.3 ± 0.1	0.25 ± 0.02
5.0 at.%	45.6	54.4	0.84	2.7 ± 0.1	0.46 ± 0.015
7.5 at.%	61.0	39.1	1.56	3.0 ± 0.15	0.61 ± 0.01

The above results indicated that doping with Sn stimulated formation of the crystalline phase in Si:Sn alloy at the Sn content starting around 1.25–2.5 at.%. This finding was in good agreement with the early results on Sn concentration dependence of the hopping conductivity mechanism onset in Si:Sn alloys [30, 31]. A

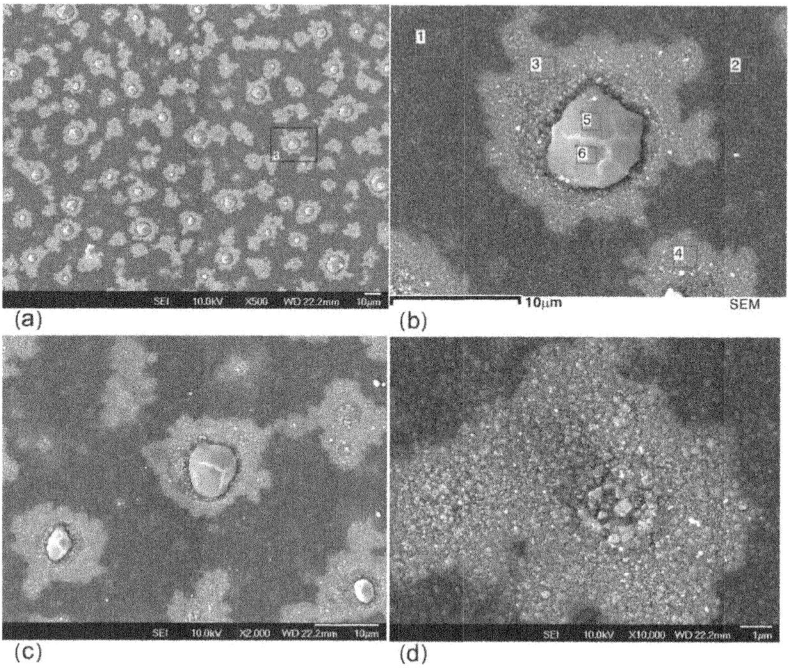

Figure 15.4 SEM images of Si:Sn film with 5.0 at.% of Sn. Panel (a) is a low resolution image of the film. Panel (b) shows the blow-up of the rectangular-outlined region "a" of panel (a). Panel (c) is a higher resolution image of several dendrites. Panel (d) shows a dendrite structure with a fully dissolved droplet (see the text below).

further increase of the Sn content from 2.5 to 7.5 at.% caused an increase of the crystalline phase partial volume from 25% to 61% with only minor change in the dominant size of nanocrystals, from ~2.3 to ~3.0 nm.

The quasi-spherical droplets surrounded with the dendrite-like aureoles were observed on the surface of the films with the Sn content of 2.5 at.% and higher. The size and density distribution of these features correlated well with the Sn content of the films. Figure 15.4 shows representative SEM images of the film containing 5.0 at.% of Sn.

The EDX micro-probing of six different sites outlined with the squares in the panel (b) of Fig. 15.4 confirmed that a quasi-spherical

Table 15.2 Results of EDX analysis at different probing sites of the sample shown in Fig. 15.4b. All data are in at.% units

Probing site	C	O	Si	Sn
1	3.57	15.59	76.79	4.06
2	3.99	14.75	77.00	4.25
3	3.21	10.80	78.24	7.74
4	3.35	9.56	78.77	8.32
5	4.95	51.45	8.96	34.65
6	5.58	49.62	7.54	37.27

formation in the middle of the image of panel (b) is a droplet of metallic Sn covered with a thin oxidation layer (see Table 15.2 for a summary). The aureoles around the droplets (light gray areas in the images of Fig. 15.4) contained Sn in the concentration twice of that outside the aureoles (dark gray areas in the images of Fig. 15.4).

To determine the phase composition (crystalline vs. amorphous Si) of the film around Sn droplets, confocal Raman micro spectroscopy was used with the microprobe delivering excitation light to a 1 μm diameter spot. Figure 15.5a shows the optical image of the same sample as in Fig. 15.4. A bright yellow spot in the center of the image is a Sn droplet. The droplet is surrounded by a dark aureole, which is the same material that shows up with light gray aureoles in the SEM images of Fig. 15.4.

The Raman spectra were taken at different distances from the Sn droplet boundary—both within the aureole area (site 1 in the image of Fig. 15.5a) and outside the aureole (site 3 in Fig. 15.5a). Site 2 in Fig. 15.5a was located within a transition area separating the aureole and outside material.

The Raman spectra (Fig. 15.5b) taken at different sites were markedly different from each other. The characteristic crystalline band around 500–525 cm^{-1} was barely seen in the spectrum from the site 3 and increased in intensity while laser excitation approached the boundary of the Sn droplet. This band became dominant showing a sharp peak at $\sim$525 cm^{-1} for the spectrum from the site 1. Table 15.3 presents the deconvolution results of the spectra from Fig. 15.5b into Gaussian and asymmetric (Eq. 15.3)

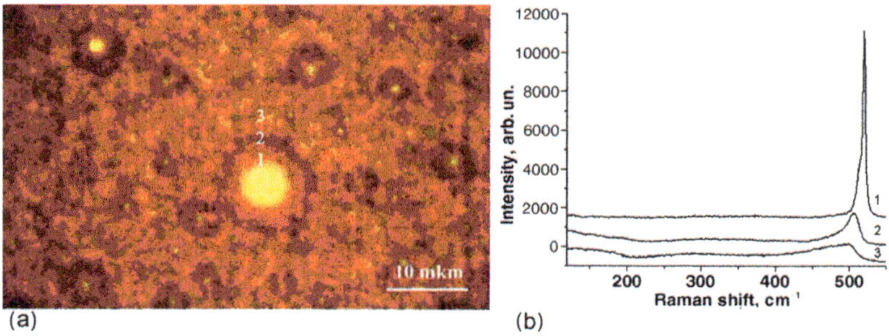

Figure 15.5 (a) Image of the film taken with an optical microscope shows quasi-spherical formations with dark aureoles on the surface of Si:Sn film. The numbers indicate the sites of Raman micro spectra probing. (b) Raman spectra taken from the three different sites of the sample shown in the panel (a).

Table 15.3 The crystal average (dominant) size L and crystalline phase partial volume X_c for the probing sites of the film shown in Fig. 15.5a

Probing Site	I_c	I_a	I_c/I_a	L	X_c
1	23.7	5.8	4.09	4.5 ± 0.2	0.8 ± 0.01
2	59.8	40.21	1.49	2.7 ± 0.1	0.6 ± 0.01
3	28.8	71.2	0.40	2.3 ± 0.1	0.28 ± 0.02

components as described above, using the nonlinear least square deconvolution algorithm.

The partial volume of the crystalline phase of Si:Sn alloy was ~80% for the probing site 1, close to the metallic Sn droplet. The partial volume of this phase dropped down to ~28% outside the aureole surrounding the droplet (site 3). The dominant size of nanocrystals was largest close to the Sn droplet decreasing with the distance from the droplet. At the periphery of the aureole structure, the nanocrystals' size was close to that measured in the experiment with a 3 mm excitation spot (see results in Table 15.1).

For the samples with Sn content lower than 2.5 at.%, we did not observe formation of metallic Sn droplets on the sample surface. No signature of the crystalline Si phase was observed in Raman spectra of these samples Crystallization of Si occurred only in the samples where the metallic Sn droplets were formed.

Analyzing metallic Sn droplets of different sizes, as well as the phase composition inside and outside the aureoles surrounding the droplets, we concluded that crystallization of Si starts at the metallic Sn droplet boundary and propagates outside the droplet. This finding agrees with recent reports [14, 15] showing that annealing of thin planar structures of amorphous Si–metallic Sn at temperatures of 300–500°C leads to crystallization of Si within the eutectic layer close to the Si–Sn boundary. Formation of polycrystalline silicon films in the bi-layered structure of α-Si film deposited on top of the metal film was discussed also in terms of the "layer exchange" process [17]. During the isothermal annealing of the structure, a continuous polycrystalline silicon film was formed within the initial metal layer and displaced it completely. Based on experimental observations, the authors concluded that Si atoms diffuse from the α-Si film into the metal film forming the eutectic layer, where the nucleation of Si nanocrystals occurs. Electron screening at the interface with metals weakens Si covalent bonds, facilitating inter-diffusion of metal atoms and silicon [38], which triggers formation of a eutectic layer. Due to supply of Si atoms from the amorphous phase and fast diffusion constant of Si, crystallization in the "layer exchange" model proceeds through protrusion of Si nanocrystalline formations inside the Sn film [20].

The layer exchange mechanism could not be applied directly to our results simply because the films in our studies were deposited as a uniform mixture of Si and Sn atoms. Moreover, the formation of dendrite-shaped polycrystalline structures around Sn droplets and existence of a gradient of the partial volume X_c extending from the Sn droplet toward a periphery of the surrounding dendrite aureole requires alternative explanations. The nucleation of crystals as treated within the layer exchange mechanism [17] assumes diffusion-controlled reactions invoking the thermodynamics of fluctuations to surpass the free-energy barrier. Further crystal grows is facilitated by crystallization heat release to the neighboring regions. For the samples studied in our work, such processes would produce most likely symmetric, oval or even circular-shaped polycrystalline formations. The protrusions of crystalline Si would be observed toward the interface with the Sn metallic droplets and the expected X_c gradient would be pointing from the periphery of the aureole toward the interface with the Sn droplet. However, one could hardly find a single symmetric-shaped aureole structure in our samples (see, e.g., Fig. 15.4). Moreover, all the crystalline protrusions in our experiments propagated from the inside of aureoles toward the α-Si matrix outside the aureoles. Therefore, different crystallization mechanism should be proposed to explain our results.

Following deposition of Si:Sn films at the temperature above the Sn melting point (231.9°C), Sn atoms may coagulate forming the clusters of liquid Sn (droplet). Si atoms diffuse promptly into the formed droplets, creating the molten Si:Sn layer, which could be considered as a solution of silicon in tin. Amorphous Si continues dissolving in this phase until saturation is reached, followed by a rapid super-saturation and local decrease in temperature caused by the dissolution of amorphous Si. The formation of the super-saturated molten Si:Sn layer triggers Si crystallization through either a nucleation process or spinodal decomposition effects [39–42]. In the latter case, within the spinodal region, the homogeneous saturated solution Si:Sn is unstable against fluctuations in density or composition, and there is no thermodynamic barrier to the growth of a new phase, it is purely diffusion-controlled. The shape of a newly formed phase component is essentially irregular as Si crystallization occurs simultaneously along the most favorable paths. Our results did not allow for explicit distinguishing between the two mechanisms (nucleation vs. spinodal decomposition) of a new phase formation. This should not look surprising since in accordance with recent studies the scenario of the initial stages of nucleation-growth processes in the two-component system resembles widely the picture of spinodal decomposition [43]. In addition, the basic mechanism of nucleation—surpass of a potential barrier in the evolution to the newly evolving phase—was shown to retain its importance for unstable initial states near the spinodal curve as well.

Si crystallization within the molten Si:Sn layer activates two other processes tending to return the system into the pre-saturated state. The first is Sn atoms segregation on the surface of formed Si nanocrystals. Since the solubility of Sn in crystalline Si is very low ($\sim 5 \times 10^{19}$ cm^{-3}) [44] the excess Sn atoms return to the molten Si:Sn layer facilitating its under-saturation. The second process is the crystallization heat release that increases the temperature of the adjacent molten Si:Sn layer pushing it below the saturation threshold. Both these processes stimulate further dissolving of amorphous Si. The overall cycle repeats again, creating dendritic, irregular-shaped formations of polycrystalline Si. Similar effects of the dendritic-shaped crystallization inside a thin molten layer of Si:Sn ahead of the advancing crystallization front were briefly reported previously [20]. In our model, the size of formed Si nanocrystals is limited by the thickness of the saturated molten Si:Sn layer. With accumulation of nanocrystals around the Sn droplet, the molten Si:Sn layer is pushed away from the droplet. Assuming that the additional supply of Sn atoms from the droplet is limited by Sn diffusion, the thickness of the saturated molten Si:Sn layer decreases with propagation of the crystallization front away from the droplet. As a result, the average size of formed nanocrystals decreases toward the aureole periphery, which agrees with our observations (see Table 15.3).

The width of the polycrystalline areas (aureoles) around the Sn droplets did not depend on the droplet size. This observation is well understood within the idea of a diffusion-limited supply of Sn atoms from the droplet to the molten Si:Sn layer. While supplying Sn atoms to the crystallization front, the droplet gradually dissolves. Some droplets with very small initial size could be dissolved completely during formation of the polycrystalline phase (see, e.g., Fig. 15.4d). It is also reasonable to suggest that a portion of Sn atoms remains between the Si grains (nanocrystals) of the polycrystalline phase, giving rise to a twice-higher Sn content inside the polycrystalline aureoles as compared to the α-Si matrix as shown with the data of Table 15.2.

The size of formed nanocrystals depends on the temperature, which is well understood within any diffusion-controlled reaction model. Recent studies have demonstrated a monotonous increase in the dominant size of Si nanocrystals with the anneal temperature increase [17]. Those studies suggested that for the anneal temperature of 200–300°C, the dominant size of the nanocrystals would be within the range of 2.3–4.5 nm. The results of our studies agree fully with the above estimates (see Tables 15.1 and 15.3).

In our experiments, the crystallization did not occur in the films containing less than $\sim$2.5 at.% of Sn. Similar results were reported previously [45], showing that an amorphous to polycrystalline phase transformation occurs during annealing of Sn-implanted Si when the peak Sn concentration exceeds about 2 at.%. Apparently, the concentrations of Sn below $\sim$2 at.% does not favor coagulation of Sn into droplets and molten Si:Sn layer formation within those Sn droplets. This agrees with the other studies suggesting that the crystallization process of amorphous Si is enhanced by the formation of the liquid Sn phase [14, 17].

The most important conclusions of this part are the following: (1) Doping with Sn allows reducing the α-Si crystallization temperature to below 300°C and decreasing the dominant size of nanocrystallites to $\sim$2 nm. (2) Sn stimulates α-Si crystallization in a threshold-type manner above the concentrations of approximately 2 at.%. (3) Formation of metallic Sn droplets is a pre-requisite for crystallization of α-Si. Crystallization propagates in the form of dendritic formations from the droplet/α-Si interface toward the bulk of the α-Si matrix. (4) Sn-induced crystallization of α-Si could be explained by formation of a molten Si:Sn (eutectic) layer at the droplet/α-Si matrix interface with a spontaneous nucleation and spinodal decomposition within this molten Si:Sn layer.

15.3.2 The Mechanism of Tin-Induced Crystallization of Amorphous Silicon

For experimental verification of the hypothesis, mentioned above, a macroscopic planar tin drop in the bulk of an amorphous silicon film was formed as planar structures (see Section 15.3.1) for simulating of microscopic droplets and studying its influence on the structure of surrounding Si during of heat treatments. The scheme of structures is shown on Fig. 15.6.

The central portion (direction A) of the engineered samples contained three deposited layers (α-Si/Sn/α-Si), whereas the peripheral parts of the samples (direction B) contained α-Si only. The samples were characterized with Raman, Auger, and electron

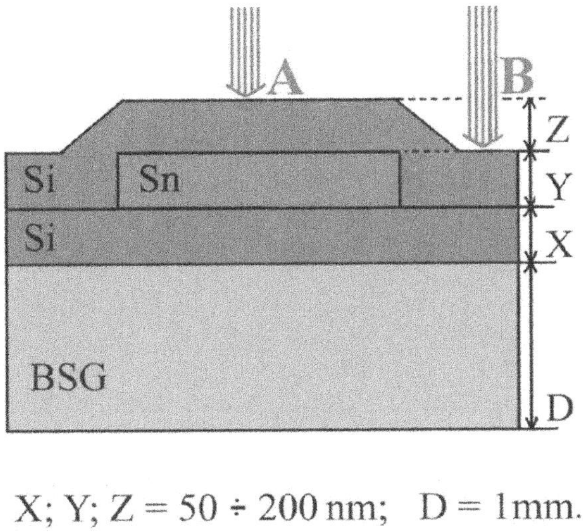

Figure 15.6 Scheme cross section of layered structures Si–Sn–Si, studied in this work.

microscopy techniques prior and after the 30 min anneal at 300 and 400°C.

Figure 15.7 demonstrates the Raman spectra measured at various sites of the examined BSG/Si/Sn/Si structure: as-fabricated (1 and 2) and after its thermal treatment in vacuum for 30 min at 300 (3 and 4) and 400°C (5). Spectra 1 and 4 were registered on the peripheral part of the surface, where the tin layer is absent; and spectra 2, 3, and 5 on the central part of the structure above the tin layer. One can see that spectra 1 and 4 with a wide band peaked at about 470 cm^{-1} almost coincide. They correspond to a purely amorphous Si state. This means that in the absence of Sn in the Raman measurement zone, the phase state of a-Si does not change after the thermal treatment at temperatures up to 300°C inclusive. On the contrary, spectra 2, 3, and 5, which were measured above the tin layer in the same specimen, contain an additional sharp peak at 505, 512, and 515 cm^{-1}, respectively, which corresponds to the crystalline Si phase. In other words, the material structure in this section of the specimen is an amorphous-crystalline composite.

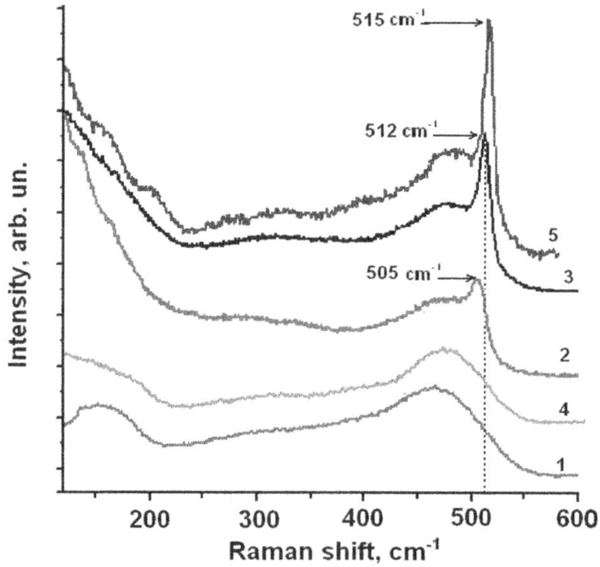

Figure 15.7 Raman spectra of BSG/Si/Si (1 and 4) and BSG/Si/Sn/Si (2, 3, and 5) structures: before (1 and 2) and after TT at 300 (3 and 4) or 400°C (5) for 30 min.

The resolution of Raman spectra into two components, as is shown in Fig. 15.3, and the application of the algorithm presented in work [20] allowed us to estimate the dominant sizes of silicon crystals according to the model of spatial phonon correlation [35–37] and the volume ratios between the crystalline and amorphous phases in the composite. From Fig. 15.7 (spectrum 2), one can see that metallic Sn induces the crystallization in amorphous Si already in the course of BSG/Si/Sn/Si structure fabrication. The analysis of Raman spectra showed that the dominant size of crystals increased from 2.7 nm in the as-fabricated specimen to 3.2 nm after the HT of the specimen at 300°C and to 4.5 nm after its HT at 400°C. Correspondingly, the content of the nanocrystalline Si phase (nc-Si) grew more than two times.

In Fig. 15.8, the concentration distributions of chemical elements across the Si/Sn/Si structure thickness are plotted. They were obtained by measuring the Auger spectra, while sputtering the layers using Ar^+ ions with an energy of 3 keV. The sputtering rate

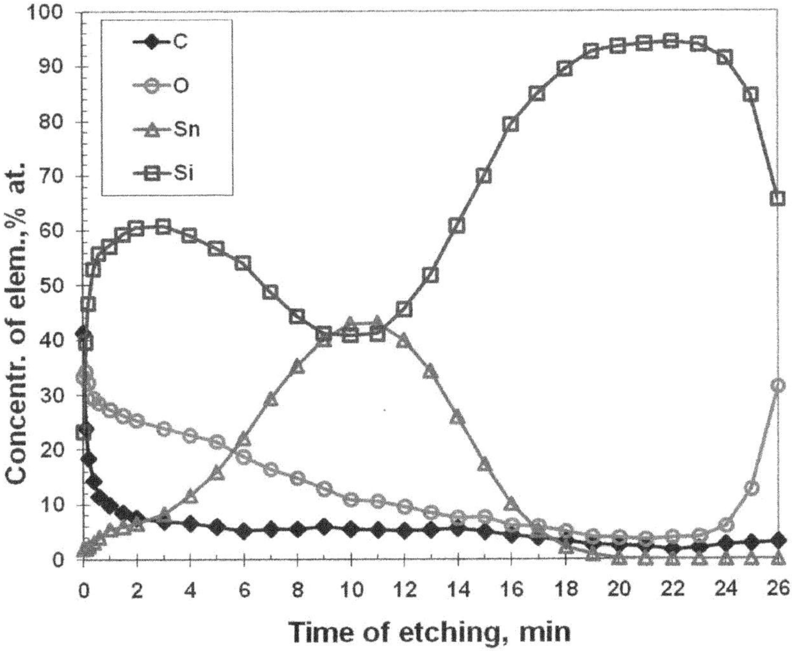

Figure 15.8 Concentration profiles of elements across the BSG/Si/Sn/Si structure thickness before a thermal treatment.

amounted to 12–14 nm/min. The left edges of the distributions shown in Fig. 15.8 correspond to the external surface of the central section of the structure exhibited in Fig. 15.6. The right edges correspond to the BSG substrate. High contents of oxygen and carbon technological impurities in the formed films resulted from a low level of vacuum in the course of impurity deposition and a subsequent atmospheric oxidation of the surface. In Fig. 15.8, one can clearly distinguish two concentration maxima of Si with an Sn maximum between them, which correspond to three layers deposited on the substrate. At the same time, the attention is attracted to an unusual concentration ratio between tin and silicon in each of three layers of the structure. First, the maximum concentration of silicon in the external layer (about 60 at.%) is 1.5 times lower than the maximum Si concentration in the internal layer (about 90 at.%). Second, the maximum tin concentration does not

Figure 15.9 SEM images of the BSG/Si/Sn/Si structure surface before a thermal treatment.

exceed 45 at.%, being almost equal to the Si concentration. At first sight, this looks strange, because the layers in the Si–Sn–Si structure were formed sequentially and under conditions that would provide an identical thickness of about 100 nm for every layer, should it be deposited onto the glass substrate alone.

Those results can be explained with regard for the results of electron microscopy researches of the Si/Sn/Si structure carried out in the secondary electron mode. They are shown in Fig. 15.9. It is evident that the surface of the as-fabricated Si–Sn–Si structure consists of microscopic balls from 120 to 470 nm in diameter. The majority of balls have diameters of about 300 nm. The reflection of visible light from this surface is very low, and the latter looks black. The balls are covered with a characteristic crack-like ornament similar to that on tin balls in the Si:Sn film alloy (Fig. 15.4). Those cracks might probably arise at cooling owing to the difference between the coefficients of thermal expansion of tin in the ball core (the light areas of the ornament) and silicon in the ball shell (the gray areas).

Of course, Fig. 15.8 exhibits the distributions of elements averaged over the thickness of the layer composed of tin balls covered with a silicon shell (Fig. 15.9). Every ball consists of a

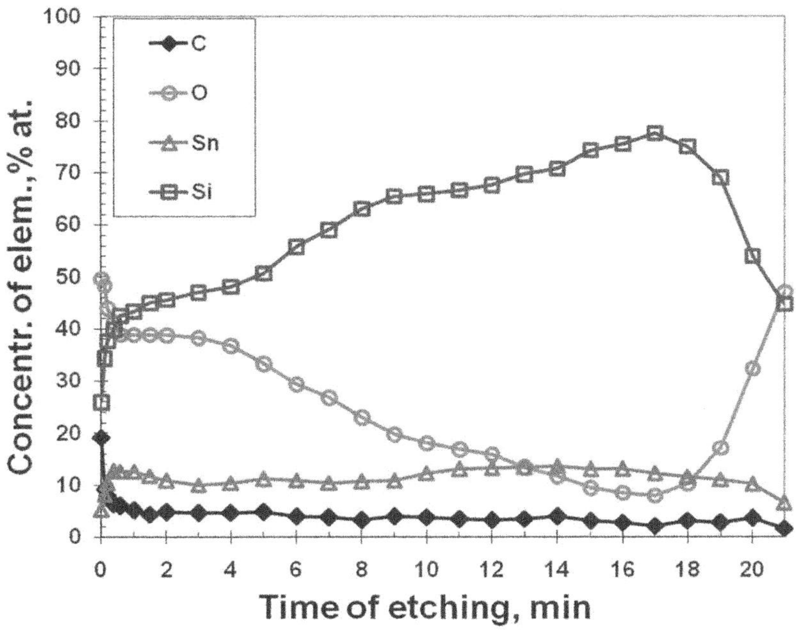

Figure 15.10 Same as in Fig. 15.8, but after TT at 300°C.

tin core covered with a silicon shell. Therefore, the integral Auger analysis registers a smeared maximum of the tin concentration at the depth equal to the average ball radius. This structure was probably formed when depositing the second Si layer on the droplets of liquid Sn. The Sn droplets may be formed as a result of the surface tension forces when the tin film was fused by the atomic and radiation flows from the silicon evaporator. Figure 15.10 demonstrates the element distributions across the Si/Sn/Si structure thickness after its 30 min annealing at 300°C. In comparison with Fig. 15.8, it is evident that the distributions of C and O technological impurities were not changed substantially. Only the oxygen content increased a little. This fact can be related to the additional oxidation of the specimen surface when the specimens were in the air atmosphere. Attention is attracted by the disappearance of the tin concentration maximum and the silicon concentration minimum in the etching time interval of 10–12 min observed in Fig. 15.8. Tin became uniformly distributed across the

Figure 15.11 The same as in Fig. 15.9, but after TT at 300°C.

whole structure thickness from the surface to the substrate. Silicon also became distributed more uniformly in comparison with Fig. 15.8. The Raman spectra show that the corresponding content of the crystalline Si phase increased by approximately a factor of two. In other words, during the TT at 300°C, silicon is mixed with tin and partially transformed from the amorphous into the crystalline state.

Figure 15.11 illustrates how the surface of the BSG/Si/Sn/Si structure changed after this process. One can see that instead of a crack-like ornament, there emerged light bumps on the balls and smaller white balls between initial gray balls. The contrast of the SEM image characterizes the electron concentration at the reflecting surface. Therefore, it is possible to suppose that white spots on the image correspond to metallic tin.

Hence, our model macrodroplet prepared in the form of a tin film embedded in a Si–Sn–Si sandwich structure has fused and decayed into microdroplets as early as in the course of the second Si layer deposition. This layer was accumulated on microdroplets of liquid Sn. It formed a shell of balls shown in Fig. 15.9 and the corresponding distribution of chemical elements across the thickness (Fig. 15.8). Within the time interval when tin remained liquid in the balls, Si might probably crystallize

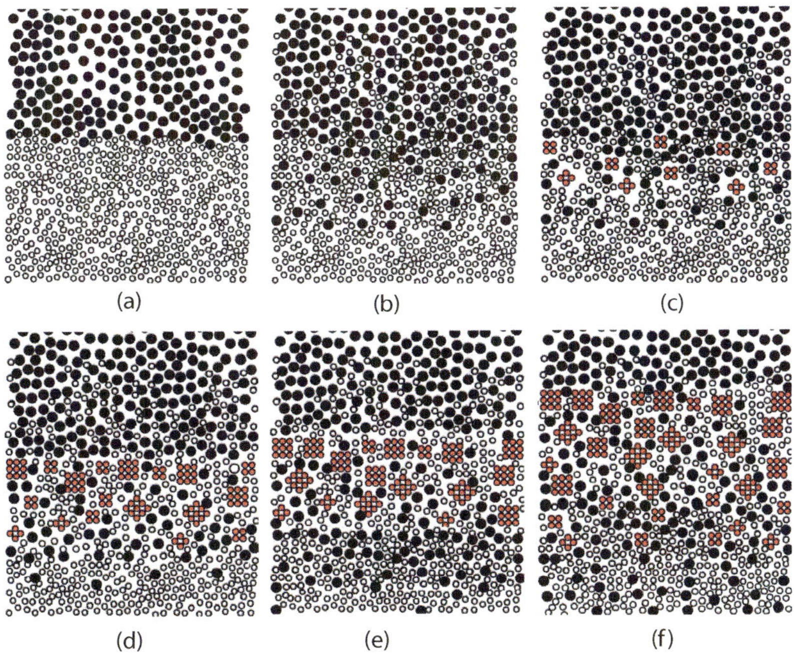

Figure 15.12 Schematic illustration of the time-dependent changes at the interface between amorphous silicon and liquid tin, which explains the process of a-Si transformation into nc-Si through its dissolution in tin.

partially, which is evidenced by curve 2 in Fig. 15.7. The silicon crystallization continued during the additional TT, when the tin cores of balls became liquid again (curves 3 and 5 in Fig. 15.7). The Auger researches testify that, at this stage, tin becomes uniformly redistributed across the whole thickness of the structure (Fig. 15.10). Let us consider, in detail, a physical model that explains the process of partial crystallization of amorphous silicon with the participation of liquid metallic Sn proposed by us in work [20]. In Fig. 15.12, light circles stand for the atoms of amorphous silicon and dark ones for the atoms of liquid tin. Owing to the screening of Si bonds by free electrons in the metal [38] and to the diffusion, the silicon and tin atoms "cross" the Si/Sn interface and partially mix together (Fig. 15.12b). As a result, the layer of a eutectic silicon solution in tin is formed. The dissolution of amorphous silicon is known to occur, being accompanied by the heat absorption. Therefore, as the solution enrichment with silicon continues, the solution temperature decreases, and, at a certain stage, the solution becomes oversaturated. As a result, the dissolved silicon atoms precipitate, i.e., they agglomerate into groups with the crystalline structure (ordered groups of gray or red circles in Fig. 15.12c). The number of such groups increases, and they become bigger (Figs. 15.12d–f). In real structures, their analog is nanocrystals, which were registered in Raman spectra (Fig. 15.7, spectra 2, 3, and 5).

The silicon crystallization in liquid tin initiates two processes, which bring the system into the previous state. One of them is the tin segregation on the surface of growing silicon crystals. The threshold solubility of tin in crystalline silicon is known to be low (about 5×10^{19} cm^{-3} [39, 44]). Therefore, the dissolved silicon atoms, which group into clusters in the form of nanocrystals, squeeze out tin atoms back to the eutectics, making the solution unsaturated. The other process is the release of a crystallization heat during the nanocrystal formation. It elevates the temperature and also reduces the saturation of a silicon solution in tin, so that the process of dissolution of amorphous silicon is resumed. Those two processes cyclically repeat, by resulting in the growth of new nanocrystals and the expansion of the crystallization zone in amorphous silicon around the Sn droplet. The accumulation of nanocrystals during the dissolution of amorphous silicon is provided by the fact that their solubility in tin is much lower than the solubility of amorphous silicon. Really, the absence of a crystal symmetry in a-Si and a significant number of dangling bonds give rise to a reduction in the total energy of interatomic bonds required for the a-Si dissolution in comparison with crystalline Si. In Fig. 15.13, the phase diagram of the Sn-Si system is plotted according to the data of work [46].

The eutectic point for the Sn:Si alloy, i.e., the minimum temperature, at which the components coexist in the liquid phase, almost coincides with the melting temperature of Sn (232°C). From this diagram, one can see that, as the temperature of Sn:Si solution decreases to the eutectic point, the solubility of Si diminishes to very low values. In other words, for the Si crystallization to start at temperatures of 300–400°C, even an insignificant dissolution of amorphous Si is enough. The smallest violations of

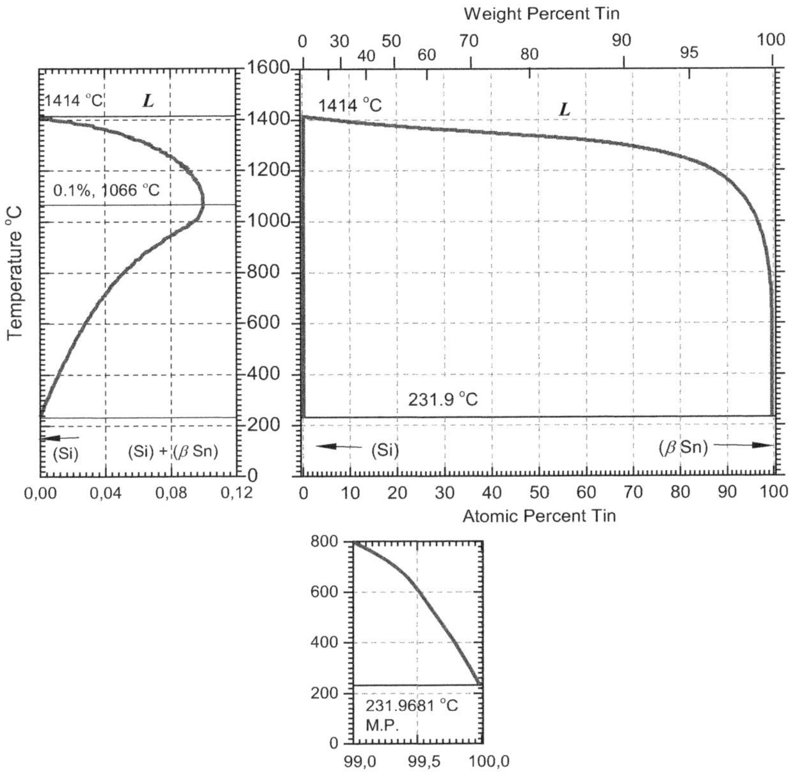

Figure 15.13 Phase diagram for the system Si–Sn according to the data of work [46].

the "temperature–concentration" equilibrium in the eutectic layer at the Sn–Si interface stimulate either the clustering of dissolved Si atoms into nanocrystals (if the temperature decreases or the oversaturation with respect to Si takes place) or the additional dissolution of a-Si (if the temperature or the tin concentration in the solution increases). Crystallization and dissolution change the temperature of eutectics in the direction opposite to that, which caused them, i.e., they return the system to the equilibrium. This factor governs the recurrence of the processes of a-Si dissolution and nc-Si formation in the eutectic layer. Therefore, the system "a-Si– Sn" can be described with the use of the adiabatic approximation, in which there is no heat exchange with the environment. In the framework of this approximation, the layer of Sn–Si eutectics together with nonequilibrium flows of silicon and tin atoms from the adjacent regions acts as a self-oscillating system until a complete exhaustion of either component. As a result, Si transforms from the amorphous state into the crystalline one. A driving force of those processes is not only the difference between the internal energies of the crystalline and amorphous Si states, but also the difference between the Gibbs potentials of a-Si and nc-Si dissolutions in liquid Sn. The thermodynamics of the described mechanism deserves a separate and more detailed analysis.

An additional argument in favor of the considered mechanism is the fact that the silicon recrystallization in the tin melt is applied to purify Si from impurities [47, 48]. For this purpose, high temperatures are used, at which the solubility of Si in Sn and the dominant size of Si crystals are much larger in comparison with nanocrystals formed in our specimens. For example, in work [48], metallurgical silicon was dissolved in the Sn melt at 1200°C (the Si solubility is approximately 15 at.%). Then the alloy temperature was lowered down to 700°C in order to create a solution oversaturated with silicon (the Si solubility is approximately 1 at.%). As a result of the solution decay at 700°C, there appeared polycrystalline Si flakes composed of crystals 100–200 μm in dimensions. According to the data of work [17], the maximum size of Si crystallites in liquid Sn at 300°C was 5 nm. Nanocrystals on just this scale were observed in our experiments: in vicinities of Sn microdroplets in the Si:Sn film alloy [20], and in a vicinity of a Sn macrodroplet in this work. Even the low solubility of Si in Sn at 300°C turns out enough for the formation of an appreciable number of such nanocrystals in Raman spectra. When silicon is deposited from the gas phase on the surface of liquid tin, the eutectic layer emerges, and the subsequent nanocrystals are formed more rapidly than if amorphous Si is dissolved. It is so because, in this case, there are no energy losses spent to break interatomic bonds of a-Si for its dissolution: free Si atoms in vapor transit immediately into liquid Sn. This mechanism explains the high formation rate of Si nanocrystals in the course of Si:Sn alloy film deposition [18–20] and when the second Si layer was deposited on the Sn layer in this work. The results obtained in this work confirm the model proposed by us in works [20, 49, 50] for the description of the mechanism of amorphous Si crystallization induced by the doping with tin. The tin-doping effect consists in (1) the emergence of microscopic droplets of metallic Sn in the bulk of amorphous Si, (2) the dissolution of amorphous silicon by liquid tin at temperatures above the Sn melting point and the formation of a Si–Sn eutectic layer, and (3) the cyclic processes of Si solution decay and recreation in Sn in the eutectic layer, which result in the dissolution of amorphous Si and the formation of nanocrystalline Si. Evidently, processes 2 and 3 compose a general mechanism of tin-induced silicon transformation from the amorphous into the crystalline state. This mechanism does not depend on the shape and the size of the interface between a-Si and liquid Sn metal; it can be either a macroscopic sandwich of Si and Sn films [14, 49, 50], or microscopic Sn inclusions in Si films [18–20], or silicon balls with a tin core, as in this work. With certain specifications, the mechanism is also applicable to the description of the Si crystallization induced by other metals, which can form eutectic solutions with silicon. The application of the considered mechanism of Si crystallization allows one, in principle, to smoothly control the number and the size of crystallites in the range of several nanometers. This possibility can serve as a ground for the development of new technologies aimed at manufacturing the cheap solar cells on the basis of nanosilicon films.

15.4 Tin-Induced Crystallization of Si at Laser Irradiation

The above mechanism induced tin MIC realized even at low temperatures near the melting point of tin ($T = 232^{\circ}$C). This allows us the measurement of the Raman to control of MIC in mode "on line."

We investigated Raman spectra of the structures with three layers of different thickness as shown in Fig. 15.6, at excitation laser at a wavelength of 633 nm with a maximum power light flux $I_0 = 10$ mW. The diameter of the laser beam was 1 mkm. That ensured flux density light to 3×10^5 W/cm^2. Raman spectra were measured at room temperature in two specific areas of the surface

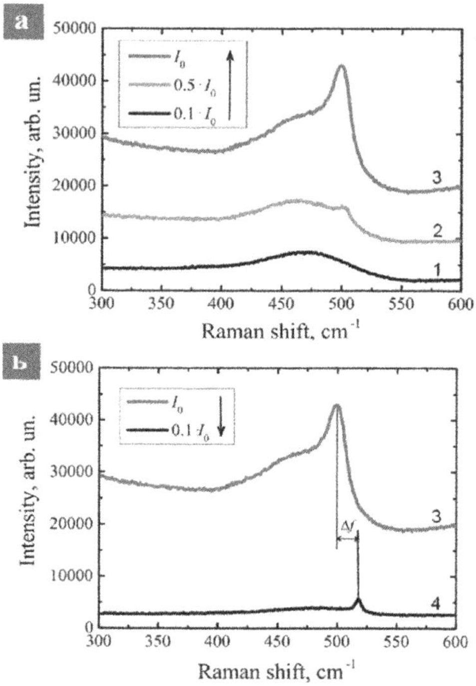

Figure 15.14 Raman spectra of the surface area A, registered in the sequence: growth (a) 1—I_1; 2—I_2; 3—I_3 and reducing (b) 4—I_4 of light intensity. A sample X:Y:Z = 100:50:100.

of the sample: region A in Fig. 15.6 (above Sn layer) and region B in Fig. 15.6 (only a-Si film). The scan time of spectra and therefore the laser beam on the sample was 30 s. Raman spectra in the same place sample were recorded at various excitation of laser power in the following sequence: $I_1 = 0{,}1 \times I_0$; $I_2 = 0{,}5 \times I_0$; $I_3 = I_0$ and again $I_4 = 0{,}1 \times I_0$.

It was found that the increase in power of the excitation light stream in zone A at some point leads to heating the sample sufficient to MIC. This is illustrated by Fig. 15.14, which shows the evolution of the Raman spectrum by increasing (a) and subsequent reduction (in) power laser excitation.

One can see that at laser power I_2 and I_3 appears typical narrow strip of Si crystalline phase in range 500–520 cm^{-1}. That is the formation of silicon nanocrystals takes place. Crystallization of amorphous Si under the influence of laser radiation was observed by many researchers [51–57]. It is believed that the crystallization after local heating Si by the laser beam to a temperature above 1000°C [51–53] or after the so-called "cold melting" Si under short but powerful laser pulses (10^{11}–10^{12} W/cm^2) [54–57] is caused by thermal crystallization mechanism. However, in our case, the laser light power is not enough for thermal crystallization. This is evidenced by the absence of Raman bands of crystalline silicon in zone B of samples even after 3-time scanning spectra with $I = I_0$.

The emergence of such bands in the A-zone apparently caused by the presence of a layer of tin, which significantly lowers the temperature of crystallization and Si [18, 19] through the mechanism of the MIC. At the interface of Si/Sn this mechanism is implemented at temperatures above the melting point of Sn at 232°C [20, 21].

It is known that Raman band of crystalline silicon (the maximum of which at room temperature has a frequency $f_c = 520$ cm^{-1}) shifted toward lower energies by the amount Δf_T proportional to the temperature of the crystal [57–59]. We also experimentally have observed low-frequency Raman shift peak of crystalline Si phase with increasing intensity laser radiation. Newly formed crystalline phase peak shifts toward more energy at reducing power laser light from the I_0 to $0.1 \times I_0$ (Fig. 15.14b).

Figure 15.15 shows the Raman spectra of the sample with X:Y:Z = 100:200:100 nm initially scanned at $I = I_0$, and then when $I = 0, 1 \times I_0$. Further reduction in the intensity of laser excitation is the order in the range $(0.01–0.1) \times I_0$ is not noticeably affect on the provisions and width of band. Such light heats the sample little compared to room temperature. Therefore Δf_T at $I > 0, 1 \times I_0$ characterizes the temperature increase in the sample site, where measuring Raman spectrum, compared with potted as a result of local heating by light.

Experimental dependence of crystalline phase peak shift on temperature [59] may be used to evaluate local temperature in the place of laser radiation effect. This shift has variations in range from 350 to 960°C (Δf_T 8–22 cm^{-1}) in samples with different ratios of X:Y:Z. However, there is about the same within

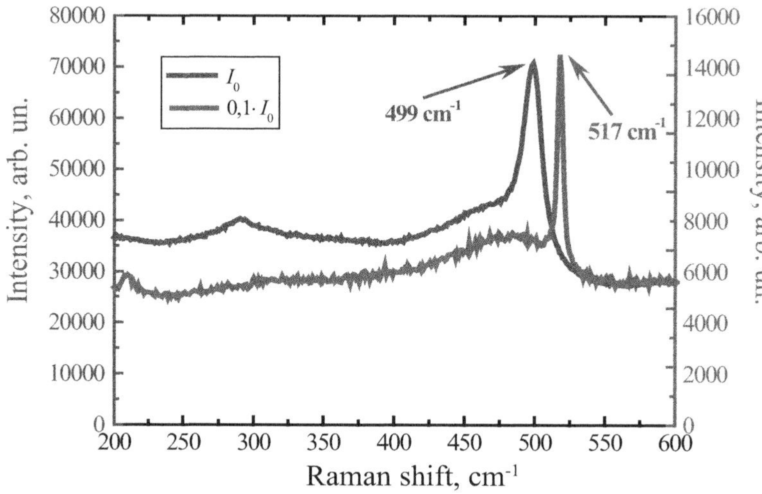

Figure 15.15 The temperature shift of the crystalline phase band. Raman spectra recorded at $I = I_0$ (red) and at $I = 0,1 \times I_0$ (green). X:Y:Z = 100:200:100.

each sample. Obviously, Si and Sn layer thickness and thicknesses relation determine the conditions of laser light absorption and heat dissipation.

Raman spectroscopy is used to determine the size of nanocrystals very long [35–37]. In particular, when Si crystal size less than 20 nm peak bandwidth LO-phonon shifts the position of the 520 cm^{-1} toward lower energies by the amount Δf_R. Figure 15.16 shows a graphical view of this shift depends on the size of crystals $\Delta f_R(\Delta R)$ at room temperature according to [60].

The share of volume of amorphous-crystalline composite occupied nanocrystals (X_C) can be determined from the ratio of integrated intensities of the Raman bands of the crystalline and amorphous phases [35]. In Fig. 15.17 shows the dependence of the amplitude band of crystalline phase on the number of 30 s scans Raman spectrum in the same place the sample.

It can be seen amplitude increases at each scan, i.e., the accumulation of nanocrystals in the process tin-induced crystallization of Si as a result of heating by laser light. In particular, in the sample 50:100:200 mkm X_C stabilizes at 92% at the fifth scanning. That

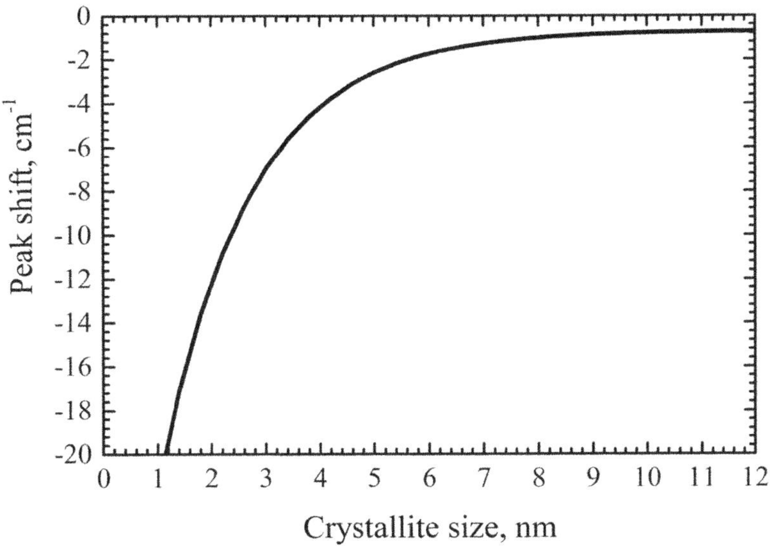

Figure 15.16 Dependence $\Delta f_R(\Delta R)$ of shift nanocrystalline phase peak compared to monocrystalline Si peak at 520 cm^{-1} on the size of crystallites [60].

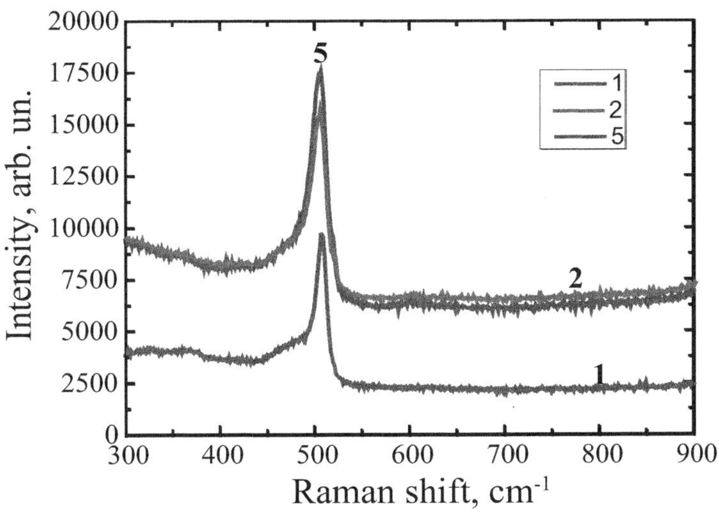

Figure 15.17 Raman spectra of sample at $X{:}Y{:}Z = 50\text{-}100\text{-}200$ mkm. $I = I_0$. (1) After first scanning of spectra; (2) after second scanning; (5) after fifth scanning at laser beam location on the same place of sample.

is 2.5 min through laser treatment. In samples with outer Si layer thickness 100 nm or less X_C reaching the level of 90% already at the first scan, i.e., 30 s. This is an order of magnitude higher crystallization rate than we have seen in similar structures as a result of heat treatments at 300–400°C in the dark [20]. If it was only due to a high temperature in the area of the laser beam, according [17] size crystallites (R) would be significantly more than 5 nm. However, this is not observed. We can assume that the MIC in the outer layer-Si is accelerated as a result of strong photoionization silicon by power laser light. Dissolving of the amorphous silicon in liquid tin on the interface layer is easier due screening of silicon covalent bonds by free electrons and due formation the additional dangling bonds [38]. Verification of this hypothesis requires further research. However, regardless of the mechanism of acceleration MIC laser light it can be used for the controlled formation of nanocrystals in the film-layered structures of Si–Sn.

Measurement and analysis of the Raman spectrum excited by laser light during the tin-induced transformation of amorphous silicon in to crystalline state allows to define (at modulating the laser and to control) in mode "on line" of crystallite size and proportion of the volume occupied by them.

The most important conclusions of this part are the following: The use of laser light to create conditions for tin-induced crystallization of amorphous silicon and to measure the size and concentration of nanocrystals using Raman spectroscopy may be the principal basis for the new technology of quality control in the manufacturing process of nanocrystalline silicon film. A significant acceleration of tin-induced crystallization of amorphous Si at heating by laser light versus dark opening act may indicate a significant role of ionization in the MIC processes.

15.5 Conclusions

Tin-induced crystallization of amorphous silicon occurs at temperatures above 232°C. The equilibrium size of silicon crystals can be several nanometers. Relative part of the volume occupied by the nanocrystals can reach 90%. Their formation occurs in a narrow layer of eutectic at the interface amorphous silicon - liquid tin. The transformation of silicon from amorphous into the nanocrystalline state is a result of repetitive cycles of formation and decay of a solution of silicon in the tin. These processes can be organized in a Si–Sn alloys with tin content greater than 2 at.% or in layer structures (Si–Sn–Si) with assistance method physical vapor deposition. The use of laser annealing to melt tin greatly accelerates the MIC. Measurement and analysis of Raman spectra during laser annealing allows us to control the formation of nanocrystals.

References

1. M. C. Beard, J. M. Luther, and A. J. Nozik, *Nat Nano* **9**, 951 (2014).
2. Z. I. Alferov, V. M. Andreev, and V. D. Rumyantsev, *Semiconductors* **38**, 899 (2004).
3. B. Yan, G. Yue, X. Xu, J. Yang, and S. Guha, *Phys. Status Solidi* **207**, 671 (2010).
4. N. S. Lewis, *Science* **315**, 798 (2007).
5. R. Søndergaard, M. Hösel, D. Angmo, T. T. Larsen-Olsen, and F. C. Krebs, *Mater. Today* **15**, 36 (2012).
6. M. Birkholz, B. Selle, E. Conrad, K. Lips, and W. Fuhs, *J. Appl. Phys.* **88**, 4376 (2000).
7. B. Rech, T. Roschek, J. Müller, S. Wieder, and H. Wagner, *Sol. Energy Mater. Sol. Cells* **66**, 267 (2001).
8. M. K. van Veen, C. H. M. van der Werf, and R. E. I. Schropp, *J. Non. Cryst. Solids* **338–340**, 655 (2004).
9. Y. Mai, S. Klein, R. Carius, H. Stiebig, L. Houben, X. Geng, and F. Finger, *J. Non. Cryst. Solids* **352**, 1859 (2006).
10. H. Li, R. H. Franken, R. L. Stolk, C. H. M. van der Werf, J. K. Rath, and R. E. I. Schropp, *J. Non. Cryst. Solids* **354**, 2087 (2008).
11. R. Amrani, F. Pichot, L. Chahed, and Y. Cuminal, *Cryst. Struct. Theory Appl.* **1**, 57 (2012).
12. G. Fugallo and A. Mattoni, *Phys. Rev. B* **89**, 045301 (2014).
13. O. Nast and A. J. Hartmann, *J. Appl. Phys.* **88**, 716 (2000).
14. M. Jeon, C. Jeong, and K. Kamisako, *Mater. Sci. Technol.* **26**, 875 (2010).
15. M. A. Mohiddon and M. G. Krishna, *J. Mater. Sci.* **47**, 6972 (2012).

16. D. Van Gestel, I. Gordon, and J. Poortmans, *Sol. Energy Mater. Sol. Cells* **119**, 261 (2013).
17. A. Mohiddon and G. Krishna, in *Cryst.-Sci. Technol.*, ed. A. Marcello (InTech, 2012), pp. 461–480.
18. V. V Voitovych, V. B. Neimash, N. N. Krasko, A. G. Kolosiuk, V. Y. Povarchuk, R. M. Rudenko, V. A. Makara, R. V Petrunya, V. O. Juhimchuk, and V. V Strelchuk, *Semiconductors* 45, 1281 (2011).
19. V. B. Neimash, V. M. Poroshin, A. M. Kabaldin, V. O. Yukhymchuk, P. E. Shepelyavyi, V. A. Makara, and S. Y. Larkin, *Ukr. J. Phys.* **58**, 865 (2013).
20. V. Neimash, V. Poroshin, P. Shepeliavyi, V. Yukhymchuk, V. Melnyk, A. Kuzmich, V. Makara, and A. O. Goushcha, *J. Appl. Phys.* **114**, 213104 (2013).
21. A. Chandra and B. M. Clemens, *J. Appl. Phys.* **96**, 6776 (2004).
22. F. Lin and M. K. Hatalis, *MRS Proc.*, **279**, 553 (1992).
23. V. B. Neimash, A. Kraitchinskii, M. Kras'ko, O. Puzenko, C. Claeys, E. Simoen, B. Svensson, and A. Kuznetsov, *J. Electrochem. Soc.* **147**, 2727 (2000).
24. V. B. Neimash, A. Kraitchinskii, M. Kras'ko, et al., *Ukr. J. Phys.* **45**, 342 (2000).
25. C. Claeys, E. Simoen, V. Neimash, A. Kraitchinskii, M. Kras'ko, O. Puzenko, A. Blondeel, and P. Clauws, *J. Electrochem. Soc.* **148**, G738 (2001).
26. E. Simoen, C. Claeys, V. B. Neimash, A. Kraitchinskii, N. Krasko, O. Puzenko, A. Blondeel, P. Clauws, *Appl. Phys. Lett.* **76**, 2838 (2000).
27. E. Simoen, C. Claeys, A. M. Kraitchinskii, M. M. Kras'ko, V. B. Neimash, and L. I. Shpinar, *Solid State Phenomena*, **82**, 425 (2002).
28. M. L. David, E. Simoen, C. Claeys, V. Neimash, M. Kras'ko, A. Kraitchinskii, V. Voytovych, A. Kabaldin, and J. F. Barbot, *J. Phys. Condens. Matter.* **17**, S2255 (2005).
29. D. Girginoudi, N. Georgoulas, and F. J. Thanailakis, *J. Appl. Phys.* **66**, 354 (1989).
30. G. N. Parsons, J. W. Cook, G. Lucovsky, S. Y. Lin, and M. J. Mantini, *J. Vac. Sci. Technol. A* **4**, 470 (1986).
31. R. Ragan, K. S. Min, and H. A. Atwater, *Mater. Sci. Eng. B* **87**, 204 (2001).
32. K. A. Johnson and N. W. Ashcroft, *Phys. Rev. B* **54**, 14480 (1996).
33. M. Vergnat, M. Piecuch, G. Marchal, and M. Gerl, *Philos. Mag. Part B* **51**, 327 (1985).
34. S. Yu. Shiryaev, J. L. Hansen, P. Kringhyj, and A. N. Larsen, *Appl. Phys. Lett.* **67**, 2287 (1995).

35. E. Bustarret, M. A. Hachicha, and M. Brunel, *Appl. Phys. Lett.* **52**, 1675 (1988).
36. H. Richter, Z. P. Wang, L. Ley, *Solid State Commun.* **39**, 625 (1981).
37. H. Campbell, P. M. Fauchet, *Solid State Commun.* **58,** 739 (1986).
38. A. (Akio) Hiraki, *Low Temperature Reactions at Si/metal Interfaces; What Is Going on at the Interfaces?* (Amsterdam: North-Holland Pub. Co., 1984).
39. M. Hillert, *Acta Metallurgica* **9**, 525 (1961).
40. G. Horvay, J. W. Cahn, *Acta Metallurgica* **9**, 695 (1961).
41. J. W. Cahn, *J. Chem. Phys.* **42**, 93 (1965).
42. L. R. Gomez and D. A. Vega, *Phys. Rev. E* **83**, 021501 (2011).
43. J. Schmelzer, A. Abyzov, and J. Moller, *J. Chem. Phys.* **121**(14), 6900–6917 (2004).
44. A. G. Milnes, *Deep Impurities in Semiconductors* (Wiley, New York) p. 526 (1973).
45. R. P. Thornton, R. G. Elliman, and J. S. Williams, *J. Mater. Res.* **5**, 1003 (1990).
46. R. W. Olesinski, and G. J. Abbaschian, *Bull. Alloy Phase Diagrams* **5**, 273 (1984).
47. G. F. Wakefield, H. S. Nagaraja Setty. Tin-lead purification of silicon. Patent US 3933981 A. 20.01.1976.
48. E. Maronchuk, T. F. Kulyutkina, and I. I. Maronchuk. Method for purification of technical purity silicon. Patent UA84653 (Ukraine). 16.02.2010.
49. V. B. Neimash, A. O. Goushcha, P. E. Shepeliavyi, V. O. Yukhymchuk, V. A. Dan'ko, V. Melnyk, and A. G. Kuzmich, *Ukr. J. Phys.* **59**, 1168 (2014).
50. V. B. Neimash, A. O. Goushcha, P. Y. Shepeliavyi, V. O. Yukhymchuk, V. A. Danko, V. V. Melnyk and A. G. Kuzmich *J. Mater. Res.* **30**(20), 3116–3124 (2015).
51. S. Chen and I. C. Hsleh, *Solid State Technol.* **39**, 113 (1996).
52. A. A. D. T. Adikaari and S. R. P. Silva, *J. Appl. Phys.* **97**, (2005).
53. T. Y. Choi, D. J. Hwang, and C. P. Grigoropoulos, *Opt. Eng.* **42**, 3383 (2003).
54. J.-M. Shieh, Z.-H. Chen, B.-T. Dai, Y.-C. Wang, A. Zaitsev, and C.-L. Pan, *Appl. Phys. Lett.* **85**, 1232 (2004).
55. V. A. Volodin and A. S. Kachko, *Semiconductors* **45**, 265 (2011).
56. A. V. Emelyanov, A. G. Kazanskii, P. K. Kashkarov, O. I. Konkov, E. I. Terukov, P. A. Forsh, M. V Khenkin, A. V Kukin, M. Beresna, and P. Kazansky, *Semiconductors* **46**, 749 (2012).

57. M. Balkanski, R. F. Wallis, and E. Haro, *Phys. Rev. B* **28**, 1928 (1983).
58. B. Stoib, S. Filser, N. Petermann, H. Wiggers, M. Stutzmann, and M. S. Brandt, *Appl. Phys. Lett.* **104**, 161907 (2014).
59. S. Périchon, V. Lysenko, B. Remaki, D. Barbier, and B. Champagnon, *J. Appl. Phys.* **86**, 4700 (1999).
60. W. Cheng and S.-F. Ren, *Phys. Rev. B* **65**, 205305 (2002).

Chapter 16

Nanoparticle Formation via Magnetron Sputtering with Inert Gas Aggregation

Panagiotis Grammatikopoulos[a] **and Mukhles Sowwan**[a,b]
[a]*Nanoparticles by Design Unit, Okinawa Institute of Science and Technology (OIST) Graduate University, Okinawa, Japan*
[b]*Nanotechnology Research Laboratory, Al-Quds University, East Jerusalem, Palestine*
pgrammatikopoulos@oist.jp

During the last part of the 20th century, the technological revolution of microelectronics meant that modern world industry and, consequently, global economy, would be increasingly dominated by silicon and its utilisation in electronic devices and integrated circuits. Therefore, it comes as no surprise that an overwhelming number of research studies published within the last decades had Si as a focal point, with both a basic science and an application perspective. With the advent of the nano era providing new opportunities to materials design, all challenges that the Si research community faced now have to be scaled down to the nanoscale.

Nanostructured Semiconductors: Amorphization and Thermal Properties
Edited by Konstantinos Termentzidis

ISBN 978-981-4745-64-2 (Hardcover), 978-1-315-36445-2 (eBook)
www.panstanford.com

16.1 Introduction

Nanoparticles (NPs) are particles of any shape, with dimensions ranging between 1 and 100 nanometres [1], which behave as whole units with respect to their transport and properties. Possessing an extra degree of freedom, i.e. their size, these "artificial atoms" have tunable properties that can lead to novel materials with bespoke structure, properties and potential applications [2]. NPs differ from their bulk counterparts in many respects. They possess a very high atomic surface-to-volume ratio, affecting both physical and chemical properties, such as low melting temperature or enhanced catalytic activity. Semiconductor NPs, in particular, are also characterised by a property known as quantum confinement: as the size of a NP diminishes close to its bulk exciton Bohr diameter, its corresponding energy gap increases, shifting its fluorescence emission towards the blue end of the spectrum [3]. In effect, this means that the whole visible light spectrum can be generated just by fine-tuning the NP diameters [4].

Due to the ubiquity of Si in a multitude of technological applications, various physical and chemical methods have been utilised for the fabrication of Si NPs. Electrochemical etching or high-energy mechanical ball milling of bulk Si are but two examples of the top-down approach to Si NP synthesis, whereas bottom-up techniques include, among many others, numerous wet chemistry methodologies [5–7]. Si NP synthesis has opened new paths in the microelectronics technology that had never been accessible to conventional methods in the past. However, additional desirable characteristics, such as biocompatibility, mean that possible uses are not confined to electronics. With potential applications ranging from integrated circuits to semiconducting devices, from photovoltaics to optoelectronics and from insulations to bioimaging and disease diagnostics, it is safe to assume that the industrial interest in Si NP technologies is only going to increase in the years to come [3].

One of the key challenges in these technologies is to control simultaneously the size and crystallinity of Si NPs with as high accuracy as possible, since these two parameters can have a beneficial or detrimental impact on uses for specific applications. For example, crystallisation of Si NPs is of particular interest for the Si-based photovoltaic industry, since it can affect the efficiency of solar cells [8–10]. Nevertheless, conventional thin-film synthesis systems suffer from inadequate concurrent control of both particle size and crystallinity, especially in the smaller diameter regime. The recently optimised magnetron sputtering inert gas condensation technique, however, has managed to overcome these limitations. Herein, we will present an overview of this method, and show how it allows for a highly desirable flexible design and synthesis of Si NPs of controlled size and structure.

16.2 The Magnetron Sputtering Method

Physical vapour deposition (PVD) has long been one of the most common bottom-up thin film production methods, which offers an attractive combination of speed, high-quality and low-cost [2, 11, 12]. During PVD, atoms from a source material target evaporate by means of some physical method, forming a highly super-saturated vapour, and are deposited on a suitable substrate. Depending on the physical method, several variants of PVD exist, namely (i) cathodic arc deposition, (ii) electron-beam PVD, (iii) evaporative deposition, (iv) pulsed laser deposition, and (v) sputter deposition [13]. In the latter method, which is of interest in this chapter, it is a glow plasma discharge that bombards the target, sputtering material away. To enhance yield, the electrically charged plasma is usually localised close to the target surface by a magnet; hence the name magnetron sputtering.

During recent years, magnetron sputtering has been widely utilised to synthesise single- and multi-component metal and semiconductor NPs, via subsequent inert-gas cooling and gas-phase condensation of the sputtered atomic vapour. Following the pioneering work of Haberland et al. [14], rapid advances in magnetron sputtering have been made; in-depth theoretical studies elucidated the main principles of particle formation [15, 16] and numerous technologically advanced variations of the NP synthesis systems facilitated higher cluster yields, post-deposition thermal treatment, oxidation control, etc. [17–23]. The current chapter does not aspire to offer an exhaustive review on the magnetron sputtering

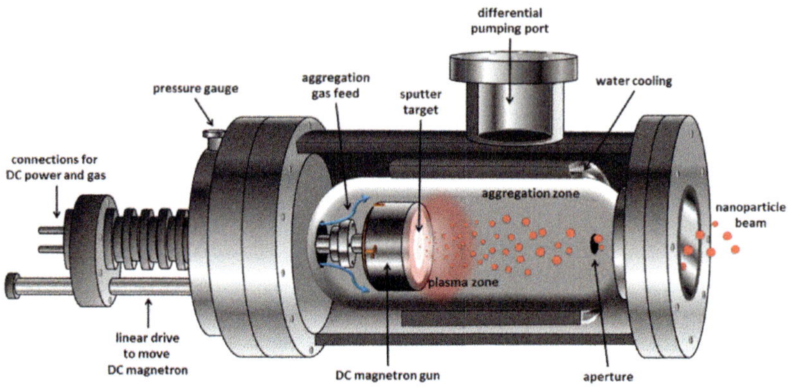

Figure 16.1 Schematic representation of a modified magnetron sputtering inert-gas condensation system.

method; for this, the interested reader can resort to dedicated literature [24–27]. However, an in-depth but rather qualitative description of the NP synthesis process is deemed expedient, to showcase the flexibility of the method and to prove how it can offer good control on the size and crystallinity of Si NPs.

A schematic representation of an ultra-high vacuum magnetron sputtering inert-gas aggregation reactor is shown in Fig. 16.1. A pure Si sputter target is placed on a DC magnetron gun, inside a water-cooled sputtering reactor. Inert gas, usually Ar, is fed into the reactor from behind the magnetron and is heated into forming ionised plasma which contains a mixture of Ar^+ ions, fast electrons and still-cool Ar atoms. Other gasses may also be introduced, such as He, or mixtures of gasses. The newly formed Ar^+ ions are directed towards the target; they bombard its surface and vaporise it. As energetic Si atoms leave the target, they collide with Ar atoms and their kinetic energy is decreased. Upon cooling down, they also collide with each other, forming bi-atoms, tri-atoms etc., which act as nuclei for cluster formation. After leaving the hot plasma zone, further growth occurs within the cool aggregation zone, first by more Si atoms landing on the nascent nanoclusters and subsequently by collisions and coalescence of the nanoclusters. A pressure differential is maintained between the aggregation zone (high pressure) and the sample deposition chamber (low pressure), which, along with the

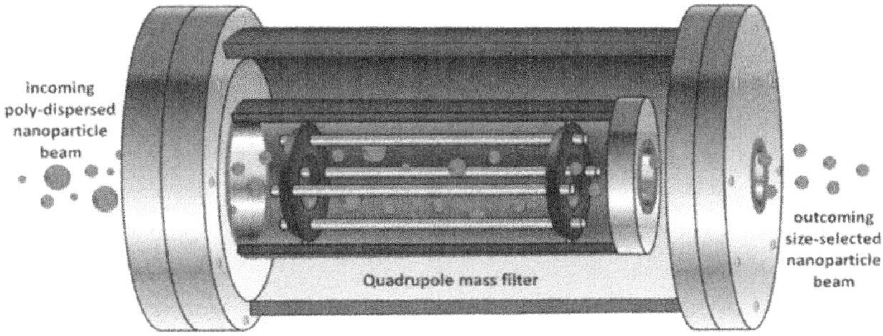

Figure 16.2 Schematic representation of a quadrupole mass filter. NPs lying outside a mass selection window are trapped by electrical fields inside the filter, allowing only a size-selected NP beam to go through.

inert gas drift velocity, transports the NPs away from the target. Finally, the NPs exit the aggregation zone through an aperture in the reactor, forming a NP beam towards the sample deposition chamber [28, 29].

An inline quadrupole mass filter (QMF) positioned immediately after the sputtering reactor further ensures the size-selected monodisperse nature of the deposition (see Fig. 16.2). The underlying principle behind its function is based on the fact that the beam contains single-charged NPs. The QMF consists of four parallel cylindrical rods; when oscillating electric fields are applied to them, NPs passing through maintain particular trajectories based on their mass-to-charge ratio. Fine-tuning the fields so that NPs lying outside a certain mass window are diverted out of the beam enables one to select only NPs of specific sizes, filtering out all the rest.

The main advantage of the magnetron sputtering method is its inherent flexibility. It allows control over the main growth parameters such as magnetron power (and, therefore, sputtering yield and Si atom number density), inert gas composition and pressure (affecting cooling and growth rates), NP drift velocity (determining residence time and effectively defining NP size) and aggregation zone length (the magnetron gun being mounted on a linear translation system enables movement along the reactor's axis with the resultant increase or decrease of the aggregation zone length, with potential effect on NP sizes). The use of alloy targets, or additional sputtering sources, either within the main reactor or beyond it, allow for the design of multi-component NPs with extra degrees of freedom, those of composition and structure. As will be shown later on, clever usage of such tools also enables control over the crystalline structure of the NPs. Finally, further control techniques employed after NP growth, such as in situ mass filtration, and deposition temperature and energy control, offer a plethora of possibilities for the design and facile, one-step synthesis of NPs of a desired size, composition, structure, and, what is of interest here, crystallinity.

16.3 Computer Simulation of Gas-Phase Synthesis

16.3.1 Why Is Computer Modelling Necessary?

In the macroscopic scale, the most decisive parameter that defines the crystalline structure of bulk materials is temperature evolution during or after their formation. Various thermal treatments, such as annealing or quenching, can completely alter the phase of a material, with grave effects on its physical and chemical properties. Therefore, in order to understand the growth procedure of particles in the nanoscale and the crystalline state that it results in, it is reasonable to analyse NP formation with a focus on temperature and its fluctuations during the growth procedure.

Actual in situ temperature measurements, however, are extremely difficult. In fact, since the nascent NPs and the inert gas around them are only weakly and sporadically coupled, a temperature reading at a certain point inside the growth chamber may be significantly different than the temperature of each specific NP at any given moment in time. As such, it can provide little useful information on the thermodynamic processes of NP growth, coalescence, or crystallisation. To make things worse, the high fraction of surface atoms in NPs means that such processes are even more sensitive to temperature, since surface atoms behave differently than inner atoms; one needs to have literally atomic temperature resolution to be able to predict NP growth governed by adatom adsorption, NP interfacial interactions, or surface pre-melting.

To elucidate the atomistic nature of the growth and crystallisation mechanisms, computer simulation methods have been employed instead, since the temporal and spatial scales involved are well within their resolution capabilities. Using the classical molecular dynamics (MD) simulation method, one can track atomic motion in a multiphase system at the intrinsic lattice vibration rates, around 10^{13} Hz. The greatest challenge is to design a simulation model that corresponds to the conditions in the experimental setup as closely as possible. Computational time and power limitations mean that certain approximations are necessary. As long as these approximations maintain the physics of the simulated system without creating artefacts, MD can be an indispensable tool in order to explain experimental results. Methodological details on how the simulations presented below mimicked the experimental conditions can be found in the respective articles [28, 30, 31].

16.3.2 MD Simulation of NP Growth with Inert Gas Temperature Control

Let us first imagine a cubic simulation box 40 nm in edge size, representing the growth chamber; to avoid spatial limitations, let it also have periodic boundaries along all directions. Moreover, let a number of sputtered and inert gas atoms (e.g. 500 Si and 1500 Ar atoms) be introduced in the box, in random positions (Fig. 16.3a). Even though the 1/3 ratio of sputtered-to-inert gas atoms is exaggerated compared to experimental values, which are estimated around ten times lower, this will not affect our analysis significantly, as will be shown later on [31].

The initial temperature of the energetic Si atoms is very high (e.g. 3000 K), in order to correspond to their initial velocities after being sputtered from the target. For the sputtered species only inert gas temperature control is used, i.e. by collisions with cool Ar atoms. In the real physical process, black-body radiation emitted from the surfaces of the nascent nanoclusters would also lower their temperature. However, this process is significant only in coarser

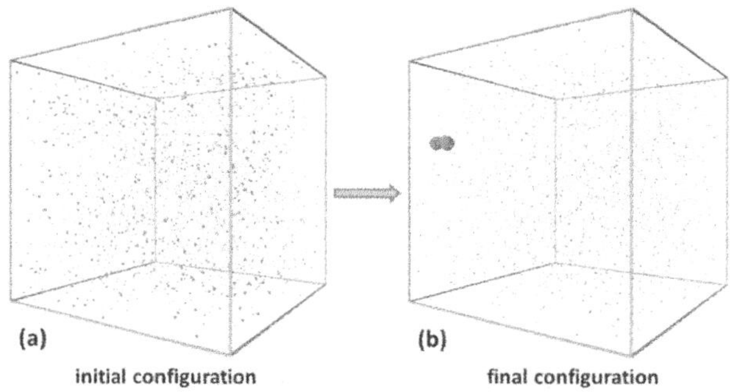

Figure 16.3 Initial (a) and final (b) configuration of MD simulation of Si NP growth. The box contains 500 Si (red) and 1500 Ar (blue) atoms.

timescales, and, hence, for our current purposes, atomic collision cooling is sufficient [32].

The temperature of Ar atoms is set at 300 K. Upon collisions with hot Si atoms, Ar atoms temporarily heat up. However, to mimic the flow of the aggregation gas, every time an Ar atom passes through the simulation box boundaries, its velocity is rescaled to correspond to room temperature; this way each simulated Ar atom represents a new Ar atom after every passage through the boundaries. Utilising a thermostat only at the simulation box boundaries sets the Ar atoms back to room temperature with a proper frequency. This should be neither too fast, to keep the collisions between Si and Ar elastic, nor too slow, so that not many collisions occur with Ar atoms possessing high velocities (i.e. colliding more than once before they had a chance to cool down going through the box boundaries). This corresponds to an Ar flow where each Ar atom would, most of the times, collide with a sputtered atom not more than once (there is no problem if more collisions happen on a few occasions, but that should not happen too often, either).

The first question that needs to be answered is whether inert gas cooling is really necessary for nucleation to take place. To help formulate a qualitative yet intuitive answer, a 2D schematic representation of our 3D simulation box is shown in Fig. 16.4. The z-coordinate axis is replaced by one displaying potential energy. Si and Ar atoms are roughly approximated by red and blue 3D curves representing their respective potential functions (indicated

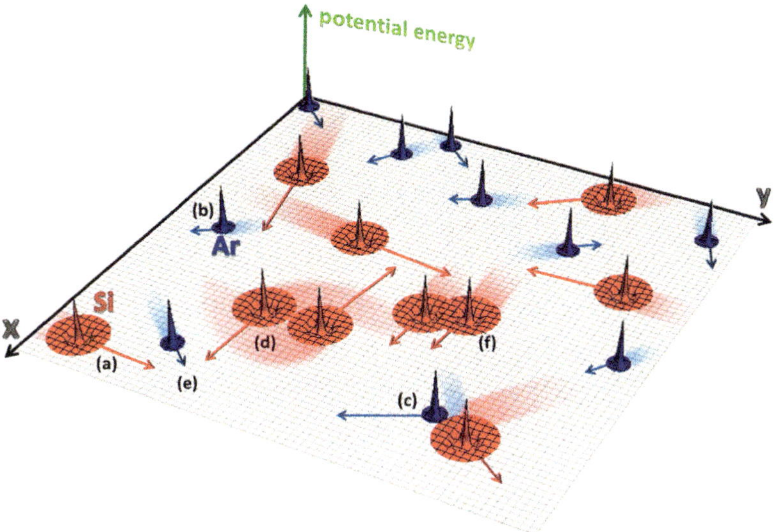

Figure 16.4 2D schematic representation of a 3D simulation box, containing Si (red, e.g. (a)) and Ar (blue, e.g. (b)) atoms. Atom velocities are indicated by arrows. Instance (c) depicts a Si and Ar system post-collision, with the resultant velocity exchange, whereas instance (d) shows temporary mutual attraction of two Si atoms which did not end up in bond formation. In contrast, instance (e) indicates an imminent three-body collision that can result in bond formation, such as the one shown in instance (f), where a Si bi-atom was formed.

by (a) and (b) in Fig. 16.4); note that the inert gas potential is almost purely positive, indicating repulsive forces only, whereas the Si potential has both negative (long distances) and positive (short distances) values and can therefore produce both attractive and repulsive forces, depending on inter-atomic distance. The nature of the long-distance attraction depends on the type of bonding that exists between particular atoms, which, for Si atoms, can be covalent or van der Waals. The strong repulsion at short distances is Pauli-type, dictated by the overlap of outer electron orbitals that occurs when atoms get too close to each other. Arrows of various lengths indicate the velocities of the atoms. Since stable bonds are always associated with a decrease in potential energy, only the negative components of the potentials can contribute to nucleation. Ar atoms repel all other atoms; therefore they cannot form any bonds. Instance (c) in Fig. 16.4 represents a Si and an Ar atom immediately after collision. Note the exchange of velocities: now the Ar atom is fast, whereas Si has slowed down. Si atoms are attracted to each other when they feel their attractive potentials. However, no stable bond can be created when two Si atoms collide; the atoms will either get too close and crash on one another, eventually bouncing back after feeling the repulsive parts of the potentials at short distances, or just deviate from their original course, not unlike cosmological objects when feeling the gravitational fields of stars (instance (d)). If, however, their velocities are low enough, a third atom may have just enough time to collide with them while they are still interacting, and carry away some energy and momentum. When this happens, e.g. in instance (e), the Si atoms can be trapped by the attractive component of their respective potentials and, thus, form stable bonds (such as in instance (f)). This mechanism is known as three-body collision [33–37] and explains why cooling by inert gas is an indispensable mechanism for the nucleation of NP seeds. Indeed, in MD simulation runs where no cool inert gas atoms were introduced, the temperature of the Si atoms always remained too high for any bonding to occur, and no nanocluster formation was ever observed. Note that three-body collisions are only necessary for the first step of clustering. Once a dimer is formed, it can later grow into a cluster seed without the need of a simultaneous double collision; it can just adsorb extra atoms, with the already existing bonds taking care of energy and momentum conservation.

Temperature and potential energy evolution during the growth procedure are shown in Fig. 16.5. MD simulation snapshots designate various instances of the growth mechanism. The first stage of the process involves the initial inert gas cooling of the sputtered atoms described above (snaphsot (a)). Since this stage is very fast, the inset zooms in at the initial 2.5 ns for clarity. It is evident that the temperature curve takes an exponential form as it approaches asymptotically a value that allows for bond formation.

When the temperature becomes sufficiently low, the second stage begins: dimers, trimers etc. start forming (snapshot (b)). Once atoms start bonding to each other, their potential energy decreases (see blue curve) and the released energy is transformed into kinetic energy (according to the equipartition theorem), leading

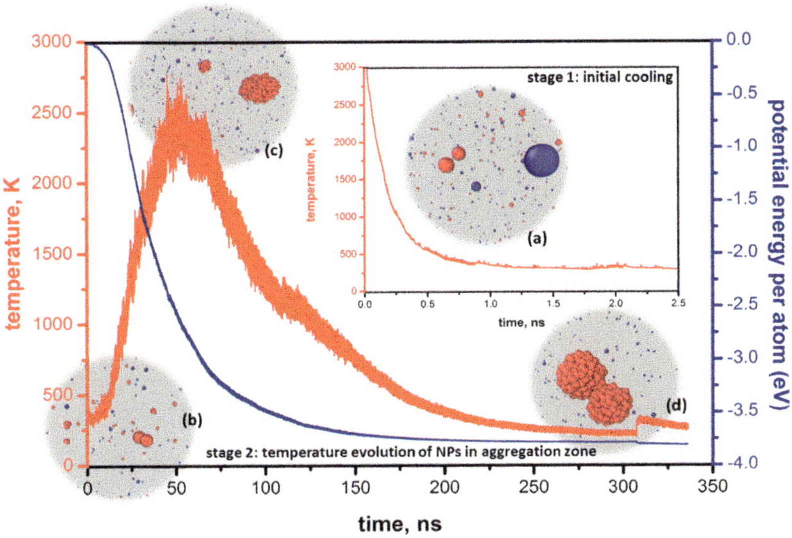

Figure 16.5 Temperature and potential energy evolution during growth of a Si NP containing 500 atoms. The inset zooms in at the initial 2.5 ns of the process. MD simulation snapshots designate the various stages of the growth mechanism.

to a temperature rise; this explains the minimum in the temperature curve at about 5 ns. As more atoms bond into small primary particles, the temperature keeps rising. How high the overall temperature reaches depends on how fast bonds form; collisions with Ar atoms, not leading to bond formation, still cool the system down, counterbalancing the heating.

Eventually, the simulation box gets almost depleted of Si monomers, and a new maximum temperature is reached, when the heating due to bond formations is just as significant as the cooling due to collisions with Ar atoms (snapshot (c)). At a first glance this looks like a simulation artefact; in the real world, newly sputtered Si atoms would come to replace the ones already in nanoclusters. However, one should bear in mind that the nascent nanoclusters drift away from the target. Therefore, they soon reach a distance where the number of available free Si atoms is not enough to keep heating them up, corresponding to the simulation setup.

Cooling ensues until the system reaches an equilibrium temperature, which is near that of the Ar atoms. Obviously, temperature decrease is not smooth. Temperature is due to oscillate between values that favour either bond formation or inert-gas cooling at each moment in time; these values change as the system evolves and the concentrations of free and bonded atoms change. Violent events also do happen, when nanoclusters of considerable sizes collide with each other, with the resultant upward jump in temperature and downward jump in potential energy. Such an event happened at ~310 ns (see snapshot (d)). Since this collision occurred after the two nanoclusters had already cooled down, it did not lead to their full fusion into a spherical NP, but rather to partial coalescence (see also Fig. 16.3b) with the resultant necking between the two NPs [38].

During the growth process, there can also be collisions which lead to breakups of bonds instead of creating ones. In such a scenario, there are three contemporary rates which affect the temperature of the system: those of bond formation, bond breaking and elastic collisions with Ar atoms. The balance of the effects of these rates can be very delicate. As mentioned above, the sputtered-to-inert gas atom ratio in this discussion was exaggerated; a lower ratio would yield lower overall temperatures, as the balance would be offset towards a faster cooling rate. However, for individual nanoclusters during atom absorption that would not be very significant. In fact, since the cohesive energy of Si is of the order of a few eV [39], the amounts of heat locally released are vast; the temperature of a cluster containing a small number of Si atoms, initially at 300 K, which takes in one atom more, can rise up to a few thousand Kelvin. Such local temperature upsurges are inevitable during the early stages of NP growth, keeping the NP at a liquid state. Nevertheless, it is the subsequent cooling (and, most importantly, its rate) that dictates the order of the phase that the NPs end up with.

16.4 Analysis of Temperature Evolution in NP Formation

An analytical derivation of the discussed temperature profile is the subject of the following subsections. It is decomposed into the aforementioned two stages. Even though the former describes the initial cooling before condensation starts taking place, and, therefore, does not seem to be directly relevant to our following discussion on NP crystallinity, a simplified explanation is presented here for completion. As will be shown later on, initial cooling rate is of interest in the case of simultaneous co-sputtering of more than one species.

16.4.1 Stage 1: Initial Cooling of Sputtered Atoms

In order to make our analysis as general as possible, we assume atoms of some sputtered material of a certain mass, m, possessing a high initial velocity, u. The high initial velocity is the average velocity of the sputtered atoms, and corresponds to a high initial temperature, through the kinetic/thermal energy relation ($E_{\text{kin}} = E_{\text{therm}}$, or $\frac{1}{2}\mu^2 = \frac{3}{2}k_{\text{B}}T$) for individual atoms. This is realistic considering that sputtered materials are initially very hot. Inert gas atoms of mass μ are introduced in the aggregation zone with a velocity β. This velocity corresponds to the average velocity of the inert gas atoms, and is assumed lower than u, since the inert gas is cooler than the sputtered material.

We assume that the initial cooling of the sputtered material is achieved by elastic collisions between its atoms and those of the inert gas. For simplicity, we also assume that these collisions occur in one dimension. A 3D treatment is possible but rather more complicated; yet, it would produce similar results [40, 41]. Here, we take β to have a positive value always; therefore u can be either negative, when the sputtered atom moves in the opposite direction from the inert gas atom, yielding head-on collisions, or positive, when the sputtered atom moves in the same direction, yielding smoother collisions (since $u > \beta$, same-direction collisions do occur).

The resultant velocity of a sputtered atom after collision with an inert gas atom is

$$v(u, m) = \frac{m - \mu}{m + \mu}u + \frac{2\mu\beta}{m + \mu}, \qquad (16.1)$$

which is of the linear form $y = ax + b$. Since $\left|\frac{m-\mu}{m+\mu}\right| < 1$ for all values of m, the slope of this line will always be lower than that of $y = x$, which is the limit for velocity conservation of the sputtered atom before and after a collision. Therefore, cooling does take place.

According to the mass of the sputtered atom, there are three distinct cases.

(1) $m = \mu$
This case is unrealistic, since the sputtered atoms are always of different atomic mass than the inert gas ones, but it is considered for completion reasons, and because it will help for a better understanding of the other two cases.
If $m = \mu$, i.e. if the two atoms have identical masses, Eq. (16.1) becomes $v(u) = \beta =$ constant. In other words, the final velocity of the sputtered atom is equal to the initial velocity of the inert gas atom; the two species have exchanged velocities (and momenta, since their masses are equal). This fact (well established amongst billiard players) explains why the sputtered species does not cool itself by collisions of its atoms with each other.

(2) $m > \mu$
This is the most common case for metals, since usually the sputtered species is heavier than the inert gas. When $m > \mu$, it is always $0 < \frac{m-\mu}{m+\mu} < 1$. The larger m becomes, the more the slope of the linear expression (16.1) approaches 1. For example, when $m = 2\mu$, $\frac{m-\mu}{m+\mu} = \frac{1}{3}$, whereas when $m = 10\mu$, $\frac{m-\mu}{m+\mu} = \frac{9}{11}$. Therefore, rather intuitively, the heavier the sputtered atom, and the bigger its inertia, the more it retains its original velocity. With $m \to \infty$, Eq. (16.1) becomes $v = u$, i.e. there is no change in velocity (a tennis ball colliding with the Great China Wall hardly changes the Wall's momentum).

(3) $m < \mu$
When $m < \mu$, such as in the case of Si studied here, it is always $-1 < \frac{m-\mu}{m+\mu} < 0$. The negative sign indicates a downward slope; therefore, for small values of m, $\frac{m-\mu}{m+\mu}$ becomes small too (but its absolute value increases), approaching -1 in the limiting case when $m \to 0$. This means that a hypothetical infinitesimally

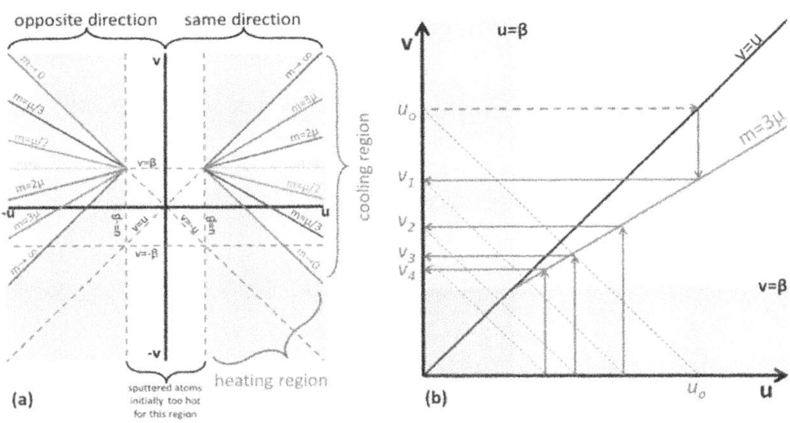

Figure 16.6 (a) Examples of resultant velocities of sputtered atoms after a single elastic collision for various mass differences between sputtered and inert atom species. (b) Schematic representation of iterative mechanism of cooling after a series of elastic collisions between a sputtered and an inert gas atom. A sputtered atom with an initial velocity u_0 cools down to velocities, v_1, v_2, v_3, etc., asymptotically reaching towards the set velocity of the inert gas atoms, thus producing an exponential velocity curve.

small mass would maintain its velocity (by measure) upon head-on collision with an inert gas atom, and would only change direction. Thermally speaking, it would be exactly equivalent to an infinitely large mass.

All these cases are summarised in Fig. 16.6a. No matter the mass ratio between species, cooling always takes place, as the final velocity is within the cooling region, designated by blue colour in the figure. It is clear that the more similar the mass of the sputtered atom is to the mass of the inert gas atom, the more its final velocity, v, will be affected by the elastic collisions. In the equal mass extreme scenario, the two species exchange velocities. When the mass difference between the species is pronounced, the sputtered atoms are more prone to preserve their initial velocities (with or without direction inversion). Therefore, if two species are co-sputtered, the one whose atomic mass is more similar to that of the inert gas will be, in principle, cooled faster.

To conclude, a couple of more observations about the plots in Fig. 16.6a could be interesting. It was mentioned that for all values of m, $v(u, m)$ is linear, i.e. of the form $y = ax + b$, but it should also be stressed that for all m values, $u = \beta$ means that $v = \beta$, too. This means that all lines go through the point (β, β) (or $(-\beta, \beta)$, if we also include negative velocities). Our initial assumption that the sputtered species is hotter than the inert-gas atom means that what happens when $|u| < \beta$ is of no importance, i.e. the definition field for $v(u)$ is $(-\infty, -\beta] \cup [\beta, \infty)$, see Fig. 16.6a, where the forbidden velocity region where sputtered species atoms cannot enter is designated.

Since all straight lines start from the same point on the $v = u$ line (i.e. (β, β)) and their slopes are always $\neq 1$, they deviate from $v = u$ more with higher u values. This means that for higher temperatures, the velocity (or equivalently, the temperature) decrease with each collision is more pronounced. For example, in the case where $m = 3\mu$ (Fig. 16.6b), the sputtered atom has an initial velocity u_0. After colliding with an inert gas atom for the first time, its velocity drops to v_1. Upon further collisions, its velocity drops to v_2, v_3, v_4, etc., approaching β asymptotically. In other words, the hotter the sputtered gas is, the faster it initially cools down, reaching equilibrium when it acquires the same velocity as the inert gas.

This cooling behaviour is intuitive and in complete agreement with the exponential-type curves of cooling vs. time graphs from MD simulations of inert-gas condensation (see inset of Fig. 16.5 or, for example, in Singh et al. [30]). Obviously, in these graphs when condensation begins heat is released in the system, and deviations from a perfect exponential-curve occur, but still, all deviations are superimposed on a exponential underline.

16.4.2 Stage 2: Temperature Evolution in Aggregation Zone

Moving to stage 2, let us consider a small, nascent Si nanocluster which flies in a chamber with both inert gas and free Si atoms (Fig. 16.7). Frequent collisions with Ar atoms lower its temperature. Simultaneously, the size of the nanocluster increases as it absorbs Si atoms landing on its surface, with the corresponding heat release.

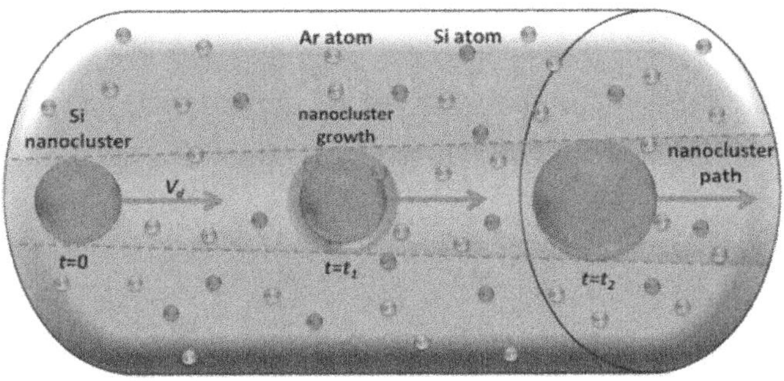

Figure 16.7 Schematic representation of a nascent Si nanocluster flying in a chamber with both inert gas and free Si atoms. The cluster grows absorbing Si atoms and cools by collisions with Ar atoms that it finds on its path.

The number of atoms in the nanocluster N can be solved as a function of flying time:

$$N(t) = N_0 + \int_0^t n_{\text{co-Si}}(t)dt = N_0 + \int_0^t \pi r(t)^2 \rho_{\text{Si}} v_{\text{d}} dt, \qquad (16.2)$$

where N_0 is the initial number of Si atoms in the cluster, ρ_{Si} is the number density of Si atoms in the gas phase, $n_{\text{co-Si}}$ is the collision rate between the nanocluster and Si free atoms, and v_{d} is the drift velocity of the nanocluster. Let $\overline{V_0}$ be the average volume for a single Si atom; then, the relation between the radius, $r_{\text{c}}(t)$, of the nanocluster and the number of atoms it contains, is

$$r_{\text{c}}(t) = \left(\frac{3\overline{V_0}N(t)}{4\pi}\right)^{1/3}. \qquad (16.3)$$

Assuming the initial condition $N(t=0) = N_0$, $N(t)$ can be solved as:

$$N(t) = \frac{1}{27}\left[\pi v_{\text{d}} \rho_{\text{Si}} \left(\frac{3\overline{V_0}}{4\pi}\right)^{2/3} t + 3N_0^{1/3}\right]^3 \qquad (16.4)$$

If each Si atom which collides with the nanocluster forms a bond, thus releasing potential energy equal to the bond cohesive energy, E_{co}, the kinetic energy change of the cluster $\frac{3}{2}k_{\text{B}}N(t)dT$ over small dt is:

$$\frac{3}{2}k_{\text{B}}N(t)dT(t) = \frac{k}{2}\left(T_{\text{Ar}} - T(t)\right)n_{\text{co-Ar}}dt + n_{\text{co-Si}}E_{\text{co}}dt, \qquad (16.5)$$

where $n_{\text{co-Ar}}$ is the collision rate between the nanocluster and Ar atoms, k_{B} is Boltzmann's constant and k is an energy exchange constant, which is a function of the gas atomic mass, cluster mass and interaction strength [42]. Substituting $n_{\text{co}-\text{Si}} = \pi r_{\text{c}}^2 v_{\text{d}} \rho_{\text{Si}}$ and $n_{\text{co-Ar}} = \pi r_{\text{c}}^2 v_{\text{d}} \rho_{\text{Ar}}$, respectively, we obtain

$$\frac{dT(t)}{dt} = \frac{k\pi r_{\text{c}}^2 v_{\text{d}} \rho_{\text{Ar}} \left(T_{\text{Ar}} - T(t)\right) + 2\pi r_{\text{c}}^2 v_{\text{d}} \rho_{\text{Si}} E_{\text{co}}}{3k_{\text{B}} N(t)}. \qquad (16.6)$$

One can further substitute expressions (16.3) and (16.4) into (16.6). With the initial condition $T(t = 0) = T_0$, the last equation can be solved analytically, giving the temperature of the nanocluster as a function of time, controlled by ρ_{Ar}, ρ_{Si}, and v_{d}.

Various exemplary solutions to Eq. (16.6) are shown in Fig. 16.8, for a 4 nm NP with an initial temperature of 3000 K. It is evident from these solutions that the equilibrium temperature depends solely on the ratio of the number densities $\rho_{\text{Si}}/\rho_{\text{Ar}}$; more importantly, this dependence is a very sensitive one. Of course, Eq. (16.6) makes the assumption of a constant ρ_{Si}. As discussed above, however, the number density of free Si atoms decreases with time (or else, with distance from the target). This means that at the very early growth stages, when there is a super-saturation of Si atoms, the temperature can, in principle, increase (following, for example, the $\rho_{\text{Si}}/\rho_{\text{Ar}} = 1/3$ curve of Fig. 16.8) only to decrease later on, when the abundance of Si atoms ceases to exist (following progressively lower curves). A combination of such behaviours (and graphs) would result in a temperature profile very similar to that shown in Fig. 16.5. A temperature evolution following this scheme is implied by the yellow points in Fig. 16.8. Note that the right-most point corresponds to room temperature, following a complete depletion of Si free atoms far away from the sputtering target. The time intervals between these points are dictated by the drift velocity of the NPs, which is, in turn, determined by the pressure differential. It should be stressed, however, that the ones shown in Fig. 16.8 were chosen arbitrarily and do not necessarily correspond to real growth times; this is just an attempt to indicate the cooling curve qualitatively.

In closing this discussion, it is important to emphasise that the parameters that determine the cooling rate can be experimentally controlled, varying the inert gas pressure, sputter atom density or

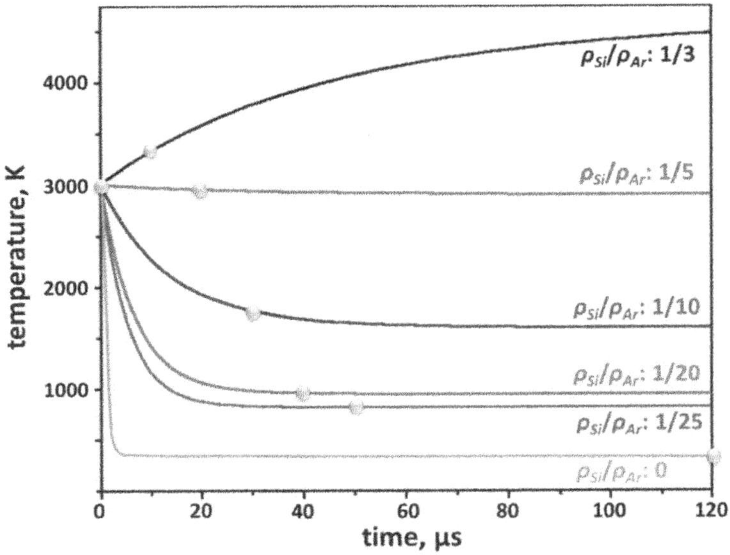

Figure 16.8 Various exemplary solutions to Eq. (16.6), for different atomic ratios between Si and Ar. As the nascent nanocluster drifts away from the sputtering target and the abundance of Si atoms decreases, the solution that best describes cooling at any given moment changes with time, as implied by the yellow points. A curve connecting them would be very similar to that of Fig. 16.5, describing the overall temperature evolution of the growth process. Adapted with permission from Ref. [31]. Copyright 2015, American Physical Society.

pressure differential. In other words, one can effectively control the cooling rate of the nascent nanoclusters within the aggregation zone. How this defines the crystalline state of the NPs will be the subject of the next section.

16.5 Case Study (i): Crystallisation of Si NPS Using Inert Gas Temperature Control

Utilising the magnetron sputtering inert gas condensation cluster beam source of Fig. 16.1, Si NPs can be synthesised with excellent control over their size. Growth can be controlled by varying deposition parameters such as carrier gas pressure, composition and flow rate, magnetron power, and aggregation length, resulting in NPs of different size distributions. Of these parameters, the most significant factors determining NP magnitude are the inert gas pressure and composition. Commonly, Ar gas is used both to ignite the plasma and act as the aggregation gas. However, other inert gasses, such as He, can also be introduced in the growth chamber. Variations in a mix of Ar/He gas flow can result in NPs of different corresponding sizes. High flow rates of He gas result in shorter residence times within the aggregation zone, with the resultant reduction in NP sizes. Ar gas of the same temperature is much slower, due to its greater mass; hence, it produces a slower drift velocity, allowing the NPs to grow larger. Simultaneously, as discussed in the previous section, slower drift velocity translates into slower cooling.

Atomic force microscopy (AFM) topography images of three Si NP depositions on a Si substrate are shown in Fig. 16.9a [31]. Keeping all other growth parameters unchanged, the three samples correspond to different Ar/He gas flow conditions, namely, Si-1 (Ar: 80 sccm and He: 40 sccm), Si-2 (Ar: 80 sccm and He: 10 sccm) and Si-3 (Ar: 100 sccm and He: 0 sccm). Their respective size distributions calculated from the AFM images are shown in Fig. 16.9b, and verified by QMF readings (insets of Fig. 16.9b). The main observation of this figure is that with a decreasing He gas flow rate (i.e. increasing residence time), the depositions shift from having a large number density of small-size NPs to having a lower number density of larger-size NPs. Along with the actual NP sizes, the size distribution width increases, too; however, stricter QMF settings could narrow down these distributions further.

More importantly, transmission electron microscopy (TEM) analysis shows that NPs of the Si-1 and Si-2 samples are amorphous, whereas those of Si-3 obtain polycrystalline structures. Figure 16.10 shows exemplary high-resolution TEM (HRTEM) images of a 10-nm (a) and a 16-nm (c) NP synthesised under the Si-2 and Si-3 deposition conditions, respectively. The corresponding fast Fourier transform (FFT) analysis clearly indicates amorphous structure in the former case (b) and polycrystalline structure in the latter (d); the main reflections (111), (220), and (311) are indexed in the

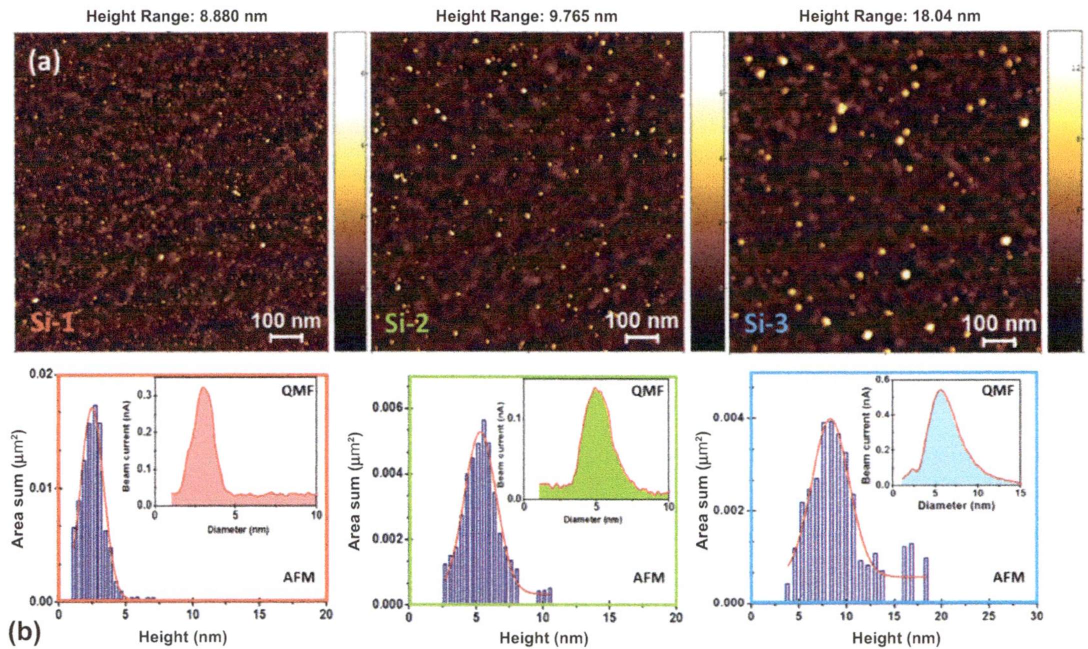

Figure 16.9 (a) High-resolution AFM topography images and (b) AFM height distributions of Si NPs deposited using varying Ar and He gas flow (Si-1, Si-2, and Si-3) at a magnetron discharge power of 90 W and time of 10 min. Insets of (b) show corresponding QMF profiles, in agreement with the AFM readings. Adapted with permission from Ref. [31]. Copyright 2015, American Physical Society.

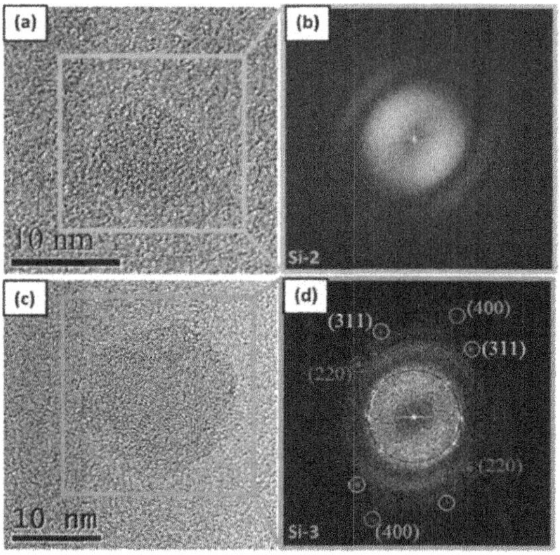

Figure 16.10 HRTEM images with corresponding FFTs from a 10-nm (a) and a 16-nm (c) NP synthesised under the Si-2 and Si-3 deposition conditions, respectively. In the latter case, the FFT with clear spot pattern shows a lattice structure characteristic of crystalline Si. Adapted with permission from Ref. [31]. Copyright 2015, American Physical Society.

figures (JCPDS card No. 27-1402), and correspond to the Si diamond structure.

In order to confirm that crystallisation occurs in the gas phase and not after landing, and to elucidate the underlying physical mechanism, numerous MD simulations of Si condensation from the gas phase were run. A spherical Si NP was placed in the centre of a simulation box, surrounded by an atmosphere of room temperature Ar atoms, as shown in Fig. 16.11. This means that, to minimise simulation time, the initial cooling and nucleation stages were omitted; the NP was assumed already formed and its initial temperature was set at 3000 K following an implicit heating during growth. As a result, the NP was in a hot liquid phase, but no evaporation yet occurred. As explained earlier, this does not affect the cooling process in any way, since, even if initial condensation was taken into account, it would result in the same, molten phase. To cover as broad a configuration space as possible, parameters such as

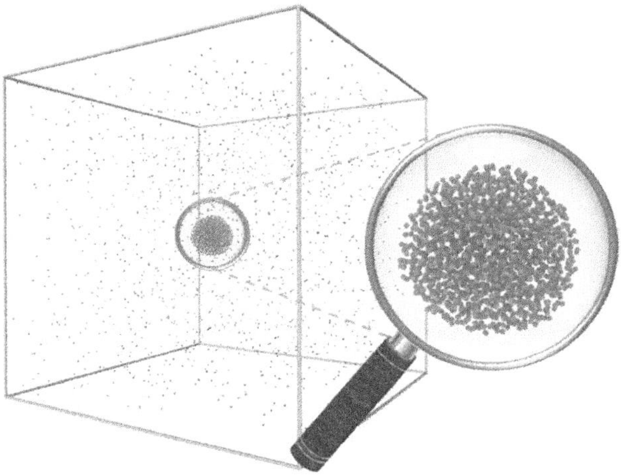

Figure 16.11 Simulation setup for a 4 nm Si nanocluster in a 40 nm box. The cluster is placed at the centre, surrounded by Ar atoms. A zoom in on the cluster shows its initial liquid state.

NP size, Ar atom number density and simulation box size were given various values.

It is well known that phase transformations can be difficult to detect. Even the most common ones, such as melting, can have indistinct transition points, which are subject to different definitions and methods of identification [43]. Therefore, to detect crystallisation unambiguously three analysis methods were utilised at the same time: the first, obvious one, was by tracking temperature evolution. Simultaneously, the angular structure factor, P_{st}, was calculated for each Si atom, which reveals the fraction of Si atoms in a disordered environment [31, 44]; a change in this fraction can be the fingerprint of crystallisation or amorphisation. Finally, to detect potential crystallisation seeds, the ring analysis algorithm was used [31, 45]. According to this method, primitive Si six-fold rings in wurtzite and diamond basic blocks can be detected, which can be considered as nuclei for the amorphous-to-crystalline transition.

Obviously, since the cooling process is inherently stochastic, numerous simulation runs were required for each configuration in order to collect reliable statistics. The number of cases where crystallisation occurred depended on NP size and cooling rate;

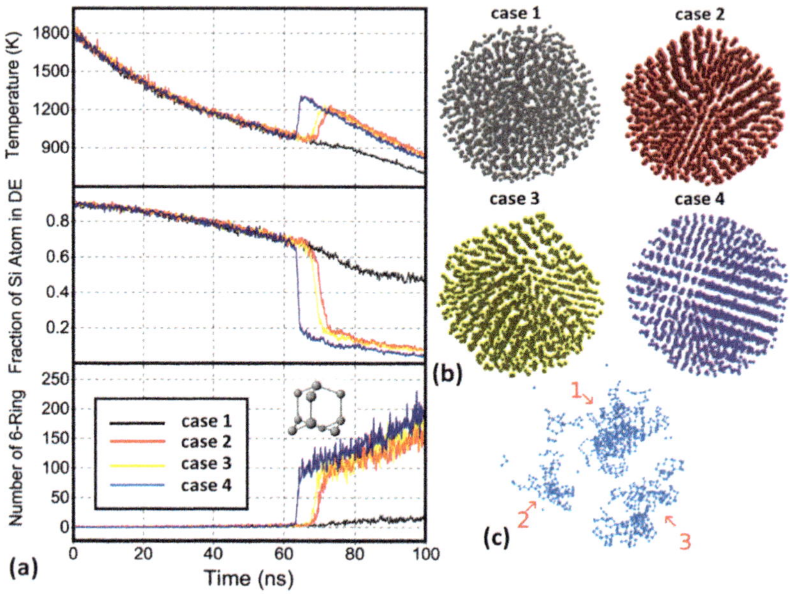

Figure 16.12 (a) Evolution of temperature of Si NPs obtained in four different simulations (top), the fraction of Si atoms in a disordered environment (middle) and the number of primitive six-fold rings (bottom). Simulation cases 2–4 resulted in crystallisation of the Si NP. Case 1 represents results of runs (including ten more cases, not shown here) where the Si NP did not crystallise upon cooling. The inset in the bottom figure shows a six-fold ring. (b) Final structures obtained from the simulations presented in (a); the NP colours agree with the plot colours. Evidently, case 1 resulted in an amorphous structure, cases 2 and 3 in polycrystalline NPs and case 4 in a single crystal. (c) Visualisation of crystallised regions after 24.3 ns of simulation of a 9 nm Si NP. Three separate crystallites are identified (red arrows), indicating that a polycrystalline structure is grown from several nuclei at random shapes. Adapted with permission from Ref. [31]. Copyright 2015, American Physical Society.

for example, out of eleven runs with NPs 9 nm in diameter, four crystallisation events were observed at low cooling rate (20 K/ns), but none at high cooling rate (30 K/ns). Smaller NPs, 4 nm in diameter, at low cooling rates presented a lower crystallisation probability: three out of fourteen cases led to crystalline NPs. Figure 16.12a contains results of the three analysis methods mentioned above for all the 4 nm cases where crystallisation was observed (cases 2, 3, and 4) and one typical case of a non-crystallised NP (case 1).

Whenever crystallisation occurred, the phase transition took place relatively quickly, within a few nanoseconds, at temperatures around 950–1000 K (note that the melting point for 4 nm Si NPs was calculated at 1350 K, far above these transition temperatures). A sharp increase in the temperature curves (topmost graph at Fig. 16.12a) at around 65 ns for cases 2–4 indicates a clear latent heat release associated with crystallisation. This is consistent with the structure curve (middle graph of Fig. 16.12a), which displays a simultaneous sharp decrease in the fraction of Si atoms in a disordered environment.

Remarkably, there is a clear difference in sharpness between case 4 (where the transition happens earlier and more abruptly) and cases 2 and 3 (where the transition happens a little later and more smoothly). This is reflected in the final structures of the NPs, shown in Fig. 16.12b; it is evident that the NP of case 4 is more ordered than those of cases 2 and 3, whereas the NP of case 1 remained amorphous. Clearly, larger amounts of released latent heat lead to more complete transition (in case 4, the temperature rise was the highest observed one, from around 1000 K to 1300 K). The NP structures of cases 2 and 3 have several nanograins growing along different crystallographic directions. Topological analysis (bottom graph, Fig. 16.12a) shows that multiple 6-membered rings (such as the one in the inset) are formed simultaneously during the transition, acting as crystallisation seeds and leading to the resultant polycrystalline morphology. This is easier to illustrate in larger NPs; thus, Fig. 16.12c shows at least three individual crystallites growing separately upon crystallisation of a 9 nm NP.

The aforementioned dependence of crystallisation probability on NP size can, once more, translate into a dependence on cooling and cooling rate. The association is twofold. First, large NPs are more likely to start cooling down from a higher initial temperature, since their size is directly linked to the number of atoms they absorbed, with the ensuing heating. Moreover, it can be calculated that larger-sized NPs cool down more slowly than smaller ones during inert gas condensation [31]. Since crystallisation is a stochastic process, higher initial temperatures and slower cooling provide longer times (and therefore more possibilities) for the phase transition to take place, as the NPs prolong their stay above a certain crystallisation temperature associated with their size (e.g. 1250 K for 4 nm NPs). This agrees not only with the MD results, which showed a higher crystallisation probability for 9 nm NPs compared to 4 nm ones, but also with experimental findings, where crystallisation was observed for the Si-3 sample alone.

From the discussion above, it appears that within the framework of inert gas condensation there is a definitive correlation between NP size and probability for crystallisation. However, this raises an important issue: suppose one wants to fabricate small size crystalline Si NPs for specific applications. Is there a minimum NP size below which quenching while still at a disordered state cannot be prevented? Is there any alternative method that enables the growth of crystalline Si NPs beyond the limitations of inert gas temperature control?

The answer to these questions is the subject of the next section.

16.6 Case Study (ii): Heterogeneous Crystallisation of Si NPS Using Ag Atoms

At the beginning of our discussion, we resorted to previous experience in crystallisation of bulk materials to propose a key parameter that governs the crystallinity of Si NPs, and we suggested that, much like their bulk counterparts, phase transitions of NPs are primarily dictated by their temperature evolution. In doing so, we implicitly assumed homogeneous crystallisation; after all, the end product was supposed to be pure Si NPs. However, as is also well known from the macroscopic world, heterogeneous crystal nucleation is much easier and takes place more quickly, without the need for significant undercooling. The reason is that when a foreign substance acts as a scaffold for a crystal to grow on, it eliminates the necessity of creating an altogether new surface with the corresponding surface energy requirements. In an analogous fashion, foreign NPs can, in principle, provide nucleation sites and facilitate crystallisation of Si NPs of relatively smaller sizes. Subsequent treatment can ensure the selective removal of the foreign NPs, finally retaining only the intact Si NPs of desired crystallinity. To allow for feasible segregation, immiscible systems (at least in bulk) appear to be the most promising candidates. The Ag-Si metal-semiconductor binary system, with extensive differences in affinity, mass and size between the atoms, is such a highly immiscible system; obtaining crystalline Si NPs via magnetron sputtering of Ag and Si will be the focus of the following sections.

16.6.1 Simultaneous Co-Sputtering of Si and Ag Nanoclusters

While introducing the magnetron sputtering method, it was mentioned that one of its most desirable characteristics was the ability to sputter alloys, in order to expand NP composition from single-component to hybrid multi-component heterostructures [12, 29]. However, utilising alloy sputter targets suffers from certain disadvantages. The most important is reduced versatility in compositional ratios; alloy targets of set compositions are only due to produce NPs of set compositions, too, partly offsetting the flexibility of the method. Unworkable single target preparation due to bulk immiscibility in the solid state can be an additional impediment. The Ag-Si system is an example of an immiscible system not apt for alloy single-target sputtering.

However, recent advances in magnetron sputtering enable the co-deposition of NPs fabricated via gas-aggregation from two neighbouring single-element targets. Pure Si and Ag sputtering targets located side by side on an integrated magnetron sputtering head are shown in Fig. 16.13. The two magnetron guns are completely independent; different powers can be applied to the two targets, enabling one to engineer NPs with a tailored composition. This is a big advantage, considering that the two materials have very different sputtering yields; magnetron powers sufficient to sputter Ag efficiently may have practically no effect on Si. In other words, independence of magnetron guns is not just a desirable feature; it is an imperative prerequisite for the method's efficiency.

Owing to the plasma zone for each species being distinct from one another and to the low affinity between the two atom types,

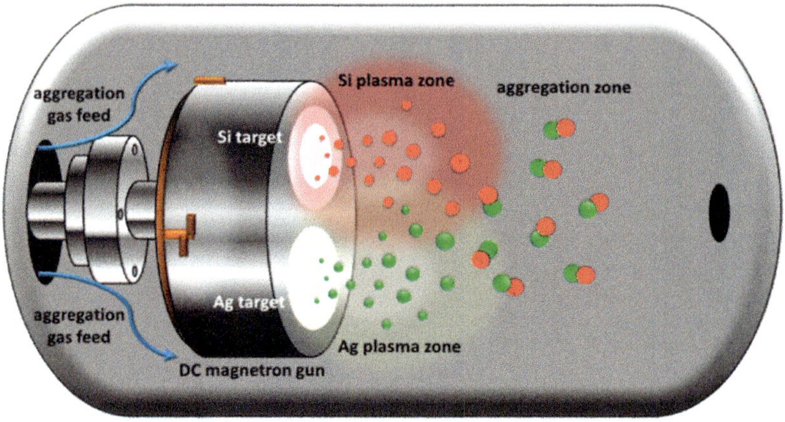

Figure 16.13 Schematic representation of simultaneous but independent co-sputtering of adjacent Si and Ag targets. Only later on in the aggregation process, do mixed NPs form.

Si and Ag nascent nanoclusters grow, in principle, separately. Only later on in the process, when the NPs mix in the cool aggregation zone, can hybrid NPs potentially form. MD simulations performed using the same scheme described above for a pure Si system put this assumption to test. A number of different configurations containing several percentages of the two sputtered materials were investigated, for various inert gas concentrations, all yielding similar results [30]. For a clear, coherent demonstration of the synthesis process, a simulation run containing 500 atoms of each species is chosen as an example case, since it features all the significant events of the aggregation mechanism in just one sequence of events. Temperature evolution of each species, along with visualisation snapshots are shown in Fig. 16.14.

The mechanism is very reminiscent of that for pure Si (Fig. 16.5). Starting with a super-saturated hot vapour of Si and Ag atoms (snapshot (a)), an initial cooling stage for both atom types reduced their temperature to a minimum value before condensation commenced (snapshot (b)). As shown in Section 16.4.1, the cooling rates rely on the mass difference between inert gas and sputtered species atoms. This means that Si should, in principle, lose heat faster. Nevertheless, the large number of Ar atoms and the resulting

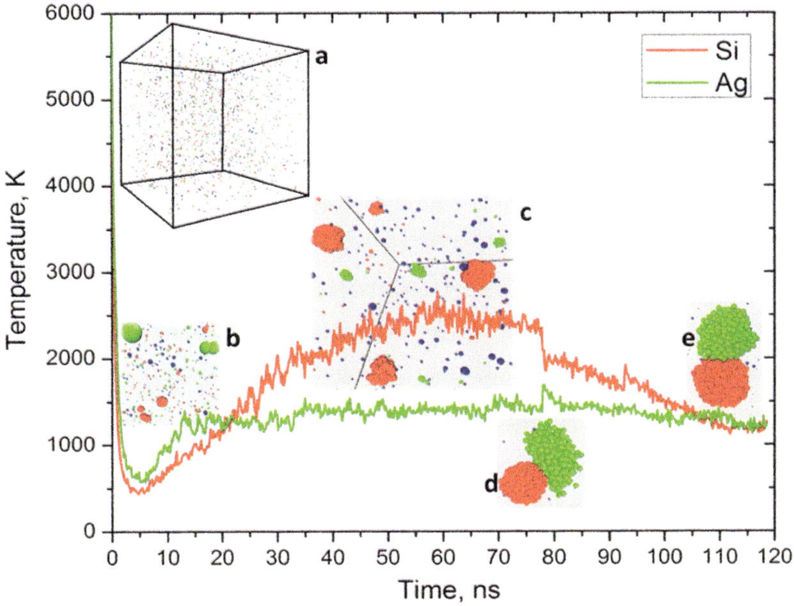

Figure 16.14 Time development of temperature during MD computer simulation of simultaneous aggregation of Si and Ag. Insets show snapshots at points of interest: (a) Initial configuration. (b) Beginning of aggregation by formation of dimers of both species. (c) Separate formation of primary particles. (d) Formation of unstable Janus cluster by partial coalescence of particles of different species. These clusters separated again after approximately 4 ns. (e) Upon subsequent collisions, the particles are sufficiently cool to form a stable Janus structure. Adapted with permission from Ref. [30]. Copyright 2014, American Chemical Society.

large number of collision events made this difference insignificant, and both species cooled down at almost the same rate. Interestingly, despite the independent condensation processes for the two species, their nucleation begun almost simultaneously. This is due to the equal numbers of Si and Ag atoms present in the box, resulting in equal probability of bond-forming collisions, once the necessary low temperature had been reached.

The two species initially nucleated and grew individually; no mixed-species dimers, trimers or small primary particles were formed. Subsequent growth was smooth, with nascent nanoclusters of both types absorbing atoms of the same type alone. The evaporation temperature of Si is higher than that of Ag; this is why its temperature soon grew higher. Several spikes in the Si temperature correspond to events where Si primary particles collided and fully coalesced, forming increasingly larger nanoclusters. Similar spikes are rarer and less pronounced in Ag, because its boiling point is much lower than that of Si (hence the temperature plateau), and condensation into stable clusters requires a longer cooling time. As the number of free Si atoms in the simulation box decreased, Si growth rate decreased, too, until it reached a maximum temperature at around 60 ns (snapshot (c)).

With a substantial number of primary particles of both types present in the box, violent collisions between NPs of all possible combinations were due to occur. While such collisions did happen, analysis of the MD simulations demonstrated that only certain cases resulted in stable conjoined structures, while in most cases the dissimilar clusters rapidly decoupled to isolated single-species clusters again. Such an incident is captured in snapshot (d); note the sudden energy exchange, with the Si temperature dropping by around 500 K and that of Ag rising by around 300 K. The two clusters partially coalesced but did not stay together for long, due to their vast temperature difference prior to the collision leading the Ag to heat up above its boiling point. In contrast, later on, at around 110 ns, the collision captured by snapshot (e) led to a stable Ag-Si hybrid NP. This time, both clusters shared the same temperature before the collision and no boiling point was exceeded. As expected, no intermixing of atoms was observed; the system remained fully segregated.

The segregated nature of Ag-Si NPs was also confirmed experimentally [30]. Pure Ag and Si targets were simultaneously co-sputtered and hybrid NPs synthesised from the gas-phase using the setup of Fig. 16.13. To obtain results comparable to those for Si NPs alone, all experimental parameters were identical to those of Section 16.5. The Ar/He gas flow ratio was kept at Ar: 70 sccm and He: 5 sccm, closely resembling the conditions of sample Si-2. The size of the NPs as well as the volume fraction of each component can be controlled by tuning the magnetron discharge power applied on each target independently while co-sputtering. We opted to hold the Si-target power fixed at 90 W (just like all samples of Section 16.5);

the tuning of the particle composition was achieved by varying the power of the Ag-target from a minimum of 30 W to a maximum of 60 W.

Depending on the ratio of magnetron power values, two types of structures were attained; namely core-satellite and Janus hybrid NPs, for low and high Ag discharge power, respectively. A core-satellite NP refers to a configuration where a number of small clusters of one element reside on the surface of a large core NP of the other element [46]. Janus NPs are, in principle, similar; only in this case the hybrid NP has a roughly spherical external morphology consisting of approximately equal-size hemispherical volumes of each constituent element [47]. It is obvious that Janus structures are a result of initially higher Ag-atom number densities, leading to larger Ag primary particles. Representative bright field HRTEM and high angle annular dark field scanning TEM (HAADF-STEM) images of both structures are shown in Fig. 16.15. In the former, the crystalline Ag domains appear dark, resulting from diffraction of the electron beam by the ordered lattice, whereas in the latter, the Ag regions appear bright due to the higher atomic number of Ag resulting in a stronger detected signal compared to Si. The core-satellite structure (Fig. 16.15a,b) corresponds to a 40 W Ag-target power and the Janus structure (Fig. 16.15c,d) corresponds to a 50 W Ag-target power.

Most importantly, though, high-resolution imaging revealed that, despite Ag domains being consistently crystalline, Si domains remained consistently amorphous, both for the core-satellite and the Janus structure, just like the Si-2 sample earlier on. In other words, utilising the co-sputtering method under conditions which, in the pure Si case, led to amorphous Si NPs, failed to enhance their crystallisation probability. The most probable explanation lies in the condensation mechanism: since the Si and Ag primary particles grew separately, each type obtained the phase it would obtain anyway, in the absence of the other type. When mixed NPs were eventually formed via collisions at room temperature in the aggregation zone, the heat and stress released during coalescence were not enough to trigger crystallisation of the Si NPs.

It is evident that co-sputtering Ag and Si simultaneously does not solve the problem. Clearly, another method has to be introduced.

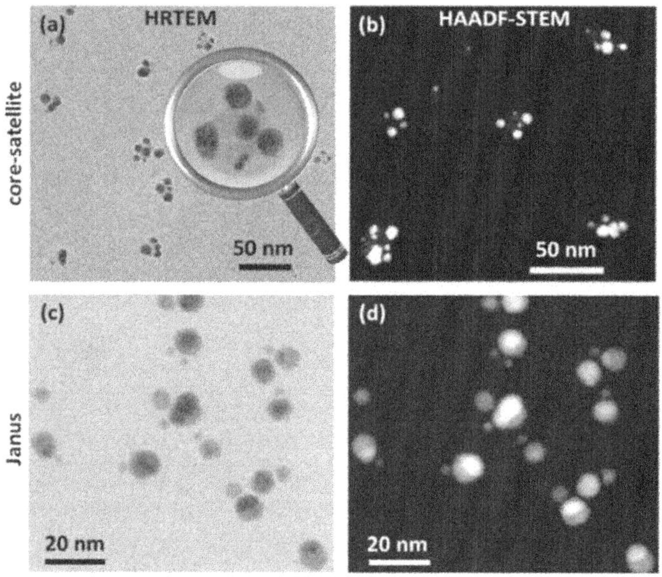

Figure 16.15 Transmission electron micrographs of hybrid Si-Ag NPs (a, c), bright field HRTEM and (b, d) HAADF-STEM images of core-satellite and Janus structures. Inset of figure (a) zooms in on one hybrid NP, to highlight its core-multiple satellite structure. Adapted with permission from Ref. [30]. Copyright 2014, American Chemical Society.

16.6.2 Si NP Crystallisation via Inoculation with Ag Atoms

Sequential sputtering of Si and Ag can be an alternative to the simultaneous co-sputtering presented in the previous section. According to this method, pure Si NPs were synthesised using the experimental setup of Fig. 16.1, under the exact conditions of depositions described in Section 16.6.1. Without further intervention, uncoated Si clusters would ultimately cool and vitrify into solid, amorphous Si NPs, just like sample Si-2, above. This assumption was indeed verified by control depositions. Instead, an incident flux of Ag atoms was applied to the Si clusters post-growth, coating them *after* they exited the aggregation zone. Ag was sputtered from a linear magnetron sputter target in a shell-coater apparatus (Fig. 16.16) positioned parallel to the flight direction of the Si NPs, en route from the aggregation zone chamber to the sample deposition chamber. Ag

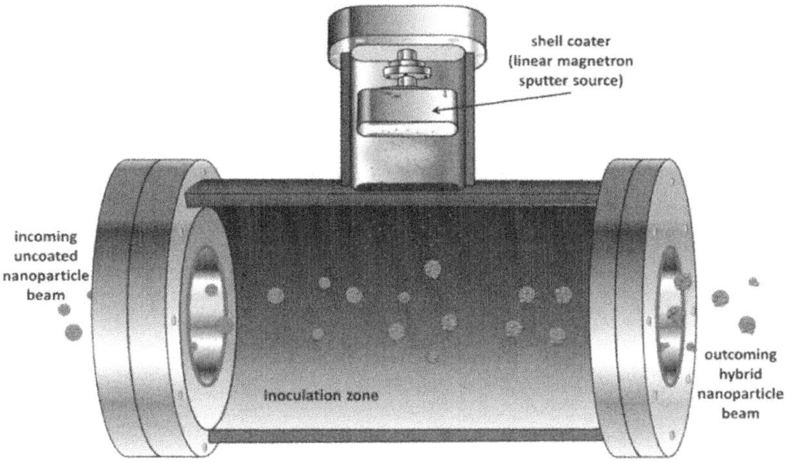

Figure 16.16 Schematic representation of shell coater. The Si NPs are inoculated with Ag atoms, forming hybrid NPs.

atoms were sputtered with an Ar gas flow of 10 sccm, whilst the DC magnetron power was varied over the range 60–120 W, to explore the impact of Ag flux on the Si NPs upon coating [28].

Just like the case of Ag-Si co-sputtering, the resultant structure was of a Si-core/Ag-satellite (or part-shell) form. However, there was a significant difference: in this case, it was not only the Ag satellites, but also the Si core that displayed crystallinity. Figure 16.17 shows bright field HRTEM images and the corresponding FFT patterns of two example NPs, an amorphous one obtained from a control sample without Ag sputtering (a-b) and a crystalline one obtained from a sample with Ag sputtering (c-d), thus confirming the phase transition achieved through inoculation of Si NPs with Ag. In the case of Fig. 16.7c, the Si NP appears to be mono-crystalline; simultaneously, only a single Ag satellite can be observed on its surface.

Remarkably, as the number of Ag satellites on the surface increases, the number of crystalline domains in the underlying Si core increases accordingly. Figure 16.18a shows a high-resolution HAADF-STEM micrograph of another exemplary Si NP, this time decorated by four Ag clusters (designated by numerals) and encapsulated with a ~2 nm oxide shell. The bright field HRTEM

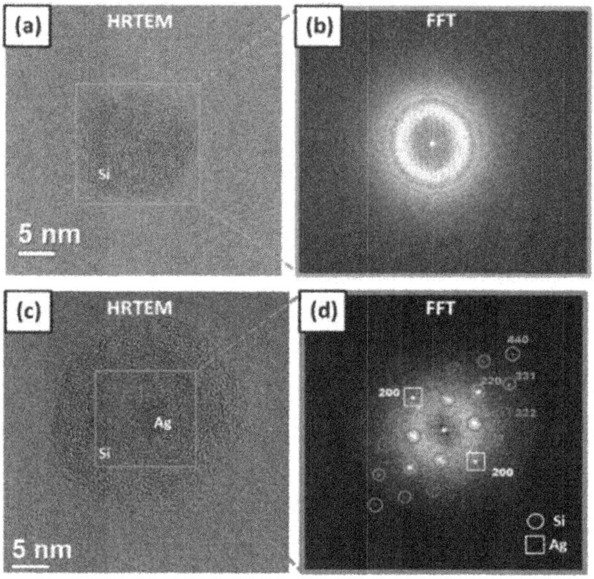

Figure 16.17 Bright field HRTEM images and the corresponding FFT patterns of two example NPs, an amorphous one obtained from a control sample without Ag sputtering (a–b) and a crystalline one obtained from a sample with Ag sputtering (c–d). A phase transition, achieved through inoculation of Si NPs with Ag, is evident. Adapted with permission from Ref. [28]. Copyright 2013, Nature Publishing Group.

image of the same NP, taken with unchanged orientation with respect to the electron beam (Fig. 16.18b), indicates the presence of three domains in the Si core of different crystallographic orientations (indicated by the arrows), separated by multiple twin boundaries [48, 49]. However, by tilting the sample by -15° relative to the incident electron beam (Fig. 16.18c) a fourth grain was confirmed, matching the number of Ag satellites on the surface of the NP.

This correlation was confirmed, as close inspection of numerous other NPs revealed a similar trend. This means that, in principle, it is possible not only to fabricate Si NPs with a high probability of exhibiting crystalline structure, but also to control (within limits, obviously) the number of crystallographic domains by controlling the number of Ag nanoclusters decorating the surface. To this

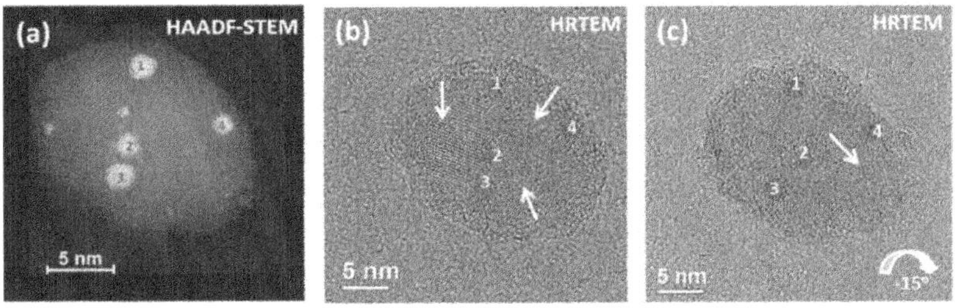

Figure 16.18 (a) HAADF-STEM micrograph showing the microstructure of a Ag-coated Si NP, decorated with several Ag nanoclusters, marked by numerals (1–4). (b) Corresponding HRTEM image of the same NP with unchanged orientation with respect to the electron beam. The Ag clusters numbered 1–4 in the HAADF mode image appear as dark spots. Three crystal grains are indicated by arrows. (c) A fourth crystal grain is confirmed by tilting the sample by −15° relative to the incident electron beam. Qualitative correlation between the number of Ag clusters and the number of grains in the Si core is evident. Adapted with permission from Ref. [28]. Copyright 2013, Nature Publishing Group.

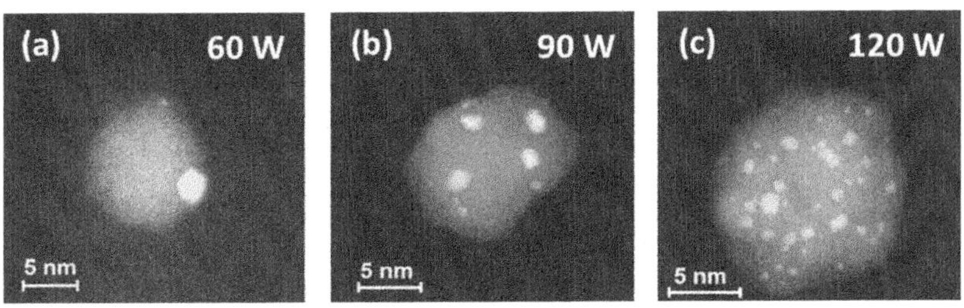

Figure 16.19 HAADF-STEM micrographs demonstrating control over the number of Ag nanoclusters decorating the surface of the Si NPs. Adapted with permission from Ref. [28]. Copyright 2013, Nature Publishing Group.

end, one merely needs to adjust the discharge power supplied to the Ag magnetron sputter source. Figure 16.19 presents typical NPs obtained with increasing power (60, 90, and 120 W for a, b, and c, respectively). Clearly, increasing power leads to higher NP coverage and to a shift from a core-satellite to a core-partial shell configuration, with the resultant change from a monocrystalline to a polycrystalline structure.

So, what is it that differentiates this method from the co-sputtering one described in the previous section, if mixing of Si and Ag happens after the plasma zone in both cases? The answer to this question is reasonably straight forward; the difference lies in the very low Ar gas flow of 10 sccm at the shell-coater. Under such low carrier gas pressure, Ag vapour cannot coalesce into nanoclusters in free space. Unlike the co-sputtering method, where nanoclusters of both species formed prior to their mixing, in the sequential sputtering method, Ag landed on the Si NPs as individual atoms. Therefore, the Ag nanoclusters observed in Figs. 16.17–16.19 actually formed *after* landing on the surface of the Si NPs, not before. This fact was verified experimentally by sputtering the Ag target under the same conditions without a Si NP flow; no Ag nanoclusters were observed on substrates in the sample deposition chamber. Most importantly, though, it leads to a completely different interaction mechanism between Si and Ag; one that can lead to Si NP crystallisation.

In order to elucidate this subtle mechanism, MD computer simulations were employed once more. Various configurations were examined. A spherical Si NP (of variable size) was introduced in the middle of a cubic box (of variable size, too, to understand the effect of pressure), along with a vapour of Ar and Ag atoms (of variable concentrations, naturally). The initial temperature was set at 3000 K, since the Ag atoms were very energetic after sputtering. Inert gas cooling was used when Ar atoms were present in the box, whereas radiative cooling (at various cooling rates) during flight was utilised for the Si NP when no Ar atoms were introduced [28, 50]. Both cooling methods eventually brought the system to equilibrium at room temperature.

Figure 16.20 shows the time evolution of temperature and fraction of disordered Si atoms, along with MD visualisation snapshots of an exemplary simulation of a Si NP (containing 1000 atoms) and Ag vapour of 243 atoms, initially isolated from each other, under radiative cooling of 20 K/ns. The inset to this figure depicts the variation in the environment of Ag, simultaneously showing the fraction of Ag atoms in vacuum, or in Ag-rich (Ag>Si), Si-rich (Ag<Si), or pure Ag surroundings. The mechanism is divided

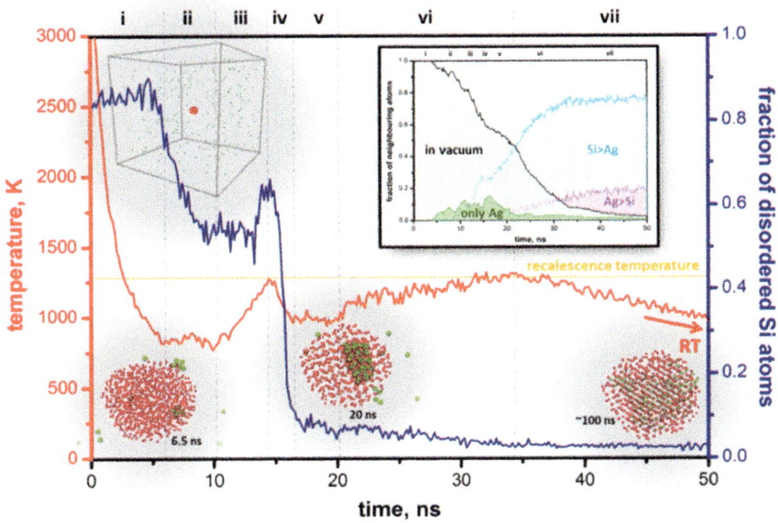

Figure 16.20 Time development of temperature and fraction of disordered atoms of a Si NP of 1000 atoms, in the centre of a MD cell surrounded by a randomly dispersed Ag gas containing 243 atoms. Snapshots depict characteristic instants of MD simulation run. The process is decomposed in stages, denoted by Latin numerals. Inset shows the time evolution of the local bonding environment of Ag atoms. Adapted with permission from Ref. [28]. Copyright 2013, Nature Publishing Group.

into seven stages, indicated by the Latin numerals at the top of the figure.

During the initial stage (stage (i)) the Si NP is quenched to around 800 K. In doing so, it remains in the liquid phase, as evident from its fraction of disordered atoms, which shows no phase transition. Therefore, at 800 K, the Si NP is seriously undercooled, since its melting temperature was calculated at around 1300–1400 K. Limited Ag nucleation also occurs at around 4 ns (see "only Ag" plot in inset figure).

Eventually, solidification commences (stage (ii)). Thermal arrest can be observed between 6.5 and 10 ns, associated with a sharp decrease in the fraction of disordered Si atoms, from around 85% to 55%. Therefore, limited crystallisation occurs upon *homogeneous* solidification. During this stage some Ag atoms end up on the

NP, forming Ag precipitates on the surface (see "Si>Ag" plot in inset figure). According to the general screening model [51], the covalent bonds in amorphous Si may become weaker at a metal-semiconductor interface, allowing for relatively high mobility of the interface atoms, which could assist crystallisation of amorphous Si at relatively low temperature.

Such is the undercooling of the Si NP, that recalescence occurring after the thermal arrest (stage (iii)) raises the temperature up to 1275 K. This value is very near the calculated melting temperature of the pure Si NP, especially considering surface pre-melting, and is the melting temperature of the hybrid NP, since Ag has a lower melting point than Si. Indeed, towards the end of this stage, some re-melting happens, signified by the sudden increase in the fraction of disordered Si atoms. As a result, Ag solubility in the Si NP increases, since Si's capacity to retain metal dopants in its matrix increases with temperature [52]. This allows the Ag precipitates on the surface to diffuse inside the Si NP, as indicated by the sharp increase in the fraction of Ag atoms in a Si-rich environment.

Consequently, after around 14 ns, further crystallisation ensues (stage (iv)). This time, however, unlike stage (ii), it is complete, *heterogeneous* crystallisation that occurs. Due to the Ag precipitates acting as an initiation seed, the fraction of Si disordered atoms drops almost instantaneously to less than 10%. If the NP consisted of a single, homogeneous phase, a second thermal arrest would be expected during stage (iv), until full crystallisation was reached. However, the temperature decreases instead; the heat generated by the crystallisation of the Si-rich phase is absorbed by the Ag precipitates, which remain liquid still.

Next, a phase transition takes place in the Ag precipitates. Assisted by the energy absorbed, Ag atoms and their surrounding Si atoms restructure themselves, forming a single, locally super-saturated, Ag-rich metastable silicide phase. This is verified by the simultaneous rise of both curves showing a mixed Si-Ag phase ("Si>Ag" and "Ag>Si") and the decline of the curve indicating pure Ag precipitation ("only Ag"). Stages (v) and (vi) are similar to (ii) and (iii), showing thermal arrest and recalescence, respectively, except now they are governed by the Ag-rich phase alone. During recalescence, the fraction of Si disordered atoms is further reduced to almost zero, since now even the Si atoms in the Ag-rich phase crystallise.

Once recalescence temperature (1275 K) is reached once more after 34 ns, the NP temperature linearly drops towards room temperature (stage (vii)). By the beginning of this stage, the formation of the silicide is completed, and its structure remains stable, as indicated by the constant values of all the graphs in the inset. However, since Ag solubility in the Si matrix gradually decreases with declining temperature, Ag is expected to segregate as a satellite onto the Si NP's surface at times exceeding the simulation time.

Due to the Ag surface precipitates connecting to form a single Ag-rich phase before crystallisation, the mechanism described above explains the formation of a single satellite-single domain configuration. In a similar fashion, structures containing more satellites (and corresponding crystal grains) could emerge, especially in larger NPs. Computational limitations make the simulations of such systems impractical; however, this analysis provides valuable insights. It should be stressed, nonetheless, that the process is highly stochastic and that very sensitive kinetics and thermodynamics are involved; therefore, other potential interactions are by no means ruled out, either as alternative or as complementary mechanisms.

So far, we showed that it is possible to obtain multicomponent heterostructured Ag-Si NPs with controllable crystallinity. Naturally, one may retain the hybrid structure if it is advantageous to utilise combined functionalities of both the semiconductor core NP and the metallic clusters decorating the surface. For example, Si is photoluminescent while Ag is plasmonic; in cell biology, NPs with a combination of such properties are attractive for multi-modal bioimaging, biosensors and 3D optical sectioning [53–55]. However, it is also possible to remove the Ag clusters selectively via plasma etching the sample, leaving the Si core of desired crystallinity intact. For example, a HAADF-STEM and a bright field HRTEM image of a NP after plasma etching is shown in Fig. 16.21, where it is evident that the Ag satellites were either removed altogether, or their content was drastically reduced to negligible values.

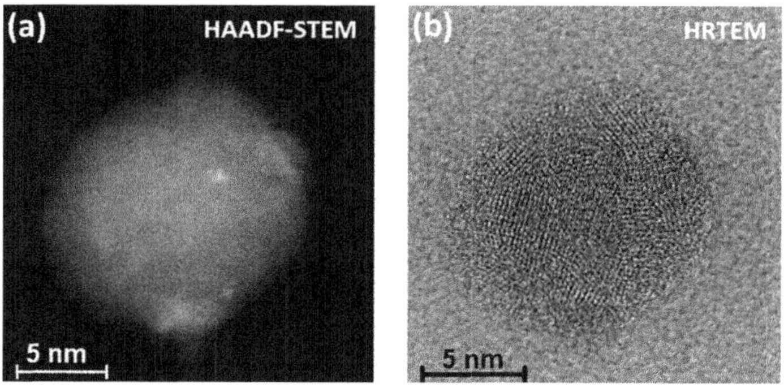

Figure 16.21 Removal of Ag nanoclusters by plasma etching: (a) HAADF-STEM and (b) bright field HRTEM images of a hybrid NP after plasma etching process. Ag nanoclusters on the surface were removed or drastically reduced, and the nanocrystalline Si core was left intact. Adapted with permission from Ref. [28]. Copyright 2013, Nature Publishing Group.

16.7 Summary

Magnetron sputtering inert gas aggregation is a powerful and flexible method to design and synthesise Si NPs with good size and crystallinity control. The cooling rate of the NPs during their growth is the dominant factor determining their structure. Inoculation with foreign atoms, such as Ag, enables further control, where synthesis of Si NPs alone fails; later on, pure Si NPs can be segregated and harvested by techniques such as plasma etching, leaving their structure intact. The ability to explicitly and straightforwardly tune the atomic order from amorphous to polycrystalline to single crystalline offers tenability of the dependent optical, electrical and chemical properties, as well as a direct insight into the fundamental physical and chemical processes at work inside such NP systems.

Acknowledgements

The authors would gratefully like to thank their colleagues in Okinawa and in Helsinki for their work in the journal articles this chapter is based on. Special thanks go to Ms Evropi Toulkeridou, Dr Jerome Vernieres, Dr Vidyadhar Singh, Mr Junlei Zhao and Mr Sakis

Varkas for their extra help and feedback during the preparation of this manuscript. This work was supported by funding from the Okinawa Institute of Science and Technology Graduate University. We are also grateful to the Finnish IT Centre for Science CSC and the Finnish Grid Infrastructure (FGI) for grants of computation time.

References

1. Poole CP, Jr, Owens FJ (2003) *Introduction to Nanotechnology*, John Wiley Sons Inc, New Jersey.
2. Palmer RE, Pratontep S, Boyen H-G (2003) Nanostructured surfaces from size-selected clusters, *Nature Mater*, **2**, 443–448.
3. Fan J, Chu PK (2010) Group IV nanoparticles: synthesis, properties, and biological applications, *Small*, **6**(19), 2080–2098.
4. Smyder JA, Krauss TD (2011) Coming attractions for semiconductor quantum dots, *Mater Today*, **14**(9), 382–387.
5. Baldwin RK, Pettigrew KA, Ratai E, Augustine MP, Kauzlarich SM (2002) Solution reduction synthesis of surface stabilized silicon nanoparticles, *Chem Commun*, 1822–1823.
6. Neiner D, Chiu HW, Kauzlarich SM (2006) Low-temperature solution route to macroscopic amounts of hydrogen terminated silicon nanoparticles, *J Am Chem Soc*, **128**(34), 11016–11017.
7. Lin SW, Chen DH (2009) Synthesis of water-soluble blue photoluminescent silicon nanocrystals with oxide surface passivation, *Small*, **5**(1), 72–76.
8. Schropp REI, Carius R, Beaucarne G (2007) Amorphous silicon, microcrystalline silicon, and thin-film polycrystalline silicon solar cells, *MRS Bulletin*, **32**(3), 219–224.
9. Choi JH, Cheon JH, Kim SK, Jang J (2005) Giant-grain silicon (GGS) and its application to stable thin-film transistor, *Displays*, **26**(3), 137–142.
10. Mahan AH, Ahrenkiel SP, Schropp REI, Li H, Ginley DS (2008) A comparison of grain nucleation and grain growth during crystallization of HWCVD and PECVD a-Si:H films, *Thin Solid Films*, **516**(5), 529–532.
11. Bromann K, Felix C, Brune H, et al. (1996) Controlled deposition of size-selected silver nanoclusters, *Science*, **274**(5289), 956–958.

12. Yin F, Wang ZW, Palmer RE (2011) Controlled formation of mass-selected Cu-Au core-shell cluster beams, *J Am Chem Soc*, **133**(27), 10325–10327.
13. Benelmekki M (2015) *Designing Hybrid Nanoparticles*, Morgan & Claypool Publishers, San Rafael, CA.
14. Haberland H, Mall M, Moseler M, et al. (1994) Filling of micron-sized contact holes with copper by energetic cluster impact, *J Vac Sci Technol A*, **12**(5), 2925–2930.
15. Kashtanov PV, Smirnov BM, Hippler R (2007) Magnetron plasma and nanotechnology, *Phys-Usp*, **50**(5), 455–488.
16. Smirnov BM (2011) Processes involving clusters and small particles in a buffer gas, *Phys-Usp*, **54**(7), 691–721.
17. Imanaka M, Katayama T, Ohshiro Y, Watanabe S, Arai H, Nakagawa T (2004) Nanocluster ion source by plasma-gas aggregation, *Rev Sci Instrum*, **75**(5), 1907–1909.
18. Iles GN, Baker SH, Thornton SC, Binns C, (2009) Enhanced capability in a gas aggregation source for magnetic nanoparticles, *J Appl Phys*, **105**, 024306.
19. Momin T, Bhowmick A (2010) A new magnetron based gas aggregation source of metal nanoclusters coupled to a double time-of-flight mass spectrometer system, *Rev Sci Instrum*, **81**, 075110.
20. Stoyanov S, Huang Y, Zhang Y, Skumryev V, Hadjipanayis GC, Weller D (2003) Fabrication of ordered FePt nanoparticles with a cluster gun, *J Appl Phys*, **93**(10), 7190–7192.
21. Kennedy MK, Kruis FE, Fissan H, Mehta BR, Stappert S, Dumpich G (2003) Tailored nanoparticle films from monosized tin oxide nanocrystals: particle synthesis, film formation, and size-dependent gas-sensing properties, *J Appl Phys*, **93**(1), 551–560.
22. Goldby IM, Issendorff BV, Kuipers L, Palmer RE (1997) Gas condensation source for production and deposition of size-selected metal clusters, *Rev Sci Instrum*, **68**(9), 3327–3334.
23. Majumdar A, Köpp D, Ganeva M, Datta D, Bhattacharyya S, Hippler R (2009) Development of metal nanocluster ion source based on dc magnetron plasma sputtering at room temperature, *Rev Sci Instrum*, **80**(9), 095103.
24. Xirouchaki C, Palmer RE (2004) Deposition of size-selected metal clusters generated by magnetron sputtering and gas condensation: a progress review, *Philos Trans R Soc A*, **362**, 117–124.

25. Bansmann J, Baker SH, Binns C, et al. (2005) Magnetic and structural properties of isolated and assembled clusters, *Surf Sci Rep*, **56**, 189–275.
26. Wegner K, Piseri P, Vahedi-Tafreshi H, Milani P (2006) Cluster beam deposition: a tool for nanoscale science and technology, *J Phys D: Appl Phys*, **39**, R439–R459.
27. Binns C, Trohidou KN, Bansmann J, et al. (2005) The behaviour of nanostructured magnetic materials produced by depositing gas-phase nanoparticles, *J Phys D: Appl Phys*, **38**, R357–R379.
28. Cassidy C, Singh V, Grammatikopoulos P, Djurabekova F, Nordlund K, Sowwan M (2013) Inoculation of silicon nanoparticles with silver atoms, *Sci Rep*, **3**, 3083.
29. Xu Y-H, Wang J-P (2008) Direct gas-phase synthesis of heterostructured nanoparticles through phase separation and surface segregation, *Adv Mater*, **20**, 994–999.
30. Singh V, Cassidy C, Grammatikopoulos P, Djurabekova F, Nordlund K, Sowwan M (2014) Heterogeneous gas-phase synthesis and molecular dynamics modeling of Janus and core-satellite Si-Ag nanoparticles, *J Phys Chem C*, **118**, 13869–13875.
31. Zhao J, Singh V, Grammatikopoulos P, Cassidy C, Aranishi K, Sowwan M, Nordlund K, Djurabekova F (2015) Crystallization of silicon nanoclusters with inert gas temperature control, *Phys Rev B*, **91**, 035419.
32. Elihn K, Landström L, Heszler P (2002) Emission spectroscopy of carbon-covered iron nanoparticles in different gas atmospheres, *Appl Surf Sci*, **186**, 573–577.
33. Haberland H, Karrais M, Mall M, Thurner Y (1992) Thin films from energetic cluster impact: A feasibility study, *J Vac Sci Technol A*, **10**(5), 3266–3271.
34. Nielsen RM, Murphy S, Strebel C, Johansson M, Chorkendorff I, Nielsen JH (2010) The morphology of mass selected ruthenium nanoparticles from a magnetron-sputter gas-aggregation source, *J Nanopart Res*, **12**, 1249–1262.
35. Krashnochtchekov P, Albe K, Averback RS (2003) Simulations of the inert gas condensation processes, *Int J Mater Res*, **94**, 1098–1105.
36. Pack RT, Walker RB, Kendrick BK (1998) Three-body collision contributions to recombination and collision-induced dissociation. I. Cross sections, *J Chem Phys*, **109**, 6701–6713.
37. Pack RT, Walker RB, Kendrick BK (1998) Three-body collision contributions to recombination and collision-induced dissociation. II. Kinetics, *J Chem Phys*, 109, 6701–6713.

38. Grammatikopoulos P, Toulkeridou E, Nordlund K, Sowwan M (2015) Simple analytical model of nanocluster coalescence for porous thin film design, *Modelling Simul Mater Sci Eng*, **23**, 015008.
39. Farid B, Godby RW (1991) Cohesive energies of crystals, *Phys Rev B*, **43**, 14248.
40. Kesälä E, Kuronen A, Nordlund K (2007) Molecular dynamics simulation of pressure dependence of cluster growth in inert gas condensation, *Phys Rev B*, **75**, 174121.
41. Westergren J, Grönbeck H, Rosén A, Nordholm S (1998) Statistical theory of cluster cooling in rare gas. I. Energy transfer analysis for palladium clusters in helium, *J Chem Phys*, **109**(22), 9848–9858.
42. Westergren J, Grönbeck H, Kim S-G, Tománek D (1997) Noble gas temperature control of metal clusters: a molecular dynamics study, *J Chem Phys*, **107**(8), 3071–3079.
43. Mei QS, Lu K (2007) Melting and superheating of crystalline solids: from bulk to nanocrystals, *Progr Mater Sci*, **52**, 1175–1262.
44. Nordlund K, Averback RS (1997) Point defect movement and annealing in collision cascades, *Phys Rev B*, **56**, 2421–2431.
45. Yuan X, Cormack A (2002) Efficient algorithm for primitive ring statistics in topological networks, *Comp Mater Sci*, **24**, 343–360.
46. Chen TH, Jean R-D, Chiu K-C, Chen C-H, Liu D-M (2012) Transparent SiO_2-Ag core-satellite nanoparticle assembled layer for plasmonic-based chemical sensors, *Appl Phys Lett*, **100**, 223101.
47. Hu J, Zhou S, Sun Y, Fang X, Wu L (2012) Fabrication, properties and applications of Janus particles, *Chem Soc Rev*, **41**, 4356–4378.
48. Hofmeister H, Tan GL, Dubiel M (2005) Shape and internal structure of silver nanoparticles embedded in glass, *J Mater Res*, **20**(6), 1551–1562.
49. Hofmeister H, Dutta J, Hofmann H (1996) Atomic structure of amorphous nanosized silicon powders upon thermal treatment, *Phys Rev B*, **54**(4), 2856–2862.
50. Zhao J (2013) *Molecular Dynamics Simulations of Silver-Induced Crystallization in Silicon Nanocluster*, Master's Thesis, University of Helsinki, Helsinki.
51. Hiraki, A (1980) A model on the mechanism of room-temperature interfacial intermixing reaction in various metal-semiconductor couples—what triggers the reaction, *J Electrochem Soc*, **127**(12), 2662–2665.

52. Hudelson S, Newman BK, Bernardis S, et al. (2010) Retrograde melting and internal liquid gettering in silicon, *Adv Mater*, **22**(35), 3948–3953.
53. Jin Y, Gao X (2009) Plasmonic fluorescent quantum dots, *Nat Nanotechnol*, **4**, 571–576.
54. Michalet X, Pinaud FF, Bentolila LA, et al. (2005) Quantum dots for live cells, in vivo imaging, and diagnostics, *Science*, **307**(5709), 538–544.
55. Erogbogbo F, Liu X, May JL, et al. (2013) Plasmonic gold and luminescent silicon nanoplatforms for multimode imaging of cancer cells, *Integr Biol*, **5**(1), 144–150.

Chapter 17

Scanning Thermal Microscopy

Séverine Gomès and Ali Assy
Université de Lyon, CNRS, INSA-Lyon, CETHIL, UMR5008, F-69621 Villeurbanne, France
Université Lyon 1, CETHIL, UMR5008, F-69622 Villeurbanne, France
severine.gomes@insa-lyon.fr

Scanning thermal microscopy (SThM) enables nanometer-scale heat flow measurements and materials' thermal characterization. This chapter gives and reviews first the principles and the main resistive scanning thermal microscopy techniques operating in active mode (heated probe) for thermal imaging with sub-microscale spatial resolution. After recalling the approaches currently used to calibrate SThM probes for thermal property measurement, it presents selected SThM applications.

17.1 Introduction

Scanning thermal microscopy (SThM) techniques use miniaturized thermal probes and are based on scanning probe microscopy (SPM) methods. Their spatial resolutions depend on the probe tip size and the characteristic lengths associated to the heat transfer between the tip and the sample to be characterized.

Nanostructured Semiconductors: Amorphization and Thermal Properties
Edited by Konstantinos Termentzidis

ISBN 978-981-4745-64-2 (Hardcover), 978-1-315-36445-2 (eBook)
www.panstanford.com

Contrary to far-field optical techniques, SThM is not limited in lateral resolution by optical diffraction at a few hundreds of nanometers: It can perform thermal imaging and measurements far beyond the micron-scale. SThM probes can indeed be tailored with tips of curvature radii in the range of few tens of nanometers.

Since the 1990s, SThM has been developed actively. Because of its high spatial resolution, it has found a wide range of applications to diverse areas such as microelectronics, optoelectronics, polymers, and carbon nanotubes.

The first section of this chapter reviews research on the design of SThM experiments for localized thermal property imaging and characterization. Among the various SThM methods nowadays proposed, resistive techniques are mainly used for thermal conductivity analysis. This chapter will then focus on them. The second section then presents the approaches currently used to model and calibrate SThM probes for the measurement of thermal conductivity or thermal conductance of materials. The third section gives some examples of selected applications of SThM with the Wollaston wire probe.

17.2 Experimental Setup

17.2.1 General Principle

Conventional SThM systems are nowadays mainly based on atomic force microscopy (AFM) systems. The usual AFM probe is replaced with a cantilever equipped with a thermosensitive element at the tip. In such configuration, surface thermal imaging in force constant mode, and thermal measurements as a function of the tip-sample force and distance, can be performed. Figure 17.1 describes the setup of an AFM-based SThM system. This comprises the following main components:

- the probe operating as thermal sensor
- a source of heat that can be external or internal to the sample and/or is incorporated into the probe itself when SThM operates in active mode

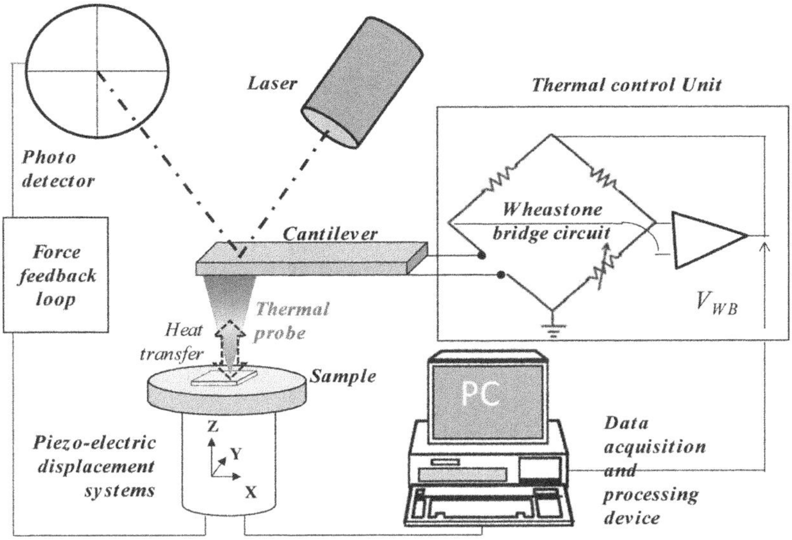

Figure 17.1 General layout of a SThM AFM-based system.

- a cantilever on which the probe is mounted and whose deflection senses the force that acts between the probe and the sample
- a piezo-ceramic system associated with an electronic circuit that servo-controls the probe-sample near-field interaction intensity
- a second piezo-ceramic system enabling the scanning of the sample surface by the probe

Measurement consists in determining the temperature or the temperature-dependent characteristics of the sample surface through the analysis of the heat exchanged between the probe and the sample. This may be performed when the probe and the sample are in contact or separated by a distance of a few nanometers to several μm.

Cantilever deflections are probed by reflecting a laser beam off a reflective part of the probe, such as the cantilever itself or a mirror appropriately glued on its back, towards a photodiode. The deflection generates an electrical signal that is detected. Usually real-time thermal signal measurement is performed with the help of a thermal control unit.

In the imaging mode, the deflection signal is used in a feedback control loop to maintain a constant tip-sample contact force while the tip scans laterally the sample surface. Piezoelectric scanners are used to move the sample vertically (Z direction in Fig. 17.1) and to scan the sample surface laterally (in X–Y directions). The combination of the X–Y scan position data, the force feedback signal and the thermal signal measured by the sensor located either at the tip or on the cantilever gives the raw data for both the topography image and the "thermal" image of the sample surface.

The force feedback control system operates simultaneously and independently of the process of the thermal measurement, enabling to obtain topography and thermal contrast images simultaneously. As an example, topography and thermal images and profiles obtained with a SThM probe used in "thermal conductivity contrast" mode (see Section 17.2.2.1) are given in Fig. 17.2. The contrast of the thermal image and profile reflects the change in the amount of heat locally exchanged between the tip and the sample while the tip scans the surface. This contrast clearly shows the lost in thermal

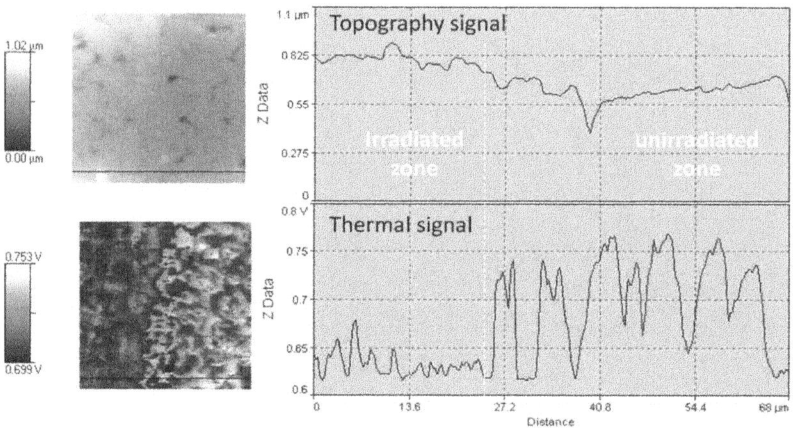

Figure 17.2 Topography (top) and thermal (bottom) images and profiles obtained at the level of an interface between two zones of a zirconium carbide (ZrC) sample, locally irradiated with 86 MeV krypton (Kr) ions at a 5×10^{15} ions.cm^{-2} dose.

conductivity of a ZrC sample, made amorphous due to a large individual point defects generated within the material subsurface by an irradiation with 86 MeV Kr ions [3]. Furtheremore, it highlights that many other parameters, such as surface topography and then roughness, also contribute to SThM image and measurement. It is then recommended to identify and analyze them before any interpretation of a SThM image.

17.2.2 Resistive SThM techniques

Various kinds of SThM probes, based on resistance thermometry, in particular metallic probes and doped silicon probes, have been implemented for SThM in active mode.

17.2.2.1 Metallic probes: description

Proposed in 1994 by Pylkki et al. [4], one of the first SThM metallic resistive probes was the Wollaston probe. The cantilever arms of this probe are made of a Wollaston wire that encloses a core of diameter 5 μm made of platinum/10% rhodium. As shown in Fig. 17.3a, the wire is etched by electrochemically removing the outer silver layer of the Wollaston wire over a length of 200 μm. The liberated core constitutes the thermal sensitive element. Its bending enables to form a probe of V shape. A mirror is stacked on the cantilever

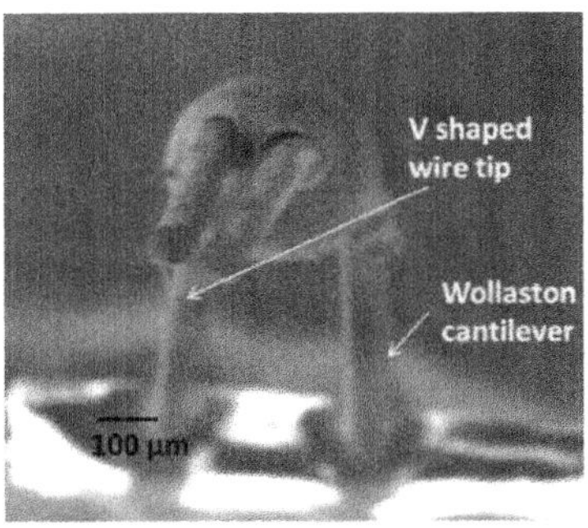

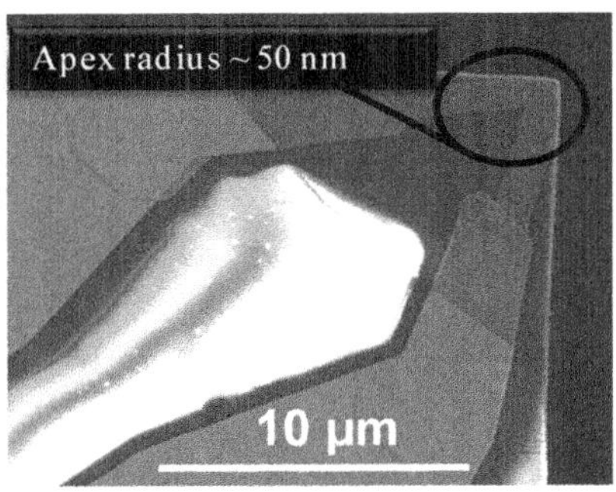

Figure 17.3 Scanning electron microscopy (SEM) images of SThM resistive probes: (a) Wollaston microprobe [2], (b) Palladium probe [1].

arms for the control of the cantilever deflection by optical method. Cantilever has a spring constant of 5 N/m.

This probe has a high endurance and a time response estimated to 200 μs in air [5]. The high temperature coefficient of platinum/10% rhodium is about 0.00166 K^{-1}. As a consequence, the Wollaston probe is attractive. It had been commercialized for more than 15 years, and demonstrated applications encompasses microsystem diagnostics [6–9] and local thermophysical characterization of various materials [10–20].

Smaller resistive metallic probes have been proposed to improve the spatial resolution of the technique. As an example, Fig. 17.3b shows a probe (KNT probe) that combines a thin Pd film resistor positioned at the very end of a flat tip at the end of a thin cantilever. The cantilever is made of silicon oxide SiO_2, or silicon nitride Si_3N_4 [21]. It has a low spring constant (0.35 N/m). The tip has a curvature radius of around 50 nm [22] and its electrical resistance has a temperature coefficient about 0.0012 K^{-1} [23]. Its time response has been estimated of a few tens of μs [24].

17.2.2.2 Metallic probe: operating modes

All the resistive metallic probes can operate in passive and active modes. The passive mode is used for thermometry. In this mode, a very small electrical current is passed through the probe. This results in minimal Joule self-heating of the tip and enables the measurement of its electrical resistance. During a scan, heat flows from the hot sample to the probe and changes the electrical resistance of the probe. Indeed at first order,

$$R_{\mathrm{el,p}}(T) = R_{\mathrm{el,p0}}(1 + \alpha(T - T_0)), \tag{17.1}$$

where $R_{\mathrm{el,p}}(T)$ is the electrical resistance of the probe thermosensitive element at a reference temperature T, $R_{\mathrm{el,p0}}$ is the electrical resistance of the element at temperature T_0, and α is the temperature coefficient of its electrical resistivity.

The active mode, which is mainly discussed in this chapter, is used to measure thermophysical properties of materials such as thermal conductivity. In this case, a larger electrical current I is passed through the probe, resulting in a significant Joule heating.

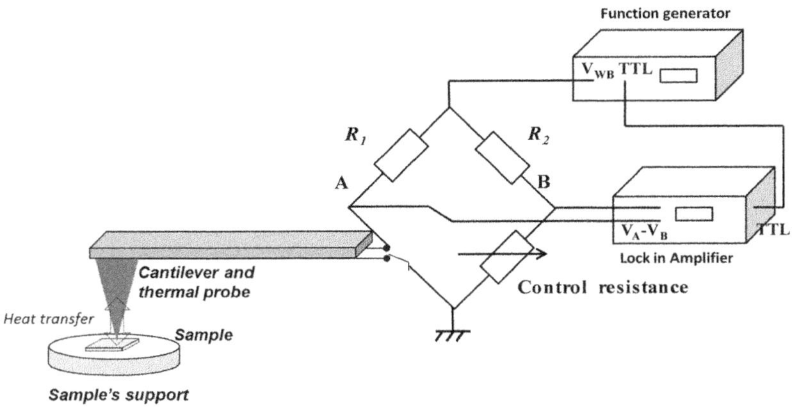

Figure 17.4 Schematic diagram of a typical setup operating in ac regime.

Part of the Joule power flows into the sample, depending on its thermal conductivity. The tip acts as a resistive thermometer and delivers heat to the sample simultaneously. The investigation of a sample by use of a very-localized heat source then enables to probe materials at the level of very small subsurface volumes.

For thermal conductivity measurements, the probe temperature that is related to the thermal conductivity of the sample can be monitored by measuring the probe voltage $V_{\text{el,p}} = R_{\text{el,p}}(T)I$. Dc and ac or both measurements can be performed (see Fig. 17.4).

Exciting the probe with an ac electrical current can be useful for an improved signal-to-noise ratio, since lock-in detection is possible. The 3ω method can be used [25]. This consists in measuring the third-harmonic voltage $V_{3\omega}$ of the resistor. The $V_{3\omega}$ amplitude is then directly proportional to an increase of mean temperature of the probe due to Joule heating:

$$V_{3\omega} = I_0 R_{\text{el,p0}} \alpha \frac{\overline{\theta_{2\omega}}}{2}, \tag{17.2}$$

where I_0 is the amplitude of the exciting current and $\overline{\theta_{2\omega}}$ is the amplitude of second-harmonic of the probe mean temperature.

Another method is to use a Wheatstone bridge equipped with a feedback loop as represented in Fig. 17.1. The method is currently called "constant temperature mode" or "thermal conductivity contrast mode." In this mode, the balanced Wheatstone bridge

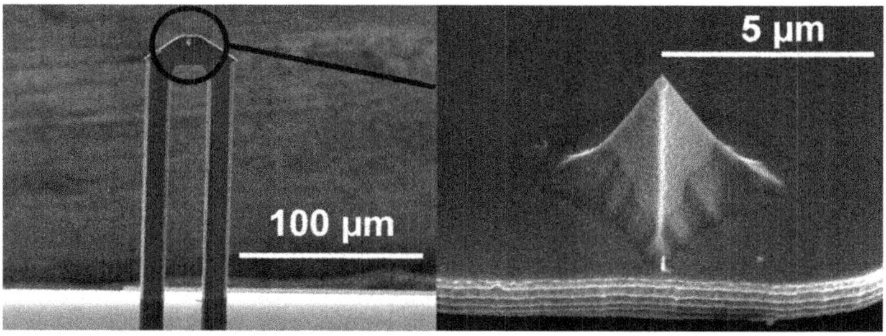

Figure 17.5 Doped silicon probe (DS probes) [1].

enables setting a constant value of the probe electrical resistance. When the probe contacts the sample, a part of the heat is lost and then the probe temperature decreases. The feedback loop increases the probe current to reset the initial electrical resistance of the probe and then its temperature.

17.2.2.3 Other probes used in active mode

Thermocouple probes, initially mainly used for thermometry in SThM, are also metallic probes. Their use in active mode was extended in 2006 to the investigation of material thermal conductivity with the 2ω method, which consists in monitoring the amplitude of the 2ω signal from the thermocouple junction of the probe that is heated by joule effect by an ac current. The first measurements were performed in contact mode [26]. The null-point SThM (NP SThM) method, also based on thermocouple probe used in ac mode [27], has recently been shown promising for quantitative thermal conductivity profiling for experiments under ambient condition. However, two temperature profiles must be obtained by scanning the probe at two different heights above the sample at each position on the sample, adding to the complexity of the technique.

SThM doped Si resistor SPM probes also operate in active mode. As represented in Fig. 17.5, they are based on U-shaped cantilever. They have a heater element at the cantilever end, placed above a sharp silicon tip of micrometric height. The tip curvature radius can reach 10 nm. The electric current flowing through the resistive element causes its resistive heating and a temperature rise of the tip. This kind of probe enabled the imaging of thermal conductivity contrast of biological materials with a spatial resolution of 10 nm and thermal resolution of 50 nW [28] and an estimation of the thermal conductivity of 3 nm thin films of HfO_2 [29].

17.3 Approach of the Measurement

Purely experimental calibration and calibration based on the determination of modeling parameters have been used for characterizing the thermal conductivity of materials in SThM. In the second case, a thermal modeling of the probe-sample system is required. This modeling involves the heat transfer

(1) through the probe,
(2) through the sample,
(3) between the probe and the sample,

while considering the probe and sample surrounding environment.

17.3.1 Modeling of the Probe-Sample System

17.3.1.1 Description of the probe

The SThM probes that are the most used for thermal conductivity determination are the Wollaston and KNT probes. Although the resistive element constitutes the AFM tip for the Wollaston probe and is deposited at the AFM tip for the KNT probe, almost similar descriptions of the probe-sample thermal system can be used for both probes. In these descriptions, the tip is described as two thermal fins [30] of length L_t (see Fig. 17.6 for the Wollaston probe).

For experiments performed under ambient conditions, the heat losses along the resistive element to the environment through radiation, convection and conduction are described through a loss coefficient h (W.m^{-2}.K^{-1}). Solving the heat equation for the resistive element heated by Joule effect enables to determine the temperature along the tip ($T(x)$, $x = 0$ to $x = L_t$) and the heat amount Q

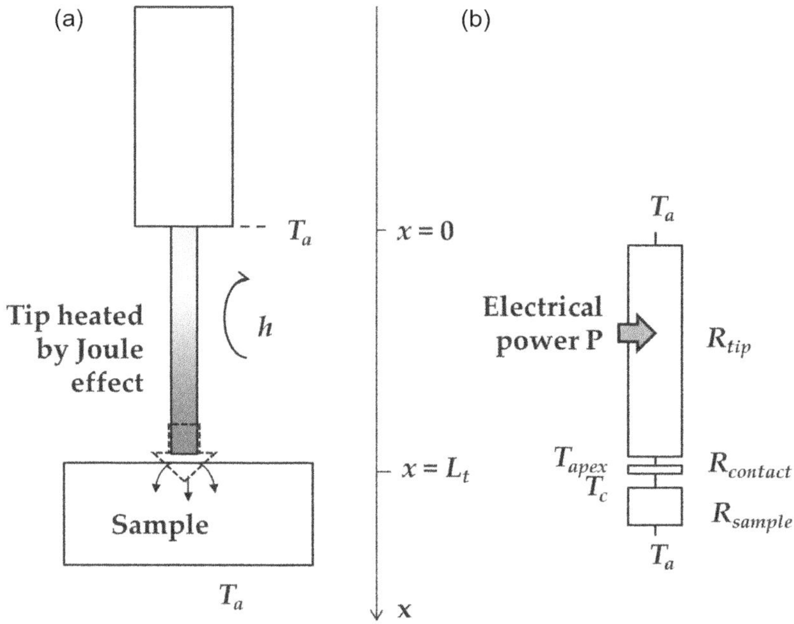

Figure 17.6 Usual description of the probe-sample thermal system (a) and associated network of thermal resistances (b) for the Wollaston probe.

transferred from the probe to the sample in $x = L_t$. Cantilever is then usually assumed to operate as a thermal bath at ambient temperature ($T(x = 0) = T_a$) and sample is described through the expression of its spreading thermal resistance R_{sample}.

17.3.1.2 Description of sample

For homogenous bulk samples, R_{sample} is given as

$$R_{\text{sample}} = \frac{1}{C_S K_S b}, \tag{17.3}$$

where C_s is a constant [31] and b is the radius of an isothermal circular area A that is taken equivalent to the real exchange surface at the tip-sample thermal contact. Let us notice that such expression only applies if the thermal contact area is sufficiently large compared with the energy carrier mean free path within the sample

In the case of a thin film deposited on a semi-infinite substrate, R_{sample} can be given as [32]

$$R_{\text{sample}} = \frac{1}{\pi b^2} \int_0^\infty \frac{2(1+\eta) J_1^2(qb)}{K_f q^2 (1-\eta)} dq, \tag{17.4}$$

where k_f is the film thermal conductivity and η is given as

$$\eta = \frac{(K_f - K_{sb})}{(K_f + K_{sb})} e^{-2qe}. \tag{17.5}$$

Here, K_{sb} and e denote respectively the substrate thermal conductivity and the film thickness. No interfacial thermal resistance is taken into account at the film-substrate interface. When b is larger than e, the intrinsic thermal conductivity of the film can be calculated independently of the film thickness and the values of interfacial thermal resistance by use of a one-dimensional terminology [33] for thin films. K_f can be expressed as a function of an intrinsic film thermal conductivity K_{fi} and an interfacial thermal resistance R_{int} through the following relation:

$$\frac{e}{K_f} = \frac{e}{K_{fi}} + R_{\text{int}} \tag{17.6}$$

This modeling was successfully applied to study the thermal conductivity of homogeneous in depth mesoporous silicon in form of thin films on mono-crystalline silicon substrate through SThM measurements [17].

17.3.1.3 Description of the probe-sample thermal contact

The tip-sample thermal contact is usually taken into account through an effective thermal resistance R_{contact}, which is associated with the effective thermal contact area taken equal to A so that

$$Q = \frac{T_{\text{apex}} - T_c}{R_{\text{contact}}} = \frac{T_{\text{apex}} - T_a}{R_{\text{contact}} + R_{\text{sample}}} \tag{17.7}$$

Under ambient conditions R_{contact} is associated with the tip-sample heat transfer through conduction through the air, solid–solid contact and water meniscus that is formed between the tip apex and the sample due to capillary condensation. Heat transfer through radiation is negligible.

Let us notice that Assy et al. [1, 34] have recently demonstrated that for the Wollaston probe the probe-sample heat exchange is mainly due to thermal conduction through the air. They have also shown that heat transfer through water meniscus can be neglected for various SThM resistive probes (Wollaston, KNT and doped Si probes) whatever the probe temperature is [35].

17.3.1.4 Measurement modeling

From such descriptions, an expression relating the mean temperature of the tip (temperature averaged on its length L_t) to the thermal resistance of the sample R_{sample}, and then to its thermal conductivity can be found. Using Eq. (17.1), the link between the mean temperature and the electrical resistance of the tip can then be established.

In practice, the measurement for a probe free (without interaction with a sample) in air is used as reference. In this configuration, $Q = 0$ in the modeling. For a given sample, the experimental data compared with simulated results are

- either the deviation of the tip electrical resistance $R_{\text{el,tip}}$:

$$\Delta R_{\text{el,tip}} = R_{\text{el,tipc}} - R_{\text{el,tipi}}, \tag{17.8}$$

  when measurements are performed with a constant current,
- or the relative deviation $(\Delta P / P)$ of the Joule power P in the tip:

$$\frac{\Delta P}{P_{\text{c}}} = \frac{R_{\text{el,tip}} I_{\text{c}}^2 - R_{\text{el,tip}} I_{\text{i}}^2}{R_{\text{el,tip}} I_{\text{c}}^2}, \tag{17.9}$$

  when a balanced Wheatstone bridge is used for measurement and $\boldsymbol{R}_{\text{el,tip}}$ is kept constant.

In Eq. (17.9), I is the amplitude of the current through the tip. In Eqs. (17.8) and (17.9), i and c indexes indicate respectively a free tip and a tip in contact with sample.

The expression (17.8) or (17.9) depends in the end on various parameters. It includes the thermal resistance of the tip R_{tip} that depends on its dimensions and components, the loss coefficient h, the effective contact thermal resistance R_{contact}, and the effective thermal contact radius b. All these parameters can be determined through complementary experiments with other measurement methods. As an example, the estimation of the real dimensions of the probe can be made from SEM observations. R_{tip} and h can be determined through the fitting of simulated measurement in ac mode with experimental data obtained under the environmental conditions used for measurements and a free probe [36]. $R_{contact}$ and b are obtained through calibration with reference samples.

17.3.2 Calibration

As introduced before, purely experimental calibration and calibration based on the determination of modeling parameters have been used. In both cases, calibration is performed using bulk samples as references of well-known thermal conductivity that lie in the range of the expected value of the thermal conductivity K_s of the sample of interest. The usual approach is here given for the Wollaston probe used in a balanced Wheatstone bridge (probe electrical resistance and temperature are kept constant; see Section 17.2.2.2).

In purely experimental calibration, the variation of the curve $\Delta R_{el,tip} = f(K_s)$ or $\Delta P/P_c = g(K_s)$ in the range of the expected value of K_s of interest is graphically identified (see Fig. 17.7).

In calibration based on modeling parameters determination, once h and probe parameters (R_{tip}) are known, $R_{contact}$ and b are identified by fitting values of $\Delta P/P_c$, which are simulated with modeling previously presented, with experimental data of $\Delta P/P_c$ obtained on reference samples. All the modeling parameters being known, the thermal conductivity of the sample to be characterized can be derived from the value of $\Delta P/P_c$ measured for this sample.

In principle, these calibrations are possible only for materials with comparable surface states (roughness, mechanical properties). Fortunately, most of the heat exchanged between probe and sample is transferred through the air, not through the mechanical contact for the Wollaston probe operating in ambient conditions. As a consequence, the measurement is robust against probe and/or surface such as defects and irregularities in the main range of sensitivity to K_s of the Wollaston probe; that means for insulating materials (K_s <10 W.m^{-1}.K^{-1} [1]).

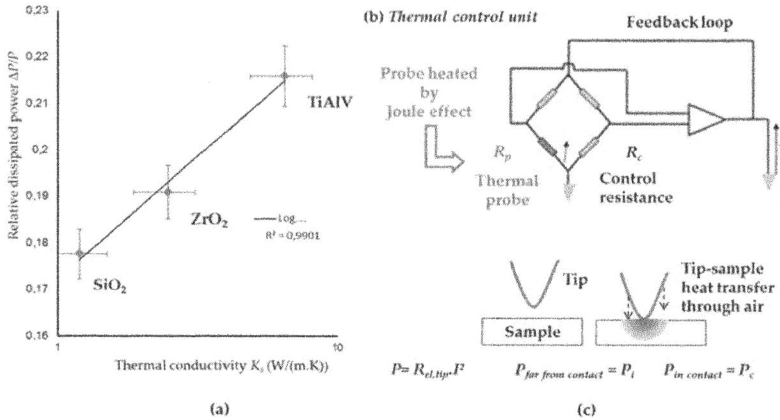

Figure 17.7 (a) Calibration curve. Three samples of known thermal conductivity in a range that covers the expected value of K_S to be measured were used. (b) Schematic diagram of a typical balanced Wheatstone bridge used for measurement with the Wollaston probe. (c) Principle of measurements with a probe free in air and in contact with the sample.

The method could furthermore be performed with the Wollaston probe from measurements in ac. mode. In this mode, the 3ω voltage $V_{3\omega}$ of the probe is measured as a function of the current frequency [23, 25, 37]. When the probe is out of contact, the measurements permits the identification of the probe geometrical parameters and h to be used in the modeling [25, 38]. In contact, R_{contact} and b can also be derived by fitting $V_{3\omega}$ values modeled with $V_{3\omega}$ measurements on reference samples. Such calibration was successfully used to measure the thermal conductivity of single silicon nanowires with the palladium probe [23].

17.4 Selected Application

This part concerns applications of the Wollaston wire probe, which was the leading probe during a long time.

This probe was applied to the characterization of various bulk materials and thin films. The studied materials were mainly insulating due to the narrow range of thermal conductivity sensitivity in SThM [36]. Its applications comprise materials such as bulk ZnO

(0001) [11], and Ba_8Si_{46} that is a simple binary representative of intermetallic clathrates, guest-host systems with a high potential for thermoelectric applications due to their ultralow thermal conductivity even their perfect crystal structure [39]. The study of the thermal conductivity of porous [14, 18, 40, 41] and mesoporous [17] bulks or thin films of submicrometric thickness [99, 152] was also possible using this probe. Some results obtained with SThM were compared with values measured on the same material samples by Raman thermometry and were in good accordance [17, 32, 40, 42], validating our approach.

Here we discuss the investigation with the Wollaston probe of the thermal conductivity degradation of mesoporous silicon (PSi) made amorphous through swift (high-energy) heavy ion-irradiation. Combined with structural analysis, SThM results have been shown useful in the framework of development of this new method for reducing the thermal conductivity of PSi.

PSi film samples were formed by electrochemical etching of p+ monocrystalline (100)-oriented Si wafers ($\rho = 0.01$–0.02 cm). Film thickness was about 10 μm and their porosity value, which was controlled via the anodization current density, was 41%. A pulsed current was used, with zero-current etch-stop periods to improve morphological uniformity of the PS layers [43]. PS fabricated in this way consists of dendritic silicon quasi-columns, with an average diameter on the order of 10 nm, perpendicular to the wafer surface and separated by pores. These quasi-columns remain monocrystalline and retain the crystalline orientation of the original Si substrate. Irradiation of the PS samples was carried out with 110 MeV uranium ions. Six different fluences were used, between 10^{12} and 7×10^{13} cm^{-2}.

The evolution of the structural properties was studied by scanning electron microscopy (SEM) and micro-Raman spectroscopy. This structural study has shown that swift heavy ion irradiation causes the creation of an amorphous phase in PSi but without suppressing its porous structure. This amorphization has been shown caused by electronic-regime interactions: swift heavy ion irradiation induces local melting of the PS dendrites. Indeed, according to the thermal spike model [14], the target electrons, previously excited by the incoming ions in the first 10–15 s following

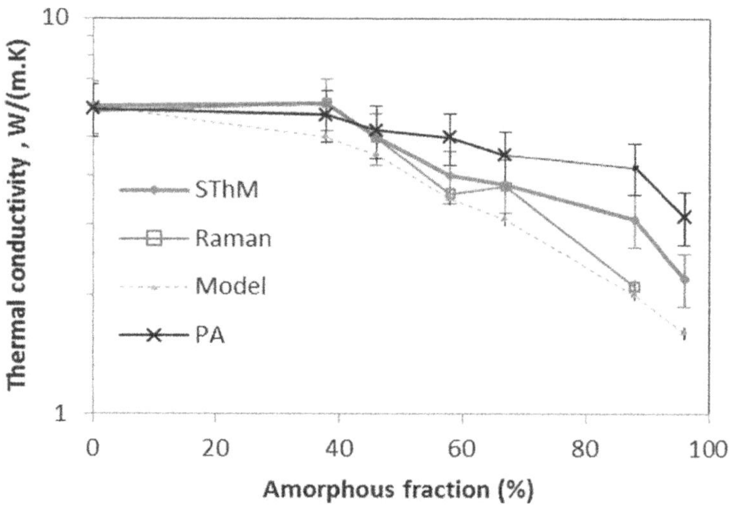

Figure 17.8 Thermal conductivity versus amorphous fraction measured by SThM [40], Raman spectroscopy [40], Photoacoustic (PA) technique [44], and compared to a simple model [42] for a PSi porosity of 41%.

their passage, transfer their energy to the crystal lattice via electron–phonon coupling (10–12 s). The electronic interaction between the ions and lattice is accompanied by the formation of an amorphous Si phase in the PS layer. Raman spectroscopy was moreover used in particular to estimate the dependence of the amorphous fraction on the ion fluence. The amorphous fraction was found to be increased when the irradiation fluence increases. All details related to this measurement can be found in [40].

The dependence of thermal conductivity on the amorphous fraction of the PS layer measured with SThM is shown in Fig. 17.8. We can observe that an increase in amorphous fraction causes a reduction of thermal conductivity. The lowest thermal conductivity is more than 3 times lower than thermal conductivity of the non-irradiated sample (about 6 W m^{-1} K^{-1}).

The thermal conductivity values obtained from the SThM measurements are compared to those obtained by micro-Raman spectroscopy [40] and by photoacoustic technique [44]. Good agreement can be stated. All data show a decrease in thermal conductivity as amorphous fraction rises. The difference found between photothermal method results, i.e., Raman and photoacoustic data, is mainly due to specificities of these methods. Indeed, in the case of Raman technique, the studied samples are locally heated by the exciting laser irradiation to ensure temperature induced Raman peak shifts. This leads to an increase of its surface temperature by more than 100°. As a result, the laser-induced local thermal elastic stresses can significantly influence the experimentally obtained data, especially in the case of amorphous samples. As for the SThM and PA techniques, the temperature increase is less than 1° and, therefore, the obtained thermal conductivity values are estimated much more precisely. Let us remark that the thickness (of 10 μm) of the PSi samples does not affect SThM results, as SThM probes here a depth of around 1 μm [17].

More recently Massoud et al. [45] studied three different mass to energy ratios (^{238}U at 110 MeV and ^{130}Xe at 91 MeV and 29 MeV) with fluences ranging from 10^{12} cm^{-2} to 7×10^{13} cm^{-2}. Thermal conductivity was measured using SThM. Although, bulk Si is insensitive to ion interaction in the electronic regime, the amorphization of a PSi of 56% porosity has been shown to result in a thermal conductivity reduction even for the low dose and energy. For the highest irradiation dose a very important reduction factor of four was obtained. Consequently swift heavy ion irradiation seems an effective new method for improving the thermal-insulation properties of PSi layers, as well as a good alternative to already known approaches, such as partial oxidation or increasing porosity.

17.5 Conclusions

This chapter presents the main principle, tools (probes and modeling), and experimental data approaches that are currently used in resistive SThM operating in active mode for local thermal imaging and thermal conductivity measurements. It highlights the possibility to characterize bulk, thin films and nanostructures, mainly in insulating materials with SThM. This is of great interest for the development and the measurement of the new nanostructured materials and structures viewed for applications such as insulation in integrated component circuits or thermoelectricity.

Acknowledgments

The research leading to this chapter has received funding from the European Union Seventh Framework Programme FP7-NMP-2013-LARGE-7 under grant agreement n°604668.

References

1. Assy, A., Development of two techniques for thermal characterization of materials: Scanning Thermal Microscopy (SThM) and 2ω method, 2015 PhD Thesis, INSA de Lyon.
2. Gomes, S., Contribution théorique et expérimentale à la microscopie thermique à sonde locale: calibration d'une pointe thermorésistive, analyse des divers couplages thermiques, 1999.
3. David, L., Développement de la microscopie thermique à sonde locale pour la détermination de la conductivité thermique de films minces Application aux céramiques pour le nucléaire, 2006, INSA de Lyon.
4. Tong, T. W., *Thermal Conductivity 22*, vol. 22. 1994: CRC Press.
5. Buzin, A., et al., Application of a Wollaston wire probe for quantitative thermal analysis. *Thermochimica Acta*, 2002. **381**(1): 9–18.
6. Lopez, L. D. P., et al., Characterization of the thermal behavior of PN thermoelectric couples by scanning thermal microscope. *Microelectronics Journal*, 2004. **35**(10): 797–803.
7. Gomès, S., et al., Temperature study of sub-micrometric ICs by scanning thermal microscopy. *Components and Packaging Technologies, IEEE Transactions on*, 2007. **30**(3): 424–431.
8. Altet, J., et al., Four different approaches for the measurement of IC surface temperature: application to thermal testing. *Microelectronics Journal*, 2002. **33**(9): 689–696.
9. Altet, J., et al., Dynamic surface temperature measurements in ICs. *Proceedings of the IEEE*, 2006. **94**(8): 1519–1533.
10. Pollock, H. and A. Hammiche, Micro-thermal analysis: techniques and applications. *Journal of Physics D: Applied Physics*, 2001. **34**(9): R23.
11. Florescu, D. I., et al., High spatial resolution thermal conductivity of bulk ZnO (0001). *Journal of Applied Physics*, 2002. **91**(2): 890–892.

12. Florescu, D., et al., Thermal conductivity of fully and partially coalesced lateral epitaxial overgrown GaN/sapphire (0001) by scanning thermal microscopy. *Applied Physics Letters*, 2000. **77**(10): 1464–1466.
13. Meinders, E. R., Measurement of the thermal conductivity of thin layers using a scanning thermal microscope. *Journal of Materials Research*, 2001. **16**(09): 2530–2543.
14. Volz, S., et al., Thermal conductivity measurements of thin amorphous silicon films by scanning thermal microscopy. *International Journal of Thermophysics*, 2002. **23**(6): 1645–1657.
15. Gorbunov, V., et al., Probing surface microthermal properties by scanning thermal microscopy. *Langmuir*, 1999. **15**(24): 8340–8343.
16. Hammiche, A., et al., Localized thermal analysis using a miniaturized resistive probe. *Review of Scientific Instruments*, 1996. **67**(12): 4268–4274.
17. Gomes, S., et al., Application of scanning thermal microscopy for thermal conductivity measurements on meso-porous silicon thin films. *Journal of Physics D: Applied Physics*, 2007. **40**(21): 6677.
18. Zhang, Y., et al., A noncontact thermal microprobe for local thermal conductivity measurement. *Review of Scientific Instruments*, 2011. **82**(2): 024902.
19. Callard, S., et al., Thermal conductivity of SiO_2 films by scanning thermal microscopy. *Journal of Non-Crystalline Solids*, 1999. **245**(1): 203–209.
20. Pelzl, J., M. Chirtoc, and R. Meckenstock, Thermal Wave-Based Scanning Probe Microscopy and Its Applications. *International Journal of Thermophysics*, 2013. **34**(8–9): 1353–1366.
21. Dobson, P. S., J. M. Weaver, and G. Mills. New Methods for Calibrated Scanning Thermal Microscopy (SThM). in *Sensors, 2007 IEEE*. 2007. IEEE.
22. Tovee, P., et al., Nanoscale spatial resolution probes for scanning thermal microscopy of solid state materials. *Journal of Applied Physics*, 2012. **112**(11): 114317.
23. Puyoo, E., et al., Scanning thermal microscopy of individual silicon nanowires. *Journal of Applied Physics*, 2011. **109**(2): 024302.
24. Puyoo, E., et al., Thermal exchange radius measurement: Application to nanowire thermal imaging. *Review of Scientific Instruments*, 2010. **81**(7): 073701.
25. Lefèvre, S. and S. Volz, 3ω-scanning thermal microscope. *Review of Scientific Instruments*, 2005. **76**(3): 033701-033701–6.

26. Roh, H. H., et al., Novel nanoscale thermal property imaging technique: the 2ω method. I. principle and the 2ω signal measurement. *Journal of Vacuum Science & Technology B*, 2006. **24**(5): 2398–2404.
27. Chung, J., et al., Quantitative temperature profiling through null-point scanning thermal microscopy. *International Journal of Thermal Sciences*, 2012. **62**: 109–113.
28. Haeberle, W., M. Pantea, and J. Hoerber, Nanometer-scale heat-conductivity measurements on biological samples. *Ultramicroscopy*, 2006. **106**(8): 678–686.
29. Hinz, M., et al., High resolution vacuum scanning thermal microscopy of HfO_2 and SiO_2. *Applied Physics Letters*, 2008. **92**(4): 043122-043122–3.
30. Assy, A. and S. Gomès, Heat transfer at nanoscale contacts investigated with scanning thermal microscopy. *Applied Physics Letters*, 2015. **107**(4): 043105.
31. Yovanovich, M. and E. Marotta, Thermal spreading and contact resistances. *Heat Transfer Handbook*, 2003. **1**: 261–394.
32. David, L., S. Gomes, and M. Raynaud, Modelling for the thermal characterization of solid materials by dc scanning thermal microscopy. *Journal of Physics D: Applied Physics*, 2007. **40**(14): 4337.
33. Lambropoulos, J. C., et al., Thermal conductivity of dielectric thin films. *Journal of Applied Physics*, 1989. **66**(9): 4230–4242.
34. Assy, A., et al., Analysis of heat transfer in the water meniscus at the tip-sample contact in scanning thermal microscopy. *Journal of Physics D: Applied Physics*, 2014. **47**(44): 442001.
35. Assy, A. and S. Gomès, Temperature-dependent capillary forces at nano-contacts for estimating the heat conduction through water meniscus. *Nanotechnology*, 2015. **26**(35): 355401.
36. Gomès, S., A. Assy, and P. O. Chapuis, Scanning thermal microscopy: A review. *Physica status solidi (a)*, 2015. **212**(3): 477–494.
37. Grauby, S. P., et al., Si and SiGe nanowires: fabrication process and thermal conductivity measurement by 3ω-scanning thermal microscopy. *The Journal of Physical Chemistry C*, 2013. **117**(17): 9025–9034.
38. Chapuis, P.-O., et al. Heat transfer between a hot AFM tip and a cold sample: impact of the air pressure. In *MRS Proceedings*. 2013. Cambridge University Press.
39. Pailhès, S., et al., Localization of propagative phonons in a perfectly crystalline solid. *Physical Review Letters*, 2014. **113**(2): 025506.

40. Newby, P. J., et al., Amorphization and reduction of thermal conductivity in porous silicon by irradiation with swift heavy ions. *Journal of Applied Physics*, 2013. **114**(1): 014903.

41. Borca-Tasciuc, T., Scanning probe methods for thermal and thermo-electric property measurements. *Annual Review of Heat Transfer*, 2013. **16**(16): 211–258.

42. Gomès, S., et al., Characterization of the thermal conductivity of insulating thin films by scanning thermal microscopy. *Microelectronics Journal*, 2013. **44**(11): 1029–1034.

43. Perichon, S., et al., Technology and micro-Raman characterization of thick meso-porous silicon layers for thermal effect microsystems. *Sensors and Actuators A: Physical*, 2000. **85**(1): 335–339.

44. Isaiev, M., et al., Thermal conductivity of partially amorphous porous silicon by photoacoustic technique. *Materials Letters*, 2014. **128**: 71–74.

45. Massoud, M., et al., Swift heavy ion irradiation reduces porous silicon thermal conductivity. *Nuclear Instruments and Methods in Physics Research Section B: Beam Interactions with Materials and Atoms*, 2014. **341**: 27–31.

Chapter 18

Thermal Wave Methods

Ali Assy,[a,b] Séverine Gomès,[a,b] Pavlo Lishchuk,[c] and Mykola Isaiev[c,d]

[a] *Université de Lyon, CNRS INSA-Lyon, CETHIL, UMR5008, F-69621 Villeurbanne, France*
[b] *Université Lyon 1, CETHIL, UMR5008, F-69622 Villeurbanne, France*
[c] *Faculty of Physics, Taras Shevchenko National University of Kyiv, 64/13, Volodymyrska Str., Kiev, 01601, Ukraine*
[d] *Laboratoire LEMTA, Faculté des Sciences et Technologies, CNRS UMR 7563, Université de Lorraine, BP 70239, 54506, Vandoeuvre les Nancy Cedex, France*
aliassy01@gmail.com, mykola.isaiev@knu.ua

This chapter presents and discusses the methods involving thermal waves for the characterization of thermal conductivity or diffusivity of crystalline and amorphous nanostructured materials. It analyzes the influence of the temporal characteristic of the exciting heat source on the temperature evolution within the sample. We focus on the methods with periodic heating sources. The chapter overviews the application of electrothermal techniques (2-omega and 3-omega) for the evaluation of in-plane and cross-plane thermal conductivities of anisotropic systems. The last part demonstrates the use of photothermal and photoacoustic techniques as powerful tools for the study of crystalline/amorphous materials.

Nanostructured Semiconductors: Amorphization and Thermal Properties
Edited by Konstantinos Termentzidis

ISBN 978-981-4745-64-2 (Hardcover), 978-1-315-36445-2 (eBook)
www.panstanford.com

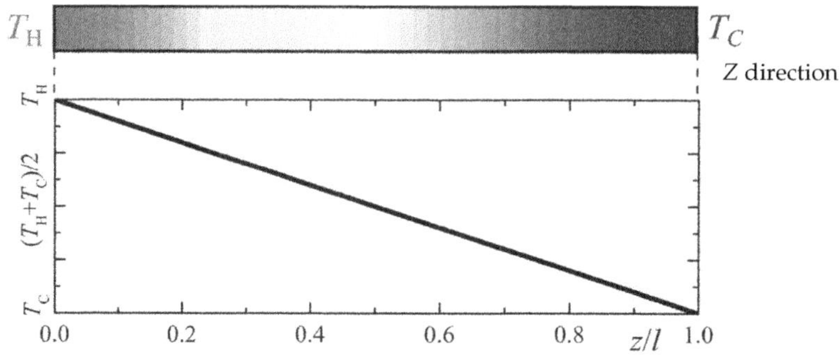

Figure 18.1 Temperature distribution across a long rod with a temperature gradient applied between its two ends ($T_H > T_C$).

18.1 Introduction

The most used methods for the thermal characterization of materials are based on the creation of a heat flux in the sample to be studied. Thus, the evaluation of the thermal conductivity from the Fourier law is given as

$$\vec{J} = -k.\nabla T, \tag{18.1}$$

where $\vec{J}$ is the heat flux, k is thermal conductivity of the sample, and ∇T is the temperature gradient within the sample.

The thermal conductivity evaluation of the one-dimensional object could be mentioned as a simple example. Let us consider the situation where the ends of thin (X and Y directions) and long (Z direction) rod are kept under constant temperatures T_H and T_C ($T_H > T_C$). In the case of a steady-state regime, the temperature distribution across Z-axis is linear as shown in Fig. 18.1. The temperature distribution could be presented as follows:

$$T(z) = T_H - \frac{(T_H - T_C).z}{l}, \tag{18.2}$$

where l is a length of the rod.

In this case, the thermal conductivity k could be easily evaluated from the following equation:

$$k = \frac{J_z.l}{T_H - T_C}, \tag{18.3}$$

where J_z is the projection of the vector $\vec{J}$ on the Z direction.

J_z could be estimated from the heat powers applied to the hot (P_H) and cold (P_C) ends of the rod:

$$J_z = \frac{P_H - P_C}{A}, \quad (18.4)$$

where A is the cross-sectional area of the rod (XY plane). Combining Eqs. (18.3) and (18.4) enables us to find the final expression of thermal conductivity.

In the practical implementation of this technique, a constant power is applied to the ends of the rod and their respective temperatures are measured (using thermocouples as an example).

Hereinabove, we considered a stationary method for thermal conductivity measurements. This means that the heating does not depend on time. Thus, one should be careful during the measurements and check that the temperatures are also independent of time.

18.2 Thermal Wave Fundamentals

18.2.1 Concept

Other popular methods for thermal conductivity measurements are non-stationary techniques, such as thermal waves and transient techniques.

Let us consider a one-dimensional system of thickness d. Its top surface is heated with a time-dependent power $P(t)$. We consider that the back-surface of the sample surface is kept also at temperature T_0. The heat equation that defines the time-evolution of the temperature rise ($\theta = T - T_0$) in the sample is written as follows:

$$c\rho \frac{\partial \theta}{\partial t} = k.\frac{\partial^2 \theta}{\partial z^2} \quad (18.5)$$

where c and ρ are the heat capacity and the density of the sample, respectively.

The boundary conditions for the Eq. (18.5) are

$$-k.\frac{\partial \theta}{\partial z}(z = 0) = P(t)/A$$

and

$$\theta(z = d) = 0,$$

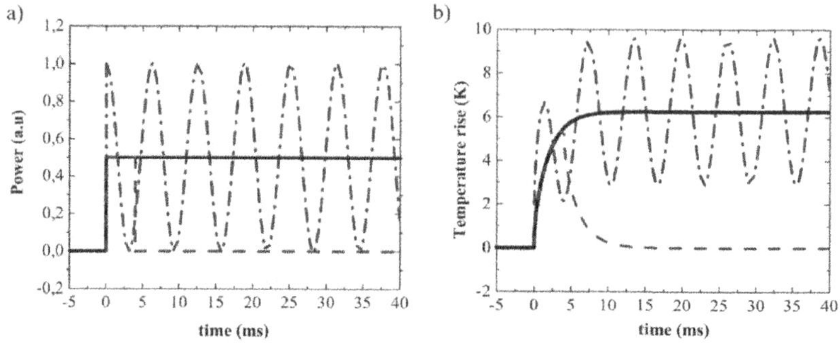

Figure 18.2 Time dependence of the power of excitation sources (a) and induced surface temperature (b) in the different regimes of heating.

where A is the heated area at the top surface. The character of the dependence $P(t)$ defines the temporal behavior of the $\theta(z, t)$. When $P(t)$ is constant (solid line at in Fig. 18.2a), the temperature on the top surface ($T(0, t)$) will increase with time (solid line in Fig. 18.2b) until a certain time $t > l^2/\alpha$ beyond which it becomes constant. Here $\alpha = \kappa/(c\rho)$ is the thermal diffusivity of the sample material.

The simple case of the pulse excitation source is also presented in Fig. 18.2a through the dashed line. In this case, the temporal behavior of the surface temperature in the "active period" ($t<\tau$) is similar as for constant source and afterwards it decreases to zero (see dashed line in Fig. 18.2b). Let us note that the rate of increasing and decreasing of the temperature strongly depend on the thermal diffusivity of the investigated sample. Therefore, the techniques with such sources are often implemented for the evaluation of this parameter.

The methods with periodic heating are also used for the determination of thermal parameters. An example of periodic heating source with harmonic law could be presented as follows:

$$P(t) \cong P_0.\,(1 + \cos(\omega t)), \tag{18.6}$$

where P_0 and ω are, respectively, the amplitude and angular frequency of the source power.

The time dependence of the periodic source is presented in Fig. 18.2a by the dashed dots line. In this case, one can see two regions in the time dependence of the surface temperature:

(i) transient regime and (ii) steady-state regime. Using these methods, the measurements of thermophysical properties are mostly performed in the regime (ii). In this regime, the temperature has two components: stationary component (which is often called "dc-component" based on electrical analogy) and non-stationary oscillations (ac-component). Thermal wave techniques are based on the exploitation of frequency-dependency of the amplitude and phase of the ac-component of the temperature.

Equation (18.5) can be used for describing the behavior of the alternating temperature. For this, one should take into account only non-stationary term from the boundary conditions (18.7), which could be presented as follows:

$$T(z, t) = \mathrm{Re}(\theta(z).\exp(i\omega t)), \qquad (18.7)$$

where $\theta(z)$ is a function that defines a spatial behavior of the ac-component of the temperature. The equation for this function can be found in the result of substitution of Eq. (18.7) into Eq. (18.5):

$$\frac{\partial^2\theta}{\partial z^2} - \frac{i\omega c\rho}{k}\frac{\partial\theta}{\partial t} = 0 \qquad (18.8)$$

The solution of this equation could be presented as follows:

$$\theta(z) = B_1.\exp\left(\sqrt{\frac{i\omega}{\alpha}}z\right) + B_2.\exp\left(-\sqrt{\frac{i\omega}{\alpha}}z\right), \qquad (18.9)$$

where B_1 and B_2 are complex constants that can be determined from the boundary conditions.

As one can see, the factor in the exponent of Eq. (18.9) has real and imaginary parts. Particularly, the term "thermal-wave" appeared in the result of the specific behavior of the $\theta(z)$ as a decaying wave—the wavelength of this wave μ is the same as the characteristic distance of the decaying (see Fig. 18.3):

$$\mu = -\mathrm{Im}\left(\sqrt{\frac{\alpha}{i\omega}}\right) = \sqrt{\frac{2\alpha}{\omega}} \qquad (18.10)$$

18.2.2 Various Methods

There are different methods of thermal wave excitation. In particular in this chapter, we will discuss electrical (2ω and 3ω techniques) and optical (photoacoustic) thermal sources.

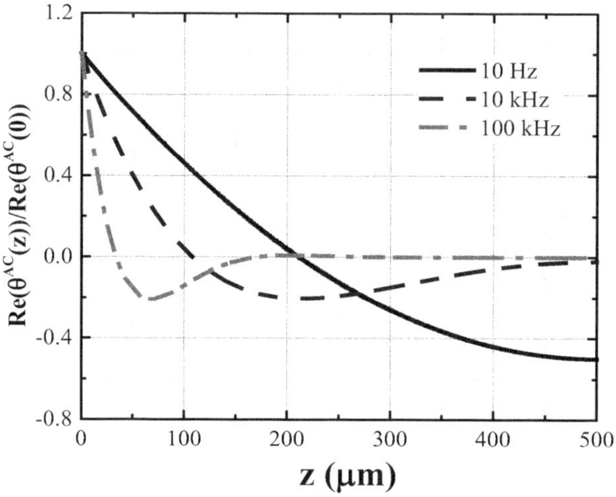

Figure 18.3 View of normalized real part of $\theta(z)$ for different frequencies of the excitation source.

In electrothermal methods, the heating element consists of a long metallic wire deposited on the sample surface. An alternating current with magnitude I_{ac} and angular frequency ω is applied across the wire. As a result, the Joule power dissipated in the wire is

$$P_p(t) \cong \frac{R_W I_{ac}^2}{2} \cdot (1 + \cos(2\omega t)), \tag{18.11}$$

where R_W is the electrical resistance of the wire.

In the material, the conductive diffusion of such heating produces a dc-temperature component (T_{DC}) and ac-temperature component at angular frequency 2ω (T_{2_ω}):

$$T(t) = T_{DC} + T_{2\omega} \cos(2\omega t + \Phi) \tag{18.12}$$

Then, the electrical resistance of the heating wire $R(t)$ is

$$R_W(t) = R_0 (1 + \beta T(t)) = R_0 + \Delta R_{DC} + \Delta R_{2\omega}, \tag{18.13}$$

where β and R_0 are, respectively, the temperature coefficient of the electrical resistivity and the electrical resistance of the wire at ambient temperature T_0.

In the photothermal and photoacoustic techniques, the surface of the sample is illuminated with the light beam. In this case, the periodical heating can be performed with different ways. Among them, one can mention, the modulation of a light source power, or direct interruption of the light beam with a mechanical chopper or optoacoustic modulator.

18.3 Methods

18.3.1 3ω Method

The 3ω method was developed by Cahill [1] for determination of the thermal properties of bulk materials. In this method, a metallic wire is deposited on the sample surface and used simultaneously as a heater and a thermometer. In this case, the determination of $T_{2\omega}$ can be achieved by the measurement of the voltage $V(t)$ across the heating wire that can be presented as follows:

$$\begin{aligned} V(t) &= R(t)I(t) = R_0(1+\beta\theta_{\mathrm{DC}})I_{\mathrm{ac}}\cos(\omega t) \\ &\quad + \frac{R_0\beta I_{\mathrm{ac}}T_{2\omega}}{2}\cos(\omega t+\phi) + \frac{R_0\beta I_{\mathrm{ac}}T_{2\omega}}{2}\cos(3\omega t+\phi) \\ &= V_\omega + V_{3\omega} \end{aligned} \tag{18.14}$$

Through electronic circuits (Wheatstone bridge as an example), the first harmonic voltage V_ω can be suppressed. Therefore, $T_{2\omega}$ is determined through the measurement of the third harmonic voltage $V_{3\omega}$.

Bulk materials: Based on the solution of 2D heat equation in the case of an isotropic semi-infinite sample, the alternating temperature $T_{2\omega}$ across the heater can be expressed as [1]:

$$T_{2\omega,\mathrm{s}} = -\frac{P_{\mathrm{rms}}}{2\pi k_{\mathrm{s}}}\left(\ln(2\omega) + \ln\left(\frac{w^2}{\alpha}\right) - 2c\right) - i\frac{P_{\mathrm{rms}}}{4k_{\mathrm{s}}}, \tag{18.15}$$

where c is a constant equal to 0.9228 [2] and P_{rms} is the power per unit heater length.

The current frequency f can be tuned to ensure the assumptions of a semi-infinite substrate and a line heating source. In these cases, μ has to be, respectively, larger than $5w$ and smaller than $e/5$ [1].

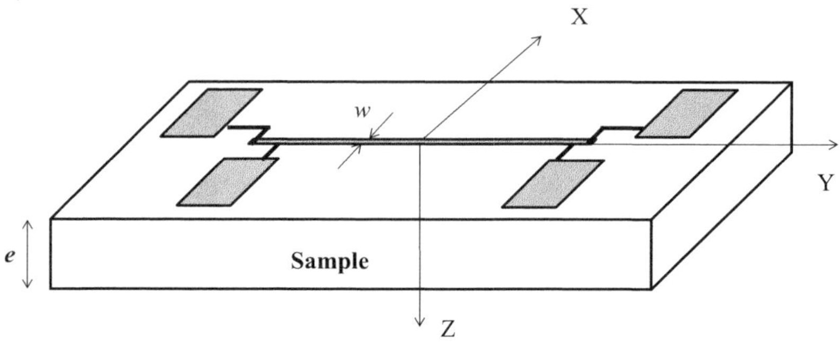

Figure 18.4 Schematic of a sample instrumented for 3ω method measurements.

Here, w and e denote, respectively, the width of the line heating source and the thickness of the sample (see Fig. 18.4).

Moreover, the thickness of the heating wire has to be small in order to avoid any additive errors as studied by Wang et al. [3]. Additionally, the solution (Eq. 18.15) of a 2D heat equation requires a uniform temperature distribution along the heating wire. Based on numerical simulations, Jacquot et al. [4] determined the domain of validity of Eq. 18.15. The temperature is always uniform along the wire when its length l is 15 times larger than μ. In such technique, the measurements are performed as a function of the heating frequency 2ω and the thermal conductivity of the substrate is usually determined from the slope of the real part in Eq. (18.15) [1, 5].

In the case of a semi-infinite anisotropic substrate, $T_{2\omega}$ is given as [5]

$$T_{2\omega,s} = \frac{-P_{\text{rms}}}{2\pi k_{s,z}\sqrt{k_{s,xz}}}\left[\ln(2\omega) + \ln\left(\frac{w^2}{\alpha_{s,z}k_{s,xz}}\right) - 2c\right] -i\left(\frac{P_{\text{rms}}}{4k_{s,z}\sqrt{k_{s,xz}}}\right), \quad (18.16)$$

where $k_{s,xz} = k_{s,x}/k_{s,z}$ and x, z denote, respectively, the in-plane and cross-plane components of the thermal conductivity.

We have to note that Eq. (18.16) is valid only in the case where the ratio $e(k_{s,xz})^{0.5}/w<25$ [5]. Otherwise, one has to refer to the exact solution of $T_{2\omega}$ [5]:

$$T_{2\omega} = \frac{P_{\text{rms}}}{\pi k_{s,z}} \int_0^\infty \frac{1}{A} \frac{\sin^2(w\theta)}{w^2\theta^2} d\theta, \tag{18.17}$$

where A is given as

$$A = \sqrt{k_{xz}\theta^2 + \frac{i2\omega}{\alpha_z}}. \tag{18.18}$$

Anisotropic thermal properties of nanochanneled alumina templates were determined using measurements of $T_{2\omega} = f(\omega)$ and Eq. (18.17) [6].

In the 3ω techniques, the thermal properties of the sample are deter-mined at temperatures that are usually controlled through cryostats or ovens. Moreover, the amplitude of the alternating current I_{ac} is sufficiently small to avoid any additive heating of the sample.

Thin films: The 3ω method was extended to determine the thermal properties of thin films. Thin films have basically different thermal properties from their bulk materials due to reduction in size. This reduction induces many processes especially phonon scattering at the films boundaries that decrease the thermal properties of the films. Moreover, the reduction in size acts differently on the in-plane and cross-plane components of the thermal properties and, therefore, thin films are expected to have basically anisotropic thermal properties.

Cahill et al. [7] used the 3ω method to measure the cross-plane thermal conductivity of thin films with a thickness of a few tens of nanometers. In this case, the heater width w is much larger than the film thickness d_{f} and the film is deposited on a substrate of high thermal conductivity. Then, the film can be described as a 1D thermal resistance in the cross-plane direction. It can be shown that this thermal resistance e_{f}/k_z, adds an additional second harmonic temperature to Eq. (18.15) or (18.16):

$$\Delta T_{2\omega,\text{f}} = \frac{P_{\text{rms}}}{k_{\text{f},z}} \frac{e_{\text{f}}}{2w} \tag{18.19}$$

The thermal conductivity of the substrate was obtained from the slope of the curve Real $(\Delta T_{2\omega}) = f(\omega)$ (see Eq. (18.16)) and the film cross-plane thermal conductivity is calculated from the temperature drop through the film (Eq. (18.19)) [7]. The obtained thermal conductivity of the film is an average along the frequency interval. In this approach, the series of thermal resistances must be known between the heater and the substrate. Kim et al. [8] noticed a variation of the estimated value of the substrate thermal conductivity that was linked to the thermal resistance between the heater and the film. Even though Eq. (18.19) gives a simple solution and has been more to determine the cross-plane thermal conductivity of thin films, one should be careful to many factors that introduce significant errors to the identification of the thermal properties.

In order to examine an eventual lateral spreading in the thin film and the film/substrate thermal interface, the thermal conductivity of the film deposited on an isotropic substrate can be calculated from the temperature variations as [5]

$$\Delta T_{2\omega,\mathrm{f}} = \Delta T_{2\omega,\mathrm{f+S}} - \Delta T_{2\omega,\mathrm{S}} = \frac{P_{\mathrm{rms}} e_{\mathrm{f}}}{2wk_{\mathrm{f},z}} C_1 C_2, \tag{18.20}$$

where C_1 and C_2 are defined through

$$C_1 = 1 - \frac{\lambda_{\mathrm{f},x}\lambda_{\mathrm{f},z}}{\lambda_{\mathrm{s}}^2} \tag{18.21}$$

and

$$C_2 = \frac{2}{\pi}\int_0^\infty \frac{\sin^2\theta}{\theta} \frac{\tan(\theta C_3)}{[1 + (\sqrt{k_{\mathrm{f},z}k_{\mathrm{f},x}}/k_{\mathrm{S}})\tanh(\theta C_3)]C_3} d\theta \tag{18.22}$$

with

$$C_3 = \sqrt{k_{\mathrm{f},xz}}(e_{\mathrm{f}}/w). \tag{18.23}$$

It appears that the thermal conductivity of the film has to be much smaller than the one of the substrate [5]. When this condition is satisfied, the 1D cross-plane heat condition (Eq. 18.19) can be applied by replacing the heater width $2w$ by $(2w + 76d_{\mathrm{f}}k_{\mathrm{f},xz}^{0.5})$ [5].

In order to minimize the effect of thermal resistances at interfaces, the 3ω measurements can be performed in two steps as shown in Ref. [5]:

(i) with the heater directly deposited on the substrate (S)
(ii) with the heater deposited on the film-substrate (f+S) system

This technique has been known as the differential technique. Even though Eq. (18.20) gives the exact solution, Eq. (18.19) remains the mostly used due to its simplicity and, therefore, the cross-plane thermal properties can be directly determined. Equation (18.19) can be used when the heat spreading inside the film is almost eliminated. In the case of anisotropic thin film, the error induced by Eq. (18.19) is of 5% when the ratio $e_{\mathrm{f}}\left(k_{\mathrm{f},xz}^{0.5}\right)/b$ is 0.1 [5].

Therefore, the 3ω method has been usually used to determine the cross-plane thermal conductivity in the case of thin films. An alternative solution for an anisotropic determination of the thermal properties lies in depositing different heaters of different widths on the sample surface [9]. The cross-plane thermal conductivity is first determined in the case of a heater width that is much larger than the film thickness (Eq. 18.19). The in-plane component can be determined in the case of a narrow heater as seen in the 3ω differential technique (Eq. 18.20) [9].

In the case of electrically conducting wires, an alternative method with the same principle as the 3ω method can be employed. 1D heat equation can be used to extract the thermal properties of the wires of interest [10]. The samples are the rod or filament-like specimens and suspended between the points of voltage contacts and the heat is sunk to the substrate through electrical contacts [10]. An alternating current is generated through the wire and the voltage $V_{3\omega}$ is measured. In this method, the measurements performed at low-frequency limits [10] ($\omega\tau \rightarrow 0$, where τ is the thermal constant time of the specimen) enables the determination of the thermal conductivity/thermal conductance of the sample. At high-frequency limits ($\omega\tau \rightarrow \infty$), the method enables the determination of the heat capacity of the sample [10–12]. This method has been used widely in the measurements of the thermal properties of different wires of different geometries [13–15].

For the determination of the in-plane component of thin films, different methods have been developed. Other AC techniques developed to determine the in-plane thermal properties of material will be discussed in the next part.

18.3.2 2ω Method

The 2ω method has been used to determine the in-plane thermal conductivity of thin films. As in the 3ω method, the 2ω method is based on AC heating of the sample through a long metallic wire deposited on the sample surface. However, the temperature detection is not the same where the heating and temperature detection are not performed through the same wire in the 2ω method.

The 2ω method principle consists of two or several parallel and long resistive wires that are deposited on the front surface of the sample of finite thickness d as shown in Fig. 18.5. While one of these wires is used to heat the sample with an alternative current, the other wire(s) detect the second harmonic temperature $T_{2\omega}$.

By measuring the electrical resistance change of the probing wires $\Delta R_{2\omega}(x)$, the surface temperature $T_{2\omega}(x)$ is detected:

$$\Delta R_{2\omega}(x) = R_0(x).\beta.T_{2\omega}, \tag{18.24}$$

where x is the distance from the heating wire, $R_0(x)$ is the electrical resistance of the probing wire at ambient temperature T_a and β is the temperature coefficient of the electrical resistivity.

A direct current i_0 of small magnitude is generated through the probing wires so that the $\Delta R_{2\omega}(x)$ measurements and the second harmonic voltage at the probing wires $\Delta V_{2\omega}$ can be measured:

$$\Delta V_{2\omega}(x) = \Delta R_{2\omega}(x)\, i_0 = R_0(x).\beta.T_{2\omega}(x).i_0 \tag{18.25}$$

Based on these measurements, this technique is called the 2ω method. The probing wires should be shorter than the heating one to detect mainly the lateral heat spreading from the heating wire (see Fig. 18.5) [16, 17].

In order to describe the heat propagation using 1D heat equation (along the X-axis), the thermal diffusion length μ_2 should be larger than the thickness of the film ($\mu_2 \gg d$) [16, 17]. In this case, the temperature can be considered uniform along z, and μ_2 is given as

$$\mu_2 = \sqrt{\frac{\alpha_f}{2\pi f}}, \tag{18.26}$$

where α_f is the film thermal diffusivity.

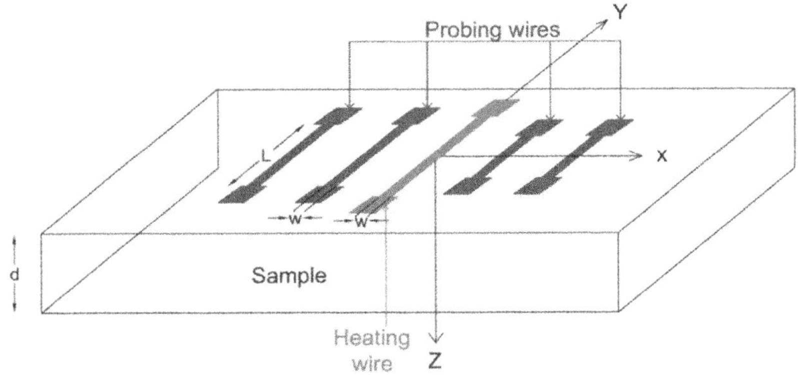

Figure 18.5 Schematic of the 2ω method principle.

The second harmonic temperature $T_{2\omega}(x)$ at a distance x from the heating sensor is

$$T_{2\omega}(x) = A_1\sqrt{\frac{\alpha_f}{2\omega}}e^{-\frac{x}{\mu_2}}e^{-i\frac{x}{\mu_2}-i\frac{\pi}{4}} \tag{18.27}$$

and

$$A_1 = \frac{P_d}{2d\lambda_f w_f}, \tag{18.28}$$

where w_f is the width of the film and P_d denotes the power dissipated in the heater.

As can be seen in Eq. (18.27), the thermal diffusivity and thermal conductivity of the film can be determined simultaneously. The measurements can be performed either as a function of the current frequency f through one probing wire (located at a distance x from the heating wire) or as a function of the distance x through multiple probing wires at a considered frequency f. In both cases, the distance x has to be at least larger than the thickness of the sample [17, 18]. For distances smaller than the sample thickness d, the Z part of the heat flux cannot be neglected and, as a result, the one-dimensional conduction model (Eq. 18.27) is not valid anymore.

The measurements are usually performed at different controlled levels of temperature using cryostats or ovens while generating small current amplitude I_{ac} through the heating wire. However, it appears that an alternative solution can be used to control the sample heating at temperatures larger than the ambient. In fact, the sample temperature at any instant is the sum of the two components T_{DC} and $T_{2\omega}$. When an alternating current passes across the heater, this latter is expected to act as a low-pass filter. The alternative solution consists of adjusting the frequency of the current in order to get a condition where $T_{DC} \gg T_{2\omega}$. Satisfying this condition enables to determine the thermal properties at the temperature T_{DC} of the considered materials through the variation of $T_{2\omega}$. The temperature T_{DC} is controlled by the amplitude of the alternating current generated within the heater, and the thermal properties of the sample at this T_{DC} temperature are determined following the variations of $T_{2\omega}$ (see Eq. 18.27). In this case, the determination of the heat losses to the environment (convection, radiation) is not required as the sample temperature $T_{sample} = T_{DC}$ is measured experimentally through the probing wires:

$$T_{DC,exp} = (1/N)\sum_{i=1} (R(x_i) - R_0(x_i))/(R_0(x_i).\beta) + T_a, \tag{18.29}$$

where N is the number of the probing wires and i is an index for a considered probing wire.

Moreover, the 2ω method appears to be less sensitive to the thermal resistance R_{th} between the wires and the film (sample) [17, 18].

We note that characterizing the thermal properties of different films can nevertheless induce the re-dimensioning of the system (sample thickness, resistive wire dimensions and the distances between the resistive wires) in order to ensure the validity of Eq. (18.27).

Experimentally, the 2ω method has been always used to determine the in-plane thermal properties of thin films. It appears that with a resolution of the 2D heat equation (*XZ* regime) in the Fourier space, the cross-plane thermal properties can be extracted [15]. $T_{2\omega}$ can be measured at the sample back surface through one probing wire as a function of the current frequency as shown in Fig. 18.6.

The comparison between 2D analytical and numerical simulations showed the validity of the method for the determination of cross-plane thermal properties of silicon thin films [15]. Yet, no experimental data are available in the literature for such determination as experimental difficulties are encountered especially in the

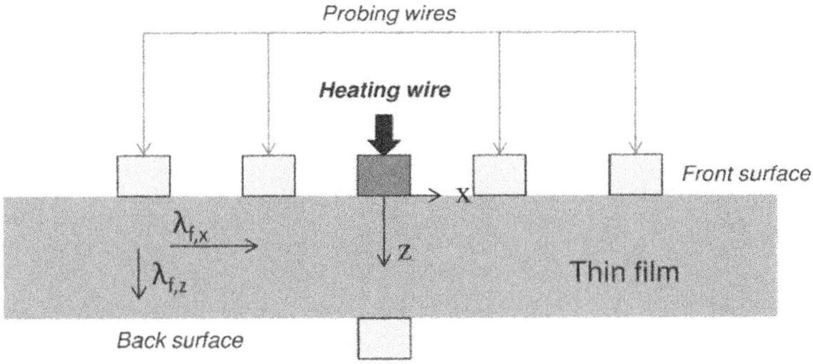

Figure 18.6 Schematic of the 2ω method principle for simultaneous characterization of in-plane and cross-plane thermal properties of a material shaped as a suspended membrane.

alignment of the heating wire and the probing wire located at the sample back surface.

18.4 Photothermal and Photoacoustic Techniques

Photothermal effect consists of heating the media by the absorbed light as the result of non-radiation decaying of the photoinduced excitation [19]. The main advantage of these methods is the possibility to heat the sample without any pretreatment of the surface.

The photothermal processes are accompanied by heating of the gas adjacent to the surface of the structure, emitting of infrared (IR) radiation, appearance of thermoelastic deformations in the material (Fig. 18.7), etc.

The registration of the IR radiation from the top [20] or bottom [21] surfaces of the sample is the basis of the laser-flash methods. On the other hand, the detection of the thermally induced change of the optical properties of the sample or adjusted gas is used in the thermal lens techniques [22].

In this section, we will discuss ones of the widely used photothermal methods: the photoacoustic techniques. The photoacoustic effect is the transformation of absorbed radiation into acoustic

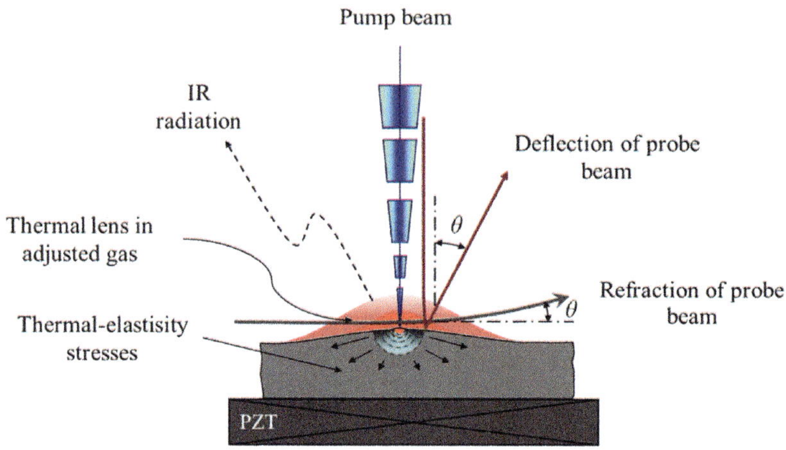

Figure 18.7 Schematic representation of the thermo-induced processes that occur during photothermal conversion: θ is an angle of test beam.

pressure as a result of media heating. There are numerous methods for the registration of the photoinduced acoustic perturbation. Nowadays, the most widely used photoacoustic techniques are gas-microphone [19, 23] and piezoelectric [24, 25] techniques. In this section, we will discuss most popular gas-microphone technique application for the study of thermal properties of amorphous semiconductor structures.

In the gas-microphone PA techniques, the light is periodically modulated and focused on the sample surface. The modulation is carried out using mechanical interruption of light beam [26] or modulation of the light source power [27].

There are several mechanisms of acoustic response formation in the PA cell. Among them, the main mechanism is the so-called "gas-piston" [28]. The irradiation of the sample surface leads to the expansion of the adjusted gas layer. Therefore, the periodic heating of the sample surface causes the periodic oscillation of the pressure in the PA cell. The pressure oscillation can be detected with the microphone (see Fig. 18.8).

In PA experiments, the dependence of the amplitude and phase shift of the pressure oscillation on the modulation frequency is

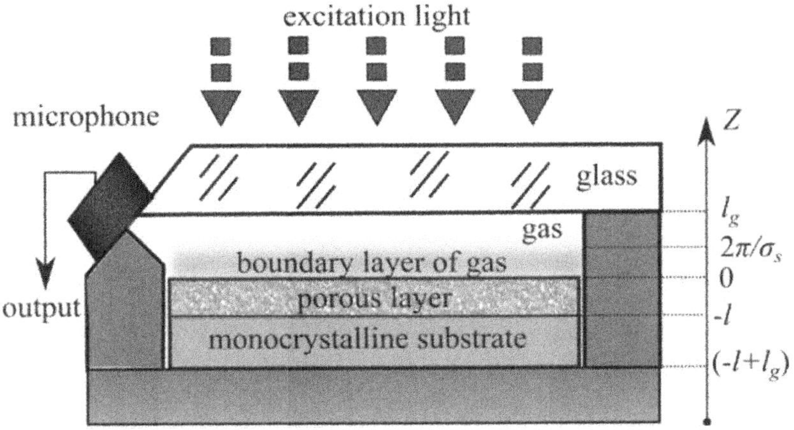

Figure 18.8 Schematic sketch of an experimental setup for photoacoustic signal measurement by the gas-microphone method in classical configuration.

usually measured. It is clear that this dependence is sensitive to the thermal and optical properties of the sample.

Due to the "gas piston" model, the temperature distribution in the gas layer defines the pressure oscillation inside the PA cell. The detected voltage and the pressure oscillation can be presented in the following form:

$$U(\omega) \sim p(\omega) \sim \int_0^{-\mathrm{Abs}(2\pi/\sigma_g)} \theta(0)\exp(\sigma_g z)\, dz \tag{18.30}$$

with

$$\sigma_g = \sqrt{(i\omega\rho_g c_g)/\chi_g}, \tag{18.31}$$

where $\theta(0)$ is the temperature on the surface of the sample and Abs $(1/\sigma_g)$ is the heat diffusion length in the gas. From Eq. (18.30), it can be seen that for the simulation of photoacoustic response, one should resolve a heat conduction problem in the considered sample.

In most cases, it is enough to apply the Fourier-based heat conduction equation that can be presented in the following form:

$$\frac{d}{dz}\left(\chi(z)\frac{d\theta}{dz}\right) - ic\rho\omega\theta = f(z), \tag{18.32}$$

where c is the specific heat, ρ is the density, ω is the modulated frequency and χ is the thermal conductivity. The function $f(z)$ describes the spatial distribution of the heat sources in the sample. Based on the Beer–Lambert–Bouguer law $f(z)$ can be presented in the following form:

$$f(z) = I_0(1 - R)\alpha \exp\left(-\int_0^z \alpha(z')dz'\right), \tag{18.33}$$

where I_0 is the intensity of radiation source and α is an absorption coefficient.

In the case of strong light absorption, the PA technique in classical configuration could be adapted for evaluation of thermal conductivity of different materials.

Let us consider the application of this technique for the thermal conductivity evaluation of the partially amorphous porous silicon (PSi) [29]. The samples of the partially amorphous silicon were obtained as a result of the irradiation of the porous silicon grown on the Si substrate with swift heavy ions.

The frequency region of the light modulation was chosen to satisfy the condition that the thermal diffusion length PSi is larger than the porous layer thickness. In this case, the pressure variation in the PA cell can be presented as

$$p(\omega) \sim \int_0^{-\infty} T(z,\omega)dz = -\frac{l_{\mathrm{por}}}{\chi_{\mathrm{por}}}\sqrt{\frac{\chi_{\mathrm{g}}}{i\omega c_{\mathrm{g}}\rho_{\mathrm{g}}}}. \tag{18.34}$$

The theoretically obtained linear dependence of the pressure in the PA cell on modulation frequency range is in good agreement with the experimental results as shown in Fig. 18.9. Furthermore, the absolute value of the signal amplitude is inversely proportional to the value of thermal conductivity. Thus, with a sample of well-known thermal conductivity, these values can be evaluated. In the considered case, the non-irradiated part of the sample was used as a reference sample (See the inset in Fig. 18.9):

$$\frac{p^{\mathrm{I}}(\omega)}{p^{\mathrm{NI}}(\omega)} = \frac{\chi_{\mathrm{por}}^{\mathrm{NI}}}{\chi_{\mathrm{por}}^{\mathrm{I}}}, \tag{18.35}$$

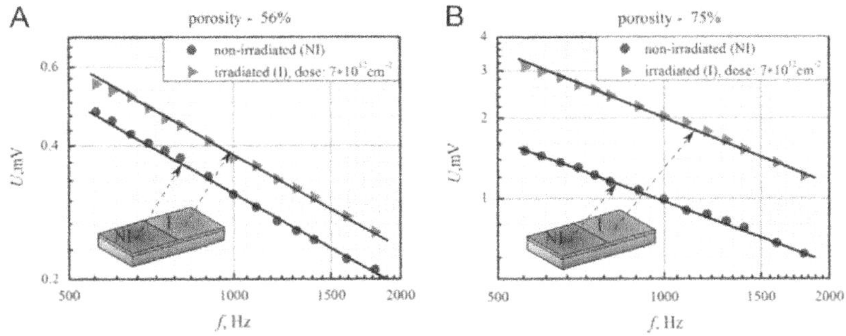

Figure 18.9 The amplitude-frequency dependences of the PA signal for the irradiated (by triangles) and non-irradiated (by dots) PSi samples with porosities 56% (A) and 75 % (B).

where indexes I and NI denote, respectively, the irradiated and non-irradiated parts of Psi sample.

This approach was applied for the evaluation of thermal conductivity of PSi samples with three different porosities and with different amorphous fractions as shown in Fig. 18.10. As we can see, the increase of amorphous fraction causes a reduction of thermal conductivity for all samples. The data obtained by Raman technique and Scanning Thermal Microscopy (SThM) [30] are also represented Fig. 18.10. The amorphization leads to the decrease of the thermal conductivity of porous silicon up to three times. As can be seen, the PA technique gives the possibility to evaluate thermal conductivity with important amorphous fraction and can be applied for samples with a big porosity.

Hereinabove, the simplest case of the strong optical absorption was considered. In the case of a finite value of the optical absorption coefficient, the analysis of the PA response formation is more complicated. However, the difference between the PA signals excited with sources of different wavelengths can be useful for more precise evaluation of the thermal properties of a partially amorphous nanostructured material. It is especially useful for the study of the thermal conductivity of the thin porous film grown on monocrystalline substrate. Particularly, it was successfully applied for the studies of the thermal properties of microporous silicon [31] and mesoporous germanium [32].

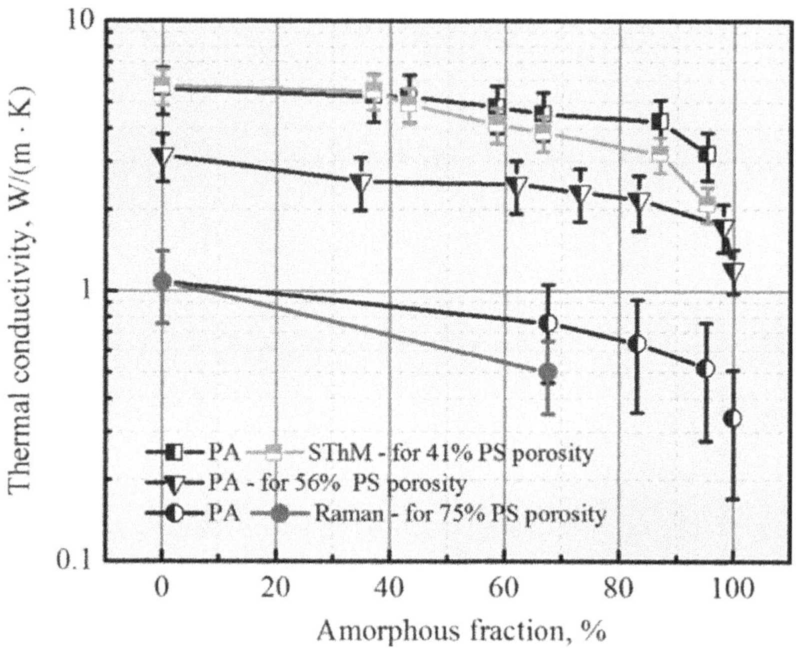

Figure 18.10 Dependences of thermal conductivity on amorphous fraction for PSi structures with different porosity, obtained by scanning thermal microscopy, Raman and PA techniques.

Let us consider the typical case for the semiconductors where increasing the wavelength of the incident radiation induces the decrease the optical absorption coefficient of the material. In the case of two-layered system, using two different wavelengths (for example, red and blue light) makes the top layer optically opaque for one wavelength and transparent for the other (see Fig. 18.11). Therefore, the heat sources located mainly near the top (case of blue light) or near the bottom (case of red light) film surfaces. As a result, the parameters of the PA signal will be different for the different sources. The amplitude-frequency responses of the PA signal excited by the light sources with wavelengths 470 and 670 nm in mesoporous germanium are shown in Fig. 18.12. The simulation of these responses allows evaluating the thermal conductivity of the sample.

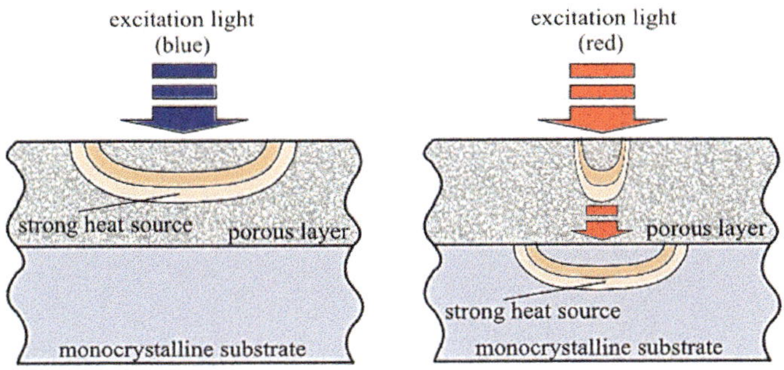

Figure 18.11 Schematic view of creating a heat source in the sample by its irradiation with blue and red light.

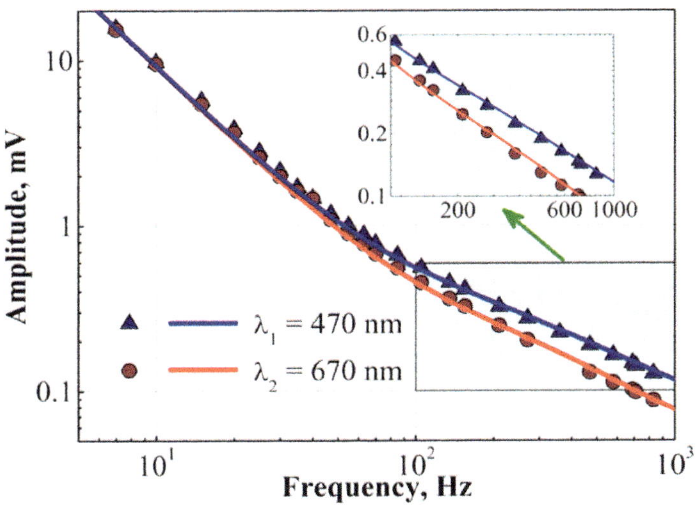

Figure 18.12 Experimental values of the amplitude-frequency dependences (markers) for the meso-PGe layers at different wavelengths of excitation light (470 and 670 nm) and the theoretical fitting of the experimental results (continuous lines).

18.5 Conclusions

The methods involving a periodic heating to characterize the thermal properties of crystalline/amorphous nanostructured materials were detailed in this chapter.

The first part was dedicated to present the electrothermal techniques that involve electrical means as heating source. These techniques consist of depositing metallic wires on the surface of the samples of interest. The electrothermal methods can be classified into two classes: 3ω method and 2ω method. The 3ω method is usually used to determine the cross-plane component of the thermal properties. The wire deposited on the sample surface is used as heater and temperature sensor. This method enables the evaluation of the in-plane component while modifying the heater width. However, more attention should be paid to the series of thermal resistances between the heating wire and the sample and/or between the film of interest and the substrate. This parameter appears to be crucial in the interpretation of the measurements.

The 2ω method is mostly used for the characterization of in-plane components of the sample thermal properties. A network of resistive wires is deposited on the sample surface. One of these wires acts as a heater while the other wires are temperature sensors. The 2ω method appears promising in the determination of the cross-plane thermal properties by depositing a probing wire on the back-surface of the sample. However, experimental difficulties prevent this type of characterization using such method.

Many factors can influence the thermal characterization of samples using electrothermal methods. We mention the dimensions of the wires and sample, frequency of the heating wire and distance between the wires (case of 2ω method). These parameters must be adjusted to ensure the validity of the physical model that is used to interpret the experimental measurements.

The photothermal and photoacoustic techniques for the study of the thermal properties of the crystalline/amorphous nanostructured material were discussed. Particularly, the gas-microphone photoacoustic technique was described in details. Moreover, the usage of the double wavelength photoacoustic technique for the study of the thermal conductivity of crystalline/amorphous nanostructured materials was demonstrated.

References

1. D. G. Cahill, *Rev. Sci. Instrum.* 61, 802 (1990).
2. J. Y. Duquesne, D. Fournier, and C. Frétigny, *J. Appl. Phys.* 108, 086104 (2010).
3. H. Wang and M. Sen, *Int. J. Heat Mass Transf.* 52, 2102 (2009).
4. A. Jacquot, B. Lenoir, A. Dauscher, M. Stölzer, and J. Meusel, *J. Appl. Phys.* 91, 4733 (2002).
5. T. Borca-Tasciuc, A. R. Kumar, and G. Chen, *Rev. Sci. Instrum.* 72, 2139 (2001).
6. D. A. Borca-Tasciuc and G. Chen, *J. Appl. Phys.* 97, 084303 (2005).
7. J. R. A. David G. Cahill, and M. Katiyar, *Phys. Rev. B* 50, 6077 (1994).
8. J. H. Kim, A. Feldman, and D. Novotny, *J. Appl. Phys.* 86, 3959 (1999).
9. Y. S. Ju, K. Kurabayashi, and K. E. Goodson, *Thin Solid Films* 339, 160 (1999).
10. L. Lu, W. Yi, and D. L. Zhang, *Rev. Sci. Instrum.* 72, 2996 (2001).
11. C. Dames and G. Chen, *Rev. Sci. Instrum.* 76, 1 (2005).
12. O. Bourgeois, T. Fournier, and J. Chaussy, arXiv preprint cond-mat/0608705, 2006.
13. H. Ftouni, C. Blanc, D. Tainoff, A. D. Fefferman, M. Defoort, K. J. Lulla, J. Richard, E. Collin, O. Bourgeois, *Physical Review B* 92, 125439 (2015).
14. L. Li, C. Jin, S. Xu, J. Yang, H. Du, and G. Li, *Nanotechnology* 25, 415704 (2014).
15. J. S. Heron, T. Fournier, N. Mingo, and O. Bourgeois, *Nano Lett.* 10, 2288 (2010).
16. A. Assy, Development of Two Techniques for Thermal Characterization of Materials: Scanning Thermal Microscopy (SThM) and 2ω Method, PhD Thesis at INSA Lyon (2015).
17. A. Assy, S. Gomès, P. Chantrenne, N. Pavye, J. Parasuraman, X. Kleber, and P. Basset, *Microelectronics J.* 45, 508 (2014).
18. A. Bejan, *Heat Transfer* (John Wiley & Sons, New York, 1993).
19. A. Rosencwaig, *Photoacoustics and Photoacoustic Spectroscopy* (Wiley and Sons, Chichester, 1980).
20. M. Chirtoc, N. Horny, J. F. Henry, A. Turgut, I. Kökey, I. Tavman, and M. Omastová, *Int. J. Thermophys.* 33, 2110 (2012).
21. M. Ruoho, K. Valset, T. Finstad, and I. Tittonen, *Nanotechnology* 26, 195706 (2015).

22. U. Bernini, R. Bernini, P. Maddalena, E. Massera, and P. Rucco, *Appl. Phys. A* 81, 399 (2005).
23. P. Lishchuk, D. Andrusenko, M. Isaiev, V. Lysenko, and R. Burbelo, *Int. J. Thermophys.* 36, 2428 (2015).
24. W. Jackson, *J. Appl. Phys.* 52, 1137 (1981).
25. M. Isaiev, D. Andrusenko, A. Tytarenko, A. Kuzmich, V. Lysenko, and R. Burbelo, *Int. J. Thermophys.* 35, 2341 (2014).
26. L. A. Skvortsov, *Quantum Electron.* 43, 1 (2013).
27. S. Alekseev, D. Andrusenko, R. Burbelo, M. Isaiev, and A. Kuzmich, *J. Phys. Conf. Ser.* 278, 012003 (2011).
28. A. Rosencwaig and A. Gersho, *J. Appl. Phys.* 47, 64 (1976).
29. M. Isaiev, P. J. Newby, B. Canut, A. Tytarenko, P. Lishchuk, D. Andrusenko, S. Gomès, J.-M. Bluet, L. G. Fréchette, V. Lysenko, and R. Burbelo, *Mater. Lett.* 128, 71 (2014).
30. P. J. Newby, B. Canut, J.-M. Bluet, S. Gomes, M. Isaiev, R. Burbelo, K. Termentzidis, P. Chantrenne, L. G. Frechette, and V. Lysenko, *J. Appl. Phys.* 114, 014903 (2013).
31. A. I. Tytarenko, D. A. Andrusenko, A. G. Kuzmich, I. V Gavril, V. A. Skryshevskii, M. V Isaiev, and R. M. Burbelo, *TPL* 40, 188 (2014).
32. M. Isaiev, S. Tutashkonko, V. Jean, K. Termentzidis, T. Nychyporuk, D. Andrusenko, O. Marty, R. M. Burbelo, D. Lacroix, and V. Lysenko, *Appl. Phys. Lett.* 105, 031912(1) (2014).

Chapter 19

X-Rays and Neutrons Spectroscopy for the Investigation of Individual Phonons Properties in Crystalline and Amorphous Solids

Stéphane Pailhès,[a] Valentina M. Giordano,[a] Pierre-François Lory,[b] Marc De Boissieu,[b] and Holger Euchner[c]

[a] *Institute of Light and Matter, CNRS, UCBL, UMR-5586, F-69622 Villeurbanne, Lyon, France*
[b] *Institut Laue-Langevin 6, rue Jules Horowitz 38042 Grenoble, France*
[c] *Institute of Materials Science and Technology, TU Vienna, Vienna, Austria*
stephane.pailhes@univ-lyon1.fr

19.1 Introduction

Today, the active global concern about energy issues has brought thermal engineering of materials at the forefront of current research, as thermal processes dominate the production and use of energy. Materials that can enhance or inhibit heat transfer can substantially improve thermal efficiency and energy sustainability. In this context, dielectric materials with ultra low thermal conductivity are of the greatest interest for a number of applications, from thermal

Nanostructured Semiconductors: Amorphization and Thermal Properties
Edited by Konstantinos Termentzidis

ISBN 978-981-4745-64-2 (Hardcover), 978-1-315-36445-2 (eBook)
www.panstanford.com

barriers to waste heat recovery through thermoelectric conversion. In the past decades, many efforts have been done for decreasing the thermal conductivity in crystalline materials by "phonon engineering," which have gone from nanostructuring and dimension reduction to introduction of disorder and crystal complexity [46, 56, 73, 82, 83].

In ordered and disordered dielectric materials, heat transport is mainly assured by atomic vibrations, described by quantized quasi-particles, called phonons. During its lifetime a phonon traveling through the material transports a quantum of heat flux, given by the product of its energy and its velocity. Generally speaking, the lattice thermal conductivity is given by the piling up of the contributions of all phonons and it is clear that a low thermal conductivity can be achieved by reducing the phonons velocity and lifetime. The number of phonon states in a material is given by the number of vibrational degrees of freedom, and typically the phonon total spectrum is spread out on a quite large energy range, whose extension is determined by the Debye energy, typically up to $\approx$50 meV. However, the relative phonon contributions are heavily energy and temperature dependent, as the number of phonons occupying each phonons state follows the Bose–Einstein distribution. Consequently, the dominant phonon energies and phonon scattering processes, which govern the thermal properties, depend on temperature. For example, in a typical thermoelectric semiconductor material, at room temperature, the majority of thermal energy is carried by phonons with energies larger than 5 meV, with lifetimes of the order of 10 ps, which in terms of energy corresponds to a ten of μeV. This would indicate that depending on the temperature range of interest phonon engineering should aim at phonons of different energy. It is thus clear that the modeling of thermal transport in crystals and glasses requires getting a full understanding of phonon dynamics at all energies and temperatures: The knowledge of the individual properties of phonons is essential and the understanding of the mechanisms underlying the single phonon lifetime fundamental for new materials engineering. Phonons spectroscopy thus turns out to be a very ambitious task, as it aims to get an insight into such properties at different energy and temperature scales.

In this chapter, we will present the experimental techniques currently used for measuring phonon velocities and lifetimes: inelastic

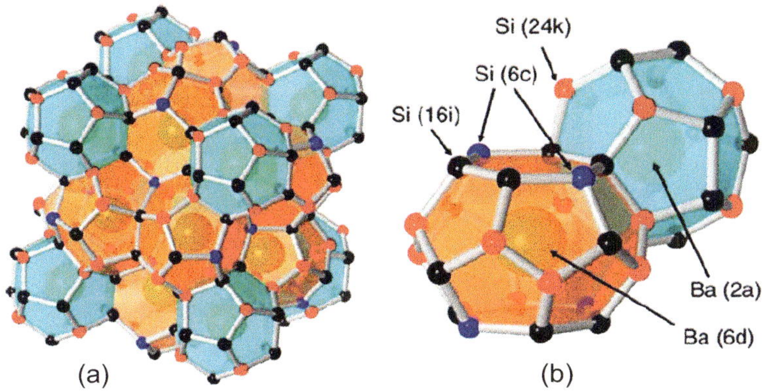

Figure 19.1 Clathrate. The clathrate-I structure of Ba_8Si_{46} with 20-atoms polyhedra (light blue) and 24-atoms polyhedra (orange), centered by Ba1 (2a) and Ba2 (6d) atoms, respectively (space group Pm-3n (223) with lattice parameter $a = 10.3$ Å) [21, 80]. It is a guest-host system in which Ba fills large polyhedral cavities in a covalently bound, fully ordered Si framework. Ba_8Si_{46} has a low thermal conductivity of about 1.7 W/m.K at ambient temperature [57]. (b) Representation of face sharing $BaSi_{20}$ and $BaSi_{24}$ cages as building units of type-I clathrates.

neutron and X-rays scattering. After a short reminder of the phonon properties (Section 19.2.1), we will briefly give the measurement principle (Sections 19.2.2 and 19.2.3) and show the experimental limitations, which currently hinder the measurement of phonon lifetimes in most crystals (Section 19.2.4). For a pedagogical aspect, we will consider an example of phonons measurement in a complex thermoelectric crystal structure—a so-called clathrate—based on the results published in [29, 57] (The crystal structure of the simplest representative clathrate of type 1, the binary Ba_8Si_{46}, is shown in Fig. 19.1a. It is a guest-host system in which Ba fills large polyhedral cavities, shown in Fig. 19.1b, in a covalently bound, fully ordered Si framework. It is a silicon based crystal with a remarkable low thermal conductivity at room temperature of 1.7 W/mK.). In the second part of this chapter, we will detail the differences in the atomic dynamics of an amorphous system with respect to crystals (Section 19.3), and what can be measured in an inelastic experiment performed on a disordered system. We will conclude with an example, and more specifically, presenting new data collected on the same metallic glass presented in Chapter 13 (Section 19.3.4).

19.2 Measurements of Phonons in Crystals

19.2.1 Phonons

We recall here concisely the concept of phonon as defined in harmonic crystals. There is an abundant literature on the topic [3, 10, 26, 49, 77] where all the formalism is deeply presented. Here the goal is to describe the quantum particle, the phonon, that we want to probe by neutrons or X-rays inelastic scattering in a crystal.

A crystal contains an extremely large number (N) of atoms, of the order of the Avogadro number ($N \sim 6 \cdot 10^{23}$), chemically bounded and arranged in an ordered microscopic structure. Each of them being allowed to move in the three-dimensional space, in order to determine the physical properties of this assembly of atoms it is necessary to solve a system of 3N equations of motion. Taking advantage of the translational symmetry in a crystal, the number of equations can be reduced to $3r$ with r the number of atoms in the primitive cell. Assuming that they vibrate with small amplitudes around their equilibrium positions (harmonic approximation), the theory of the lattice vibrations in harmonic crystals tells us that all the atomic motions can be described by considering a gas of $3r$ independent quantized quasi-particles or waves, called phonons or normal modes of vibration, whose occupation number obeys the Bose–Einstein statistics.

A phonon is a collective displacement of ions, defined by

- a wave vector, $\vec{q}$, giving the propagation direction of the lattice distortion;
- a polarization vector, $\vec{e}_{\vec{q}}$, giving the direction of the atomic motion with respect to the propagation vector. In the long-wavelength limit, i.e., $\vec{q} \rightarrow 0$, the two vectors are mutually perpendicular and parallel for shear (transverse) waves and compression (longitudinal) waves, respectively. For each wave vector, there are two possible independent transverse polarizations;

- a dispersion relation, $\omega(\vec{q})$, relating the phonon's angular frequency, ω, to its wave vector, and defining a group velocity $v(\vec{q}) = d\omega/d\vec{q}$. In the Debye approximation, the dispersion relation is linear, $\omega(\vec{q}) = v_s q$ where v_s is the average sound velocity;
- a lifetime, τ, measuring the average time a given phonon exists before colliding with defects, phonons, electrons, etc.

The phonon states are obtained by diagonalizing the $3r$ linear coupled differential equations of atomic motion whose determinant is the so-called dynamical matrix, $D(\vec{q})$, related to the spatial Fourier transform of the force constants matrix. In this basis, the equations of motion take thus the simple form:

$$\omega_j(\vec{q})^2.\vec{e}_j(\vec{q}) = D(\vec{q}).\vec{e}_j(\vec{q}), \tag{19.1}$$

where j defines one out of $3r$ phonon branches. Each Bravais lattice has always three acoustic modes (two transverses and one longitudinal) and $3r$-3 optical modes. Phonon energies depend on the wavelength of the atomic distortion, the masses of the atoms, and the binding forces that connect them. In most solids ω is of the order of a few terahertz (THz), corresponding to phonon energies of a few meV (1 THz $\sim$ 4.18 meV $\sim$ 33.7 cm^{-1}).

Over the past decade, quantum mechanical packages based on density functional theory (DFT) have been developed and are now routinely used to provide the phonon dispersion relations by calculating the harmonic force constants [9, 84]. In Fig. 19.2, we present the phonon dispersions computed with the codes VASP [45] and Phonons [58] going from the simple case of silicon to the more complex case of the silicon clathrate Ba_8Si_{46}. The dispersions are drawn along the reciprocal space direction (111) where the three acoustic branches are not degenerate so that the two transverse and longitudinal acoustic branches are separated.

Phonon lifetime is the result of all possible interactions that the phonon experiences with its environment. In a spectroscopic measurement, this lifetime leads to a finite energy width of the phonon peak which can be resolved if it is comparable or larger than the energy width of the instrumental resolution. There exist several scattering channels reducing the phonon lifetime, which are in general efficient in separated regions of the phase space

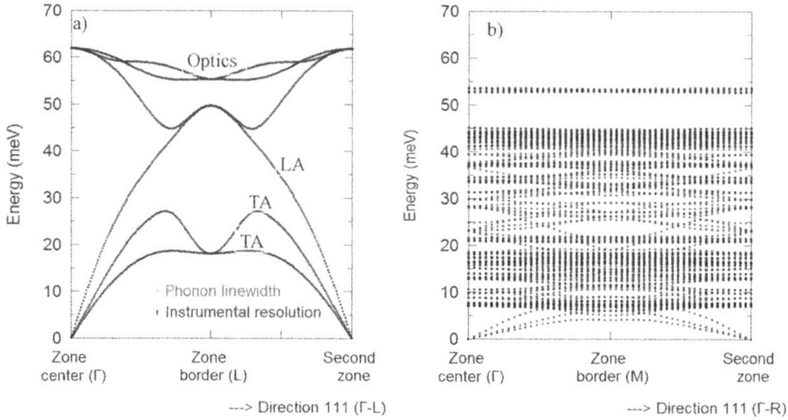

Figure 19.2 Simulated phonon dispersion relations in simple and complex crystals. Phonon dispersions along the reciprocal direction Γ-L obtained from ab initio DFT calculations in pure Si (a) and in a silicon clathrate Ba_8Si_{46} (b) [57]. The calculations were conducted using the DFT code VASP [45] and the PHONON package [58]. TA and LA stand for longitudinal and transverse acoustic phonons. In figure (a) is represented the size of a common phonon linewidth (0.1 meV, red line almost a dot at this scale) for phonons with energies of a few meV in crystals and a typical value of the instrumental energy resolution (1 meV, blue line) obtained on neutrons or X-rays spectrometers configured for phonons measurements.

and in different temperature ranges [4, 19, 40, 42]. These are mainly differentiated between two-phonons scattering processes (a phonon collides and transforms into another phonon) and three-phonons scattering processes (a phonon decays into two phonons and vice versa). The latter, supposed to dominate at high temperature, is the most relevant for applications but its calculation is extremely challenging as it requires the knowledge of the anharmonic component of the interatomic potential. Indeed, in the harmonic approximation, phonons are non-interacting and have therefore an infinite lifetime. This issue has been theoretically tackled in recent years thanks to a considerable progress in ab initio computational methods [14, 20, 28, 79].

In the general case, the order of magnitude of the lifetime of an acoustic phonon with a frequency of about 1 THz ($\hbar\omega \sim 4$ meV) at ambient temperature is between ten and a few hundreds of picoseconds, which in terms of energy corresponds to a few hundreds and a few μeV, respectively. These typical values are compared to the typical instrumental energy resolution ($\sim$1 meV) of a spectrometer (either X-rays or neutrons) configured for phonons measurements in Fig. 19.2a, where they are represented as vertical bars. At the energy scale of the phonon dispersions, the phonon linewidth appears as an almost invisible dot. Already here, one simply realizes the experimental difficulty to extract from the data the intrinsic phonon lifetime as it will be further discussed in the following. The difficulty arises from the fact that the spectrometers are optimized to map out phonon energies over a wide energy range of a few tens of meV (typically related to the Debye energy) and, simultaneously, we ask them to resolve the energy widths of acoustic phonons which are one hundred times smaller. Therefore, while phonon energies and thus velocities are experimentally mastered properties, the measurement of acoustic phonons lifetime turns out to be primitive especially in very simple crystals like Si. Of course, the instrumental resolution can be improved, more specifically in the case of neutrons, but at the price of a great reduction of intensity which can only be compensated by an increase of the sample volume. On the other hand, larger phonon linewidth in the meV range as recently found in some poor thermal conductors like in the thermo-electrics PbTe [24] or $AgSbTe_2$ [48] can more easily be detected.

19.2.2 Basics Concepts of a Scattering Experiment

Inelastic scattering either by neutrons or X-rays aims at probing the properties of individual quantized excitations in materials, such as phonons, by scanning step by step the entire phase space, i.e., in energy ($\hbar\omega$) and reciprocal space $\vec{Q}$. As shown above in the case of phonon dispersion relations in Fig. 19.2, such quantized excitations are defined by discrete values of wave vectors and energies in the phase space. They are the channels through which the material is able to exchange energy with its environment. The general idea of a scattering experiment is to measure the energy ($\hbar\omega$) and the wave vector ($\vec{Q}$) transfers between a beam of probe particles and the material. The exchange can take place only when the pair ($\vec{q}$, $\hbar\omega$) corresponds to the one of an elementary excitation in the material

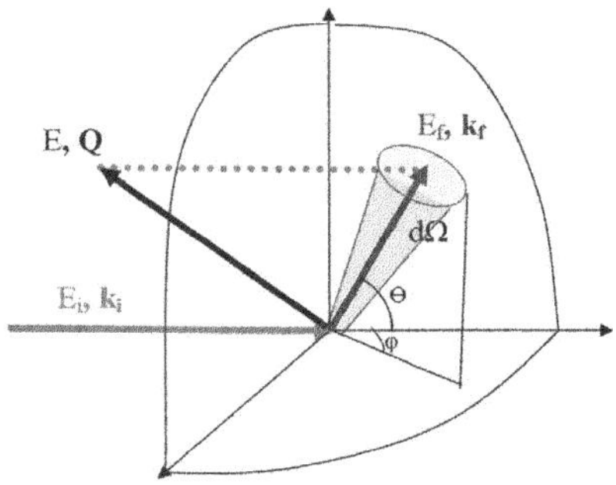

Figure 19.3 Principle of a scattering experiment. The figure illustrates the typical scattering geometry. A beam of probe particles filtered from a source with a given energy and wave vector (E_i, k_i) impinges onto a sample. The detection system collects the scattered particles traveling in the direction of k_f into a solid angle element dΩ with energy between E_f and E_f+dE_f. The ratio between the flux of detected particles and the incident flux onto the sample defines the double differential cross section (Eq. 19.2). The momentum and the energy transfer to the sample (E and $\vec{Q}$) are determined from the energy and momentum conservation laws (Eq. 19.3).

(the relation between $\vec{Q}$ and $\vec{q}$ is given below). The scheme of a typical inelastic scattering experiment is illustrated in Fig. 19.3. The sample is illuminated by an incident beam of probe particles prepared with a well-defined wave vector $\vec{k}_i$ and energy E_i. The initial flux of these particles onto the sample is measured. After interaction with the sample, the probe particles may be scattered with different final wave vectors, $\vec{k}_f$, and/or final energies, E_f. The scattering direction and the energy of the scattered probe particles are analyzed. The observable is then the flux of scattered particles emitted in the direction of $\vec{k}_f$ into a solid angle element dΩ and with energy between E_f and $E_f + dE_f$. The ratio between the scattered and the initial flux of probe particles gives the double differential cross section:*aaa bb*

$$\frac{d^2\sigma}{d\Omega dE_f} = \frac{\begin{array}{c}\text{flux of probe particles with energy}\\ \text{between } E_f \text{ and } E_f + dE_f \text{ scattered into } d\Omega\end{array}}{\begin{array}{c}\text{initial flux of probe particles}\\ \text{onto the sample with the energy } E_i\end{array}} [\mathrm{m}^2 \cdot \mathrm{eV}^{-1}] \tag{19.2}$$

The momentum and energy conservation imposes for the transferred momentum $\vec{Q}$ and energy ω:

$$\begin{aligned} \vec{Q} &= \vec{k}_f - \vec{k}_i \iff Q^2 = k_f^2 + k_i^2 - 2k_i k_f \cos(\theta) \\ \hbar\omega &= E_f - E_i, \end{aligned} \tag{19.3}$$

where θ is the scattering angle between the incident and scattered particles beams.

In the case of the elastic scattering, it is $E_f = E_i$, $|k_i| = |k_f|$, so that there is no energy transfer ($\hbar\omega = 0$) between the probe particles and the sample and Eq. (19.3) results in Bragg's law: $Q = 2k_i \sin(\theta/2)$. On the other hand, an inelastic scattering, for which $E_f \neq E_i$, is detected when the probe particles create or annihilate an excitation $(\vec{q}, \omega)$ in the material. If $\hbar\omega \geq 0$, i.e., $E_f \geq E_i$, the probe particle gains energy from the scattering process, whereas in the sample an excitation gets annihilated. If on the contrary, $\hbar\omega \leq 0$, i.e., $E_f \leq E_i$ the probe particle transfers energy to the sample, in which excitations are created. These two scenarios correspond to the so-called anti-Stokes and Stokes sides of the spectrum, respectively.

Here, we are interested in the determination of the double differential cross section in the case of a scattering process of the probe particle with a phonon in the material. Establishing the useful final formula relating the measurable observable to the crystal dynamics requires to go into the formalism of the inelastic scattering theory in the case of a harmonic crystal. However, this is beyond the scope of this chapter, the derivation of the formulas being given in detail in many textbooks and review articles [18, 37, 44, 47, 71, 76]. We only present here briefly the main concepts of the scattering theory.

The scattering of a beam of particles by a crystal is obviously complex since it involves a large number of scattering processes (of the order of the Avogadro number), which are a priori not independent, between the particles and the target atoms. The double differential cross section is expressed in terms of transition probabilities from initial to final states of the global system {probe particles + material}. The quantum mechanical formalism used to calculate these probabilities is derived from the one developed to solve the Schrödinger equation that describes a single particle diffusion in a static potential using the Born approximation. Here the particle is replaced by the target atoms and the probe particles and the number of equations is multiplied by 10^{23}. In a scattering experiment, one can distinguish three main components: the probe particles, the target material, and the interaction. The main approximations introduced to disentangle these components in the double differential cross section are the following:

- The interaction between the probe particles and the target material is assumed to be very weak such that it does not change their eigenstates. The Born approximation is then employed to express the quantum states of the global system after interaction as a product of the interacting potential and the states of the unperturbed system. The main role of the interacting potential is only to allow transitions between the unperturbed states of the global system. The transition probabilities are determined from the Fermi golden rule using the interacting potential as the transition operator and the energy conservation law. Secondly, in the context of a weak interaction, the states of the target do not depend on the position of the probe particles, so that all integrals over the spatial coordinates of the probe particles and the target can be factorized. Finally, the scattering processes are assumed to be independent and the interacting potential is decomposed into a sum of individual contributions. In this context, the interacting term is determined by the cross section of the scattering between the probe particles and an individual target atom described in [18, 37, 44, 47, 71, 76]. The atomic cross sections being tabulated for neutrons and X-rays, there is then no difficulty to switch from one to the other in the derivation of the double differential cross section.
- In a standard scattering experiment, the initial and final states of the target material are not determined. Only the transitions between them are probed, meaning that all possible transitions leading to the same changes of the probe particles states are equally detected. The above probabilities have then to be summed over the initial and final states of the target. While the former, defining the

target thermodynamical equilibrium, are known, the final excited states are not. As usual in this case, the interaction between the target and the probe particles is assumed to be an adiabatic perturbation, meaning that it builds up sufficiently slowly for the system to permanently remain in its equilibrium state. The final states are then expressed as a linear combination of the initial states.

The double differential cross section is given by a scattering function (also called dynamic structure factor), $S(\vec{Q},\hbar\omega)$, which gives the probability of a probe particle to go from the state $|k_i\rangle$ to the state $|k_f\rangle$ while exchanging the energy $\hbar\omega = E_f - E_i$ with the sample:

$$\frac{d^2\sigma_{\vec{k}_i \to \vec{k}_f}}{d\Omega dE_f} = \frac{k_f}{k_i} S(\vec{Q}, \hbar\omega) \tag{19.4}$$

The scattering function is asymmetric in terms of the frequency and wave vector as stipulated by the detailed balance equation:

$$S(-\vec{Q}, -\hbar\omega) = e^{-\hbar\omega\beta} S(\vec{Q}, \hbar\omega) \tag{19.5}$$

with $\beta = \frac{1}{K_B T}$.

If the system is invariant under inversion of the direction $\vec{Q}$, as it is the case for disordered systems like liquids and glasses and for crystals that possess an inversion center, the relation simplifies to $S(\vec{Q}, -\hbar\omega) = e^{-\hbar\omega\beta} S(\vec{Q}, \hbar\omega)$.

Although the intrinsic probabilities of the Stokes and anti-Stokes processes are identical, the factor $e^{-\hbar\omega\beta}$ introduces a difference due to the occupation of the initial states of the sample. As it is always smaller than one for $\hbar\omega \geq 0$, the collected signal for $\hbar\omega \leq 0$ will always be inferior to the one for $\hbar\omega \geq 0$. Eventually, at very low temperature, where the sample is in its ground state and no de-excitation is possible, $S(\vec{Q}, \hbar\omega) = 0$ for $\hbar\omega \leq 0$.

To compare the experiment with classical calculations obtained from molecular dynamics simulations, it is useful symmetrizing the scattering function by multiplying $S(\vec{Q}, \hbar\omega)$ by $e^{\frac{1}{2}\hbar\omega\beta}$. It can then be compared with simulations that provide functions real and even in time.

Table 19.1 Elastic and inelastic atomic scattering factors for X-rays and neutrons for Ba and Si atoms

Element	Z^2	M	σ_c	Z^2/M	σ_c/M
Ba	3136	137.1	3.23	22.8	0.023
Si	196	28.1	2.16	6.9	0.076

Note: Z is the atomic number, M is the atomic mass in g.mol^{-1} and σ_c is the neutron coherent scattering cross section in barn (10^{-24} cm^2)

19.2.3 Elastic Scattering

In the case of the elastic scattering, when there is no energy transfer between the probe particles and the material, the scattering depends only on the structural properties of the material. The partial differential cross section takes thus the form

$$\left(\frac{d^2\sigma}{d\Omega dE_f}\right)^{el}_{coh} = N_c\frac{(2\pi)^3}{V_c}|F_B(\vec{Q})|\sum_{\vec{G}}\delta(\vec{Q}-\vec{G}), \tag{19.6}$$

where N_c and V_c are, respectively, the number of unit cells in the crystal and the volume of the unit cell, and $F_B(\vec{Q})$ is the elastic form factor for the unit cell defined as

$$F_B(\vec{Q}) = \sum_j f_j e^{-W_j(\vec{Q})} e^{i\vec{Q}.\vec{d}_j}. \tag{19.7}$$

Here the sum runs over all the atoms j in the unit cell, located at positions $\vec{d}_j$, f_j is the atomic structure factor and B stands for Bragg. f_j differs between neutrons and X-rays: for neutrons it is the coherent atomic cross section, $f_j = \sigma_j$, while for X-rays it is the Fourier transform of the electron density which, in the limit of Q$\rightarrow$0, is equal to the atomic number Z, $f_j \sim Z$. Table. 19.1 reports as an example the values of the atomic structure factors in Ba and Si, constituent atoms of the binary clathrate Ba_8Si_{46}.

Equation (19.6) shows that for diffusive vectors, $\vec{Q}$, equal to reciprocal wave vectors, $\vec{G}$, Bragg peaks are measured with an intensity modulated by the elastic form factor. Figure 19.4 reports the reciprocal lattice points, whose size is modulated by the X-rays and neutrons form factors for the cubic crystal system (Pm-3n (223)) of the Ba_8Si_{46} clathrate in the plane defined by the axis ([h h 0],[0 0 1]).

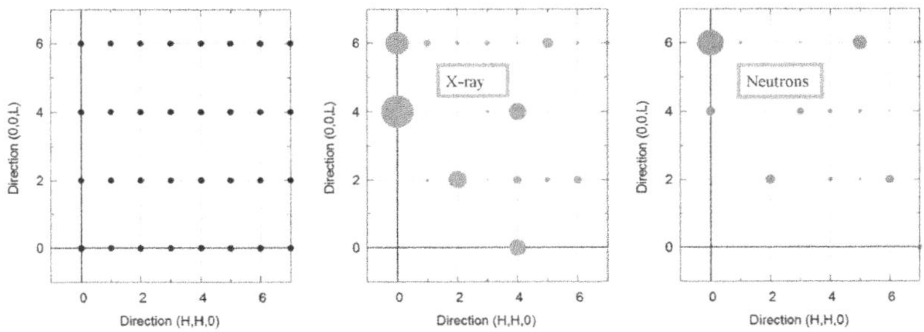

Figure 19.4 Reciprocal space. (Left) Reciprocal lattice points for the cubic crystal system of Ba_8Si_{46} in the plane ([h h 0],[0 0 l]) where the integers h and l are the Miller indices. The diameter of the points are scaled by the X-ray (middle) and neutrons (right) form factors as defined in Eq. (19.7).

19.2.4 Inelastic Scattering from One Phonon

The scattering function of a scattering process that creates or annihilates a single phonon state, $\omega_{\vec{q}}$ where $\vec{q}$ is its wave vector, is given by

$$S^{1-ph}(\vec{Q},\omega) \propto \frac{|F(\vec{Q})|^2}{\omega_{\vec{q}}}\left(\left[(1+n(\omega_{\vec{q}}))\delta(\omega-\omega_{\vec{q}})\right]+\left[n(\omega_{\vec{q}})\delta(\omega+\omega_{\vec{q}})\right]\right). \tag{19.8}$$

Here $n(\omega_{\vec{q}})$ is the thermal occupation number given by $n(\omega_{\vec{q}}) = 1/(e^{\frac{\hbar\omega_{\vec{q}}}{k_BT}} - 1)$. As we will see below, the scattering wave vector $(\vec{Q})$ is the sum of a reciprocal wave vector $(\vec{G})$ with the phonon wave vector transfer $(\vec{q})$. The two Dirac terms in $S^{1-ph}(\vec{Q},\omega)$ impose that the energy transfer, $\pm\hbar\omega$, coincides with a phonon energy, $\hbar\omega_{\vec{q}}$. Positive (negative) values of the energy transfer correspond to the creation (annihilation) of one phonon. The dependence of the scattering intensity on $\omega^{-1}n(\omega)$ ($\propto \omega^{-2}$ for $\hbar\omega < k_BT$) shows that the scattering is most sensitive to low energy modes.

The intensity of the scattering process is given by the structure factor $F(\vec{Q})$, which reflects the details of the crystal structure:

$$|F(\vec{Q})|^2 = \left|\sum_k \frac{f_k}{\sqrt{m_k}} e^{-i\vec{Q}.\vec{R}_k^0} e^{-W_k(\vec{Q})}\left(\vec{Q}.\vec{e}_k^{\vec{q}}\right)\right|^2 \tag{19.9}$$

Here, the sum runs over the atoms k within the unit cell, with m_k, f_k, $\vec{R}_k^0$ and $W_k(\vec{Q})$ their mass, atomic structure factor, equilibrium position and Debye–Waller factor, respectively.

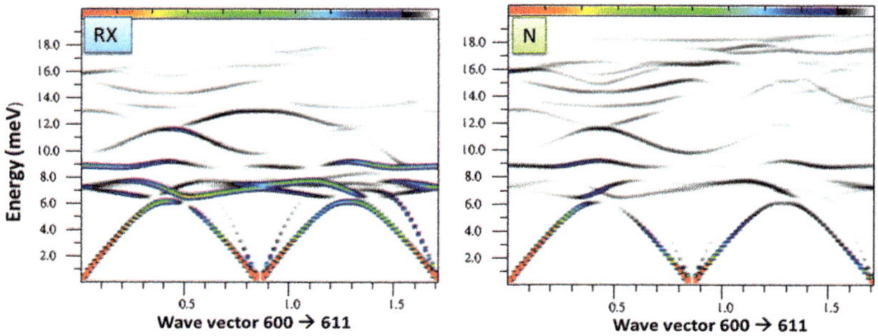

Figure 19.5 Neutrons and X-rays $S^{1-ph}(\vec{Q},\omega)$. Mapping of the (a) IXS (labeled "RX") and (b) INS (labeled "N") 1-phonon inelastic coherent diffusion function $S_{coh}^{1ph}(\vec{Q},\omega)$ simulated in Ba_8Si_{46} by ab initio density functional theory (DFT) calculations as explained in the text. It is represented in energy as a function of the reduced phonons wave vector, $\vec{q} = \vec{Q} - \vec{Q}_{600}$, in the direction (011). The color code is indicated on the top of the figures; red gives the strongest intensity.

Equation (19.8) can be simplified using the relation $\left[(1+n(\omega_{\vec{q}}))\delta(\omega-\omega_{\vec{q}})+n(\omega_{\vec{q}})\delta(\omega+\omega_{\vec{q}})\right] = \frac{1}{1-exp(\frac{-\hbar\omega_{\vec{q}}}{k_B T})}[\delta(\omega-\omega_{\vec{q}})-\delta(\omega+\omega_{\vec{q}})]$. $S^{1-ph}(\vec{Q},\omega)$ takes thus the form

$$S^{1-ph}(\vec{Q},\omega) \propto \frac{|F(\vec{Q})|^2}{\omega_{\vec{q}}^2}\frac{\omega_{\vec{q}}}{1-exp(\frac{-\hbar\omega_{\vec{q}}}{k_B T})}\left[\delta(\omega-\omega_{\vec{q}})-\delta(\omega+\omega_{\vec{q}})\right] \tag{19.10}$$

The 1-phonon scattering function can be completely simulated in the entire phase space $(\vec{Q},\omega)$, given the knowledge of the crystal structure and the phonon eigenmodes. In Fig. 19.5, we show the computed phonon dispersions along the direction (011) in Ba_8Si_{46} weighted by the dynamical structure factors for X-rays (left) and neutrons (right). The X-ray and Neutron inelastic atomic structure factors, (f_k^2/m_k), for Ba and Si are reported in Table 19.1. Here, contrary to the elastic case discussed previously, the contrast Ba/Si for phonons is opposite between neutrons and X-rays because of the division by the masses. Neutrons are roughly three times more sensitive to silicon than barium modes, while for X-ray the exact opposite is true. Both probe particles are then complementary for measuring all phonon modes, especially optical modes. The optical modes with the lowest energies in this material are mainly due to vibrations of Ba atoms encapsulated in silicon cages, for which reason they are more intense in the X-ray map than in the neutron map.

We now write the expansion of $S^{1-ph}(\vec{Q}, \omega)$ in the acoustic limit for phonons close to the center of the Brillouin zone, $\vec{G}$, which have a wave vector, $\vec{q}$, negligible with respect to $\vec{G}$ so that the scattering vector $\vec{Q} = \vec{q} + \vec{G} \sim \vec{G}$. For these modes all atoms are vibrating in phase and with equal displacements $\vec{e}_k(\vec{Q})/\sqrt{m_k}$. The structure factor becomes proportional to the elastic structure factor: $\lim_{q \ll Q} F(\vec{Q}) \sim (\vec{e}.\vec{G}) F_B(\vec{Q})$, where $F_B(\vec{Q})$ was defined in Eq. (19.7). In the acoustic limit, the 1-phonon inelastic function is thus inversely proportional to the squared energy of the mode times the squared structure factor (i.e., Bragg intensity):

$$S^{1-ph}(\vec{Q}, \omega) \propto (\vec{e}.\vec{G})^2 \frac{|F_B(\vec{Q})|^2}{\omega_j^2(\vec{q})}. \tag{19.11}$$

It is clear that the acoustic phonons intensities are directly proportional to the Bragg intensities, implying that their measurement should be performed close to strong Bragg peaks. Moreover, because of the G^2 factor, the furthest strong Bragg peaks in the Brillouin zone should be chosen for the measurement. For neutrons, since the resolution degrades by increasing the wave vector, as explained below, a compromise should be found to optimize the resolution and intensity. Equation (19.11) shows also that as long as phonons are purely acoustic $S_{ph}(\vec{Q}, \omega) \times \omega_j^2(\vec{q})$ is constant.

In the expansion of $S^{1ph}(\vec{Q}, \omega)$ in the acoustic limit, Eq. (19.11), the polarization term $(\vec{e}.\vec{G})$ imposes selection rules for detecting transverse or longitudinal phonons. Only phonons with a component of the polarization vector parallel to the scattering vector contribute to the scattering intensity. Transverse phonons have a polarization vector, $\vec{e}^T$, perpendicular to $\vec{q}$, whereas for longitudinal phonons, $\vec{e}_L$ is parallel to $\vec{q}$. Then if $\vec{q}$ is perpendicular to $\vec{G}$, the scalar product $(\vec{e}.\vec{G}) = G.e_T$ selects only the transverse modes. On the contrary, when $\vec{q}$ is parallel to $\vec{G}$, only the longitudinal phonons are measured. Figure 19.6 reports the reciprocal plane {(k k 0);(0 0 l)}, with points weighted by the elastic neutron structure factors (as already shown in Fig. 19.4) with two measurement geometries for transverse and longitudinal acoustic phonons detection. In

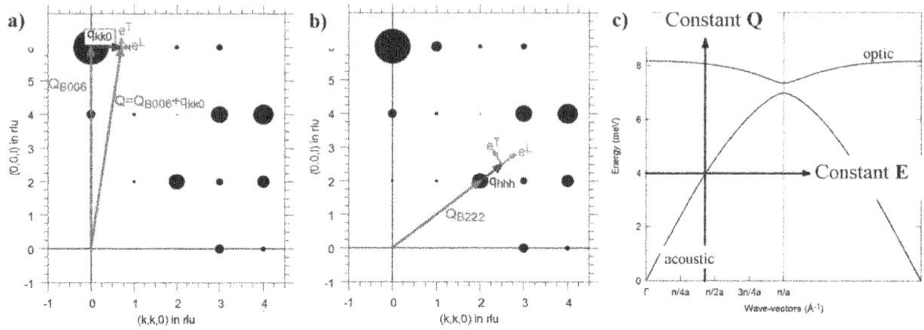

Figure 19.6 Wave vectors geometry for measurements of transverse (a) and longitudinal (b) phonons drawn in the reciprocal plane [k k 0];[0 0 l] of Ba_8Si_{46} whose points are weighted by the neutron elastic structure factor. The configuration in (a) close to the Bragg peak (006) allows for the measurement of phonons propagating along the [k k 0] direction with wave vectors q_{kk0}. The polarization term $\left(\vec{e}.\vec{Q}\right) \sim e_T.G_{006}$ selects only transverse polarization modes. On the other hand, in (b), the configuration sketched close to the Bragg peak (222) allows to measure only longitudinal phonons propagating along the direction [k k k]. (c) The two usual scanning directions, constant $\vec{Q}$-scan and constant E-scan (energy), are sketched in a simple phonon spectrum with optical and acoustic branches.

Fig. 19.6a, the measurement is performed close to the strongest Bragg peak (600) and the transferred wave vector, $\vec{Q} = \vec{Q}_{B006} + \vec{q}_{kk0}$, is chosen such that the polarization term $\left(\vec{e}.\vec{Q}\right) \sim |\vec{Q}_{B006}|e^T$ selects only the transverse phonons polarized in the direction (0 0 1) propagating in the direction (1 1 0). On the other hand, in Fig. 19.6b, the measurement is performed close to the Bragg peak (222) and the transferred wave vector, $\vec{Q} = \vec{Q}_{B222} + \vec{q}_{hhh}$, is chosen such that the polarization term $\left(\vec{e}.\vec{Q}\right) \sim |\vec{Q}_{B222}|e^L$ selects only the longitudinal phonons propagating in the direction (1 1 1).

INS (triple axis) and IXS spectrometers allow to measure the scattered intensity at a single wave vector transfer and a single energy transfer, in other words, they measure a point $(\vec{Q}, \omega)$ in the phase space. An instrumental setup made of an assembly of single crystals allows to set the probe incident wave vector and to analyze the scattered probe wave vector. To measure an inelastic peak, a sequence of measurements is needed, either at constant $\vec{Q}$, while changing the energy transfer, or at constant E while

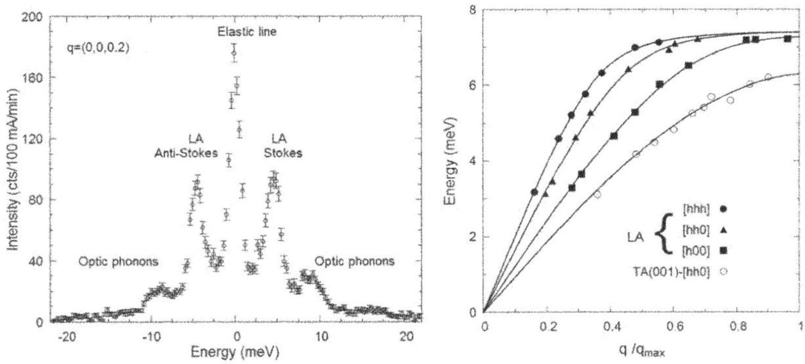

Figure 19.7 Constant Q-scan and dispersion relations. (Left) Example of IXS constant Q-scan performed across the dispersion of longitudinal acoustic (LA) phonons propagating along the [0 0 1] direction in Ba_8Si_{46} with the spectrometer ID28@ESRF. The measurement was performed at the wave vector $\vec{Q} = (0, 0, 6.2)$ around the Bragg peak $\vec{G} = (0, 0, 6)$, i.e., a phonon wave vector $q = (0, 0, 0.2)$. The transfer energies were scanned from the anti-Stokes to the Stokes parts of the spectrum. (Right) Dispersion relations of longitudinal acoustic (LA) phonons propagating along the directions (h00) (squares), (hh0) (triangles) and (hhh) (circles) and of transverse acoustic (TA) phonons polarized in the direction (001) and propagating along the direction (hh0) (empty circles). The phonon energies were obtained from fits of constant Q-scans like the one shown in the left panel.

changing the wave vector transfer, as sketched in Fig. 19.6c. This latter measurement mode is, however, rarely used. In Fig. 19.7 (left panel), we report an example of IXS raw data collected on Ba_8Si_{46} at the wave vector $\vec{Q} = \vec{Q}_{600} + \vec{q}_{0,0,0.2}$ by scanning the transfer energy from the anti-Stokes (negative energy) to the Stokes side. The geometrical configuration is such that the longitudinal acoustic phonon propagating in the direction [0 0 1] with a wave vector $\vec{q} = (0, 0, 0.2)$ is probed. This is indeed the most intense peak at the energy of about ~5 meV, while higher energy peaks are interpreted as the superposition of several optical branches. Performing this kind of scans at several $\vec{q}$ allows for drawing the whole dispersion curve for the acoustic modes as well as the optic modes, as the ones reported in the left panel of the same figure for longitudinal (LA) and transverse (TA) acoustic phonons propagating in the main symmetry directions.

Time-of-flight (TOF) neutrons technique offers an alternative way to map out large regions of the phase space $(\vec{Q}, \omega)$ by using large array detectors which simultaneously collect neutrons by timing their travel over fixed path lengths from either side of the target. The time-of-flight of a neutron is directly related to its velocity and thus its energy ($E = \frac{m_n v^2}{2}$ where m_n is the neutron mass). The philosophy of the scattering is different from the one introduced in Section 19.2.1. Indeed, here, the quantity that is determined is the double differential cross section, $\frac{d^2\sigma}{d\Omega dt}$, given by the ratio between the flux of probe particles scattered into $d\Omega$ and detected at a time between t and $t + dt$ and the initial flux of probe particles impinging onto the sample with energy E_i. The relation with the scattering function $S(\vec{Q}, \omega)$ is given by

$$\frac{d^2\sigma}{d\Omega dt} = \frac{d\omega}{dt}\frac{d^2\sigma}{d\Omega d\omega} = \frac{A}{t^4} S(\vec{Q}, \omega), \tag{19.12}$$

where A is a constant depending on the TOF geometry and incident neutrons energy. In Fig. 19.8, we show a series of constant-energy maps in the reciprocal plane (0,0,l)(1,1,0) measured by means of TOF in a Ge-based clathrate around the Bragg peak (600). The overlay of the outer and inner circles reproduces, respectively, the transverse and longitudinal cones of dispersion of acoustic phonons.

The dispersion relations of acoustic phonons, $\hbar\omega(\vec{q})$, are obtained from the fitted phonons energies. They are linear close to the center of the Brillouin zone and then bends over until they become no dispersive or flat at the border of the Brillouin zone. The dispersion is commonly modeled with a linear term and a curvature: $\hbar\omega(\vec{q}) = v_s q - cq^2$ where v_s is the sound velocity of the considered acoustic branch and c is a positive parameter describing its curvature. It is then trivial to calculate the group velocity for each phonon state: $V_g(\vec{q}) = \partial\hbar\omega/\partial q$ which, in the limit $\vec{q} \rightarrow 0$, is equal to v_s. The order of magnitude of the sound velocities in silicon-based clathrates is of about 6000 m/s for LA phonons and 3000 m/s for TA phonons.

19.2.5 Phonons Lifetimes

As shown in the previous section, INS and IXS techniques allow to scan point by point the entire phase space. However, because of

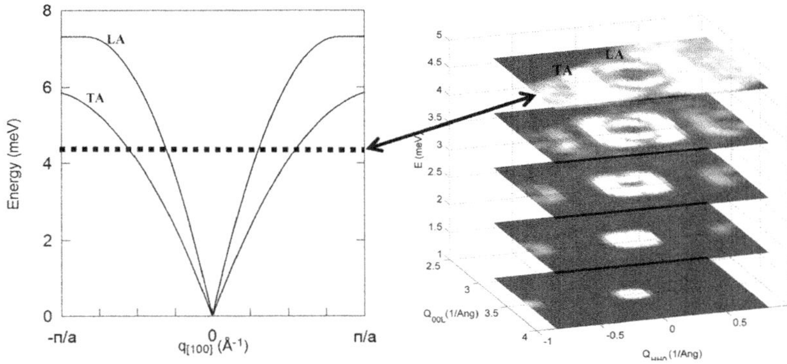

Figure 19.8 Dispersion cones of longitudinal and transverse acoustic phonons. (Left) Dispersion relations of longitudinal (LA) and transverse (TA) acoustic phonons deduced from INS measurements in a Ge-based clathrate around the Bragg peak 600. They are drawn along the propagation direction (100). (Right) We present constant-energy maps of the neutron intensity in the reciprocal plane (0,0,l)(1,1,0) of $S(\vec{Q}, \omega)$ extracted from time-of-flight measurements with IN5@ILL. White color corresponds to strong intensities and then indicate the locus of the LA and TA phonons states. As indicated by the arrow, each slices maps the TA and LA phonons states in the reciprocal plane (0,0,l)(1,1,0) at a given energy transfer. The TA phonon states appear on the outer ring while the LA states on the inner ring. The overlay of these maps gives the two dispersion cones for LA and TA phonons. Note that these maps were extracted from a larger data file in which the neutrons intensities were collected in the 3D reciprocal space and in energy.

the crystals mosaicities of monochromator and analyzer and the beam angular divergences, a whole set of values $(\vec{Q}, \hbar\omega)$ centered on $(\vec{Q}_0, \hbar\omega_0)$, where the spectrometer was positioned, is in fact probed. The function giving the weight of this distribution is called the instrumental resolution function, $\mathrm{R}(\vec{Q} - \vec{Q}_0, \hbar\omega - \hbar\omega_0)$. The measured intensity is then given by the integration over $\vec{Q}$ and $\hbar\omega$ of this function, weighted by the scattering law of the sample.

$$I(\vec{Q}_0, \hbar\omega_0) = \int S(\vec{Q}, \hbar\omega) R(\vec{Q} - \vec{Q}_0, \hbar\omega - \hbar\omega_0) \delta\vec{Q} \delta\hbar\omega \qquad (19.13)$$

The instrumental resolution is a volume defined in the four-dimensional phase space $(Q_x, Q_y, Q_z, \hbar\omega)$, whose average dimensions in inelastic X-ray or Neutron spectrometers configured for phonons measurements are $\Delta Q \sim 0.02$ Å^{-1} and $\Delta\omega \sim 1$ meV. In

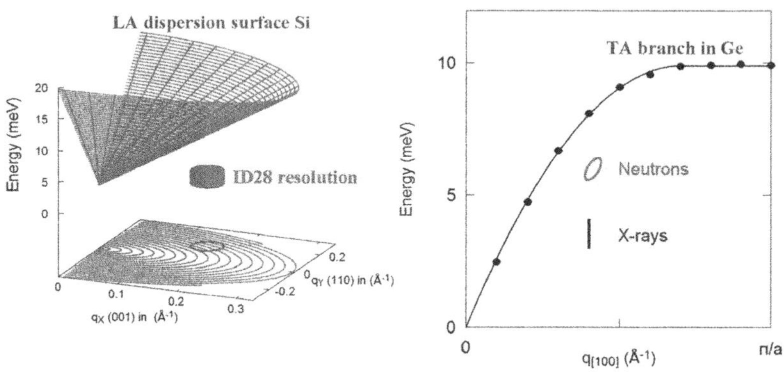

Figure 19.9 Instrumental resolution volume. (Left) The resolution volume of the spectrometer ID28@ESRF, which was positioned to measure acoustic phonons in silicon, in the scattering plane space {(001);(110)}[57]. In the same figure, we report the energy dispersion surface of the longitudinal acoustic (LA) phonons. (Right) Sketch of the X-ray and neutrons (2T@LLB) instrumental resolution functions drawn with the transverse acoustic (TA) branch in Ge.

Fig. 19.9, we represent the resolution volume of the spectrometer ID28@ESRF, which was positioned to measure acoustic phonons in silicon, in the scattering plane space {(001);(110)}. In the same figure, we report the energy dispersion surface of the longitudinal acoustic phonons. The IXS resolution function can be represented by a Q-ellipse smeared along the energy axis, with uncoupled Q and energy components. The best energy resolution achieved on a IXS spectrometer is of about 1.1 meV, clearly limiting the possibility of measuring phonons lifetimes in crystals.

The situation is different on a triple axis neutron spectrometer, where the resolution forms an ellipsoid in the four-dimensional space with a certain inclination in the $(\vec{Q}, \hbar\omega)$ space due to their coupling, as represented in Fig. 19.9 for the spectrometer 2T@LLB, which was positioned to measure acoustic phonons in germanium. Its derivation has long been established [23, 25, 55] and software is available [61, 62], that determine it either by using analytical formulas or by performing Monte Carlo ray-tracing simulations [68]. The volume of the triple-axis neutron resolution depends essentially on the choice of the final neutrons energy (in a fixed final energy mode) and can be reduced down to $\approx$10 μeV in some configurations of the spectrometer, unfortunately usually not compatible with the measurement of phonons.

The first consequence of the finite sizes in $\vec{Q}$ and in ω of the instrumental resolution and of the convolution product is that it limits the measurement of acoustic phonons close to the zone center where their energies tend to zero because a part of the Bragg peak intensity (~40,000 cts/s) is integrated and masks the phonon intensity (~200 cts/s). In the case of a dispersive excitation, the $\vec{Q}$- and energy-components of the instrumental resolution couple in the convolution product in Eq. (19.13) through the phonon dispersion relation. In a constant Q-scan, the resolution volume is moved along the energy axis through the dispersion surface $\omega(q)$ of the acoustic phonon. The observed intensity profile depends on the local curvature of $\omega(q)$ (and thus on the local group velocity) over the range of the instrumental resolution. The phonon is therefore measured with an effective energy resolution which is different from the intrinsic instrumental one along the energy axis.

We now discuss a few aspects of the issue of such four-dimensional resolution on the measurement of acoustic phonons linewidths by means of IXS or INS.

- In a neutron triple-axis spectrometer, the effective resolution can be improved without any intensity loss by focusing the orientation of the instrumental resolution ellipsoid with respect to the slope of the dispersion curve. The best focusing is obtained when the long axis of the ellipsoid is parallel to the phonon dispersion surface. Unfortunately, although the neutron resolution is perfectly determined, its inclination in $(\vec{q}, \omega)$ cannot be easily handled and then optimized for each point of the dispersion curve. Consequently, the effective energy resolution varies over the dispersion surface because of the change of the focusing effect. This is illustrated in Fig. 19.10, where we report the values of the effective energy resolution as a function of the local curvature of the phonon dispersion. It can be seen that the best effective resolution is obtained at Q values for which the group velocity is about 12 meV.Å^{-1} (~1800 m.s^{-1}).

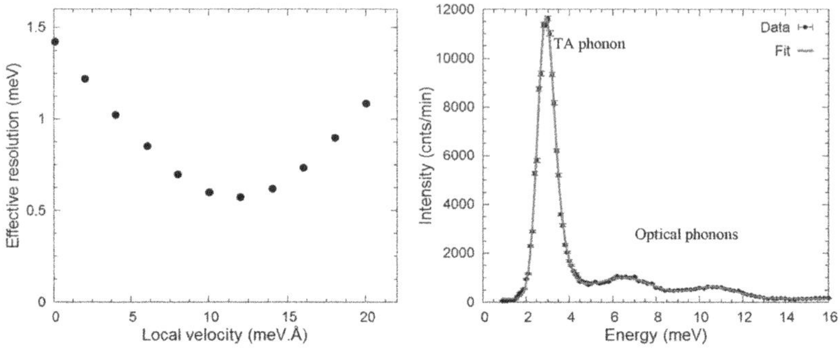

Figure 19.10 Neutrons effective resolution. (Left) We report the instrumental effective resolution as a function of the local slope, or group velocity, of the phonons dispersion relation in the standard geometry of the spectrometer 2T@LLB with optimized flux. (Right) We report the measurement of a transverse acoustic (TA) phonon on a Ge-based clathrate, made with a constant Q-scan. The red line shows the result of the fitting process with the four-dimensional convolution product, as explained in the text, used to fit the intensity profile of the acoustic phonon. Gaussians are used for the optical phonons.

- Even in the case of an ideal spectrometer with a sub-μeV energy resolution, the measured intensity profile would have an effective energy resolution in the meV range if the resolution in $\vec{Q}$ is not improved. Indeed, for acoustic phonons with a sound velocity of about $v_s \sim 3000$ m.s^{-1}, a q-resolution $\Delta Q \sim 0.02$ Å^{-1} leads to an effective energy resolution of $\delta\omega \sim v_s \delta q \sim 0.4$ meV. The effective resolution is efficiently improved by reducing only its size in the $\vec{Q}$ space in the direction of the dispersion. This can be done by using collimators and/or slits at the price of a reduction of the intensity. The situation becomes more delicate in the case of acoustic phonons close to the center of the Brillouin zone. Here, all phonon states of the dispersion surface propagating in different directions are integrated over the range of the instrumental resolution. Moreover, phonon states of other acoustic phonon branches can also be integrated. For example, the measurement of the LA phonons in Si dispersing in the direction (100) can be contaminated by a spurious contribution from TA phonons dispersing in the direction (1-10), perpendicular to the scattering plane.

- The crystal mosaicity, which is the mean angle (usually expressed in minutes) between misaligned planes in the crystal, affects the effective resolution. While a mosaicity of about 60′ is enough for acoustic dispersions measurements, the measurement of phonon linewidths in the sub-meV range requires a crystal quality of a few minutes only.
- In the best conditions on a neutron triple-axis spectrometer, with horizontal and vertical collimation of 10–20′, a perfect (crystal mosaicity less than 10′) and centimetric single crystal and at the focusing point of the instrumental resolution with the dispersion surface, an effective resolution for acoustic phonons in the sub-100 μeV range can be achieved. Still, it remains sizable with respect to the linewidth of acoustic phonons with energies of a few meV. A challenging data analysis is therefore needed in order to extract properly the intrinsic phonon linewidth, requiring the calculation of the instrumental resolution and the performance of a four-dimensional convolution. As discussed above, for phonon states close to the center of the Brillouin zone, analytical knowledge of the complete dispersion surface in all directions is needed since all directions of the ellipsoid will contribute to the broadening. Finally, the convolution has to be included in a fitting process in order to fit the phonon linewidth.

In the 1-phonon scattering function (Eq. 19.8), the Dirac terms indicate that the phonon state has an infinite lifetime, at least much larger than the time resolution of the experiment. To take into account a finite lifetime, Eq. (19.8) is generalized by introducing a phonon spectral function $\mathcal{F}_{\vec{Q}}(\omega)$ normalized to unity: $\int_{-\infty}^{+\infty} \mathcal{F}_{\vec{Q}}(\omega) d\omega = 1$. Using the relation $(1 + n(\omega)) = -n(-\omega)$, the 1-phonon coherent inelastic diffusion function (Eq. 19.8) can be rewritten:

$$S_{coh}^{1ph}(\vec{Q}, \omega) = \sum_i \frac{|F_i(\vec{Q})|^2}{\omega_i^2} \frac{\omega}{1 - exp(-\hbar\omega/k_B T)} \mathcal{F}_{\vec{Q}}(\omega) \qquad (19.14)$$

In the case where the intrinsic energy width of phonons is not sizable compared with the instrumental resolution, the spectral function is simply given by the sum of the two Dirac functions: $\mathcal{F}_{\vec{Q}}(\omega) = \left(\delta(\omega - \omega_i) + \delta(\omega + \omega_i)\right)/2$.

On the other hand, the spectral function of a single phonon with an intrinsic energy width Γ (full width at half maximum) beyond the instrumental resolution can be described by semi-phenomenological functions, like the sum of two Lorentzian functions: $\mathcal{F}_{\vec{Q}}(\omega) = \frac{1}{2\pi}\left[\frac{\Gamma}{(\omega-\omega_i)^2+\Gamma^2} + \frac{\Gamma}{(\omega+\omega_i)^2+\Gamma^2}\right]$, or the damped harmonic oscillator (DHO) function[a]: $F_{\vec{Q}}(\omega) = \frac{1}{\pi}\frac{\Gamma\omega_i^2}{\left((\omega^2-\omega_i^2)^2-\Gamma^2\omega^2\right)}$. In the latter case, the diffusion function becomes

$$S_{coh}^{1ph}(\vec{Q}, \omega) = \frac{1}{\pi}\sum_i |F_i(\vec{Q})|^2 \frac{\omega}{1 - exp(-\hbar\omega/k_B T)} \frac{\Gamma}{(\omega^2 - \omega_i^2)^2 - \Gamma^2\omega^2} \tag{19.15}$$

In Fig. 19.10, we present two examples of fit of data corresponding to the measurement of transverse acoustic phonons in a Ge-based clathrate. In this case, the fitted value for the damping is zero because all the intensity profile is dominated by the effective resolution which was of about 0.7 meV.

If Γ is the half width at half-maximum, the correct relationship between the phonon lifetime, τ and Γ is

$$\tau = \frac{1}{\Gamma}. \tag{19.16}$$

19.2.5.1 Neutron resonance spin echo technique

We discussed above the importance of the focusing effect on the effective instrumental resolution for the measurement of the phonon linewidth with a triple-axis spectrometer. A significant

---

[a]The DHO is the scattering function of a single damped harmonic oscillator, with mass m, under the influence of an external force, $f(t)$, whose classical equation of motion in the Fourier frequency domain is $m(-\omega^2 + \omega_0^2 + i\gamma\omega)x(\omega) = f(\omega)$. Here $\omega_0 = \frac{K}{m}$, with K denoting the force constant, and γ is a frictional damping. The dissipative part (or imaginary part) of the response function, here given by $\chi''(\omega) = \mathcal{I}m\chi(\omega) = \mathcal{I}m\left(\frac{1}{m}\frac{1}{-\omega^2+\omega_0^2+i\gamma\omega}\right)$, is connected to the dynamical structure factor through the fluctuation-dissipation theorem. The scattering function of the DHO is then: $S(\omega) = (1 + n(\omega))\frac{1}{\pi m}\frac{\gamma\omega}{(\omega^2-\omega_0^2)^2+(\gamma\omega)^2}$.

improvement of the measurement accuracy is obtained for the limited part of the dispersion surface where the local curvature is aligned with the largest axis of the instrumental resolution. The Neutron Resonance Spin Echo technique (NRSE) [30, 31, 41, 52], sometimes called the Zero-field Spin Echo, installed on a triple-axis spectrometer, offers the unique opportunity to optimize the focusing effect for all points of the dispersion at once by measuring the neutrons group velocity. Moreover, the effective energy resolution is brought down to about 10 μeV.

The neutron spin echo technique (NSE) is based on the Larmor precession, ω_L, of the neutron magnetic moment around the axis of an applied magnetic field, B. The Larmor frequency depends linearly on the magnetic field $\omega_L = \gamma_L B$ where γ_L is the gyromagnetic ratio. Thus, in a solenoid of length (L) with a well-defined homogeneous magnetic field, the neutrons spins acquire a phase $\phi = \omega_L t$ proportional to their traveling time. In NSE spectroscopy, the precession angle of the neutron spins accumulated in the two spectrometer arms before and after the sample is measured. The spin echo condition is fulfilled when the total Larmor precession angle of the spins is zero. The beam recovers then its full polarization, P^{out}, and an intensity is measured. The final polarization is simply given by the cosine of the total Larmor precession angle: $P^{out} = \langle cos(\phi) \rangle$. In the case of inelastic scattering with a phonon, the total Larmor precession depends linearly on the energy transferred ($\hbar\omega$) to the sample (phonon energy), $\phi = \omega\tau$ where τ is the spin echo time which depends on the instrumental parameters and on the phonon group velocity. It follows that the analysis of the amplitude of the final beam polarization, $\langle P^{out} \rangle$, provides directly the inverse lifetime of the phonon, Γ_{ph}:

$$P^{out}(\tau_{NSE}) = P_0^{out} e^{-\Gamma_{ph}\tau_{NSE}}, \tag{19.17}$$

where $P_0^{out} = P^{out}(\tau_{NSE} \to 0)$ is the intrinsic polarization of the spectrometer.

In Fig. 19.11, we show an example of a phonon lifetime measurement in a Ge-based clathrate performed on the spectrometer IN22@ILL with the ZETO option [50, 51]. The spectrometer is first positioned at the coordinate of a phonon state $(\vec{q}, \omega)$ in the phase space. The field boundaries are calculated so that the spin

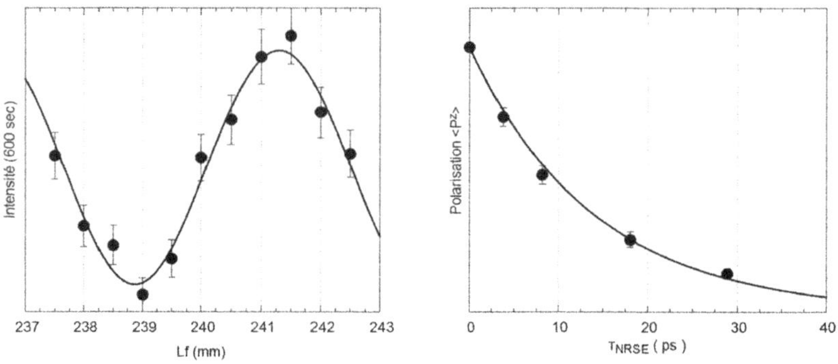

Figure 19.11 Typical measurements of phonon lifetime with the NRSE technique. (Left) One period of spin echo curve recorded by moving the distance (L_f in mm) of the coils along the neutron beam from either side of the sample. The measurement was performed on the spectrometer IN22@ILL with the ZETA-NRSE option [43, 50, 51] in a Ge-based clathrate. The field boundaries and the spectrometer settings were configured for a neutron spin echo time $\tau_{NRSE} = 8$ ps. The polarization, P^z, for this τ_{NRSE} is obtained from the fit of the neutron intensity using the expression of Eq. (19.17). The measurement of P^{out}, repeated for different values of τ_{NRSE}, is shown in the panel on the right. The fit of the exponential decay provides a direct measurement of the phonon lifetime. Here a lifetime of 14 ps was obtained.

echo resolution is adjusted to be parallel to the tangent to the dispersion surface. For this, the phonon dispersion relation is required to correctly position the spectrometer at a given $(\vec{q}_0, \omega)$-point of the phonon dispersion and to determine the needed field boundaries of the spin echo conditions from the phonons group velocity. The neutron beam polarization is measured for several spin echo times values in order to obtain $P^{out}(\tau_{NSE})$. For each τ_{NSE}, the polarization P^{out} is measured by changing the total Larmor precession angle, here by translating the magnetic coil in the beam direction, and extracting the amplitude of the resulting sinusoidal intensity modulation.

The NRSE is an advanced spin echo technique dedicated to the measurement of linewidths of dispersive excitations in solids over the entire Brillouin zone. As recent, this technique still needs to be proved on different systems and it is not yet clear that it will resolve the difficulties of a correct measurement of phonons linewidths.

19.3 Phonons in an Amorphous System

19.3.1 Dynamics in a Disordered System

The difficulty in studying the dynamics of amorphous materials stems from the absence of any possible simplification in the structural description of the system. First of all, the lack of long-range periodicity hinders the application of the theories developed for lattice dynamics in crystals; moreover, the disordered structure is not close enough to a simpler, ordered one, to allow for a perturbative approach to be used for solving the dynamics of the system. We have seen in Section 19.2.1 that, thanks to the periodicity of the structure, the description of the dynamics of a crystal is reduced to that of the elementary unit cell: For each wave vector there exist $3r$ modes (phonon branches), where r is the number of atoms in the unit cell. These modes are well-defined plane waves extending spatially through the whole material.

In an amorphous system no such unit cell can be identified; ideally the whole system is an infinite unit cell, with $r \to \infty$. Still a short-range order is kept, on the nearest-neighbors distance length scale. A first consequence of the long-range topological disorder is that atoms do vibrate but the excitation is not characterized by a well-defined wavelength λ or wave vector q, i.e., the vibrational modes are no longer plane waves and the concept of "phonon" loses its original meaning as a normal mode of the system. Indeed, it has been shown theoretically that a harmonic disordered system can support longitudinal and transverse acoustic modes, even if q cannot be used as a good quantum number to strictly classify the vibrational modes as "phonons" [81].

The first question which arises is how the dispersion relations in an amorphous system look like, compared to a crystalline one.

The first thing to understand is that in a disordered system, thanks to its isotropy, there is no preferential orientation for measuring phonons. In principle, their propagation is isotropic and the acoustic dispersion will only depend on the absolute value of the exchanged momentum: $\omega(\vec{q}) \to \omega(q)$.

Similarly to crystals, a pseudo-Brillouin zone can be identified even in disordered systems, though perfectly spherical because of isotropy. If we consider a crystal, the static structure factor $S(q)$ as measured by diffraction techniques is composed of Bragg peaks, whose q position is related, according to Bragg's law, to the interplanar distance. In an amorphous system, the $S(q)$ does not present sharp, well-separated, narrow peaks, but typically a first, large, pronounced peak (so-called First Sharp Diffraction Peak, FSDP), associated with a typical correlation length between the atoms, followed by less intense bumps at higher q. These peaks are due to the persistence of a short-range order in amorphous systems, so that the FSDP allows to identify a pseudo-Brillouin zone whose border is at $q_P/2$, with q_P the position of the FSDP. As an example, we report in Fig. 19.12 (left) the $S(q)$ as measured on a metallic glass of composition $Mg_{65}Cu_{25}Y_{10}$. Data were taken at the beamline ID11 of the European Synchrotron Radiation Facility (ESRF), in Grenoble, France, with an incident energy of 76 keV. The FSDP is located at $q_P \approx 2.55$ Å^{-1} [66], from which a Brillouin zone border $q_P/2 \approx 1.27$ Å^{-1} is found.

It is worth mentioning here that the pseudo-Brillouin zone border is much less sharply defined than in a crystal. While there indeed exists a short-range order, the atoms relative positions have a probability distribution. This can be visualized through the pair correlation function $g(r)$, which gives the probability of finding an atom at the position r if there is an atom at the position $r = 0$. $g(r)$ can be calculated by Fourier transforming $S(q)$ [18, 37, 44, 47, 71, 76]: for an isotropic disordered system, it results in $g(r) = 1 + \frac{1}{2\pi^2\rho}\int_0^{+\infty} dq q^2 (S(q) - 1)\frac{\sin qr}{qr}$, with ρ the average density. As can be seen in Fig. 19.12 (right), this function presents a first sharp peak, which corresponds to the nearest neighbor shell, and then several less intense bumps, corresponding to the next shells of neighbors. In the case of a monoatomic system, the width of the first peak gives the width of the probability distribution of the nearest neighbor's position. This directly reflects into a distribution of sizes of the pseudo-Brillouin zone, which turns out to be related to the width of the FSDP [32].

As the persisting order is only short-ranged, it is clear that there are no higher order Brillouin zones in a disordered system. Phonons are in principle defined only in the first zone. As a consequence, $Q = q$, for which reason we will always call the exchanged momentum q in this section.

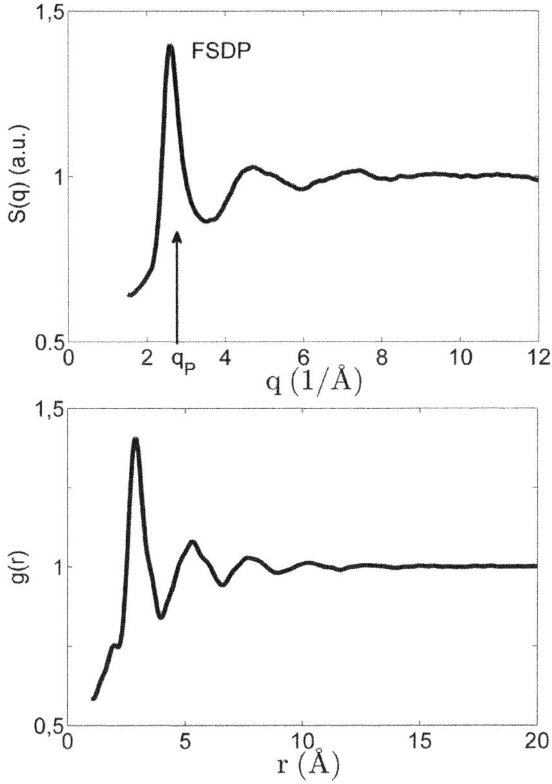

Figure 19.12 Static structure factor in disordered systems. (Left) We report here the S(q) of a metallic glass of composition $Mg_{65}Cu_{25}Y_{10}$, as measured by high-energy X-ray diffraction (incident energy $E_i = 76$ KeV) at the ID11 beamline of the ESRF. Data are presented after background subtraction and normalization, in absolute units. The arrow marks the position q_P of the FSDP (see text). (Right) The pair correlation function $g(r)$ as calculated by Fourier transforming $S(q)$.

19.3.2 A Length Scale–Dependent Dynamics

The position of the FSDP sets a length scale with respect to which we can classify the different dynamical behavior in the dispersion relation of a glass.

In the long wavelengths limit, when $q << q_P/2$, corresponding typically to frequencies of a few hundred GHz, a glass appears as a continuum elastic medium and the details of the structural disorder are irrelevant to the vibrational dynamics. The acoustic modes can still be described in terms of plane waves, as in the crystalline case, with well-defined dispersion curves, which give information on the sound velocity in the material. This is the regime where the Debye approximation works well for describing the vibrational density of states and where the acoustic dispersion is linear.

On the other hand, when $q \leq q_P/2$, i.e., for wavelengths in the nanometer range and frequencies of the order of the THz, the lack of periodicity does not allow for a plane wave description of the collective excitations and q cannot be considered any longer a well-defined quantum number. This is the mesoscopic regime, in which a breakdown of the Debye approximation (i.e., the continuum medium description) has been recently reported, with deviations of the acoustic dispersion from the linear behavior [5, 53, 67] and the appearance in the density of states of an excess of modes with respect to the Debye prediction, the so-called Boson Peak [13, 17, 22, 74, 75]. Despite that in the mesoscopic regime phonons are ill-defined, the presence of acoustic phonon-like excitations in glasses has been assessed down to wavelengths comparable to the interparticle distance [15], i.e., $q \approx q_P/2$. This is the microscopic regime, when the presence of a short range order becomes more and more important starting to affect the vibrational dynamics, and giving raise to closer similarities with the dynamics of the corresponding crystalline phase, i.e., the one with the same mass density and local order [8, 32].

This length scale analysis leaves us with three well distinct dynamical regimes, whose differences manifest not only in the acoustic dispersion, i.e., the $\omega(q)$ law, but also, and more clearly, in the width of the inelastic peaks, i.e., in the phonons lifetime. This is because different scattering mechanisms dominate at different length scales and change the energy dependence of the phonons lifetime.

A number of sources of damping exists in a disordered system. Some are the same as in a crystal, like anharmonicity (3-phonon collisions), or scattering from defects and impurities, see Section 19.2.1. Moreover, there are some glass-specific sources of phonons scattering, which play a major role in different energy ranges.

One renowned example is the scattering from two-level tunneling states characteristic of glasses, active at very low temperature and very low frequency, interpreted as scattering from structural defects which quantum-tunnel between two equilibrium configurations: more specifically, in the disordered system there exist some entities (individual atoms, groups of atoms) which can tunnel between two nearly degenerate configurations [2, 59]. This scattering process affects phonons at ultrasonic frequencies.

Another example, which is at the focus of the current research, is a strong phonon scattering which takes place at energies close to the Boson Peak energy, ω_{BP}, usually located at $\approx$1 THz (few meV), and thus in the frequency range which can be accessed by IXS or INS.

Below ω_{BP}, the linewidth Γ grows with the fourth power of the frequency ($\Gamma \propto \omega^4$), while above it the dependence is approximately quadratic ($\Gamma \propto \omega^2$) [5, 6, 53, 65]. The crossover between the two frequency dependencies takes also place close to the Ioffe–Regel limit, defined as the frequency at which phonon's mean free path and wavelength are comparable. The Ioffe–Regel frequency, $\omega_{IR} \approx \pi\Gamma$ [63] marks the energy above which the phonon is over-damped and cannot be considered anymore as a propagative mode, but more as a diffusive mode. It is, however, worth underlying that this does not mean that above the Ioffe–Regel frequency modes are localized. In fact, they are still extended waves.

Allen [1] proposed a tentative classification of vibrational modes in glasses on the basis of their propagation character and space extension. The energy space is divided first in two categories: extended and localized modes, the latter lying at the upper limit of the vibrational spectrum. The extended modes divide into "propagons" at energy lower than 1–2 THz, corresponding to propagative plane waves with a mean free path of the order of 2 to 3 inter-atomic distances (ballistic propagation), and "diffusons" at energies above ω_{IR}, which are extended modes but with ill-defined wave vector. Here the propagation is diffusive and the concept of mean free path loses any meaning. Localized modes, at high energy, are called "locons." Their wavefunction exponentially decays with a localization length of a few Angstrom. Allen found that, in amorphous silicon, only 3% of the modes are locons, and only 4% are propagons, the most part of the excitations being then diffusons.

The nature of the strong scattering regime below ω_{BP} is still matter of discussion in the scientific community. Two main explanations have been proposed: (i) Rayleigh scattering from density fluctuations [27, 34] or elastic constants heterogeneities on the nanoscale [69, 70]; (ii) a resonant interaction between vibrational soft modes and sound waves, as predicted by the soft potential model [16, 64].

Concerning the ω^2 behavior for $\omega \geq \omega_{BP}$, this has been recently explained on the basis of the close similarity in the lattice dynamics of a glass and its polycrystalline counterpart in the microscopic regime [8]: At these length scales phonons are confronted to the presence of a short-range order. Their acoustic dispersion results completely dictated by the local order, thus very similar to the one of the corresponding crystalline phase. The glass being isotropic, the comparison needs to be done with the polycrystal, whose signal is the result of an orientational average of the branches over all directions. In this q-range, the inelastic peak in the glass starts also acquiring contributions from transverse modes, which appear in the $S(q, \omega)$ at these qs (as explained in Section 19.3.3), as well as optic modes, so that it cannot be considered any more as a single, well-defined mode. In a fit with a single-mode, the ω^2 dependence will thus come from the increase of the intensity of the optic modes, also present now in the spectrum, which start to get more intense than the acoustic mode, causing thus a change in the slope of the linewidth.

19.3.3 The Typical Spectrum of a Disordered System

In disordered systems, like liquids, glasses and dense gases, very few neutron studies have been performed so far. In fact, due to the kinematics of the scattering process, thermal neutrons are not efficient for studying at very low q in the first Brillouin zone sound modes in materials with a large speed of sound, typically above 2000 m/s. This is one reason for which IXS has been developed, as it does not suffer from the same kinematic constraints. We therefore talk only of IXS from now on.

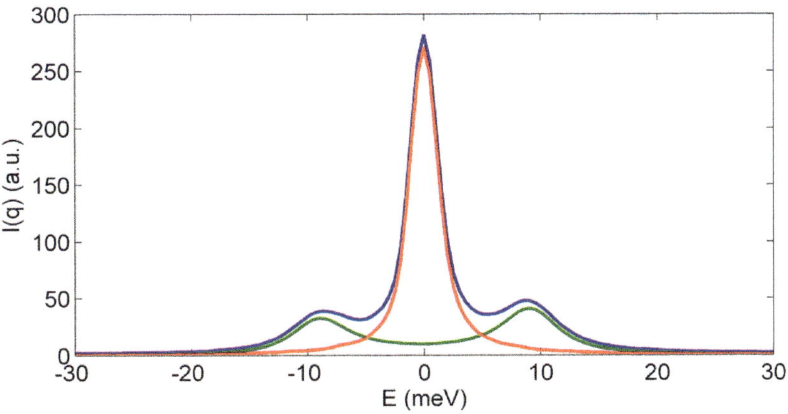

Figure 19.13 Typical inelastic spectrum in a disordered system. Sketch of an inelastic spectrum of a disordered system at low q, as can be acquired by IXS (blue line). Dotted red and dashed green lines depict the elastic line and the inelastic (Stokes and anti-Stokes) peaks.

As shown in Fig. 19.13 a typical IXS spectrum in a disordered system has a triplet shape: there is a quite important elastic/quasi-elastic peak and the inelastic Stokes and anti-Stokes excitations.

For describing the dynamics in an amorphous system it is useful to remember that $S(q, \omega)$ is actually the space-time Fourier transform of the density-density correlation function [18, 37, 44, 47, 71, 76]:

$$S(q, \omega) = S(q) \int_{-\infty}^{+\infty} dte^{-i\omega t} \Phi_q(t), \qquad (19.18)$$

where the static structure factor is related to the density fluctuations $\delta\rho$ at $t = 0$: $S(q) =< \delta\rho_q^*(0)\delta\rho_q(0) >$ and Φ_q is the density correlator, which monitors the decay of the density fluctuations:

$$\Phi_q(t) = \frac{< \delta\rho_q^*(0)\delta\rho_q(t) >}{< \delta\rho_q^*(0)\delta\rho_q(0) >} \qquad (19.19)$$

This decay is ruled by relaxation processes, similarly to liquids. It is important to remind that an amorphous system is an out-of-equilibrium state, classically obtained by fast quench from the liquid, trapped in a local minimum of the energy landscape [33]. Thanks to thermal fluctuations, with time, the system can eventually overcome the barriers of its local minimum to drop into another minimum, thus exploring the energy landscape towards the absolute minimum. This evolution happens through relaxation processes, characterized each by a different characteristic time. Two main processes are generally identified: a fast relaxation process, called β, corresponding to atomic vibrations around the disordered instantaneous positions (typical time $\tau_\beta \approx$ ps), and a slow structural one, called α, related to the cooperative rearrangements of the disordered structure (typical time $\tau_\alpha \approx$ s) [38]. If we think of a single particle as instantaneously vibrating in the cage of its nearest neighbors, the α process corresponds to the escape of the particle from this cage.

These two processes are present in both liquids and glasses, but in the latter the slow one becomes almost frozen, its typical time increasing by several orders of magnitude up to 100 s approaching the glass transition.

The activation of a relaxation process is cause of a decay in the density correlator, as can be seen in Fig. 19.14. At short time the decay is related to the free motion of the atoms and to their collisions, which lead to a first plateau. The onset of a secondary relaxation (fast process) gives rise to a step and a longtime plateau, followed by a last step leading eventually to complete decorrelation ($\Phi_q = 0$), due to structural relaxation [72].

Coming back to the IXS typical spectrum of Fig. 19.13, the central peak arises from density fluctuations related to the relaxation processes in the system. Its width is inversely proportional to the relaxation characteristic time. In the case of a glass, the active relaxation processes have characteristic times which are well above the time resolution of an IXS experiment ($\omega\tau \gg 1$), so that they appear completely frozen. As a consequence, their spectral contribution will be a Dirac function, and the observed width is only due to the finite value of the energy resolution of the technique.

The inelastic excitations correspond to the collective excitations, phonons, in the system: their energy position and width give information on the sound velocity and phonon lifetime.

It is important here to introduce the relation between the dynamic structure factor $S(q, \omega)$ and the longitudinal part of the current correlation function.

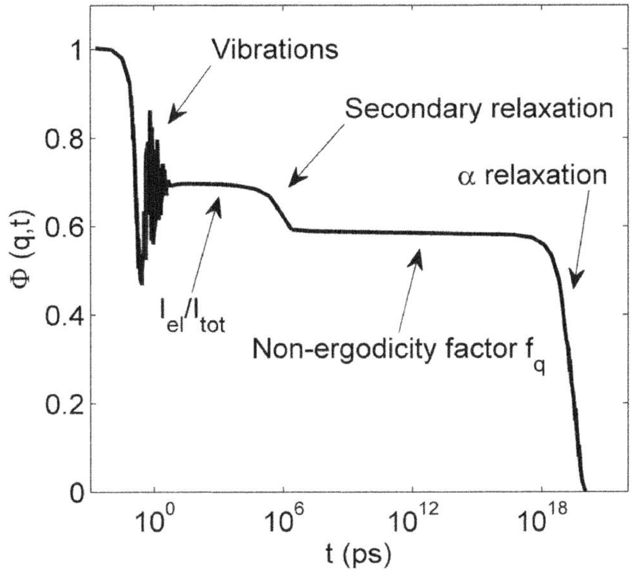

Figure 19.14 Time evolution of the density correlator. Sketch of the temporal evolution of the density correlator (Eq. 19.19). Different temporal regimes can be seen, dominated by different relaxation processes. Adapted from [72].

This latter, $J(q, \omega)$, is the space-time Fourier transform of the fluctuations of the current density $j(r, t)$: $J_{ij}(q, \omega) = < j^*_{qi}(0)\, j_{qj}(t) >$, where i and j denote spatial directions. It can be shown that the dynamic structure factor is proportional to the longitudinal current correlation function [11]:

$$S(q, \omega) = \frac{q^2}{\omega^2} J_l(q, \omega), \qquad (19.20)$$

where l stands for longitudinal. As a consequence, it has long been believed that inelastic scattering experiments could give access only to longitudinal modes in a disordered system. However, differently from the crystalline case, the lack of translational invariance make longitudinal and transverse modes to mutually intertwine, so that their symmetry is ill-defined at distances shorter than some threshold, i.e., q larger than a critical q_c. It follows that inelastic transverse modes can actually be observed together with the longitudinal ones in the $S(q, \omega)$. The L-T coupling is supposed to be stronger in network systems, although transverse modes have been observed even in simple binary metallic glasses [54, 78].

The relative intensities of elastic and inelastic peaks is related to the so-called non-ergodicity factor, $f(q)$, which corresponds to the long-time plateau of the density correlator (see Fig. 19.14) and gives a measurement of the frozen state of the glass.

The non-ergodicity factor is a central quantity in the mode-coupling theory on the glass-transition, as it is expected to go from 0 in the supercooled liquid to a finite value in the glass [60]. A zero non-ergodicity factor describes the situation where a particle eventually escapes from the nearest neighbors cage, thus the system is ergodic. Conversely, a finite value describes a frozen structural relaxation, which prevents the particle to escape from the cage, thus the system is trapped in a non-ergodic state.

The inelastic to total integrated intensity ratio in the IXS spectrum for $q \to 0$ gives an experimental measurement of an effective non-ergodicity factor, which differs from the non-ergodicity factor as it corresponds to the value of the density correlator at the finite time $t = \frac{1}{\Delta E}$ with ΔE the energy resolution.

19.3.4 Phonons in a Metallic Glass

In this subsection we present a practical example of phonons measurements in a glass. More specifically we present here data on the bulk metallic glass of composition $Zr_{52.5}Cu_{27}Al_{10}Ni_8Ti_{2.5}$, also called D", the same presented in Chapter 13.

The longitudinal and transverse sound velocities, as measured by ultrasound techniques, are $v_L = 4.7$ km/s and $v_T = 2.3$ Km/s [35], respectively, thus confirming X-rays as the probe of choice for investigating its dynamics.

The experiment was performed at the ID28 beamline of the ESRF, with an incident energy of $\approx$17.7 KeV and an energy resolution of 3 meV. The sample was cut as a thin disc with a thickness of the order of $\approx$70 μm, matching the attenuation length of X-rays at 17.7 keV, in order to maximize the diffused signal. Inelastic spectra were collected as energy scans at constant q by spanning both positive and negative energy exchanges and counting 60 s per energy point.

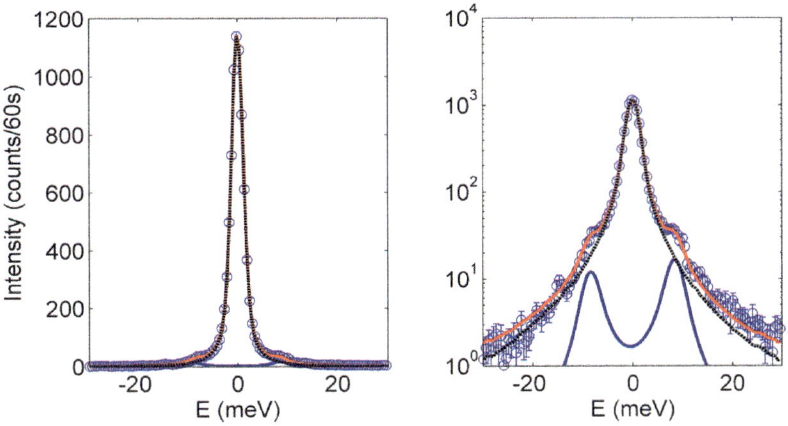

Figure 19.15 Constant q-scan spectrum in a metallic glass: Raw Data. The spectrum collected by IXS on the bulk metallic glass called D" is reported in a linear (left) and logarithmic (right) scale for the intensity at an exchanged momentum $q = 0.266$ Å^{-1}. This spectrum is the sum of three scans collected with an integration time of 60 s per energy point. The red line is the result of the fit with a damped harmonic oscillator (DHO) model, the elastic and inelastic contributions being individually reported as black dashed and solid blue lines, respectively.

Two to three spectra were collected per q point and summed up to improve statistics.

Figure 19.15 shows a spectrum collected at $q = 0.266$ Å^{-1}. While the elastic line is very intense, two inelastic peaks are still visible at an energy exchange of $\omega \approx 8.5$ meV. The same spectrum is reported on the right in a logarithmic scale for the intensity to better see the inelastic excitations.

For fitting these data, we have modeled the elastic line with a Dirac function, and the inelastic part of the spectrum with a damped harmonic oscillator (DHO):

$$S(q,\omega) = A(q)\left[f(q)\delta(\omega) + (1 - f(q))I(q)\frac{\Omega_L^2\Gamma_L}{(\omega^2 - \Omega_L^2)^2 + \omega^2\Gamma_L^2}\right], \tag{19.21}$$

where the subscript L stands for longitudinal, Ω and Γ are energy and width of the phonon, $f(q)$ the non-ergodicity factor and $A(q)$ a renormalization constant.

This model needs to be convoluted with the instrumental resolution function $R(q, \omega)$, giving thus $S(q, \omega) = f(q)R(q, \omega) + S_{in}(q, \omega) \times R(q, \omega)$. As it can be seen, the elastic line is directly given by the instrumental resolution function. We report in Fig. 19.15 the result of the fit with Eq. (19.21), as well as the individual inelastic and elastic components. In this latter one, the long-lasting tails of the pseudo-Voigt resolution function are very well visible.

As previously said, it is possible in a disordered system to measure simultaneously longitudinal and transverse modes at large q. In this system we indeed observe a transverse excitation for $q \geq 0.52$ Å^{-1}. The model is thus extended to have two DHOs, one for the longitudinal and one for the transverse phonons.

In order to enhance the phonons visibility, we report in Fig. 19.16 some spectra collected at different q values in the pseudo-Brillouin zone, subtracted of the elastic line and only for positive energies. The presence of a second, low-lying excitation, is clearly visible for $q \geq 0.52$ Å^{-1}.

We report in Fig. 19.17 the fitting results for longitudinal (LA) and transverse (TA) acoustic dispersions in the pseudo-Brillouin zone. Two branches are well distinct, both dispersive, they reach a maximum at the pseudo-Brillouin zone border, similarly to what happens in a crystal. The Debye predicted low-energy linear behavior is also reported as dashed lines based on the known speeds of sound. The agreement is very good. The TA mode is somehow at higher energies than predicted by the Debye model, which can be due to uncertainties in the measurements. Indeed, this is the mode the most affected by the subtraction of the intense elastic line.

The right panel of the same figure reports the full width at half maximum (FWHM) Γ for the longitudinal mode, together with a fit at low q with a quadratic dependence, $\Gamma \propto q^2 \propto E^2$. It is clear than in this glass we don't see the ω^4 dependence observed in many other glasses at very low q. This is most probably due to the fact that the lowest q accessible due to the intense elastic line is $q = 0.15$ Å^{-1}, corresponding to an energy of $\approx$ 4 meV, likely very close to ω_{BP}, found to be 4–6 meV in similar glasses [36, 39]. We are therefore directly in the regime $\Omega \geq \omega_{BP}$.

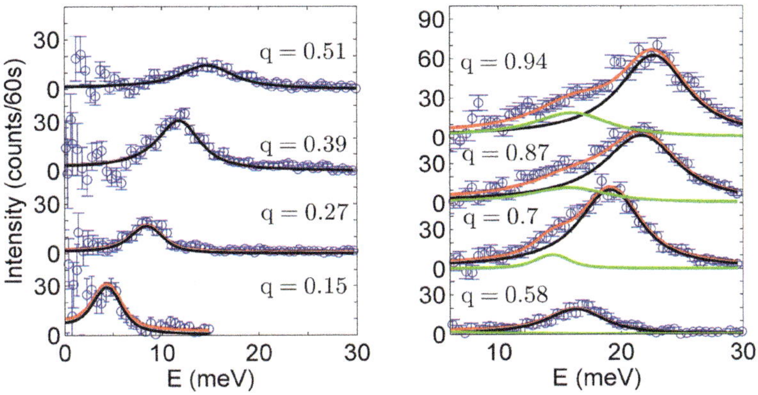

Figure 19.16 Constant q-scans in a metallic glass. We report here the inelastic part of spectra collected at several exchanged momenta from 0.1 to ≈1 Å^{-1}, i.e., the pseudo-Brillouin zone border, q_P being 2.4 Å^{-1}. The q values are reported for each spectrum in units of Å^{-1}. The fitting model (red solid line), longitudinal mode (black solid line) and transverse mode (dashed light green line) are also reported.

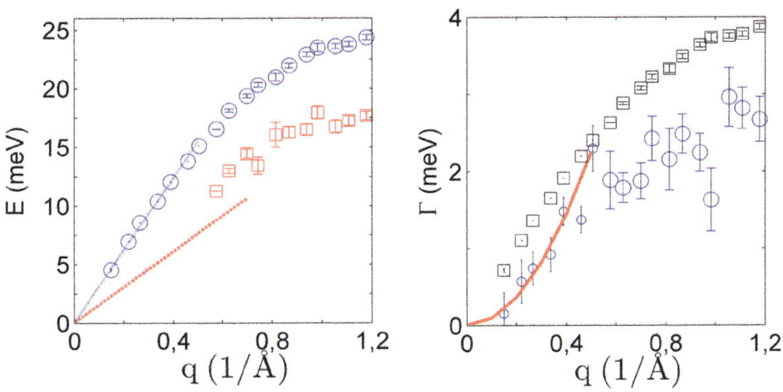

Figure 19.17 Phonon dispersion relations in a glass. (Left) Longitudinal (blue circles) and transverse (red squares) acoustic dispersion as measured in the metallic glass D" in the whole pseudo-Brillouin zone. The linear behavior expected from the sound velocities as measured with ultrasonic techniques is also reported as dashed lines for LA (blue) and dot-dashed lines for TA (red). (Right) The FWHM of the longitudinal phonons (blue circles) is reported and compared with Ω/π (black squares), showing that we are not beyond the Ioffe–Regel limit yet. At low q a quadratic dependence of Γ on q (and Ω in this linear regime) is found (red line).

Notwithstanding, we are still in the regime where the acoustic modes are well-defined, as the Ioffe–Regel limit has not been crossed yet. Indeed it results $\Omega/\pi \geq \Gamma$ as can be seen in the figure, where Ω/π points are reported as black squares.

This example addresses the main limitation of current IXS spectrometers as tools for investigating the glassy dynamics. Indeed, the currently accessible q-range is limited to 1 nm^{-1}, for the divergence of the incident beam, but also because of a still poor energy resolution, as a very intense elastic line will overwhelm low-energy low-intensity phonons lying on its tails. $q = 1$ nm^{-1} is a wave vector very close to the one corresponding to the energy of the Boson Peak in many systems. In order to investigate the role of anharmonicity and the strong scattering ($\Gamma \propto \omega^4$) regime, lower qs need to be investigated at these high frequencies. This is still a no-man land and improvement is thus needed for accessing it with a higher energy resolution.

19.4 Conclusions

Thermal properties of materials, amorphous, crystalline, or composites, are at the focus of the modern research, for a variety of applications.

As mentioned in the introduction, we are still far from getting a proper understanding of the mechanisms underlying a low thermal conductivity in materials, and developing a proper formalism for the description of heat transport.

In order to do this, microscopic measurements are essential, and especially the measurement of phonons lifetimes, which remains elusive in crystals because of experimental limitations.

In this chapter, we have shown how the dynamics of ordered as well as disordered systems can be measured and which the current limitations for the measurement of lifetimes are.

In the q-range accessible to the IXS technique these limitations do not hold for amorphous systems, the phonons lifetime being short enough to be measurable. Indeed in these systems, for $\omega \leq \omega_{IR}, \omega_{BP}$ phonons are well-defined propagative modes, and their lifetime at low q gives a typical width going from $\Gamma \geq 0.2$ meV to several tens of meV, as in the case of SiO_2, thus measurable within the IXS instrumental resolution [7]. The largest widths just below the Ioffe–Regel limit correspond to phonons mean free paths as small as 0.2 nm, i.e., of the order of the interparticle distance, which is considered to be the smallest possible.

For frequencies beyond the Boson Peak or the Ioffe–Regel limit the measured peak cannot be assimilated to a single phonon anymore, but it's an overlap of modes with different polarization and character (TA, LA, optic modes, etc.), thus the measured width looses any connection with phonons lifetime. In this regime the appropriate model for heat transport is based on a diffusive mechanism.

A similar mechanism has been recently invoked for complex crystalline materials, despite phonons lifetimes still are too large to be measured by IXS or INS and the mean free path still of tens of nm [12, 48, 57]. These recent results highlight the need of overcoming the current limitations in phonons lifetimes measurements (energy resolution and, for disordered materials, accessible q-range), in order to get an exhaustive understanding of the heat transport mechanisms in different materials, and to build a unifying theoretical description.

References

1. Allen, P. B., Feldman, J. L., Fabian, J. and Wooten, F. (1999). Diffusons, locons and propagons: character of atomic vibrations in amorphous si, *Phil. Mag. B* **79**(11/12): 1715–1731.
2. Anderson, P. W., Halperin, B. I. and Varma, C. M. (1972). Anomalous low-temperature thermal properties of glasses and spin glasses., *Philos. Mag.* **25**: 1.
3. Ashcroft, N. and Mermin, N. (2002). *Physique des Solides*, EDP Science.
4. Aynajian, P. (2011). *Electron-Phonon Interaction in Conventional and Unconventional Superconductors*, Springer Berlin Heidelberg, chapter Phonons and Their Interactions, pp. 7–13.
5. Baldi, G., Giordano, V. M. and Monaco, G. (2011). Elastic anomalies at terahertz frequencies and excess density of vibrational states in silica glass, *Phys. Rev. B* **83**: 174203.

6. Baldi, G., Giordano, V. M., Monaco, G. and Ruta, B. (2010). Sound attenuation at terahertz frequencies and the boson peak of vitreous silica., *Phys. Rev. Lett* **104**: 195501.
7. Baldi, G., Giordano, V. M., Monaco, G., Sette, F., Fabiani, E., Fontana, A. and Ruocco, G. (2008). Thermal conductivity and terahertz vibrational dynamics of vitreous silica, *PRB* **77**: 214309.
8. Baldi, G., Zanatta, M., Gilioli, E., Milman, V., Refson, K., Wehinger, B., Winkler, B., A., A. F. and Monaco, G. (2013). Emergence of crystal-like atomic dynamics in glasses at the nanometer scale, *Phys. Rev. Lett.* **110**: 185503.
9. Baroni, S., Gironcoli, S. D., Corso, A. D. and Giannozzi, P. (2001). Phonons and related crystal properties from density-functional perturbation theory, *Rev. Mod. Phys.* **73**: 515–562.
10. Birman, J. L. (1974). *Theory of Crystal Space Groups and Lattice Dynamics*, Vol. XXIV, 538, Springer-Verlag Berlin Heidelberg.
11. Boon, J. P. and Yip, S. (1991). *Molecular Hydrodynamics*, Dover Publications, Inc. New York.
12. Bouyrie, Y., Candolfi, C., Pailhès, S., Koza, M. M., Malaman, B., Dauscher, A., Tobola, J., Boisron, O., Saviot, L. and Lenoir, B. (2015). From crystal to glass-like thermal conductivity in crystalline minerals., *Phys Chem Chem Phys.* **17**: 19751.
13. Bove, L. E., Petrillo, C., Fontana, A. and Sokolov, A. P. (2008). Damping of sound waves in the terahertz range and strength of the boson peak, *J. Chem. Phys.* **128**: 184502.
14. Broido, D. A., Malorny, M., Birner, G., Mingo, N. and Stewart, D. A. (2007). Intrinsic lattice thermal conductivity of semiconductors from first principles, *Appl. Phys. Lett.* **91**: 231922.
15. Buchenau, U. (2001). *Encyclopedia of Materials: Science and Technology*, Elsevier, chapter Amorphous Materials: Low-frequency Vibrations., p. 212.
16. Buchenau, U., Galperin, Y. M., Gurevich, V. L., Parshin, D. A., Ramos, M. A. and Schober, H. R. (1992). Interaction of soft modes and sound waves in glasses, *Phys. Rev. B* **46**: 2798.
17. Buchenau, U., Nucker, N. and Dianoux, A. J. (1984). Neutron scattering study of the low-frequency vibrations in vitreous silica, *Phys. Rev. Lett.* **53**: 2316.
18. Burkel, E. (1991). *Inelastic Scattering of X-Rays with Very High Energy Resolution*, Vol. 125 of *Springer Tracts in Modern Physics*, Springer-Verlag Berlin Heidelberg.

19. Carruthers, P. (1961). Theory of thermal conductivity of solids at low temperatures, *Reviews of Modern Physics* **33**(1): 92–138.

20. Chaput, L. (2013). Direct solution to the linearized phonon boltzmann equation, *Phys. Rev. Lett.* **110**: 265506.

21. Christensen, M., Johnsen, S. and Iversen, B. B. (2010). Thermoelectric clathrates of type i, *Dalton Transactions* **39**: 978–992.

22. Chumakov, A. I., Monaco, G., Monaco, A., Crichton, W. A., Bosak, A., Rüffer, R., Meyer, A., Kargl, F., Comez, L., Fioretto, D., Giefers, H., Roitsch, S., Wortmann, G., Manghnani, M. H., Hushur, A., Williams, Q., Balogh, J., Parlinski, K., Jochym, P. and Piekarz, P. (2011). Equivalence of the boson peak in glasses to the transverse acoustic van hove singularity in crystals, *PRL* **106**: 225501.

23. Cooper, M. and Nathans, R. (1967). The resolution function of a neutron diffractometer and its application to phonon measurements, *Acta. Cryst.* pp. 357–367.

24. Delaire, O., Ma, J., Marty, K., May, A. F., McGuire, M. A., Du, M.-H., Singh, D. J., Podlesnyak, A., Ehlers, G., Lumsden, M. D. and Sales, B. C. (2011). Giant anharmonic phonon scattering in pbte, *Nature Materials* **10**: 614–619.

25. Dorner, B. (1972). The normalization of the resolution function for inelastic neutron scattering and its application, *Acta Cryst.* **A28**: 319–327.

26. Dove, M. T. (1993). *Introduction to Lattice Dynamics*, Cambridge University Press.

27. Elliott, S. R. (1992). A unified model for the low-energy vibrational behaviour of amorphous solids, *Europhys. Lett.* **19**: 201.

28. Esfarjani, K., Chen, G. and Stokes, H. T. (2011). Heat transport in silicon from first-principles calculations, *Phys. Rev. B* **84**: 85204.

29. Euchner, H., Pailhès, S., Nguyen, L. T. K., Assmus, W., Ritter, F., Haghighirad, A., Grin, Y., Paschen, S. and de Boissieu, M. (2012). Phononic filter effect of rattling phonons in the thermoelectric clathrate ba8ge40+xni6?x, *Physical Review B* **86**: 224303.

30. Gähler, R. and Golub, R. (1987). A high resolution neutron spectrometer for quasielastic scattering on the basis of spin echo and magnetic resonance, *Z. Phys. B. Condensed Matter* **65**(269).

31. Gähler, R. and Golub, R. (1988). Neutron resonance-spin-echo, bootstrap-method for increasing the effective magnetic field, *Journal de Physique* **49**: 1195–1205.

32. Giordano, V. M. and Monaco, G. (2010). Fingerprints of order and disorder on the high-frequency dynamics of liquids, *PNAS* **107**: 21985.

33. Goldstein, M. (1969). Viscous liquids and the glass transition: A potential energy barrier picture, *J. Chem. Phys.* **51**: 3728–3739.
34. Graebner, J. E., Golding, B. and Allen, L. C. (1986). Phonon localization in glasses, *PRB* **34**: 5696.
35. Gravier, S. (n.d.). Private communication.
36. Hou, L., Liu, H., Liu, Q., Dun, C., Yang, W., Huo, J. and Dou, L. (2015). Effects of crystallization on boson peak of zr52.5cu17.9ni14.6al10ti5 bulk metallic glass, *J. Low Temp. Phys.* **178**: 11–17.
37. Imae, T., Kanaya, T., Furusaka, M. and Torikai, N. (2011). *Neutrons in Soft Matter*, Wiley.
38. Jackle, J. (1986). Models of the glass transition, *Rep. Prog. Phys.* **49**: 171.
39. Jakse, N., Nassour, A. and Pasturel, A. (2012). Structural and dynamic origin of the boson peak in a cu-zr metallic glass, *PRB* **85**: 174201.
40. Kaviany, M. (2008). *Heat Transfer Physics*, Cambridge University Press.
41. Keller, T., Keimer, B., Habicht, K., Golub, R. and Mezei, F. (2003). *Neutron Spin Echo Spectroscopy*, Vol. 601 of *Lecture Notes in Physics*, Springer Berlin Heidelberg.
42. Klemens, P. G. (1958). Thermal conductivity and lattice vibrational modes, *Solid state Phys.* **7**: 1–98.
43. Klimko, S. (2003). *ZETA, A zero field spin echo method for very high resolution study of elementary excitations and first application*, PhD thesis, Technische Universität Berlin.
44. Kneller, G. R. (2000). Inelastic neutron scattering from damped collective vibrations of macromolecules, *Chemical Physics* **261**: 1–24.
45. Kresse, G. and Hafner, J. (1993). Ab initio molecular dynamics for liquid metals, *Phys. Rev. B* **47**: 558. https://www.vasp.at/
46. Li, J. F., Liu, W. S., Zhao, L.-D. and Zhou, M. (2010). High-performance nanostructured thermoelectric materials, *NPG Asia Mater* **2**: 152–158.
47. Lovesey, S. W. (1987). *The Theory of Neutron Scattering from Condensed Matter*, Vol. 2, Oxford University Press.
48. Ma, J., Delaire, O., May, A. F., Carlton, C. E., McGuire, M. A., VanBebber, L. H. and Sales, B. C. (2013). Glass-like phonon scattering from a spontaneous nanostructure in agsbte2, *Nature Nanotechnol.* **8**: 445–451.
49. Maradudin, A., Montroll, E., Weiss, G. and I. P. Ipatova (1963). *Theory of Lattice Dynamics in the Harmonic Approximation*, New York: Academic Press.

50. Martin, N., Regnault, L.-P. and Klimko, S. (2012). Nautron larmor diffraction study of the bam2(xo4)2 (m=co, ni; x=as, p) compounds, *Journal of Physics: Conference Series*.
51. Martin, N., Regnault, L.-P., Klimko, S., Lorenzo, J. E. and Gaehler, R. (2011). Larmor diffraction measurement of the temperature dependence of lattice constants in cugeo3, *Physica B: Condensed Matter* **406**: 2333–36.
52. Mezei, F. (1972). Neutron spin echo: A new concept in polarised thermal neutron techniques, *Z. Physik* **255**: 146–160.
53. Monaco, G. and Giordano, V. M. (2009). Breakdown of the debye approximation for the acoustic modes with nanometric wavelengths in glasses, *PNAS* **106**: 3659.
54. Nakashima, S., Kawakita, Y., Otomo, T., Suenaga, R., Baron, A. Q. R., Tsutsui, S., Kohara, S., Takeda, S., Itoh, K., Kato, H., Fukunaga, T. and Hasegawa, M. (2007). Phonon dispersion of metallic glass $cuzr_2$, *Journal of Physics: Conference Series* **92**: 012136.
55. Nielsen, M. and MØller, H. (1969). Resolution of triple axis spectrometer, *Acta Cryst. A* **25**: 547–550.
56. Nolas, G., Sharp, J. and Goldsmid, J. (2001). *Thermoelectrics*, Vol. 45, Springer-Verlag Berlin Heidelberg.
57. Pailhès, S., Euchner, H., Giordano, V. M., Debord, R., Assy, A., Gomès, S., Bosak, A., Machon, D., Paschen, S. and de Boissieu, M. (2014). Localization of propagative phonons in a perfectly crystalline solid, *PRL* **113**: 025506.
58. Parlinski, K. (2010). *Software phonon*. http://wolf.ifj.edu.pl/phonon/Public/phrefer.html
59. Phillips, W. A. (1972). Tunneling states in amorphous solids, *J. Low Temp. Phys.* **7**: 351.
60. Reichman, D. R. and Charbonneau, P. (2005). Mode-coupling theory., *J. Stat. Mech.* p. P05013.
61. Rescal (n.d.). https://www.ill.eu/de/instruments-support/ computing-for-science/cs-software/all-software/matlab-ill/ rescal-for-matlab/
62. Restrax (n.d.). http://neutron.ujf.cas.cz/restrax/?chap=1
63. Rufflé, B., Guimbretière, G., Courtens, E., Vacher, R. and Monaco, G. (2006). Glass-specific behavior in the damping of acousticlike vibrations, *Phys. Rev. Lett.* **96**: 045502.
64. Rufflé, B., Parshin, D. A., Courtens, E. and Vacher, R. (2008). Boson peak and its relation to acoustic attenuation in glasses, *Phys. Rev. Lett.* **100**: 015501.

65. Ruta, B., Baldi, G., Giordano, V. M., Orsingher, L., Rols, S., Scarponi, F. and Monaco, G. (2010). Communication: High-frequency acoustic excitations and boson peak in glasses: A study of their temperature dependence, *J. Chem. Phys.* **133**: 041101.

66. Ruta, B., Giordano, V. M., Erra, L., Liu, C. and Pineda, E. (2014). Structural and dynamical properties of mg65cu25y10 metallic glasses studied by in situ high energy x-ray diffraction and time resolved x-ray photon correlation spectroscopy, *Journal of Alloys and Compounds* **615**: S45–S50.

67. Ruta, B., Monaco, G., Giordano, V. M., Scarponi, F., Fioretto, D., Ruocco, G., Andrikopoulos, K. S. and Yannopoulos, S. N. (2011). Nonergodicity factor, fragility, and elastic properties of polymeric glassy sulphur, *J. Phys. Chem. B* **115**: 14052–63.

68. Saroun, J. and Kulda, J. (1997). Restrax - a program for tas resolution calculation and scan profile simulation, *Physica B* **234**: 1102–1104.

69. Schirmacher, W. (2006). Thermal conductivity of glassy materials and the "boson peak", *Europhys. Lett.* **73**: 892.

70. Schirmacher, W., Diezemann, G. and Ganter, C. (1998). Harmonic vibrational excitations in disordered solids and the "boson peak", *PRL* **81**: 136.

71. Schober, H. (2014). An introduction to the theory of nuclear neutron scattering in condensed matter, *Journal of Neutron Research* **17**: 109–357.

72. Scopigno, T., Cangialosi, D. and Ruocco, G. (2010). Universal relation between viscous flow and fast dynamics in glass-forming materials, *PRB* **81**: 100202(R).

73. Snyder, G. J. and Toberer, E. S. (2008). Complex thermoelectric materials, *Nature Materials* **7**: 105–114.

74. Sokolov, A. P., Calemczuk, R., Salce, B., Kisliuk, A., Quitmann, D. and Duval, E. (1997). Low-temperature anomalies in strong and fragile glass formers, *PRL* **78**: 2405.

75. Sokolov, A. P., Kisliuk, A., Soltwisch, M. and Quitmann, D. (1992). Medium-range order in glasses: Comparison of raman and diffraction measurements, *Phys. Rev. Lett.* **69**: 1540.

76. Squires, G. L. (2012). *Introduction to the Theory of Thermal Neutron Scattering*, Cambridge University Press.

77. Srivastava, G. P. (1990). *The Physics of Phonons*, CRC Press, Taylor & Francis Group.

78. Suck, J.-B., Rudin, H., Guntherodt, H.-J. and Beck, H. (1980). Short-wavelength transverse collective modes in a metallic glass, *J. Phys. C: Solid St. Phys.* **13**: L1045–1051.

79. Tadano, T., Gohda, Y. and Tsuneyuki, S. (2014). Anharmonic force constants extracted from first-principles molecular dynamics: applications to heat transfer simulations, *J. Phys. Condens. Matter* **26**: 225402.

80. Takabatake, T., Suekuni, K., Nakayama, T. and Kaneshita, E. (2014). Phonon-glass electron-crystal thermoelectric clathrates: Experiments and theory, *Reviews of Modern Physics* **86**(2): 669–716.

81. Takeno, S. and Goda, M. (1971). A theory of phonons in amorphous solids and its implications to collective motion in simple liquids, *Prog. Theor. Phys.* **45**: 331.

82. Toberer, E. S., Zevalkink, A. and Snyder, G. J. (2011). Phonon engineering through crystal chemistry, *J. Mater. Chem.* **21**: 15843.

83. Tritt, T. M. (ed.) (2004). *Thermal Conductivity*, Physics of Solids and Liquids, Springer US.

84. Yin, M. T. and Cohen, M. L. (1991). Theory of lattice-dynamical properties of solids: application to si and ge, *Phys. Rev. B* **26**: 3259–3272.

Index

* 9 7 8 9 8 1 4 7 4 5 6 4 2 *

#0001 - 110817 - C40 - 229/152/32 [34] - CB - 9789814745642